AF344316

HIGH ENERGY AND SHORT PULSE LASERS

Edited by **Richard Viskup**

High Energy and Short Pulse Lasers

http://dx.doi.org/10.5772/61628
Edited by Richard Viskup

Contributors

Akira Endo, Chayan Mitra, Rachit Sharma, Masaki Nakano, Heping Zeng, Qiang Hao, Tingting Liu, Hanieh Fattahi, Leonid Dmitrievich Mikheev, Losev Valery, Amirkianoosh Kiani, Mitra Radmanesh, Mendykhan Khasenov, Tobias Mey, Yuzhai Pan, Abubaker Hamad, K. Zhukovsky, Guang-Hoon Kim, Juhee Yang, Byunghak Lee, Bosu Jeong, Sergey Chizhov, Elena Sall, Vladimir Yashin, Uk Kang, Lucia Marin-Biolan, Jongmin Lee, Liang Hu, Yiqing Huang, Meng Lin, Luca Poletto, Fabio Frassetto, Paolo Miotti

CBS, Edition **2017**

Published by InTech

Janeza Trdine 9, 51000 Rijeka, Croatia

Notice

Statements and opinions expressed in the chapters are these of the individual contributors and not necessarily those of the editors or publisher. No responsibility is accepted for the accuracy of information contained in the published chapters. The publisher assumes no responsibility for any damage or injury to persons or property arising out of the use of any materials, instructions, methods or ideas contained in the book.

Publishing Process Manager Iva Lipovic

Technical Editor SPi Global

Cover Designer InTech Design Team

Additional hard copies can be obtained from orders@intechopen.com

High Energy and Short Pulse Lasers, Edited by Richard Viskup
p. cm.
Print ISBN 978-953-51-2606-5
Online ISBN 978-953-51-2607-2

Contents

Preface

The development and the application of the lasers have changed over the last almost 60 years, since the first idea of optical maser explored by Gordon Gould in 1956, and have dramatically moved forward all research.

While owning the first lasers was the sole privilege of highly advanced countries in the world up to the current time, this trend is shifting, and other universities around the globe are extending the capabilities of their departments and divisions dealing with optics, photonics, and lasers and its applications. Initially, the first lasers possessed very low gain in laser emission, low power in continuous wave (cw) or very long laser pulses. During the last half century, we have seen how these trends exponentially graduated, followed by rapid increase in laser power, starting from few milliwatts (1963), watts (1965), kilowatts (the late 1960s), megawatt (1970), gigawatt (pulsed lasers with peak power) (1975), terawatt (the 1980s), and petawatt (first peak power laser) (1996). Exawatt (EW) peak power laser is currently under development within the European project called Extreme Light Infrastructure (ELI) that should be soon ready to deliver extremely high peak EW power laser pulse.

The time duration of laser pulses also evolved from nanosecond (ns) and picosecond (ps) laser pulses in 1970 up to first femtosecond (fs) pulse from dye laser (1980) and later from solid-state laser, Ti/sapphire; the femtosecond pulses were obtained in the 1990s. The first attosecond pulses were reported in 2001 pioneered from high-harmonic generation (HHG) process.

At present we can speak about two main streams in high-power laser technology. In the first stream, we can include large-scale laser laboratories whose main task is to possess extra-high-peak power and/or ultra-short pulse lasers for study of terawatt (TW), petawatt (PW), and future exawatt (EW) laser interactions, acceleration of particles, or hot dense thermal plasma for the laser fusion.

Into the second stream, we can include the small-scale laboratories that are using for its research commercial sources of laser radiation—ns, ps, or fs laser beam.

The main task of this book is to expand the knowledge of the readers in both of these separate streams, which are often perceived as diametrically different and distinct. Why both streams? The answer is relatively easy; this is due to their common essence—the photons and light coherency.

This book is divided into six main sections dealing with short and ultrashort laser pulses, laser-produced soft X-ray sources, large-scale high-power laser systems, free-electron lasers, fiber-based sources of short optical pulse, and applications of short pulse lasers. In each chapter readers can find fascinating topics related to the high energy and/or short pulse laser technique. Naturally it is not possible to include all topics into this book, and neither was it meant like that. This book should serve as an engaging motivation for the readers to search for more scientific publications about this fascinating laser era.

It will be exciting to observe what the further development of lasers and its applications will offer within the next 60 years and how it will influence our everyday life. As the consequence of the progressive development of these technologies, it initially enabled a day-to-day usage of the LED diodes for lighting of our homes or LED displays, up to the applications in nanotechnology for production of electronic chips, in measurement diagnostics, in medicine, in biomedical technology, in mechatronics, in engineering, in automotive and aviation industry, as well as for the security and in defense technology.

In conclusion, I would like to thank all of the authors for conscientious preparation and formulation of individual book chapters that should serve the broad spectrum of readers of different expertise, layman, undergraduate and postgraduate students, scientists, and engineers, who may in this book find easily explained fundamentals as well as advanced principles of particular subjects related to high energy and short pulse laser phenomenon. Each chapter has well-compiled references of particular subject, from reviews to more advance literature, for eager readers who may find more details or further relevant work into each subtopic.

Finally, I would like to wish all the readers a pleasant experience during the reading of this exciting research and gaining new knowledge, which hopefully will be used for further explorations of this topic, with new ideas, research, and applications for our upcoming generation, in peace.

RnDr. Dr. Eng. Richard Viskup, MSc., MPhil., PhD
Johannes Kepler University Linz,
Austria

Short and Ultrashort Laser Pulses

Chapter 1

Generation of High-Intensity Laser Pulses and their Applications

Tae Moon Jeong and Jongmin Lee

Additional information is available at the end of the chapter

http://dx.doi.org/10.5772/64526

Abstract

The progress in the laser technology makes it possible to produce a laser pulse having a peak power of over PW. Focusing such high-power laser pulses enables ones to have unprecedentedly strong laser intensity. The laser intensity over 10^{19} W/cm^2, which is called the relativistic laser intensity, can accelerate electrons almost to the speed of light. The acceleration of charged particles using such a high-power laser pulse has been successfully demonstrated in many experiments. According to the recent calculation using the vector diffraction theory, it is possible, by employing a tight focusing geometry, to produce a femtosecond (fs) laser focal spot to have an intensity of over 10^{24} W/cm^2 in the focal plane. Over this laser intensity, protons can be directly accelerated almost to the speed of light. Such ultrashort and ultrastrong laser intensities will bring ones many opportunities to experimentally study ultrafast physical phenomena we have never met before. This chapter describes how to generate a high-power laser pulse. And, then the focusing characteristics of a femtosecond high-power laser pulse are discussed in the scalar and the vector diffraction limits. Finally, the applications of ultrashort high-power laser are briefly introduced.

Keywords: ultrashort laser pulse, high-intensity laser pulse, chirped pulse amplification, charged particle acceleration, tight focusing

1. Generation of ultrashort laser pulses

Femtosecond (fs) high-power laser pulses having a peak power of PW or higher are being produced for the study of laser-matter interactions in the relativistic intensity regime. An ultrashort laser pulse is generated in a mode-locked laser oscillator in the front and its energy is amplified in the following amplifiers. The mode locking is a technique to produce laser

pulses having a pulse duration in the ultrashort time scale such as picosecond (ps) or fs [1, 2]. In the technique, a gain or a loss of an oscillator is modulated in an active or a passive manner. Saturable absorber is a typical optical element modulating a loss in an oscillator. Nonlinear effect dependent on the laser intensity is used to realize a saturable absorption instantaneously responding to the intensity. Under the saturable absorption, a laser pulse experiences a lower loss at a higher intensity. As a result, a higher intensity part of a laser pulse grows much stronger and the temporal duration becomes shorter during the saturable absorption process.

As the pulse duration of a laser pulse decreases, the spectrum of the pulse becomes broader and the pulse encounters the dispersion effect in the medium. The dispersion effect frequently tends to broaden the pulse duration. Without any dispersion control device, the resultant pulse duration is determined by the balance between the pulse shortening due to the saturation absorption and the pulse broadening due to the dispersion. With a proper dispersion control device, the dispersion in a laser pulse is compensated and the pulse duration is mostly determined by the spectral bandwidth of the laser pulse. The minimum pulse duration obtainable with a spectral bandwidth is known as the transform-limited pulse duration. Up to date, sub-10 fs laser pulses from an oscillator are generated by compensating for the dispersion effect with prism pairs [3]. In this section, the basic principle of the mode-locking technique is explained for generating an ultrashort laser pulse and the formation of an ultrashort laser pulse is described.

1.1. Short pulse generation by locking phase of longitudinal mode

When a laser oscillator is formed with an optical length of L, the wavelength of standing waves inside the oscillator is determined by $2L/m$ (m is a positive integer), and alternatively, the frequency by $v_m = m \times c/2L$ (or $\omega_m = m \times \pi c/L$). The oscillating frequency in the oscillator is limited by a gain spectrum and it is called the longitudinal mode of the oscillator. A laser pulse can be decomposed into the summation of each electric field having different modes, and the resultant electric field of the pulse can be written as a superposition of oscillating modes:

$$E(z,t) = \sum_m E_m \exp\left\{i\omega_m\left(t-\frac{z}{c}\right)\right\}. \tag{1}$$

In a free running laser, the phase relation among oscillating modes is random and this is the origin of a short-timescale random intensity fluctuation. The phase relation between modes can be constant (i.e., $v_m - v_{m-1}$ = constant) under a specific condition. The situation of having the constant phase relation between modes is mentioned as "mode-locked." In this case, the intensity of the resultant electric field is given by

$$I(z,t) = |E(z,t)|^2 = |E_0(z,t)|^2 \frac{\sin^2\left\{m\omega_m\left(t-z/c\right)/2\right\}}{\sin^2\left\{\omega_m\left(t-z/c\right)/2\right\}}. \tag{2}$$

As can be seen in Eq. (2), a strong intensity peak can grow in the resonator when oscillating electric fields are added under the mode-locking condition. This is the basic principle for generating a mode-locked laser pulse (see **Figure 1**). As expected in Eq. (2), the pulse duration of a mode-locked laser pulse is determined by the number of oscillating modes. For example, a Ti:sapphire laser that typically produces 10-fs laser pulses contains several hundred-thousand modes in the spectral bandwidth. Up to date, a number of mode-locking techniques have been introduced to generate ps and fs laser pulses, but the underlying physics is basically the same and the question is how to realize locking longitudinal modes.

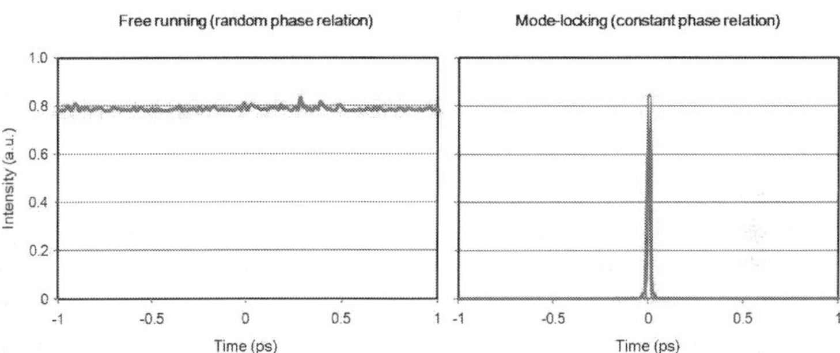

Figure 1. Power at free running and mode-locked operations. When the phase relation is random among longitudinal modes, the intensity has fluctuation because of the beating among modes (left). On the other hand, a single high peak laser pulse is formed under the constant phase relation between modes (right).

1.2. How to lock phases of longitudinal modes

In the early history of mode-locking technique, an active loss element operating at an rf-frequency was installed in an oscillator. The element periodically inducing intensity loss initiates an intensity modulation at a repetition rate corresponding to the round-trip time. A periodic loss at a round-trip time forces to form a laser pulse inside the oscillator. This is known as the active mode-locking technique. Another technique is to introduce a passive-type intensity modulation to the oscillator. Thus, in the passive mode-locking technique, an optical element that has an intensity-dependent loss is installed in the oscillator. The intensity peak in the temporal domain has higher transmittance and energy gain, but a lower intensity part has lower transmittance and energy gain. The lower intensity part is relatively suppressed by the intensity-dependent loss when an intensity fluctuation circulates in the oscillator. As the intensity peak grows, the number of oscillating modes becomes larger and larger in the spectral domain, and the phase relation between modes is automatically locked to form a laser pulse.

1.2.1. Saturable absorption

Some materials have a property that the absorption of light decreases as increasing the light intensity. This kind of material is known as the saturable absorber. In the saturable absorber, the light propagating in the medium transfers its energy to electrons in the ground level and excites them to higher energy levels. The light intensity decreases as the light propagates in the medium. The light absorption becomes very weak when the number of electrons in the ground level becomes sufficiently low, and the rest of light energy almost transmits the medium. At a time later, the excited electrons spontaneously decay into the ground level and the number of ground electrons is recovered to be ready to absorb energy from light. This phenomenon is known as the saturable absorption. The saturable absorber can be divided into slow and fast saturable absorbers, depending on the recovery time τ_r. In the slow saturable absorber, the recovery time is slower than the pulse duration τ_p and it is assumed to be shorter than the round-trip time under the mode-locking condition. Most saturable absorbers used as the form of solid state and semiconductor have the slow recovery property. In a slow saturable absorber, the intensity-dependent loss is described as follows:

$$\frac{dL}{dt} = -\frac{L - L_0}{\tau_r} - \frac{I}{F_{sat}} L \,. \tag{3}$$

Here, L_0 is the unsaturated loss and F_{sat} is the saturation fluence. Since $\tau_r \gg \tau_p$, the second term on the right-hand side of Eq. (3), is dominant and the loss exponentially decreases with respect to the pulse fluence $\int_{-\infty}^{t} I(t)dt$. For the slow absorber, two mode-locking regimes are possible depending on the soliton effect. Without the soliton effect, a slow saturable absorber absorbs the leading part of a pulse while the trailing part is less absorbed. The pulse formation is mostly determined by balancing between the net gain and losses. As a result, a pulse profile becomes shortened, and the pulse duration obtainable in this case is estimated by [4]

$$\tau_p = \frac{1.07}{\Delta v_g} \sqrt{\frac{g}{\Delta R}} \,. \tag{4}$$

Here, Δv_g is the FWHM gain bandwidth, assuming a Gaussian-shaped gain spectrum, and ΔR is the modulation depth. As will be discussed later, the laser pulse can be broadened by the dispersion. Under a proper condition, the shortening and broadening processes can be balanced. Thus, for a slow saturable absorber with the soliton effect, a short laser pulse is generated by the self-phase modulation (SPM) in combination with an appropriate amount of negative dispersion. In this case, the pulse duration can be estimated by [4]

$$\tau_p \approx 1.76 \times \frac{2|D|}{\gamma_{SPM} E_p} \,. \tag{5}$$

Here, D is the group delay dispersion (GDD) per cavity round trip, γ_{SPM} is the SPM coefficient (in rad/W) per round trip, and E_p is the pulse energy. **Figure 2(a)** and **(b)** shows the pulse formation with the slow saturable absorber. The laser pulse is formed when the loss decreases below the gain. The gain can be either unsaturated or saturated during the saturation absorption process. Under the unsaturated gain, the laser pulse gains energy quickly in the beginning of the saturation absorption process. When the gain is saturated during the saturable absorption, the decrease in the gain is slightly delayed and thus the net gain exists for the pulse formation.

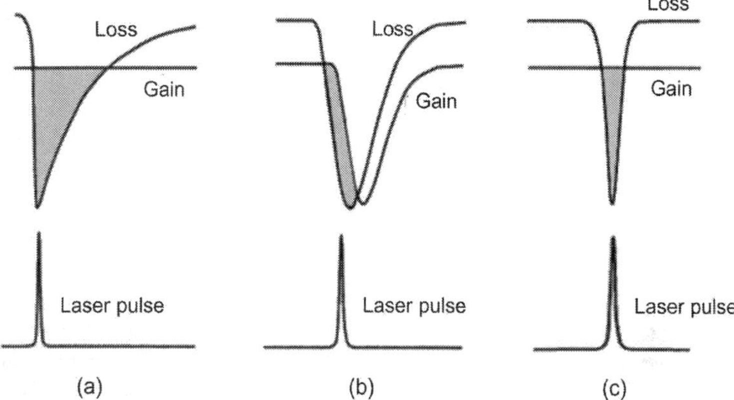

Figure 2. Laser pulse formation with saturable absorbers. The gain is not saturated in (a), but the gain is saturated in (b) during the saturation absorption process. (c) The absorption is quickly recovered in the fast saturable absorber. The laser pulse is formed when the loss decreases below the gain.

The absorption by the material is assumed to be instantaneously recovered in the fast saturable absorber (see **Figure 2(c)**). Thus, a higher intensity in the pulse center experiences a higher transmittance and a lower intensity in the side is suppressed by the saturable absorption. When a fast saturable absorber is installed in an oscillator, the intensity of a transmitted laser pulse increases in a gain medium at a growing rate of

$$\frac{dI}{dz} = \frac{g_0 I}{1 + I/I_{sat}}. \tag{6}$$

Here, I_{sat} is the saturation intensity and g_0 is the unsaturated small signal gain. Thus, the pulse profile is controlled by the intensity, and a higher gain at a higher intensity leads to the pulse shortening. The pulse duration is given by [5]

$$\tau_p \cong \frac{0.79}{\Delta v}\left(\frac{g}{L}\right)^{1/2}\left(\frac{I_{sat}}{I}\right)^{1/2}, \tag{7}$$

with the assumption of a hyperbolic secant pulse profile. Here, Δv is the gain bandwidth, g is the gain defined by $g_0/(1 + I/I_{sat})$, and L is the saturated loss. In reality, the fast saturable absorbing material operating in the femtosecond regime does not exist. Instead, there are materials having a strong nonlinear effect. These materials can possess the property of ultrafast loss modulation that is induced by the nonlinear effect. The ultrashort pulse formation by these materials can be considered as the mode locking by the fast saturable absorber. In this section, self-phase modulation as a nonlinear effect which induces ultrafast change in reflection or transmission is discussed.

(a) Self-phase modulation in time domain

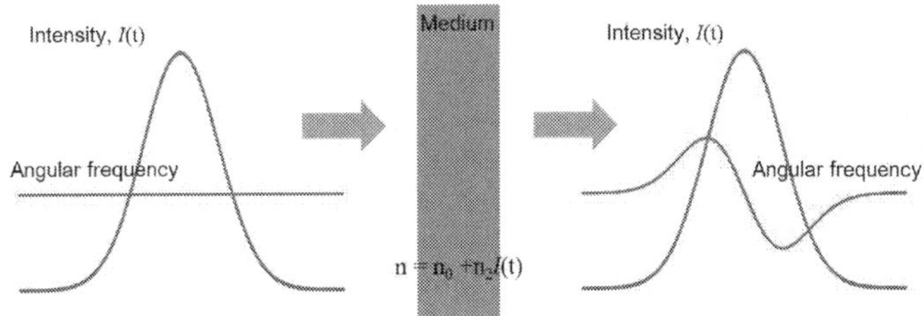

(b) Self-focusing in space domain

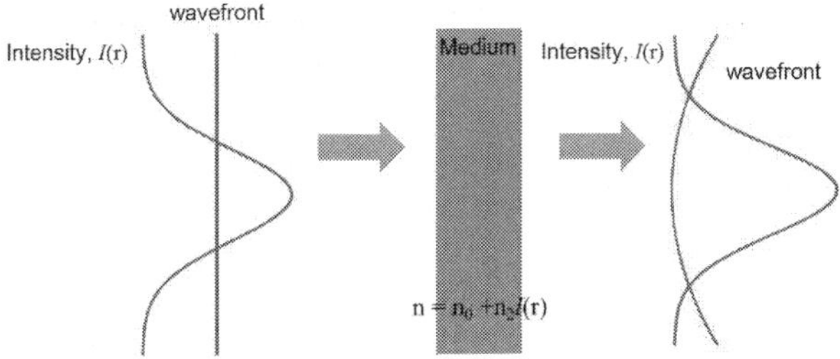

Figure 3. (a) Self-phase modulation in time induces the time-dependent phase variation. The lower angular frequency in the rising part and the higher angular frequency in the falling part are induced. (b) Self-phase modulation in space makes the wavefront quadratically curved.

When a light pulse passes through a medium, it experiences an intensity-dependent change in refractive index. This phenomenon is known as the Kerr effect. The Kerr effect can induce an instantaneous loss modulation and make a medium to act as a fast saturable absorber (see **Figure 3**). In order to derive how the Kerr effect is related with the instantaneous loss modulation, let us consider the refractive index depending on the laser intensity which is given by

$$n = n_0 + n_2 I. \tag{8}$$

Here, n_0 is the normal refractive index and n_2 is the nonlinear refractive index related with the Kerr effect. After a nonlinear medium, the phase of the laser pulse is modified by

$$\phi = nkd = n_0 kd + n_2 I kd. \tag{9}$$

With a Gaussian pulse profile, $I = I_0 \exp\left(-2t^2/\tau_p^2\right)$, in time, the second term on the right-hand side in Eq. (8) induces time-dependent phase variation, and the angular frequency is calculated as

$$\omega = -\frac{d\phi}{dt} = -\frac{d}{dt}\left(nkd\right) = -n_2 kd\frac{dI}{dt} = \frac{4n_2 kdI_0 t}{\tau_p^2}\exp\left(-2t^2/\tau_p^2\right). \tag{10}$$

Thus, after a nonlinear medium, a laser pulse has lower frequency components in the rising edge and higher frequency components in the falling edge. When a Gaussian pulse having these induced frequency components is coherently added to the original one, the constructive interference occurs at the pulse center, but the destructive interference occurs at the edge. The constructive and destructive interferences induce an instantaneous reflectance change in time. This leads to the pulse shortening effect in time. Nonlinear coupled-cavity mode-locking technique introduced as the additive-pulse mode locking (APM) uses the instantaneous reflectance change induced by the self-phase modulation [6].

A similar phenomenon happens in the spatial domain as well. With a Gaussian beam profile, $I = I_0 \exp\left(-2r^2/w_0^2\right)$, in space, the phase at a radial position, r, is given by

$$\phi(r) = n_2 kdI(r) = n_2 kdI_0 \exp\left(-2r^2/w_0^2\right) \approx n_2 kdI_0\left(1 - \frac{2r^2}{w_0^2}\right). \tag{11}$$

with an approximation of $\exp\left(-2r^2/w_0^2\right) \approx \left(1 - 2r^2/w_0^2\right)$. The phase variation induced by the nonlinear effect makes the wavefront quadratically curved in the radial direction. This means that, after the nonlinear medium with a positive nonlinear refractive index, the phase at a higher intensity becomes retarded to the phase at a lower intensity. The focal length induced by the quadratic curvature is calculated as

$$f_{nl} = -\frac{dr}{d\phi} = \frac{w_0^2}{4n_2 dI_0}. \tag{12}$$

This phenomenon is known as the self-focusing. Kerr-lens mode-locking (KLM) technique employs the self-focusing to induce an instantaneous intensity-dependent transmittance [7]. In the KLM technique, a higher intensity part can be separated by the self-focusing in combination with an aperture. A higher intensity part in time and space domain has a higher transmittance because of the self-focusing. As a result, a higher intensity grows as a laser pulse circulates in a oscillator. The KLM technique forms an ultrashort pulse using this pulse shortening process. In the technique, a gain medium in the resonator also acts as a nonlinear medium that induces the self-focusing.

1.3. Dispersion

When a laser pulse propagates in a material with a length of d, the phase is given by the refractive index, $n(\omega)$, of the medium as follows:

$$\phi(\omega) = n(\omega) \cdot k \cdot d = \frac{n(\omega) \cdot \omega \cdot d}{c}. \tag{13}$$

The refractive index of a medium is a function of the angular frequency and can be expressed as the Sellmeier's formula of wavelength as follows:

$$n^2(\lambda) = 1 + \frac{B_1 \lambda^2}{\lambda^2 - C_1} + \frac{B_2 \lambda^2}{\lambda^2 - C_2} + \frac{B_3 \lambda^2}{\lambda^2 - C_3}. \tag{14}$$

with the help of definition, $\lambda = 2\pi c/\omega$. Here, B_1, B_2, B_3, C_1, C_2, and C_3 are known as Sellmeier's coefficients for material. Because of the refractive index of material depending on the wavelength, the phase of an ultrashort laser pulse after material experiences a distortion known as the dispersion. The dispersion is responsible for the broadening of a pulse duration and the distortion of the pulse profile in time. In order to see the effect of dispersion, let us express the spectral phase depending on the angular frequency as the Taylor expansion,

$$\phi(\omega) = \sum_{m=0}^{\infty} (\omega - \omega_0)^m \left. \frac{\partial^m \phi(\omega)}{\partial \omega^m} \right|_{\omega = \omega 0} = \frac{d}{c} \sum_{m=0}^{\infty} (\omega - \omega_0)^m \left. \frac{\partial^m \{\omega \cdot n(\omega)\}}{\partial \omega^m} \right|_{\omega = \omega 0}. \tag{15}$$

Now, we define derivatives as

$$\left. \frac{\partial^n \phi(\omega)}{\partial \omega^n} \right|_{\omega = \omega 0} = \left. \frac{\partial^n \{\omega \cdot n(\omega)\}}{\partial \omega^n} \right|_{\omega = \omega 0} = D_n(\omega = \omega_0). \tag{16}$$

Because the material has a refractive index depending on the frequency, Eqs. (15) and (16) show interesting properties when a laser pulse with a broad spectrum propagates in the material. The first term, $D_0 = d \cdot \omega \cdot n/c$, in the phase relation represents a phase propagation in the material. The second term defined by $D_1(\omega = \omega_0) = \partial\phi(\omega)/\partial\omega |_{\omega = \omega 0}$ is known as the group delay (GD) that can be interpreted as d/v_g. Here, v_g is the group velocity and represents the pulse propagation in the material. The third term defined by $D_2(\omega = \omega_0) = \partial^2\phi(\omega)/\partial\omega^2 |_{\omega = \omega 0}$ is known as the group delay dispersion (GDD) that is responsible for the temporal broadening of a pulse. The temporal broadening by the group delay dispersion is sometimes known as the chirping which originally means the frequency change in time.

Two kinds of temporal broadenings are possible depending on the sign of D_2. When the sign of D_2 is positive, a long (red-like) wavelength component travels faster than a blue-like one in the pulse. On the other hand, a short (blue-like) wavelength component travels faster than a red-like one with a negative sign of D_2. A pulse is said to be positively chirped when a red-like wavelength component travels faster, or to be negatively chirped when a blue-like wavelength component travels faster (see **Figures 4** and **5**).

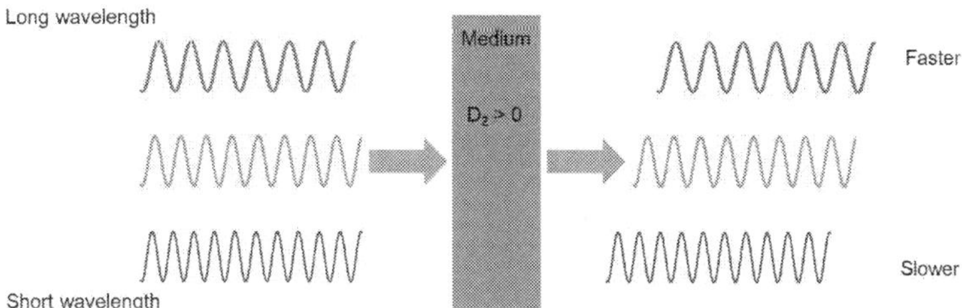

Figure 4. Refractive index depending on the wavelength induces the group delay dispersion (GDD). In the positive GDD, the long-wavelength electromagnetic field travels faster than the short-wavelength electromagnetic field in the medium.

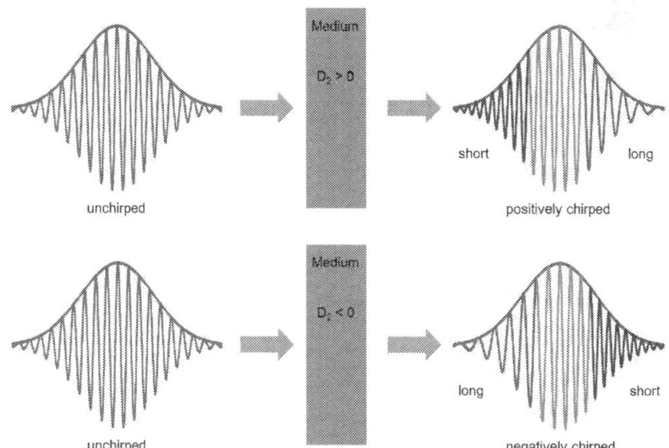

Figure 5. Frequency chirping in the laser pulse. In the upper drawing, a short laser pulse experiences the positive chirping, thus the long-wavelength (red) component arrives faster than the short-wavelength (blue) component in the laser pulse. In the lower drawing, a short laser pulse experiences the negative chirping, thus the short-wavelength (blue) component arrives faster than the long-wavelength (red) component. The pulse duration is broadened by the positive or negative chirping.

Higher-order derivatives in the Taylor expansion affect the pulse profile in time as higher-order dispersions. Even-order dispersions are responsible for the symmetric distortion of a laser pulse in time and odd-order dispersions are responsible for the antisymmetric distortion in the laser pulse. The dispersion control and compensation are key techniques to have a transform-limited laser pulse with a given spectrum. Third-order dispersion (TOD) and fourth-order dispersion (FOD) should be considered to be compensated for the generation of transform-limited pulse.

2. Amplification of ultrashort laser pulses

The energy of a mode-locked laser pulse with a pulse duration of 1 ps or below typically ranges from 10^{-12} to 10^{-10} J. The ultrashort laser pulse cannot be directly amplified in amplifiers because of damage issues in optical elements due to the nonlinear effect and the low-energy extraction efficiency. These hurdles were detoured by employing the chirped-pulse amplification (CPA) technique devised by Strickland and Mourou [8]. The key idea of the CPA technique is to temporarily stretch a laser pulse before amplification, to amplify the energy of the stretched pulse, and finally, after energy amplification, to compress the pulse duration to the original level. The CPA technique was well demonstrated in many systems around the world [9], and now it is used for producing the relativistic laser intensity ($>10^{18}$ W/cm^2).

The control of pulse duration is usually performed by an optical setup which uses the GDD induced by the grating. The stretched pulse duration ranges from few hundreds of ps to nanosecond (ns). The stretched pulse is amplified in a series of amplifier chain including regenerative and/or multipass amplifiers. The output energy can be estimated from the Frantz-Nodvik equation. In this section, the basic principles for controlling the pulse duration and for amplifying the energy are explained.

2.1. Stretching of an ultrashort laser pulse before amplification

The control of pulse duration using the dispersion was first proposed by Treacy [10]. In the proposal, two gratings with a normal separation distance of b are placed in the parallel geometry to induce a negative GDD. The total amount of GDD can be controlled by the separation distance. According to the Treacy's proposal, when a laser pulse passes through an optical setup shown in **Figure 6**, the group delay dispersion (GDD) experienced by a laser pulse is given by

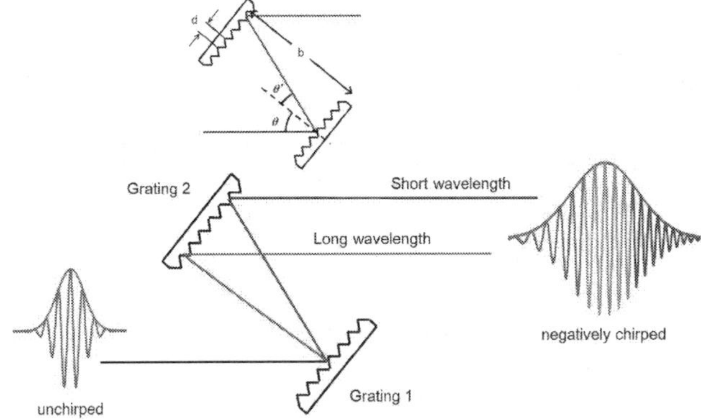

Figure 6. Parallel grating pulse stretching scheme. The parallel grating pulse stretcher introduces a negative GDD to the laser pulse.

$$\left.\frac{d^2\phi}{d\omega^2}\right|_{\omega 0} = -\frac{\lambda_0}{2\pi c^2}\left(\frac{\lambda_0}{d}\right)^2 \frac{b}{\cos^3\theta'(\lambda_0)} \cdot \qquad (17)$$

Here, d is the groove spacing of grating and θ' is the diffraction angle. The first-order diffraction is only considered in this case. The diffraction angle is calculated by the grating equation as follows:

$$d\left(\sin\theta' - \sin\theta\right) = \lambda. \qquad (18)$$

As shown in Eq. (17), the parallel grating geometry always introduces the negative GDD, and thus the blue-like wavelength component travels faster than the red-like one. The positive GDD can be either introduced by installing a telescope in the parallel grating geometry, which was proposed by Martinez [11]. A telescope is an optical device that induces an angular dispersion. The GDD induced by an angular dispersion is given by

$$\left.\frac{d^2\phi}{d\omega^2}\right|_{\omega 0} \approx -\frac{L_p \cdot \omega_0}{c}\left(\left.\frac{d\alpha}{d\omega}\right|_{\omega 0}\right)^2 \qquad (19)$$

with an approximation of $\cos\alpha \gg \sin\alpha$. In the equation, α is the deviation angle at the reference wavelength and L_p is the propagation distance after the surface of an angularly dispersive element. When a laser pulse propagates an optical setup shown in **Figure 7**, the angular dispersion is magnified by a factor of M, which is the magnification of a telescope. Then, the GDD induced by the angular dispersion after the propagation of z' is

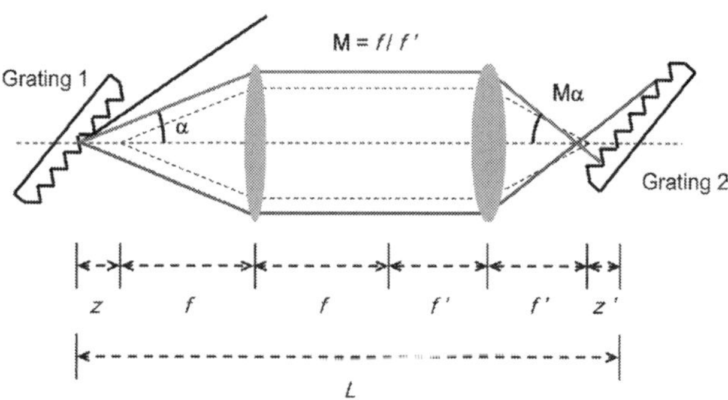

Figure 7. GDD control by the grating pair with a telescope inside. The grating pair with the telescope can induce the positive and negative GDD depending on the total length between gratings. The negative GDD is obtained by $L/2 < (f+f')$. The positive GDD is obtained by $L/2 < (f+f')$.

$$\left.\frac{d^2\phi}{d\omega^2}\right|_{\omega 0} = -\frac{\omega_0}{c}\left(\left.\frac{d\alpha}{d\omega}\right|_{\omega 0}\right)^2 z'M^2. \qquad (20)$$

As shown in **Figure 7**, the propagation distance z' is given by $L - 2(f+f')$ and the magnification by f/f'. The positive GDD can be obtained when $L - 2(f+f') < 0$ or $L/2 < (f+f')$. This condition can be met by moving a second grating before the focal point F'. In general, the first lens can be placed at a position of $z+f$. Then, an additional GDD, $\left.\frac{d^2\phi}{d\omega^2}\right|_{\omega 0} = -\frac{\omega_0}{c}\left(\left.\frac{d\theta'}{d\omega}\right|_{\omega 0}\right)^2 z$, by an angular dispersion after the propagation of z should be added to Eq. (20) to obtain

$$\left.\frac{d^2\phi}{d\omega^2}\right|_{\omega 0} = -\frac{\omega_0}{c}\left(\left.\frac{d\theta'}{d\omega}\right|_{\omega 0}\right)^2 \left(z'M^2 + z\right). \tag{21}$$

In many cases, a reflecting mirror can be put after the first lens to reduce the cost and space. The positive GDD induced by two grating geometry having a telescope can be compensated for with the parallel grating pair. This is important because the pulse duration stretched by the positive or negative GDD can be recompressed by the negative or positive GDD. This is the principle for stretching and compressing an utrashort laser pulse in the CPA technique. In a common CPA technique, a pulse stretcher introduces a positive GDD to the laser pulse and a pulse compressor introduces a negative GDD. The reason for this is that the material dispersion used in amplifier systems also produces a positive GDD. If a laser pulse has negative GDD by a stretcher, the pulse duration of a pulse is shortened as the pulse propagates in a medium having a positive GDD. This might induce damage on optical elements that the pulse propagates. The other combination that uses a pulse stretcher introducing negative GDD and a pulse compressor introducing positive GDD is also possible. This combination is known as the down-chirped pulse amplification (DCPA) technique and also demonstrated with a grating stretcher and bulk material compressor. Although the DCPA technique works for the energy amplification of an ultrashort laser pulse, the pulse duration of the pulse is somewhat broadened because higher-order dispersions, such as TOD and FOD, induced by media in the laser system remain uncompensated. As mentioned earlier, third-order dispersion (TOD), and fourth-order dispersion (FOD) should be corrected or optimized to obtain a nearly transform-limited pulse duration through the pulse compressor.

The misalignment in the parallelism of a grating induces an additional angular dispersion in the spatial domain. This is known as the spatial chirping. The spatial chirping can easily be examined by monitoring the intensity distribution of a focal spot. If there is the spatial chirping in the laser beam profile, a focal spot is elongated along the chirping direction. Sometimes, the elongation by the spatial chirping is confused with astigmatism in the beam. However, the spatial chirping can be discriminated by the through-the-focus image because the elongation by the spatial chirping is not rotated by 90 degrees while the elongation by astigmatism can be rotated.

2.2. Rate equation

When a laser pulse passes through an amplification medium, the pulse obtains energy gain from the medium. The energy gain comes from a stored energy in the medium which is provided by an external power source. Absorption by the transition between electronic energy levels is used to store an external energy. Electrons at a lower energy level are excited to a higher energy level through the pumping process. When an electromagnetic wave (photon) with a specific wavelength defined by the atomic energy transition is radiated to an excited atom, the atom emits the same electromagnetic wave (photon) as the incoming one. This means that an incoming electromagnetic wave is amplified in intensity. This dynamics can be described by the rate equation. In order to describe the situation mathematically, let us consider a four-energy-level system shown in **Figure 8**.

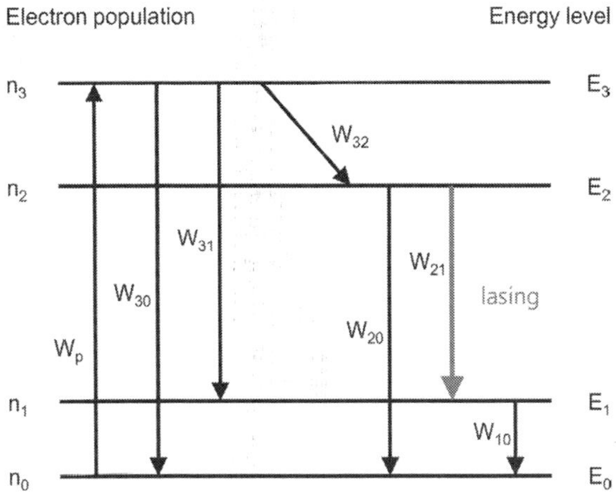

Figure 8. Diagram for energy levels, level transition rate, and the number of electrons at the energy level. In the four-level system, the storage of external energy is accomplished by the absorption due to the electronic transition from level 0 to level 3, and the lasing or energy gain is obtained by the electronic transition from level 2 to level 1.

In a four-level system shown in **Figure 8**, electrons at the lowest energy level 0 are excited to level 3 by the pumping process. The changing rate for the excited electron population increases by the electron population at level 0 and the pumping rate W_p. In a short time, electrons at level 3 lose their energy and decay into level 2 with a transition probability W_{32}. Electrons at level 3 also decay into level 1 and 0 with probabilities W_{31} and W_{30}. The changing rate for electron population at level 2 increases with the number of electrons at level 3 and the transition probability W_{32}, and it decreases with the number of electrons at level 2 and transition probabilities W_{21} and W_{20} to level 1 and level 0, respectively. The main lasing action or energy gain happens with the transition from level 2 to level 1. Under this circumstance, the rate equations for electrons at each level can be expressed as

$$\frac{dn_3}{dt} = W_p n_0 - W_{32} n_3 - W_{31} n_3 - W_{30} n_3 ,$$
(22-1)

$$\frac{dn_2}{dt} = W_{32} n_3 - W_{21} n_2 - W_{20} n_2 ,$$
(22-2)

$$\frac{dn_1}{dt} = W_{31}n_3 + W_{21}n_2 - W_{10}n_1 ,$$

(22-3)

$$\frac{dn_0}{dt} = -W_p n_0 + W_{30}n_3 + W_{20}n_2 + W_{10}n_1 .$$

(22-4)

Although rate equations for level 1 and level 0 are not explained here, those can be easily derived from **Figure 8**. In the four-level system, it is assumed that electron populations at levels 1 and 3 are very small because of the rapid transition to other levels, i.e., $n_2, n_0 \gg n_3, n_1$. The total number, n, of electrons is determined by the sum of electron numbers at levels 0 and 2, i.e., $n = n_2 + n_0$. In the steady-state condition, the change of electron populations at levels 3 and 2 are very small as well; so, we assume $\frac{dn_3}{dt} = \frac{dn_2}{dt} \approx 0$. From Eqs. (22-1) and (22-2), we obtain

$$\frac{n_2}{n_0} = \frac{W_p}{\left(W_{21} + W_{20}\right)} \frac{W_{32}}{\left(W_{32} + W_{31} + W_{30}\right)} .$$

(23)

At level 2, the approximation of $W_{21} \gg W_{20}$ is valid because the lasing action or gain is dominant. And, electron transition from level 3 to level 2 is most dominant to the other transition and thus $W_{32} \gg W_{31}, W_{30}$. Under these conditions, Eq. (23) reduces to

$$\frac{n_2}{n_0} = \frac{W_p}{W_{21}}$$

(24)

According to Eq. (24), a laser pulse can have energy gain when $n_2 > n_0$. The population inversion happens when the difference, $\Delta n = n_2 - n_0$, in electron numbers at levels 2 and 0 is positive. Under the heavily pumping condition, most electrons exist in level 2, and the number of electrons (n_2) at level 2 approximately equals n_0. The population inversion is given by

$$\Delta n = n_2 \left(1 - \frac{W_{21}}{W_p}\right) \approx \frac{n}{1 + W_{21}/W_p} .$$

(25)

By using the relation of $W_{21}/W_p = I(z)/I_{sat}$, Eq. (25) becomes $\Delta n \approx n/(1 + I/I_{sat})$. When a low-intensity laser pulse propagates in the gain medium, the intensity growing rate is linear with the product of propagation distance and population inversion as shown below:

$$\frac{dI(z)}{dz} = I(z)\sigma_{21}\Delta n .$$

(26)

Here, σ_{21} is the emission cross section. By inserting the relation of $\Delta n \approx n/(1 + I/I_{sat})$ into Eq. (26), the growing rate for the intensity becomes

$$\frac{dI(z)}{dz} = I(z)\frac{\sigma_{21}n}{1+I(z)/I_{sat}} \quad \text{or} \quad \frac{dI(z)}{dz} = g(z)I(z).$$

(27)

In Eq. (27), the gain $g(z)$ is defined by $g_0/(1+I(z)/I_{sat})$ and g_0 is defined by $\sigma_{21}n$. When the intensity of a laser pulse is small enough, the intensity exponentially grows with $I_0\exp(\int g(z)dz)$. The gain, $\exp(\int g(z)dz)$, at a small input intensity is known as the small signal gain. As the intensity becomes stronger, the growing rate for the intensity starts to be lowered and the intensity linearly grows with gL in the saturation regime, where L is the medium length.

2.3. Energy amplification

The small signal gain describes how much intensity or energy can be achieved with a given small input intensity. The small intensity means an intensity level that does not affect the population inversion. In this subsection, we will describe the energy amplification in an amplifier system. A single-pass energy gain can be measured by putting a detector before and after the amplification medium during the energy measurement experiment. The small signal and single-pass gain, G_0, at the first pass is given by $\exp(\int g(z)dz)$ or simply by

$$G_0 = \exp(g_0 L).$$

(28)

Here, g_0 is the measured gain coefficient and L is the medium length. Using the Frantz-Nodvik equation [12], the output energy of a laser pulse at the ith round trip in the amplifier can be expressed by

$$F_i = F_{sat}\ln\left[1 + G_i\left\{\exp\left(\frac{F_{i-1}}{F_{sat}}\right) - 1\right\}\right].$$

(29)

Here, F means the fluence of a laser pulse defined by $\int_{-\infty}^{\infty} I(z,\,t)dt$ and the subscripts (i and $i-1$) mean the ith and $(i-1)$th round trips. F_{sat} is the saturation fluence. In a multipass amplifier system, an amplified laser pulse is reinjected into the amplifier medium. Thus, the gain decreases as the input intensity increases. The reduced gain at the ith round trip can be calculated from the gain and the fluence at the $(i-1)$th round trip as follows:

$$G_i = \left\{1 - \left(1 - \frac{1}{G_{i-1}}\right)\exp\left(-\frac{F_{i-1}}{F_{sat}}\right)\right\}^{-1}.$$

(30)

Figure 9 shows the amplified output energy as a function of the round trip in a multipass amplifier. In the calculation, the Ti:sapphire crystal is assumed as an amplifier medium. The saturation fluence of the Ti:sapphire crystal is 1.2 J/cm² and the small signal gain of 3.5 is assumed. The energy exponentially increases in the first few round trips, but the energy linearly increases as the energy becomes comparable to the saturation energy of the amplifier medium. Finally, the output energy is saturated at a certain energy level which is close to the saturation fluence.

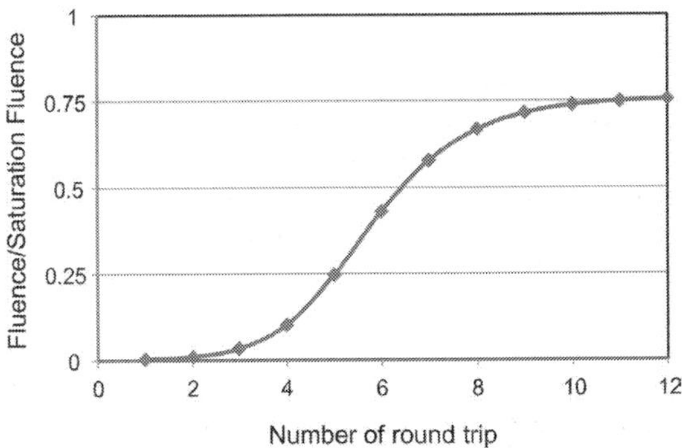

Figure 9. Fluence of the laser pulse with respect to the number of round trip. The input fluence was 1 mJ and the small signal gain was assumed to be 3.5.

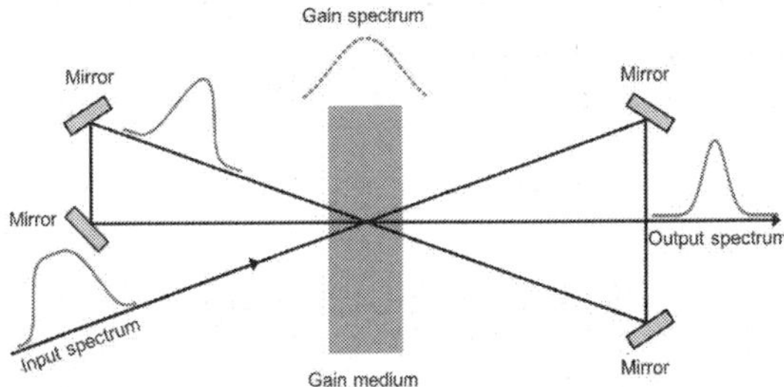

Figure 10. Diagram explaining the gain narrowing effect. The origin of the gain narrowing effect is an un-equal gain at a different wavelengths. The wavelength component at a higher gain grows faster than that at a lower gain. The gain narrowing effect broadens the pulse duration of the compressed pulse.

A series of amplifier system including a regenerative amplifier and multiple-stage amplifiers are used for energy amplification. The final output energy ranges from a couple of J to ~100 J, depending on the peak power level. The pulse energy should be amplified by a factor of ~10^{12} while keeping the pulse characteristics the same. This is not easy because of the gain narrowing effect induced by the different gains at different wavelengths. The gain narrowing phenomenon happens because a wavelength component located at a higher gain becomes stronger than a wavelength component at a lower gain as shown in **Figure 10**. The gain narrowing broadens the pulse duration of a compressed pulse. Several techniques, such as input intensity modulation, wavelength mismatch between the input and gain spectrum, gain saturation, and so on, have been developed to minimize the gain narrowing effect.

The amplified spontaneous emission (ASE) occurred in a large size gain crystal reduces an overall gain and deteriorates the spatial profile of a laser pulse as shown in **Figure 11**. A spontaneous emission traveling in the transverse direction of the gain medium has energy gain before the laser pulse arrives. When the gain and the size of gain medium are small, the ASE is negligible during the amplification process. However, as the size of gain medium is large enough with a considerable gain, the ASE becomes significant. In order to reduce the ASE, the gain medium is enclosed by the light-absorption cooling liquid having a refractive index similar to the gain medium. With the cooling liquid, the spontaneous emission transmits the boundary between the gain medium and cooling liquid, and scattered in the mount. Thus, the ASE reflected from the boundary can be suppressed. Sometimes, the spontaneous emission has enough energy gain even in a single transverse pass. In this case, a delayed pumping scheme can be useful to reduce the ASE.

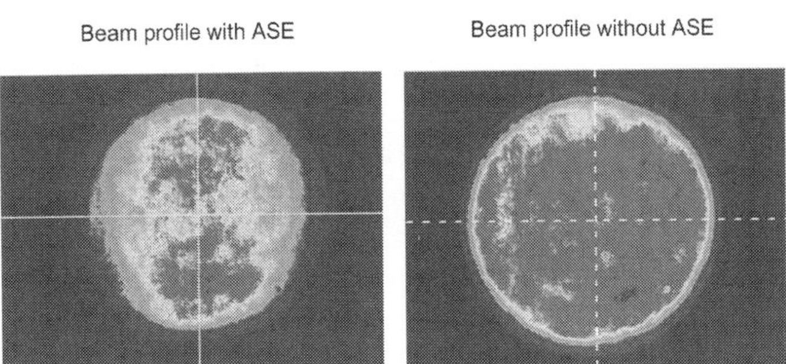

Figure 11. Laser beam profile with and without the amplified spontaneous emission (ASE). The ASE reduces energy gain and deteriorates beam profile.

Since the demonstration of laser in 1960, the laser technology has continuously advanced to build petawatt (PW) laser systems. In 1999, the first CPA PW laser has been demonstrated using a Ti:sapphire/Nd:glass hybrid system [13]. Almost a decade later, 30 fs 1 PW laser operating at 0.1 Hz repetition rate was developed [14] and more recently an amplifier for 5 PW laser system has been successfully demonstrated [15]. Now, fs and 10 PW laser systems are under construction through the European Extreme Light Infrastructure (ELI) program.

3. Focusing ultrashort laser pulses

An amplified and compressed laser pulse is focused on solid or gas target for laser-matter interaction studies. Concave mirrors are generally used and the intensity reaches at a relativistic level, $>10^{18}$ W/cm^2. The size of a focal spot is proportional to the focal length of a mirror and a shorter focal length is preferable to reach a higher intensity. Thus, one particular research interest is to tightly focus a laser pulse to reach an unprecedented intensity level. The paraxial approximation, which is commonly used in calculating focal spots under high f-number conditions, becomes invalid under tight focusing (low *f*-number) conditions. Intensities of a focal spot that have other polarization components different from an incident polarization are assumed to be negligible in the paraxial approximation. However, under tight focusing conditions, intensities at different polarizations increase and modify the overall intensity distribution of a focused laser pulse.

The intensity distributions of all polarization components of a focal spot formed under a tight focusing condition can be calculated by vector diffraction integrals developed by Stratton and Chu [16]. Recently, the intensity distributions of a focused fs high-power laser pulse under a tight focusing condition were intensively examined [17]. In this section, the intensity distributions of a tightly focused laser spot are described. The accurate assessment of the peak power and information on the intensity distribution are beneficial in simulating and predicting the motion of charged particles under a super-strong laser pulse that is provided by a tight focusing scheme.

3.1. Modeling of focusing scheme with low *f*-number parabolic mirror

The parabolic mirror is used as a focusing mirror because of its quadratic surface profile. A linearly polarized (x-polarized) laser pulse having an electric field distribution, $E_{\mathrm{inc}}(\theta_S, \phi_S)$, is incident on a parabolic mirror along the negative *z*-direction (**Figure 12**). By using Sttraton and Chu's vector diffraction integrals, the electric fields at all polarization components can be expressed as follows:

$$
\begin{aligned}
E_x\left(x_P,y_P,z_P\right) \cdot \int\!\!\!\int d\theta_s d\phi_s E_{inc}\left(\theta_s,\phi_s\right)\exp\left\{-ik\varphi\left(x_P,y_P,z_P,\theta_s,\phi_s\right)\right\}\frac{2f\sin\theta_s}{\left(1-\cos\theta_s\right)} \\
\times\left\{1-\frac{\sin\theta_s\cos\phi_s}{1-\cos\theta_s}\left(1-\frac{1-\cos\theta_s}{i2kf}\right)\frac{2f\sin\theta_s\cos\phi_s-x_P\left(1-\cos\theta_s\right)}{2f}\right\}
\end{aligned}
\tag{31-1}
$$

$$
\begin{aligned}
E_y\left(x_P,y_P,z_P\right) \sim \int_{\theta m}^{\pi}\int_0^{2\pi} d\theta_s d\phi_s E_{inc}\left(\theta_s,\phi_s\right)\exp\left\{-i\varphi\left(x_P,y_P,z_P,\theta_s,\phi_s\right)\right\}\frac{2f\sin\theta_s}{\left(1-\cos\theta_s\right)^2} \\
\times\left\{\sin\theta_s\cos\phi_s\left(1-\frac{1-\cos\theta_s}{i2kf}\right)\frac{2f\sin\theta_s\sin\phi_s-y_P\left(1-\cos\theta_s\right)}{2f}\right\}
\end{aligned}
\tag{31-2}
$$

$$
\begin{aligned}
E_z\left(x_P,y_P,z_P\right) \cdot \int\!\!\!\int d\theta_s d\phi_s E_{inc}\left(\theta_s,\phi_s\right)\exp\left\{-i\varphi\left(x_P,y_P,z_P,\theta_s,\phi_s\right)\right\}\frac{2f\sin\theta_s}{\left(1-\cos\theta_s\right)^2} \\
\times\sin\theta_s\cos\phi_s\left\{1-\left(1-\frac{1-\cos\theta_s}{i2kf}\right)\frac{2f\cos\theta_s-z_P\left(1-\cos\theta_s\right)}{2f}\right\}
\end{aligned}
\tag{31-3}
$$

and

$$
\varphi\left(x_P,y_P,z_P,\theta_s,\phi_s\right)=k\left(z_P\cos\theta_s+x_P\sin\theta_s\cos\phi_s+y_P\sin\theta_s\sin\phi_s\right).
\tag{31-4}
$$

Here, f is the focal length of a mirror. x_P, y_P, and z_P represent positions at vicinities of the focal point. θ_S is the polar angle measured from the positive z-axis and ϕ_S is the rotational angle measured from the positive x-axis. The minimum angle, θ_{min}, determines the f-number of the mirror. The distance between the source (s) and observation (p) points is expressed as $2f/(1 - \cos\theta_S)$ for the intensity of a laser spot and as $2f/(1 - \cos\theta_S) - \rho_P\{\cos\theta_S\cos\theta_P + \sin\theta_S\sin\theta_P\cos(\phi_S - \phi_P)\}$ for the phase of the spot.

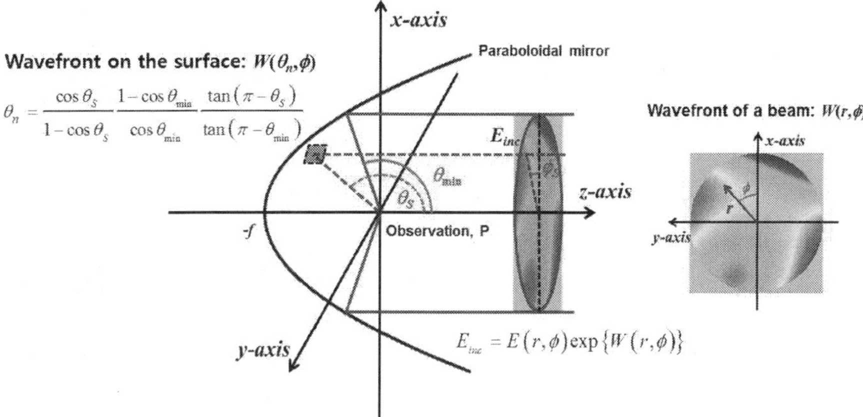

Figure 12. On-axis focusing scheme for an aberrated laser beam with a low f-number parabolic mirror.

The wavefront aberration of a laser pulse is one of the factors that determines the intensity distribution of a focal spot. The wavefront aberration is the phase delay function across the laser beam and included in the incident electromagnetic field of a laser pulse as follows:

$$E_{inc}(\theta_S,\phi_S) = E_0(\theta_S,\phi_S)\exp\{ikW_{inc}(\theta_n,\phi_S)\}. \tag{32}$$

Here, θ_n can be interpreted as a normalized radius defined by $(\pi - \theta_S)/(\pi - \theta_{min})$. The wavefront aberration is expressed by the Zernike polynomials as $W_{inc}(\theta, \phi)=\sum_{u,v} c_u^v Z_u^v(\theta_n, \phi_S)$. In this case, c_u^v means the Zernike coefficient, and $Z_u^v(\theta_n, \phi_S)$ is the Zernike polynomial for the uth radial and the vth azimuthal orders, respectively. The entire phase function on the mirror surface is modified as $k(z_P\cos\theta_S + x_P\sin\theta_S\cos\phi_S + y_P\sin\theta_S\sin\phi_S) + kW_{inc}(\theta_n, \phi_S)$. For a high f-number case, θ_S and θ_n are almost the same, and the wavefront, $W_S(\theta_n, \phi_S)$, on the mirror surface is almost equivalent to $W_{inc}(\theta_S, \phi_S)$. However, as the f-number of a parabolic mirror decreases, the wavefront on the mirror surface becomes different from the wavefront of an incident wave. In this case, the normalized radius, θ_n, on the mirror surface is modified and given by

$$\theta_n = \frac{\cos\theta_s}{1-\cos\theta_s}\,\frac{1-\cos\theta_{\min}}{\cos\theta_{\min}}\,\frac{\tan\left(\pi-\theta_s\right)}{\tan\left(\pi-\theta_{\min}\right)}. \tag{33}$$

The change in wavefront aberration due to the polarization rotation after reflection from a mirror surface should be considered for the effect of polarization. Thus, after reflection from a parabolic mirror, the normal vector to the wavefront surface is expressed by $2\hat{n}(\vec{S}\cdot\hat{n})-\vec{S}$ as shown below:

$$2\hat{n}\left(\vec{S}\cdot\hat{n}\right)-\vec{S}=\left(\hat{x}\sin\theta_s\cos\phi_s+\hat{y}\sin\theta_s\sin\phi_s-\hat{z}\cos\theta_s\right)W_s\left(\theta_n,\phi_s\right) \tag{34}$$

Expressions for normal vectors on a parabolic mirror surface which are given by

$$n_x=\frac{\sin\theta_s\cos\phi_s}{\left[2\left(1-\cos\theta_s\right)\right]^{1/2}}\,,\;\; n_z=\left(\frac{1-\cos\theta_s}{2}\right)^{1/2},\;\text{and}\;\; n_y=\frac{\sin\theta_s\sin\phi_s}{\left[2\left(1-\cos\theta_s\right)\right]^{1/2}} \tag{35}$$

are used in the calculation of Eq. (34). Finally, the wavefront component that propagates to the ρ-direction contributes to the formation of an intensity distribution near the focal plane and is given by

$$\left[2\hat{n}\left(\vec{S}\cdot\hat{n}\right)-\vec{S}\right]\cdot\vec{k}_\rho=\frac{2\pi}{\lambda}W_s\left(\theta_n,\phi_s\right)\left\{\sin^2\theta_s\cos 2\phi_s+\cos^2\theta_s\right\} \tag{36}$$

with $\hat{\rho}=\hat{x}\sin\theta_S\cos\phi_S-\hat{y}\sin\theta_S\sin\phi_S-\hat{z}\cos\theta_S$. But, as expected in Eq. (36), the contribution by the $\{\cdot\}$ term is not significant when $\sin^2\theta_S\cos 2\phi_S\ll\cos^2\theta_S$.

3.2. Coherent superposition of monochromatic fields for femtosecond focal spot

A femtosecond laser pulse typically has a broad spectrum of several tens of nm, thus the effect of broad spectrum of a femtosecond laser pulse on the focal spot should be considered in order to accurately describe the focal spot of a femtosecond laser pulse. The spatial and temporal profiles of a femtosecond focal spot can be calculated by the superposition of monochromatic electric fields near the focal point. The resultant electric fields for a femtosecond focal spot are expressed with spectral amplitude and phase as below (see **Figure 13**):

$$\begin{aligned}E_{x,y,z}\left(x_P,y_P,z_P\right)&=R_{\lambda 1}\exp\left(i\alpha_{\lambda 1}\right)E_{x,y,z}\left(\lambda_1:x_P,y_P,z_P\right)+R_{\lambda 2}\exp\left(i\alpha_{\lambda 2}\right)E_{x,y,z}\left(\lambda_2:x_P,y_P,z_P\right)+\\&\cdots+R_{\lambda n}\exp\left(i\alpha_{\lambda n}\right)E_{x,y,z}\left(\lambda_n:x_P,y_P,z_P\right)\end{aligned} \tag{37}$$

Here, R_λ defined by $\sqrt{I_\lambda/I_{\lambda,\max}}$ and α_λ are the relative amplitude and the spectral phase at a given wavelength, respectively. The subscripts (x, y, z) represent the polarization directions and $E_{x,y,z}(\lambda_n:x_P,y_P,z_P)$ induces the monochromatic electric field. Contrary to the monochromatic case, a different field oscillation period at a different wavelength induces a phase mismatch among waves at different wavelengths and reduces the intensity quickly as the observation position moves away from the origin of the focal plane. The intensity distribution along the propagation direction can be interpreted as the temporal profile of a femtosecond focal spot. Thus, the resultant electric fields, $E_{x,y,z}(x_P, y_P, z_P)$, provide the spatial and temporal (spatiotemporal) intensity distributions of a laser focal spot. The resultant electric fields are numerically calculated. In the calculation, the spectrum is sliced into n components. The monochromatic electric field distributions at all polarization components are obtained from Eqs. (31-1)–(31-4). The relative amplitude ratio and the spectral phase are obtained by the measurement of a laser pulse. After calculating the resultant electric fields, the final intensity distributions at all polarization components become

$$I_{x,y,z}\left(x_P,y_P,z_P\right) \propto \left|E_{x,y,z}\left(x_P,y_P,z_P\right)\right|^2. \tag{38}$$

This approach provides information on intensity distribution at all polarization components both in temporal and spatial domains and it is also valid under high f-number focusing conditions as well. The sum of all polarization components given by $I_x(x, y, z) + I_y(x, y, z) + I_z(x, y, z)$ is the intensity distribution measured by an image-sensing device.

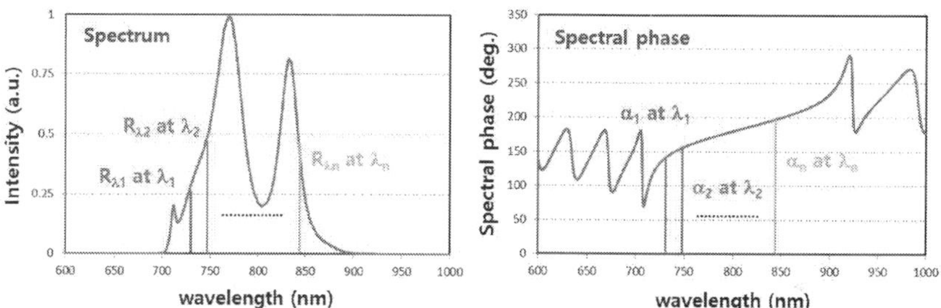

Figure 13. Spectrum and spectral phase for calculating the femtosecond focal spot.

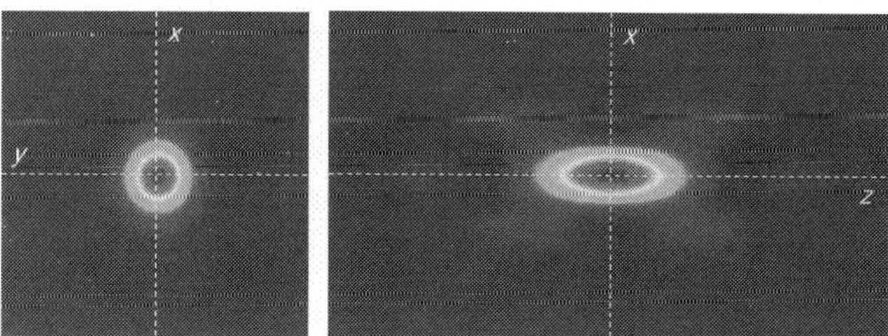

Figure 14. Three-dimensional intensity distribution of the continuous wave and spatially uniform laser beam under loose focusing condition. The x-polarized beam is assumed and the laser beam propagates along the −z direction. Under the far-field approximation, the x-polarization component is only considered to calculate the intensity distribution.

3.3. Intensity distribution in the focal plane and its vicinity

Under the loose focusing condition ($f/\# \gg 1$), the intensity distribution having the same polarization as an incoming laser pulse is only considered, and other polarization components ($I_y(x, y, z)$ and $I_z(x, y, z)$) are ignored. In this case, the far-field approximation is applied and the Fourier transform of an incoming electric field, which is derived from the scalar diffraction integral, is widely used to obtain the intensity distribution of a focal spot. **Figure 14** shows the typical intensity distributions in the x-y plane and the x-z plane.

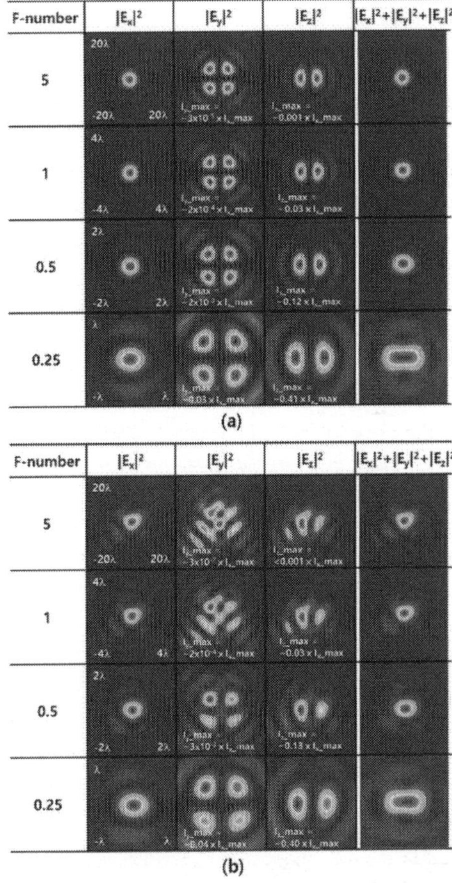

Figure 15. The change in intensity distributions as the f-number decreases. The intensity distributions for a continuous wave and uniform laser beam are calculated in the focal plane. (a) The ideal uniform laser beam profile without wavefront aberration is assumed as an input. (b) The uniform beam profile with wavefront aberration is assumed as an input.

Intensities having other polarizations different from an incoming laser pulse increase under the tight focusing condition. Typical aspects under tight focusing conditions are the increase in the intensity of a longitudinal polarization component and the elongation of a focal spot along the polarization direction. **Figure 15(a)** shows the change in the intensity distribution when the f-number of a parabolic mirror decreases from 5 to 0.25. The peak intensity of a longitudinal component, I_z, increases up to 41% of that of I_x under $f/0.25$ condition. But, compared to the x-polarization component, the peak intensity of the y-polarized component, I_y, is still negligible. Because of the increase of intensity in the longitudinal component and the deformation of x-polarized intensity, the resultant intensity is elongated in the polarization direction as shown in **Figure 15(a)**.

Figure 15(b) shows the change of a focal spot for an aberrated laser pulse as the f-number decreases. A small amount of wavefront aberration ($c_2^{-2} = 0.07$ μm, $c_3^{-3} = 0.05$ μm, $c_3^{-1} = 0.04$ μm, and $c_3^1 = 0.02$ μm) was introduced to the laser pulse to investigate the effect of wavefront aberration on the focal spot. The figure shows how the focal spot of an aberrated laser pulse is influenced by the focusing condition. Under a high f-number condition ($f/5$), the focal spot of an aberrated laser pulse is determined by the spatial characteristics of the laser pulse, such as wavefront aberration and spatial profile. The shape of the focal spot was almost same as that obtained with the Fourier transform method because the focusing condition and the amount of wavefront aberration did not violate the far-field and thin-lens approximations. Instead, under lower f-number conditions, focal spots are also influenced by the vectorial properties, resulting in the elongation along the polarization direction. With a given amount of wavefront aberration, the peak intensity of a longitudinal component, I_z, increases up to 40% of that of I_x under $f/0.25$ condition. Further calculation with a higher amount of wavefront aberration ($c_2^2 = c_2^{-2} = c_3^{-1} = 0.15$ μm) shows that intensity distribution under $f/0.5$ focusing condition was still different from the intensity distribution obtained with the Fourier transform method.

Figure 16 shows spatiotemporal intensity distributions of femtosecond focal spots for an aberrated laser pulse under two different focusing conditions ($f/3$ and $f/0.5$). The Zernike coefficient that are used again include $c_2^{-2} = 0.07$ μm, $c_3^{-3} = 0.05$ μm, $c_3^{-1} = 0.04$ μm, and $c_3^1 = 0.02$ μm. In the figure, the intensity distributions in the x-y plane provide information on spatial profiles of a femtosecond focal spot, and the intensity distributions in the x-z plane provide information on temporal profiles. By assuming a 12 fs, 10 PW, uniformly circular, and aberrated laser pulse as an input, peak intensities for x-polarized component increases up to ~8.8 × 10²² W/cm² for $f/3$ and ~2.5 × 10²⁴ W/cm² for $f/0.5$, respectively. Under same conditions, peak intensities for longitudinal component rapidly increase to ~3.1 × 10²⁰ W/cm² and ~2.4 × 10²³ W/cm². These intensities along the z-direction should be taken into account to better describe the motion of charged particles under an extremely strong EM field that is formed by tightly focusing a femtosecond high-power laser pulse.

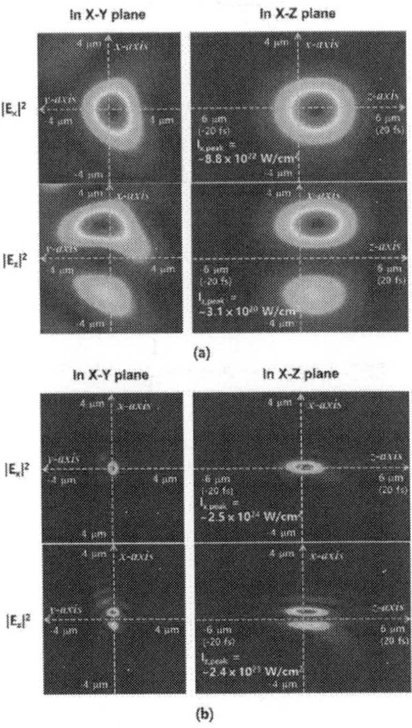

Figure 16. The spatiotemporal intensity distributions of a focal spot with (a) an $f/3$ parabolic mirror and (b) an $f/0.5$ parabolic mirror. The peak intensities of I_x reach ~8.8 × 10^{22} W/cm^2 and ~2.5 × 10^{24} W/cm^2 for $f/3$ and $f/0.5$ focusing conditions, respectively. The transverse intensity distribution is expressed in the x-y plane and the longitudinal intensity distribution is expressed in the x-z plane.

4. Interaction of an intense laser pulse with plasma

Under a strong electromagnetic field, the motion of an electron is governed by the Lorentz force as follows:

$$\frac{d\left(\gamma m_0 \vec{v}\right)}{dt} = -e\vec{E} - e\left(\frac{\vec{v}}{c} \times \vec{B}\right). \tag{39}$$

Here, m_0 is the electron rest mass, c is the speed of light, and γ is the Lorentz factor. When the electromagnetic field is not strong enough, the $\frac{\vec{v}}{c} \times \vec{B}$ term on the right-hand side is much less than the first term on the right-hand side and negligible. In this case, the Lorentz force is reduced to $d(m\vec{v})/dt = -e\vec{E}$. By assuming the sine wave for the electric field and replacing the time derivative by $-i\omega$, then the speed of an electron is

$$v = \frac{e}{m\omega} E(t).$$

(40)

The maximum speed of an electron is given by $v_{max} = eE_0/m\omega$. By comparing the maximum speed of an electron and the speed of light, we define $\beta = v/c$. In the nonrelativisitic approach, we can consider $\beta = 1$ as a reference. Then, the intensity required for an electron to have the speed of light c is given by

$$I = \frac{\varepsilon_0 c}{2}\left(\frac{2\pi mc^2}{\lambda e}\right)^2.$$

(41)

The intensity for the speed of light is ~ 2.14×10^{18} W/cm^2 for the 0.8 μm wavelength. In the nonrelativistic approach, the intensity of 10^{18} W/cm^2 is roughly estimated for electrons to have a quiver motion in which the speed is close to the speed of light. The intensity of 10^{18} W/cm^2 is known as the relativistic intensity for the electromagnetic field.

As shown in the previous section, the relativistic intensity is easily obtained by focusing a femtosecond high-power laser pulse. The femtosecond focal spot has a finite extent in the temporal and spatial domains. Let us expand the electric field of a high-power laser pulse in the Taylor series at a position of x_0, then we obtain

$$E_x(r) \simeq E_x(r)\big|_{r=x0} \cos(kz - \omega t) + x \frac{\partial E_x(r)}{\partial x}\bigg|_{r=x0} \cos(kz - \omega t) + \cdots$$

(42)

By inserting the first term on the right-hand side in Eq. (42) into the first term in Eq. (39) and solving the equation, the velocity and the position of electron are given by

$$v_x = -\frac{eE_x(r)\big|_{r=x0}}{m\omega}\sin(kz - \omega t) \text{ and } x = -\frac{eE_x(r)\big|_{r=x0}}{m\omega^2}\cos(kz - \omega t).$$

(43)

In order to see the effect of the intensity (or field) gradient of a focused intensity, let us put the expression of x in Eq. (43) into Eq. (42) and consider the Lorentz force again. Then, we obtain

$$m\frac{d}{dt}v_x = -\frac{e^2}{2m\omega^2}\frac{\partial E_x^2(r)}{\partial x}\bigg|_{r=x0}\cos^2(kz - \omega t).$$

(44)

By taking the cycle average of the force, Eq. (44) becomes

$$m\frac{d}{dt}v_x = -\frac{e^2}{4m\omega^2}\frac{\partial E_x^2(r)}{\partial x}\bigg|_{r=x0}.$$

(45)

Eq. (45) means that an electron can be pushed by the intensity or the field gradient. The force due to the intensity gradient is known as the ponderomotive force.

When a high-power laser pulse is focused in a gas target, the target immediately turns into the plasma medium. Electrons in the plasma medium feel the ponderomotive force by the laser pulse in temporal and spatial domains, and those are pushed by a focused laser field and separated from the background ions. The separation of electrons from background ions induces a strong electric field by the space charge effect. The periodic motion of oscillation for electrons occurs around heavy ions as the laser pulse propagates. The resultant pattern of alternating positive and negative charges is known as the plasma wave or laser wake. The laser wake field supports a very strong longitudinal electric field of 1 GeV/cm. Some of returning electrons can be captured into the laser wake and accelerated by the laser wake field up to GeV level. This is a short description of the laser wake field acceleration [18] (**Figure 17(a)**). Recent experiments using the laser wake field acceleration showed the quasimonoenergetic multi-GeV electron beams by focusing petawatt laser pulses [19–21]. The acceleration of electrons to 10 GeV or even 100 GeV level is now being pursued for the development of a compact electron accelerator.

Protons are also accelerated by a high-power laser pulse. In this case, a high-power laser pulse is focused onto a solid target. When a high-power laser pulse is focused on a thin metal target, the target immediately turns into plasma, and electrons in the plasma are accelerated toward the laser beam propagation direction by the ponderomotive force. Then there exists an electric field between accelerated electrons and background ions. The electric field can be used to accelerate protons existing in the metal as impurities [22, 23] (**Figure 17(b)**). At a lower laser intensity, the energy distribution for electrons is broad and the resultant proton energy distribution is also broad. As the laser intensity increases, proton energy distribution can be reduced by a narrow electron energy distribution by the radiation pressure. This is an indirect proton acceleration using electron acceleration. Protons can be directly accelerated to the light speed by an electromagnetic field as shown in Eq. (45). However, because of the proton mass, reaching to the speed of light by directly accelerating proton with an electromagnetic field requires a higher laser intensity up to $\sim 10^{24}$ W/cm^2, which is sometimes called the ultrarelativistic laser intensity. One of the ways for efficiently reaching at the ultrarelativistic laser intensity is to employ a tight focusing scheme. Based on the recent progress in the high-power laser, the demonstration of ultrarelativistic laser intensity will be possible soon.

Energetic charged particles driven by high-power laser pulses are directly used for medical applications including radiation therapy and imaging. For example, energetic proton beams having an energy range of 100–200 MeV can be used for the radiation tumor therapy [24]. When proton beams is irradiated to tumors in human body, protons dramatically lose their energy and produce x-rays in the tumor. The produced x-ray destroys DNA chains in a tumor cell and eventually kills the tumor cell. Electron beams with an energy range of 6–20 MeV can also be used for treating cancers locating at skin and lip, chest-wall and neck, respiratory and digestive-track lesions, or lymph nodes [25]. Research on stable and reliable production of energetic particles is of great interest for developing a compact particle accelerator for medical applications.

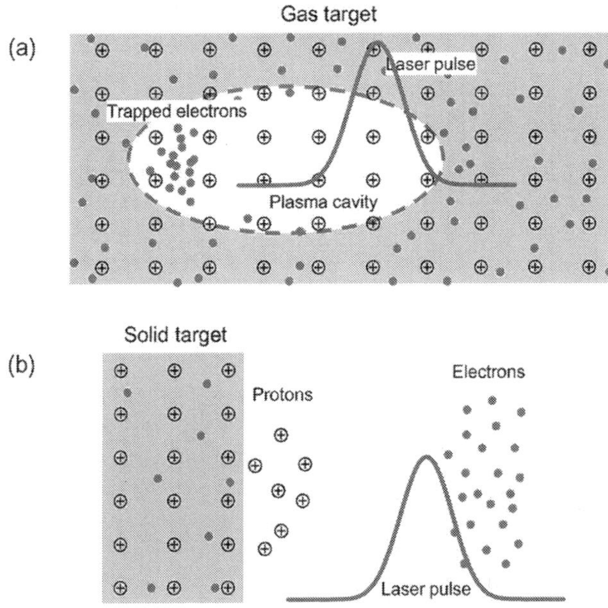

Figure 17. (a) Electron acceleration though the laser wake field. The laser pulse is focused onto the gas target. The electrons are captured in the plasma cavity and accelerated by the cavity. (b). Proton acceleration. The accelerated electrons by the laser pulse pull proton on the metal surface.

High-brightness and high-energy photons (x-ray and γ-ray) can be produced through the laser-plasma accelerator as well. By comparing to the large-scale facilities, such as synchrotron and XFEL, the laser-plasma accelerator produces high-energy photon providing an attosecond temporal resolution and subatomic spatial resolution in a small size and reasonable cost. High-energy photon can be used for research pertaining to ultrafast electron dynamics in atoms, molecules, plasmas, and solids. In the laser-plasma accelerator, many processes producing energetic photon sources, such as high harmonic generation [26], undulator radiation [27], betatron radiation [28], and Compton scattering [29], were proposed and some of them have been experimentally demonstrated. So far, basic applications for high-intensity laser pulses were described. Other interesting research topics related to fundamental physical processes are well described elsewhere [30].

5. Conclusion

The high-power laser facility is being developed for performing research on the laser-matter interaction in the relativistic and ultrarelativistic intensity regimes. The high-power laser pulse immediately ionizes solid and gas targets and makes the target medium plasma. The intensity can make use of the laser pulse as a small-scale and versatile particle accelerator. This is a primary purpose for developing high-intensity laser facilities. The interaction between an intense laser pulse and energetic charged particles produces high-energy photon as well. Many interesting schemes, such as undulator radiation, betatron radiation, and inverse Compton scattering, have been studied for producing high-energy photons. The high-energy photons can be used in many disciplines including industrial application, medical imaging, nuclear engineering, national security, and so on. As the intensity obtainable with the high-power laser pulse increases over 10^{24} W/cm^2, some of the fundamental physical processes can be investigated by light pulses with an ultrashort time scale. Since the invention of laser, the application field of laser has been dramatically expanded as the laser intensity increases. Now, the acceleration of charged particle by intense coherent light field became possible in the relativistic laser intensity regime, and new era for studying the laser-plasma interaction in the ultrarelativistic laser intensity regime will be open soon.

Author details

Tae Moon Jeong and Jongmin Lee*

*Address all correspondence to: leejm@gist.ac.kr

Handong Global University, Pohang, Republic of Korea

References

[1] L. E. Hargrove, R. L. Fork, and M. A. Pollack. Locking of He-Ne laser modes induced by synchronous intracavity modulation. Appl. Phys. Lett. 1964;5(1):4–5.

[2] C. V. Shank and E. P. Ippen. Subpicosecond kilowatt pulses from a mode-locked cw dye laser. Appl. Phys. Lett. 1974;24(8):373–375.

[3] U. Morgner, F. X. Kartner, S. H. Cho, Y. Chen, H. A. Haus, J. G. Fujimoto, E. P. Ippen, V. Scheuer, G. Angelow, and T. Tschudi. Sub-two-cycle pulses from a Kerr-lens mode-locked Ti:sapphire laser. Opt. Lett. 1999;24:411–413.

[4] R. Paschotta and U. Keller. Passive mode locking with slow saturable absorbers. Appl. Phys. B. 2001;73:653–662. DOI: 10.1007/s003400100726

[5] O. Svelto. Principles of Lasers. 4th ed. New York: Springer; 1998. 343 pp.

[6] H. A. Haus, J. G. Fujimoto, and E. P. Ippen. Structures for additive pulse mode locking. J. Opt. Soc. Am. B. 1991;8(10):2068–2076. DOI: 10.1364/JOSAB.8.002068

[7] D. E. Spence, D. E. Kean, and W. Sibbet. 60-fsec pulse generation from a self-mode-locked Ti:sapphire laser. Opt. Lett. 1991;16:42–44.

[8]   D. Strickland and G. Mourou. Compression of amplified chirped optical pulses. Opt. Commun. 1985;6:219–221.

[9]   T. M. Jeong and J. Lee. Femtosecond petawatt laser. Ann. Phys. (Berlin). 2014;526(3–4): 157–172. DOI: 10.1002/andp.201300192

[10] E. Treacy. Optical pulse compression with diffraction gratings. IEEE J. Quantum Electron. 1969;5(9):454–458. DOI: 10.1109/JQE.1969.1076303

[11] O. Martinez. 3000 times grating compressor with positive group velocity dispersion: application to fiber compensation in 1.3–1.6 μm region. IEEE J. Quantum Electron. 1987;23(1):59–64. DOI: 10.1109/JQE.1987.1073201

[12] L. M. Frantz and J. S. Nodvik. Theory of pulse propagation in a laser amplifier. J. Appl. Phys. 1963;34(8):2346–2349.

[13] M. D. Perry, D. Pennington, B. C. Stuart, G. Tietboh, J. A. Britten, C. Brown, S. Herman, B. Golick, M. Kartz, J. Miller, H. T. Powell, M. Vergino, and V. Yanovsky. Petawatt laser pulses. Opt. Lett. 1999; 24(3):160–162.

[14] J. H. Sung, S. K. Lee, T. J. Yu, T. M. Jeong, and J. Lee. 0.1 Hz 1.0 PW Ti:sapphire laser. Opt. Lett. 2010;35(18):3021–3023.

[15] Y. Chu, Z. Gan, X. Liang, L. Yu, X. Lu, C. Wang, X. Wang, L. Xu, H. Lu, D. Yin, Y. Leng, R. Li, and Z. Xu. High-energy large-aperture Ti:sapphire amplifier for 5 PW laser pulses. Opt. Lett. 2010;40(21):5011–5014.

[16] J. Stratton and L. Chu. Diffraction theory of electromagnetic waves. Phys. Rev. 1939;56(1):99–107.

[17] T. M. Jeong, S. Weber, B. Le Garrec, D. Margarone, T. Mocek, and G. Korn. Spatio-temporal modification of femtosecond focal spot under tight focusing condition. Opt. Express. 2015;23(9):11641–11656 . DOI: 10.1364/OE.23.011641

[18] T. Tajima and J. M. Dawson. Laser electron accelerator. Phys. Rev. Lett. 1979;43(4):267–270.

[19] N. M. Hafz, T. M. Jeong, I. W. Choi, S. K. Lee, K. H., Pae, V. V. Kulagin, J. H. Sung, T. J. Yu, K.-H. Hong, T. Hosokai, J. R. Cary, D.-K. Ko, and J. Lee. Stable generation of GeV-class electron beams from self-guided laser–plasma channels. Nat. Photon. 2008;2:571–577.

[20] H. T. Kim, K. H. Pae, H. J. Cha, I. J. Kim, T. J. Yu, J. H. Sung, S. K. Lee, T. M. Jeong, and J. Lee. Enhancement of electron energy to the multi-GeV regime by a dual-stage laser-Wakefield accelerator pumped by petawatt laser pulses. Phys. Rev. Lett. 2013;111(16): 165002.

[21] W. P. Leemans, A. J. Gonsalves, H.-S. Mao, K. Nakamura, C. Benedetti, C. B. Schroeder, C. Tóth, J. Daniels, D. E. Mittelberger, S. S. Bulanov, J.-L. Vay, C. G. R. Geddes, and E.

Esarey. Multi-GeV Electron beams from capillary-discharge-guided subpetawatt laser pulses in the self-trapping regime. Phys. Rev. Lett. 2014;113(24):245002.

[22] S. P. Hatchett, C. G. Brown, T. E. Cowan, E. A. Henry, J. S. Johnson, M. H. Key, J. A. Koch, A. B. Langdon, B. F. Lasinski, R. W. Lee, A. J. Mackinnon, D. M. Pennington, M. D. Perry, T. W. Phillips, M. Roth, T. C. Sangster, M. S. Singh, R. A. Snavely, M. A. Stoyer, S. C. Wilks, and K. Yasuike. Electron, photon, and ions beams from the relativistic interaction of Petawatt laser pulses with solid targets. Phys. Plasma. 2000;7(5):2076–2082.

[23] A. Macchi, F. Cattani, T. V. Liseykina, and F. Cornolti. Laser acceleration of ion bunches at the front surface of overdense plasmas. Phys. Rev. Lett. 2005;94:165003.

[24] S. V. Bulanov and V. S. Khoroshkov. Feasibility of using laser ion accelerators in proton therapy. Plasma Phys. Rep. 2002;28:453–456.

[25] K. R. Hogstrom and P. R. Almond. Review of electron beam therapy physics. Phys. Med. Biol. 2006;51:R455–R489. DOI: 10.1088/0031-9155/51/13/R25

[26] I. J. Kim, K. H. Pae, C. M. Kim, H. T. Kim, H. Yun, S. J. Yun, J. H. Sung, S. K. Lee, J. W. Yoon, T. J. Yu, T. M. Jeong, C. H. Nam, and J. Lee. Relativistic frequency upshift to the extreme ultraviolet regime using self-induced oscillatory flying mirrors. Nat. Commun. 2012;3:1231. DOI: 10.1038/ncomms2245

[27] M. Fuchs, R. Weingartner, A. Popp, Z. Major, S. Becker, J. Osterhoff, I. Cortrie, B. Zeitler, R. Hörlein, G. D. Tsakiris, U. Schramm, T. P. Rowlands-Rees, S. M. Hooker, D. Habs, F. Krausz, S. Karsch, and F. Grüner. Laser-driven soft-X-ray undulator source. Nat. Phys. 2009;5:826–829. DOI: 10.1038/NPHYS1404

[28] A. Rousse, K. Ta Phuoc, R. Shah, A. Pukhov, E. Lefebvre, V. Malka, S. Kiselev, F. Burgy, J.-P. Rousseau, D. Umstadter, and D. Hulin. Production of a keV X-ray beam from synchrotron radiation in relativistic laser-plasma interaction. Phys. Rev. Lett. 2004;93:135005.

[29] N. D. Powers, I. Ghebregziabher, G. Golovin, C. Liu, S. Chen, S. Banerjee, J. Zhang, and D. P. Umstadter. Quasi-monoenergetic and tunable X-rays from a laser-driven Compton light source. Nat. Photon. 2014;8:28–31. DOI: 10.1038/NPHOTON.2013.314

[30] G. Mourou, T. Tajima, and S. Bulanov. Optics in the relativistic regime. Rev. Mod. Phys. 2006;78:309–371.

Chapter 2

High-Power Diode-Pumped Short Pulse Lasers Based on Yb:KGW Crystals for Industrial Applications

Guang-Hoon Kim, Juhee Yang, Byunghak Lee,
Bosu Jeong, Sergey Chizhov, Elena Sall,
Vladimir Yashin and Uk Kang

Additional information is available at the end of the chapter

http://dx.doi.org/10.5772/64571

Abstract

A diode-pumped, ultrafast Yb:KYW laser system utilizing chirped-pulse amplification (CPA) in a dual-slab regenerative amplifier (RA) with spectral shaping of seeding pulse from a master oscillator (MO) has been developed. A train of compressed pulses with pulse length of 181 fs, repetition rate up to 500 kHz, and average power exceeding 15 W after compression and pulse picker was achieved.

Keywords: lasers, diode-pumped, laser amplifiers, solid-state lasers, ytterbium lasers, ultrafast lasers, mode-locked lasers, ultrafast lasers

1. Introduction

Energetic (dozens of μJ) optical pulses with femtosecond (fs) pulse lengths and hence ultrahigh focused intensities in the range of 10^{12}–10^{16} Wcm^{-2} are capable of ablating a wide range of materials including metals, semiconductors, ceramics, polymers, biological tissue, and dielectrics [1–3]. Femtosecond laser ablation has been demonstrated to be a powerful tool for various technologies [2]. Due to rapid energy delivery, the laser-plasma interaction is avoided and heat-affected zones in the irradiated targets are strongly localized with minimal residual damage. This allows generation of well-defined microstructures with high quality and reproducibility [1–3].

INTECH
open science | open minds

Femtosecond laser sources based on Yb-doped laser materials became promising tool for various technological and industrial applications including femtosecond lasers. Among Yb-media ytterbium tungstates (Yb:KYW/Yb:KGW), crystals Yb:KY(WO$_4$)$_2$ (Yb:KYW) or Yb:KGd(WO$_4$)$_2$ (Yb:KGW) exhibits an attractive set of parameters that makes it as one of the best choices for high-power fs lasers operating around 1 μm [4]. Bandwidth of Yb:KGW/Yb:KYW is sufficient for amplification of sub-200 fs pulses, but typical pulse length on the output of Yb:KYW amplifier system is limited on the level 300–400 fs [5, 6] primarily due to gain narrowing [7, 8]. A promising method to reduce the effect of gain narrowing and to increase the effective gain bandwidth is to combine laser media with separated gain maxima and to overlap broadband gain. Using Yb:KYW crystals with different orientation of crystallographic axes, e.g. N_g- and N_p-cut orientation, this approach has been realized in [9]. Another way for increasing gain bandwidth is using special spectral filters introducing controlled losses at maximum gain spectrum [7].

In this review chapter we present combination of those approaches to a double-slab regenerative amplifier (RA). Each slab is pumped separately, which enables additional possibility to control gain. Furthermore as a seed source, we used high-power master oscillator (MO) based on N_p-cut Yb:KYW crystal with output pulse length ~100 fs and central wavelength agreed well with spectral gain profile of regenerative amplifier. A highly efficient stretcher and compressor based on single transmitted diffraction grating are used for stretching and recompressing initial pulses after master oscillator.

2. Design of high-average power Yb:KGW laser system

The laser was realized as a chirped-pulse amplification (CPA) system [10]. The system shown in **Figure 1** consists of a femtosecond master oscillator (MO) based on Yb:KGW crystal, a

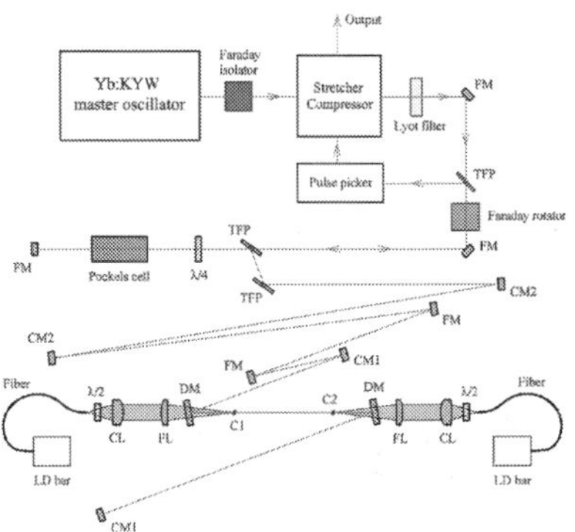

Figure 1. Schematic layout of the femtosecond laser system [15]. FM is a high reflective flat mirror; CM1 is a curved mirror with ROC = 400 mm; CM2 is a curved mirror with ROC = 600 mm; DM is a flat dichroic mirror; FL is a focusing lens; CL is a collimating lens; C1 and C2 are Yb:KYW crystals; TFP is a thin film polarizer; λ/4 is a quarter wave plate; λ/2 is a half wave plate.

common module of stretcher and compressor based on single diffraction grating, a spectral shaper based on Lyot filter, and a dual-slab regenerative amplifier (RA) with combined gain spectra. The system also includes two Faraday isolators: one for isolation of MO against leaking amplified pulses and the other for extraction of laser pulses from the system after RA. To increase the contrast ratio of output pulses according to pre-pulses and post-pulses, we used a pulse picker based on a second Pockels cell and a few thin film polarizers (TFP) with high quality.

2.1. Mode-locked Yb:KYW oscillator

A schematic design of the Yb:KYW oscillator is shown in **Figure 2**. As a gain material, we used a 3×3×3 mm at 5% Yb^{3+}-doped KGW crystal with antireflection-coated ends both for the pump radiation and for the lasing radiation. The crystal was used in an N_g-cut (N_p-cut) geometry, such that the laser polarization was aligned along the N_p axis (N_m axis), respectively. The pump beam from the 100 μm core diameter fiber output was collimated by an achromatic doublet (CL) with focal length $F = 60$ mm. Then the beam was focused into the laser crystal (LC) by another achromatic doublet (FL) with focal length $F = 60$ mm through special dichroic coated mirror (DM1) with transmittance of 95%. By using three chirped mirrors with group-velocity dispersion -1350 fs^2 per single pass, the dispersion in the laser cavity was compensated. The details of the optical layout of the laser are described in Ref. [11]. The concave mirror in the second arm of the cavity was replaced by dichroic coated mirror (DM2) with radius of curvature $R = 100$ mm and a dumper (d) is mounted additionally. Therefore, the pump beam can transmit through the dichroic mirror DM2 and laser operation can be more stabilized without heating optical mounts of mirror M3.

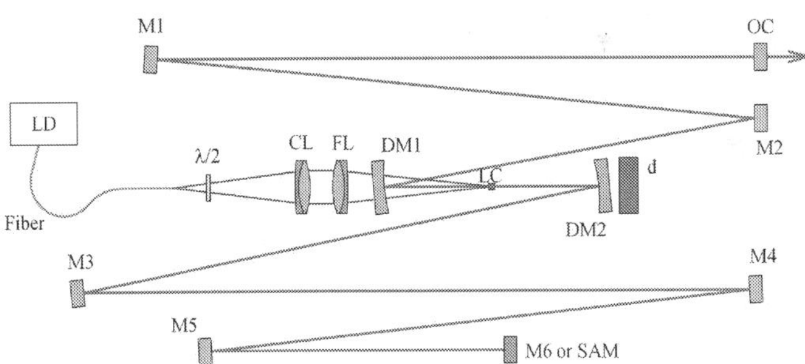

Figure 2. Schematic diagram of the Yb:KYW laser pumped by laser diode (LD) [25]. λ/2 – half wave plate; CL – colli-mating lens; FL – focusing lens; LC – Yb:KYW laser crystal; DM1, DM2 – spherical dichroic mirror; M1–M6 – cavity mirrors; OC – output coupler; SAM – semiconductor saturable-absorber mirror; d – dumper for blocking and cooling.

As is well known, the transverse size of the pump zone should match with the size of the cavity mode to achieve maximum output power. **Figure 3(a)** shows the schematic drawing to assist in understanding the mode matching between the cavity mode and pump. The pump beam is focused onto the laser crystal as shown in **Figure 3(a)** with black lines and the cavity mode is depicted as pink color. The position of laser crystal can be defined by the displacement x from the center of the caustic of the beam in the cavity. When the crystal is located in the caustic center of the cavity mode ($x = 0$ mm), the pump size is larger than the size of cavity mode. Optimal matching of cavity mode and pumped zone in the active media is important for an efficient laser operation. But it is not easy to match the pump size to cavity mode by changing the specification of pump module, because the conventional optics such as fiber and lens are limited in core size and focal length. Instead, mode matching can be easily achieved by displacing the crystal longitudinally along the optical axis. The comparison of the beam profile of pump and cavity mode for both cases of displacement $x = 0$ mm and 3.5 mm clearly shows this phenomenon as shown in **Figure 3(c)** and **(d)**. The pump profile was obtained by using ZEMAX optical system design software. The cavity mode is assumed to have Gaussian profile and the relevant factors are obtained from LASCAD simulation [12].

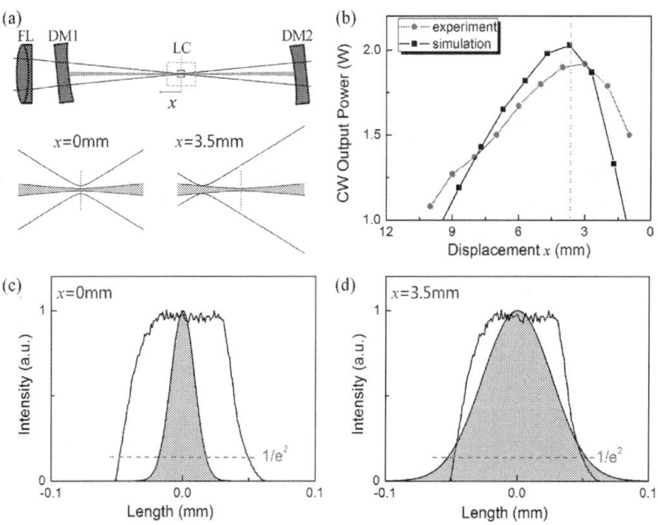

Figure 3. (a) Schematic diagram for the definition of displacement x from the center of the caustic of the beam in the cavity and for the understanding of mode matching. (b) CW output power versus position of laser crystal in the cavity for N_g-cut crystal with OC = 4% of experiment (red) and simulation (black). Comparison of beam profile between pump and cavity mode is shown in the condition of (c) $x = 0$ mm and (d) $x = 3.5$ mm [26].

To optimize the position of the crystal, we calculated the CW output power of the resonator as a function of the displacement x as depicted in **Figure 3(b)** with black dots and line by using the LASCAD software. In LASCAD simulation we approximate the top-hat pump spot in focal plane of FL as super-Gaussian distribution in xy plane with diameter 100 μm ($1/e^2$ level). Calculated cavity mode diameter between two curved mirrors DM1 and DM2 significantly changes from diameter 36 μm in a waist to 1.5 mm on the mirror. **Figure 3(b)** shows that the optimal position of the crystal is ~3.5 mm apart from the center of the beam caustic. In order to check the mode matching simulation, we performed the experiment with N_g-cut crystal oscillator, which was developed previously and in a same condition with simulation [11]. As shown in **Figure 3(b)** with red dots and line, the CW output power is highly dependent on the position of laser crystal and the experimental result coincides well with simulation result. As a note, multimode operation occurred in experiment when the size of the pump beam is larger than TEM_{00} cavity mode and it causes higher CW output power in the displacement region below 3.5 mm.

Not only the position of crystal in the cavity but also the amplification axis of crystal affects the laser operation greatly, because the absorption and emission cross sections of the crystal are quite different according to the orientation of the crystal. In previous work [11], we chose a lasing with $E//N_p$ for antireflection-coated slab crystal because of its broader emission spectrum [6]. We obtained the generator's average power exceeding 1 W at the central wavelength of 1043 nm with a pulse about 90 fs wide. But, according to the literature [13], laser can operate in the shorter wavelength range than 1043 nm by using the different amplification axis of the crystal such as $E//N_m$. For providing effective amplification as a seeding source of CPA femtosecond laser system, the oscillator with shorter wavelength such as 1030 nm is more suitable than the one with 1043 nm [14, 15].

Before going into experiments, we compare the CW output power of the oscillator between N_p- and N_m-polarization with respect to the transmission of output coupler with LASCAD simulation. The corresponding output powers for N_p- and N_m-polarization crystals are shown in **Figure 4(a)** as black and red dots and lines, respectively. From the simulated results, we can conclude that N_m-polarized crystal with 6% output coupler can be used as a candidate for the purpose of seeding source, since only the center wavelength would change with similar output power when we replace the crystal from N_p-polarization (N_g-cut) to N_m-polarization (N_p-cut). And the transmission of 6% of output coupler is chosen due to the output power saturation near 6%.

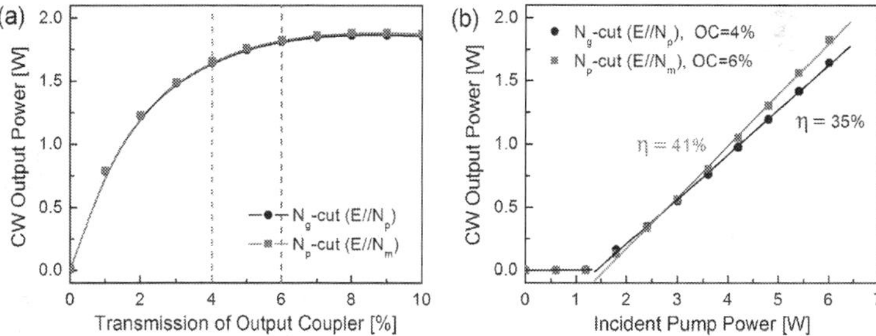

Figure 4. (a) CW output power versus transmission of output coupler for both the N_p- and N_m-polarized crystal. (b) CW output power versus incident pump power for N_p- (OC = 4%) and N_m-polarized crystal (OC = 6%) [26].

Figure 4(b) shows the simulated results according to the replacement from N_p-polarized crystal with 4% output coupler to N_m-polarized crystal with 6% output coupler, in point of the dependence of the average output power on the incident pump power. The slope efficiency with respect to the incident radiation power is 41% and 35% for each condition. From the simulation results, it is expected to get high output power and high slope efficiency with more suitable center wavelength in the new oscillator.

Based on the simulation results, we conducted the experiment for femtosecond laser based on N_m-polarization Yb:KYW with 6% output mirror and compared the result with the previously reported oscillator based on N_p-polarization Yb:KYW with 4% output mirror in Ref. [11]. An experimental setup of the laser is presented in **Figure 2** as mentioned above. An 8 W high-brightness InGaAs semiconductor injection laser module coupled into optical fiber with a 105 μm core diameter (0.11 NA) and 10 cm length was employed for optical pumping. By using the thermoelectric element (a Peltier cell), the temperature of the laser diode was adjusted for displacing its wavelength to the absorption maximum in the laser crystal. The crystal had a slab design to facilitate efficient heat removal and placed on the optimized position, 3.5 mm apart from the center of the caustic of the beam in the cavity in the longitudinal direction, for size matching of the pump beam and the cavity modes as mentioned. All the laser elements were placed in a compact one-body housing case with 424×214×84-mm size for stable laser operation.

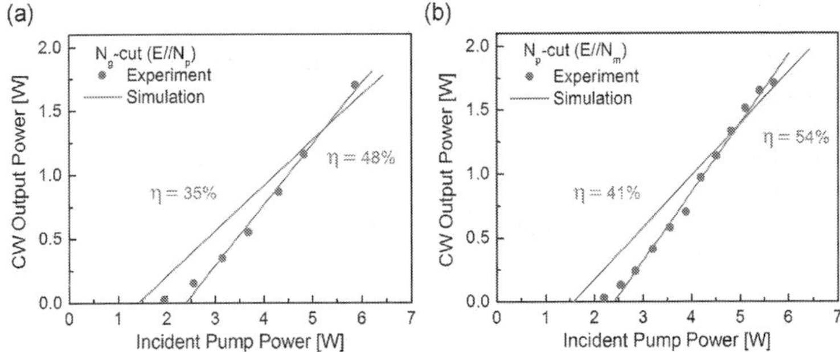

Figure 5. Experimentally observed CW output power versus incident pump power for (a) N_p-polarization (OC = 4%) and (b) N_m-polarization (OC = 6%). Gray lines indicate the calculated values from the LASCAD simulation for comparing with the experimental data [26].

The performance of the laser in the CW regime is shown in **Figure 5** for two cases of (a) the N_p-polarization crystal with 4% output coupler and (b) the N_m-polarization crystal with 6% output coupler. The high slope efficiency of 54% with respect to the incident pump power was achieved with N_m-polarization and the maximum output power exceeded 1.5 W at incident pump power of 6 W. For comparison, simulated CW output powers using LASCAD are also shown in **Figure 5** as grey lines. Computational simulations coincide well with experimental results in general, especially on high incident pump power. The discrepancy in CW output power between experiment and simulation for the low incident pump power can be explained by the characteristics of laser diodes [16]. The center wavelength of diode lasers typically depends on the injection current, operating temperature, and composition. In our case, the operating temperature of the laser diode was adjusted to 36.5°C in high injection current, corresponding to around 5 W incident pump power in **Figure 5(a)** and **(b)**, for the maximum absorption in the laser crystal. As the injected electric current of laser diode decreases, the pump light shifts toward shorter wavelengths, which results in the decreasing of pump absorption efficiency and low output power of oscillator.

We shall note that for quasi three-level laser media it is difficult to estimate the absorbed pump power experimentally because of the effect of absorption saturation appearing in such media. Therefore we considered the power of launching pump radiation for an estimation of laser efficiency instead of the absorbed power.

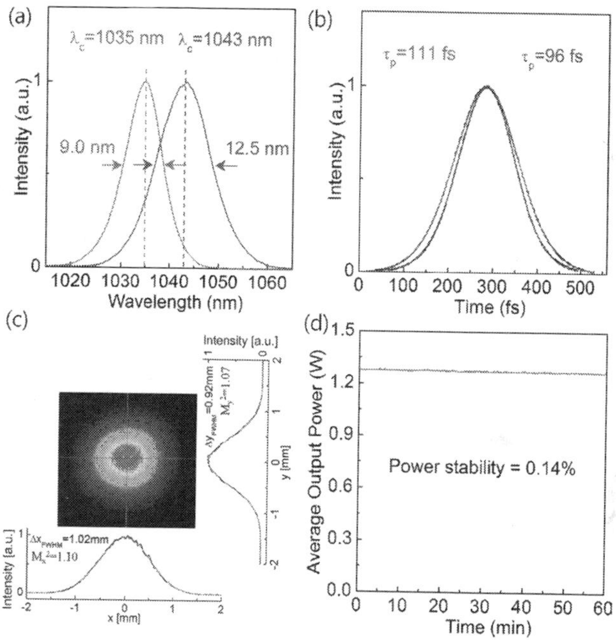

Figure 6. (a) Optical spectra and (b) autocorrelation traces of the 96 fs pulses with center wavelength 1043 nm for N_p-polarization (blue lines) and the 111 fs pulses with center wavelength 1035 nm for N_m-polarization (red lines). (c) Typical spatial pattern of the beam and the corresponding intensity profiles. (d) Measured average output power and its temporal stability [26].

The passively mode-locked regime was realized by replacing a cavity mirror M6 to semiconductor saturable-absorber mirror (SAM) at one end of the resonator in the focal region of a concave mirror (M5) with a radius of curvature R = 300 mm. In **Figure 6(a)** and **(b)**, the measured emission spectra and autocorrelation traces with the fit assuming a sech²-pulse shape are shown for the N_p- and N_m-polarization as blue and red line, respectively. For N_p-polarization (N_g-cut), 96 fs pulses were generated with the emission spectra centered at

1043 nm. Using N_m-polarization (N_p-cut) crystal, the center wavelength could be changed to 1035 nm with only somewhat longer pulse durations of 111 fs. The time-bandwidth-product of 0.308 is almost close to the Fourier-transform-limited pulse with an intensity profile described by the sech^2 function, and it indicates that dispersion is well compensated and there is no frequency modulation.

Typical spatial pattern of the beam at a distance of 1 m from the output coupler for the new oscillator based on N_m-polarization (N_p-cut) crystal is shown in **Figure 5(c)** with the vertical and horizontal intensity profiles. The cross section is nearly Gaussian with horizontal width 1.02 mm and vertical width 0.92 mm. The M^2 values of the beam are 1.10 in the horizontal plane and 1.07 in the vertical plane, which are fairly close to $M^2 = 1$ for an ideal Gaussian beam.

The thermal lens in the laser crystal did not affect the quality of output beam due to weakness of thermo-optical effects. Our calculations with LASCAD show that the focal lengths of thermal lens are 1273 and 1063 mm for horizontal and vertical directions, respectively, that are much larger than the focal length of focusing mirrors.

Figure 5(d) shows the average output power and its stability of the N_m-polarization oscillator after it was warmed up. We have achieved a maximum average power of 1.27 W at a pulse repetition rate of 87.8 MHz, and it gives a single femtosecond pulse energy greater than 14 nJ. The stability of the output power is to be 0.14% by using the definition such as twice the standard deviation of the measured output power divided by the mean output power.

As a summary, a comparison between the laser performances of the N_p- and N_m-polarization (N_g- and N_p-cut) Yb:KYW materials is given in **Table 1**. Both the relative beam angular stability and relative beam positional stability were measured by using the factors related with angular movement and beam positional movement in accordance with ISO 11670 [17]. The achieved parameters show that the new oscillator with N_m-polarization (N_p-cut) crystal is more stable and powerful with well-defined center wavelength as a seeding source.

	λ_{center} (nm)	$\Delta\lambda$ (nm)	P_{output} (W)	τ_p (fs)	M^2	Power stability (%)	Positional stability (%)	Angular stability (%)
E//N_p, OC = 4%	1043	12.5	0.64	96	1.10 (x) 1.25 (y)	0.30	0.5 (x) 3.2 (y)	0.3 (x) 1.9 (y)
E//N_m, OC = 6%	1035	9.0	1.27	111	1.10 (x) 1.07 (y)	0.14	0.1 (x) 0.15 (y)	0.2 (x) 0.3 (y)

Table 1. Comparison between N_p- and N_m-polarization Yb:KYW mode-locked oscillators with 4% and 6% output coupler, respectively.

2.2. Stretcher-compressor

Optical pulse stretcher and compressor are a key subsystem in CPA laser systems [10]. In those systems, the stretcher is used to lengthen optical pulses before the amplification, and the compressor is used to restore original short pulses. In this way, the peak power inside the amplifier cavity can be kept low enough to avoid any damage to the optical element and to suppress nonlinear distortions on the pulse temporal shape and beam spatial profile due to the self-focusing effect.

Design of stretcher and compressor module in a commercial laser system must provide maximum stretched pulse duration and good output spatial and temporal beam quality in minimum module size. To improve total efficiency and make design of stretcher and compressor more compact, we used in our stretcher-compressor module common transmission diffraction grating and folding mirrors (**Figure 7**).

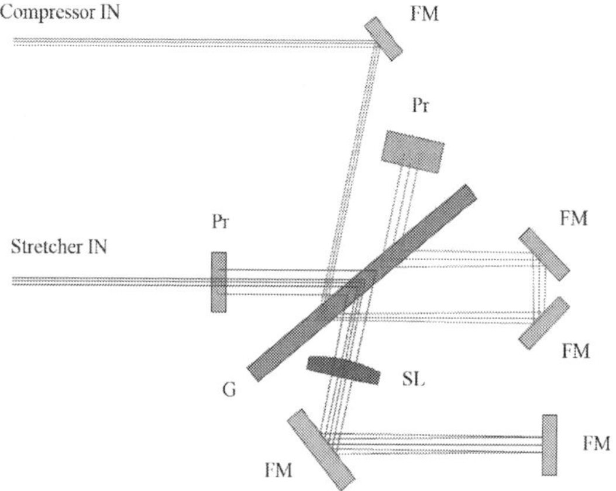

Figure 7. Optical scheme of the stretcher-compressor module based on the transmission diffraction grating (G), spherical lens (SL), right-angle prisms (Pr), and folding mirrors (FM).

In the stretcher and compressor, the beam enters and passes through transmission grating four times. In the stretcher, the beam after first and third passes through diffraction grating is focused and collimated by the spherical lens (SL). To change chirped-pulse duration, we used diffraction grating with two groove densities: $N = 1700$ mm^{-1} and $N = 1500$ mm^{-1} and two different spherical lenses with focal length $F = 100$ mm and $F = 200$ mm with NA = 0.25 and NA = 0.12, respectively. Using this stretcher-compressor module, we can supply chirped-pulse duration from 50 ps to 300 ps. The compressed pulse duration varies from 230 to 270 fs depending on the optical elements used. Increasing the chirped-pulse duration up to 300 ps using optical elements, the same aperture results in decreasing of spectral pass bandwidth of stretcher up to $\Delta\lambda = 14$ nm. Spectral clipping introduced by the aperture of optical elements in stretcher affects a recompression pulse duration and its time-bandwidth product.

Increasing the optics aperture is connected with the proportional increase of linear distances in stretcher (spherical aberration introduced by the spherical lens increase with increasing the NA number). Using the diffraction grating with higher groove density to provide more compact design is limited by increasing third-order dispersion.

To check stretcher-compressor module, we used our master oscillator as seed source. We measured initial pulse length and compressed pulse length after the beam passed through the stretcher and compressor. We also measured quality of a beam and total efficiency of the stretcher and compressor. The stretcher-compressor based on 1500 mm^{-1} grating operating with 50 ps chirped-pulse duration provides better degree of recompression and more short output pulse. It is connected with aberration avoiding due to smaller NA number. The measurements made by CCD camera shows that spatial quality of beam did not change after it passed through stretcher and compressor.

Amplification with different pulse durations shows limitation of maximum output pulse energy by Raman scattering excitation and was measured 150 µJ for chirped-pulse duration 50 ps, 300 µJ for 130 ps, and we expected more than 0.5 mJ for 300 ps. Further increase of the pulse energy needs a longer chirped-pulse duration and an increase of the stretcher-compressor module size.

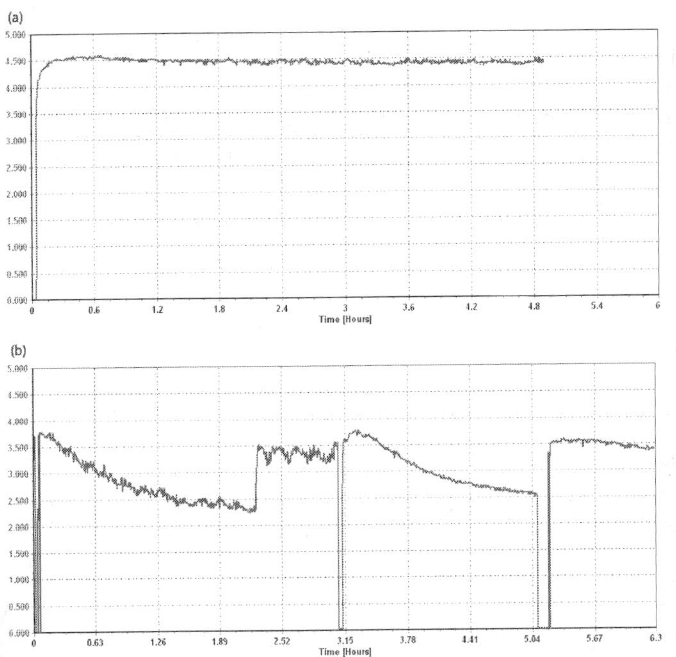

Figure 8. Second harmonic generator output power versus time using thermo stabilized compressor module (a) and nonstabilized one (b).

One of the most important parameters of stretcher-compressor is a temporal stability of compressed pulse duration. Because there is no 100% diffraction grating efficiency, approximately 30% of amplified power is dissipated inside stretcher-compressor module and causes misalignment of compressor due to thermo expansion of mechanical parts. The requirement for thermomechanical stability of stretcher-compressor elements increases with a decrease in compressed pulse duration. The thermal stabilization of the stretcher-compressor module can significantly improve temporal stability of the compressed pulse. Most dramatically, the effect of temporal detuning of compressor can by observed by the subsequent second harmonic generation. **Figure 8** shows temporal dependence of second harmonic power with water thermo stabilized stretcher-compressor module (a) and without stabilization (b).

In case of no thermo stabilized stretcher-compressor module, fundamental pulse duration increases up to two times with the gradual heating of the module resulting in significant drop of the second harmonic generation efficiency. Thermal stabilization by water cooling the module improves fundamental pulse duration stability and as a result stability of the second harmonic output power.

2.3. Spectral shaping

It is well known that regenerative amplification with high gain leads to the spectral narrowing of amplified pulses [7, 8]. To broaden spectrum in the regenerative amplifier, we tried both extra-cavity and intra-cavity spectral shaping based on the polarization-interference filter (Lyot filter). This technique is well known but it has been used generally for Ti:sapphire lasers [7]. The spectral shaper consists of a birefringent quartz plate placed between two polarizers. To realize optimal spectral shaping, a transmission minimum of the birefringent filter should coincide with maximum gain spectrum and their widths should be close. To fulfill these conditions, a quartz plate with thickness of 8 mm was cut along the optical axis and mounted to the rotation stage to rotate it in two planes for adjusting modulation depth and position of transmission minimum.

2.4. Double-slab laser and amplifier

The maximum average power of solid-state lasers is limited either by the thermal destruction of the active medium or by thermo-optical aberrations [16]. Therefore, one of a few methods of increasing this power in the case of bulk laser media is to increase the active medium length or to use several active elements in the cavity. To increase the average laser power and simultaneously to retain the high spatial quality of the output beam, we used a scheme with two active elements (see also [9, 14, 15]). The optical scheme of a laser amplifier with two active elements made of Yb:KGW crystals with different orientations of optical axes is shown in **Figure 1**. The use of N_p- and N_g-cut crystals, which have different spectral-luminescent characteristics at a corresponding orientation of the polarization vector of generated or amplified beams, makes it possible to widen the gain band and thus generate or amplify more broadband chirped or femtosecond pulses [10]. The crystals were 5 mm long and contained 3% of ytterbium ions, which ensured a good (above 70%) absorption of the pump radiation [18]. Similar to the case of the master oscillator, the pump beam and cavity mode sizes were matched by a longitudinal shift of the active elements. To eliminate possible modulation of the spectrum, the faces of the active elements were antireflection coated and inclined at an angle of ~30′. As pump sources, we used two fiber-coupled laser diode arrays with a maximal output power of 70 W each. The depolarization of the pump radiation was minimized using a short (30 cm) optical fiber 200 μm in diameter with the numerical aperture NA = 0.22. The pump beams from the fiber output were collimated and focused through dichroic mirrors into the active crystal to a spot with a diameter of about 320 μm, which is close to the diameter of the TEM_{00} cavity mode. This mode was calculated by the well-known ABCD matrix method taking into account the astigmatic thermal lenses induced in the active elements by the pump radiation. Using the LASCAD software, the focal length of these lenses was calculated to be 480–550 mm for the N_p-cut crystal at an incident power of 30 W. These values are in adequate agreement with the literature data [19, 20] and are confirmed by our measurements [18].

We used this laser as a standalone oscillator and as a regenerative amplifier. As an output mirror in this oscillator, we used a thin-film polarizer in combination with a quarter-wave plate, which was rotated to obtain the maximum output power at a given pump power. For electro-optic Q-switching, we used a Pockels cell based on a β-BaB_2O_4 (BBO) crystal and controlled by a special driver. The pulse repetition rate was smoothly changeable from single pulses to 1000 kHz.

The output laser power in the CW regime in the cases of using single active element and two elements simultaneously is shown in **Figure 9** versus the pump power incident on the crystals. The maximum output power was 12.5 and 9 W at a slope efficiency of 47% and 37% for Ng- and Np-cut single crystals, respectively. Note that the output powers for the crystal in which the polarization vector of the laser beam is parallel to the Np axis (E∥Np) are higher than for the crystal with E∥Nm despite a smaller stimulated emission cross section [9]. This behavior of power can be explained, in our opinion, by different spectral losses in the cavities with crystals emitting at different wavelengths. The maximum output power in the case of two crystals in the cavity was 18.5 W at a pump power of 72 W, which yields the total and slope efficiencies of 25% and 35%, respectively. The laser radiation spectrum was centered at a wavelength of 1035 nm, and its width did not exceed 1 nm.

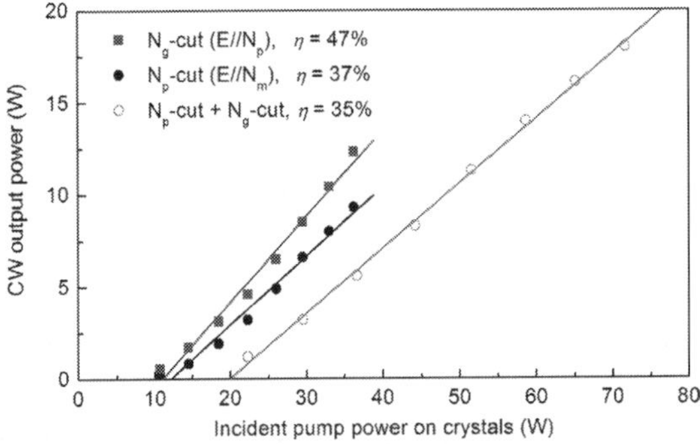

Figure 9. CW output power of laser as a function of incident pump power on crystals in CW mode operation for each single- and dual-slab configuration [15].

Note that the output laser power was maximized in measurements for each pump power by optimal adjustment of the quarter-wave plate. As can be easily shown, this measurement method leads to an underestimation of the slope efficiency. It is this fact that can explain the considerable difference in the slope efficiencies of single- and two-crystal lasers at close total efficiencies (27–30% and 25%, respectively).

The maximum output power obtained in the case of Q-switching reached 16 W at a pulse repetition rate of 100 kHz and 14 W at a repetition rate of 500 kHz (Q-switching time 800 ns). This decrease in the average power with respect to the CW regime can be explained by additional losses introduced by the electro-optic Q-switch. As the Q-switching time decreased to 400 ns, the output power decreased by three times, which obviously occurred because this time was too short for oscillation development. The output pulse duration, determined mainly by a rather long length of the cavity, was 20 ns. The width of the spectrum was 16 nm, which points to a pronouncedly multimode regime. The spectrum had two peaks, at wavelengths of 1035 and 1043 nm, which correspond to the spectral maxima of the gain coefficient for the two used crystals.

We optimized our laser by better alignment of the beam size of pump and laser modes in laser crystals at high pumping levels [21, 22]. This was achieved by moving the mirror CM1 along the optical axis. To suppress thermal losses, the thickness of the laser crystal was also reduced from 2 to 1.2 mm. The optimization results are shown in **Figure 10**. After optimization of the laser, output power increased by 32% from 18.5 to 24.5 W at incident pump power of 110 W.

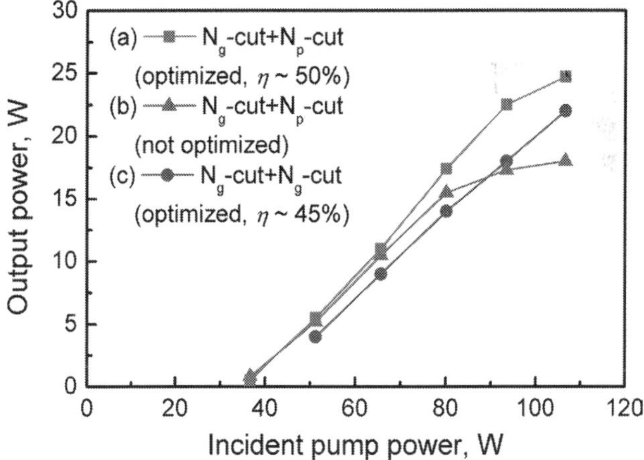

Figure 10. Output power of Q-switched laser versus incident pump power at repetition rate of 200 kHz for different dual-crystal configurations.

Output beam profiles measured using a CCD camera at distance of 1 m from output of compressor was symmetric with beam diameter of 4 mm at e^{-2} intensity level and nearly Gaussian. Increasing of the output power above 15 W resulted in gradual distortion of beam profile and transformation of beam cross section into the slightly elliptical one as shown in **Figure 5(b)**. However the beam quality parameter M^2 was measured to be below 1.5 even at the laser output power of 22 W as shown in **Figure 11**.

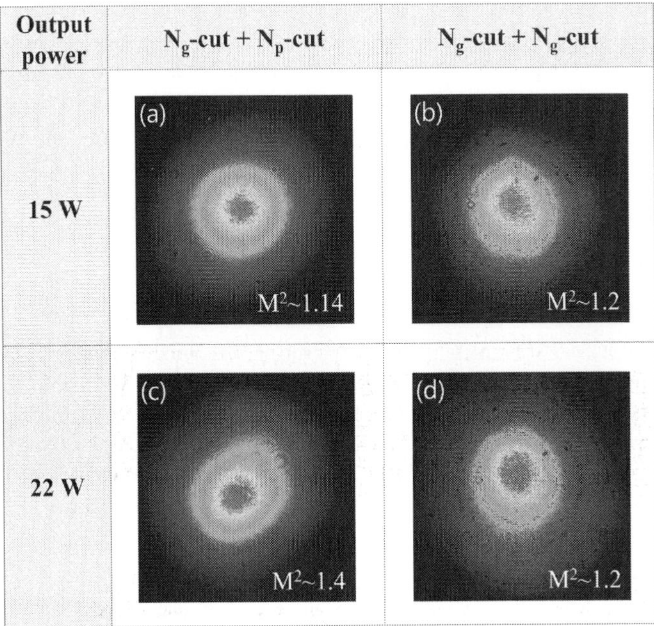

Figure 11. Near-field images and beam qualities M^2 of the laser output for different dual-crystal configurations.

3. Performance of the Yb:KGW femtosecond laser system

When the regenerative amplifier was seeded by stretched pulses, we investigated the output power, spectrum, and pulse shape after compression under different conditions. Our measurements showed that the output power after regenerative amplifier increased up to a value of 21 W when the time gate was increased up to 400 ns and the number of round trips was 24. And then the output power saturated at same value as the time gate increased up to 29 round-trips. It means that the gain is balanced by losses. **Figure 12** shows that the output power is practically linearly dependent on the incident pump power. It means the absence of such parasitic effects restricted gain as amplified spontaneous emission (ASE) and parasitic oscillations. We measured the output power with spectral shaping in addition. There was a power reduction of about 5% when the spectral shaping using a Lyot filter is applied outside the cavity. The compressor and the picker are making a loss, which reduces output power by 25% (15 W).

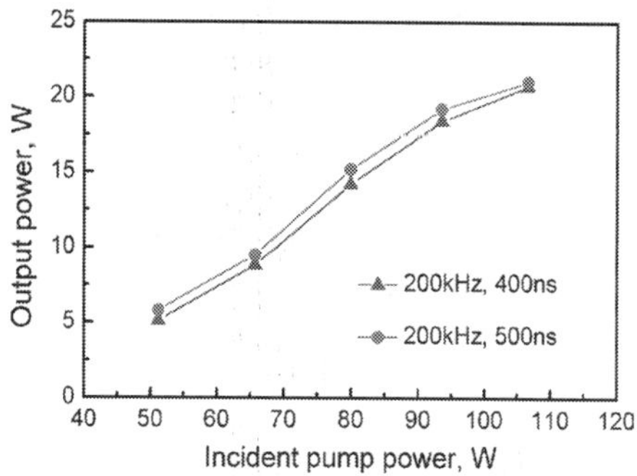

Figure 12. Average output power after RA as a function of incident pump power.

The laser system can operate in the repetition rate range of 50–500 kHz. In this range, the output power is almost not dependent on repetition rate. The single pulse energy was measured to be 300 and 30 µJ at repetition rates of 50 and 500 kHz, correspondently, that is important for microprocessing applications. Maximum pulse energy at 50 kHz was limited by Raman scattering excitation in Yb:KYW crystal that was observed in the experiment [23, 24]. Pulse shape at this repetition rate is distorted phase-modulation of the pulse at the Kerr nonlinearity [24].

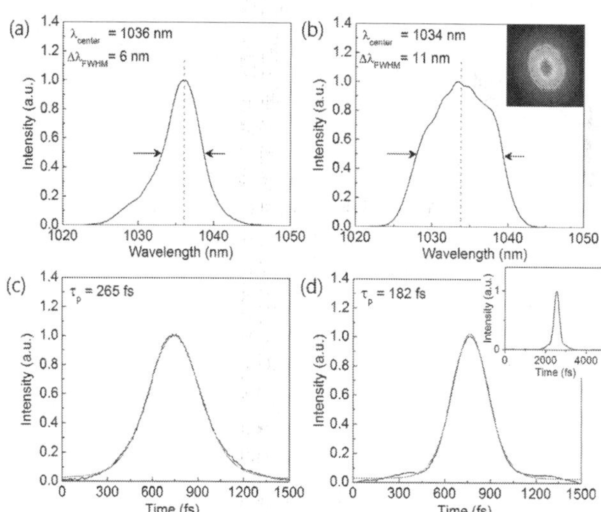

Figure 13. (a, b) Spectra and (c, d) intensity autocorrelation traces (black-experimental data, red-fitting) of output pulses at incident pump power of 67 W and repetition rate of 200 kHz without spectral shaping (a, c) and with spectral shaping (b, d). Insets show the output beam profile (b) and autocorrelation trace in the range of 5 ps (d) [15].

The shape of the output spectrum for equal pump power in both arms of pumping is shown in **Figure 13(a)**. The gain narrowing effect is noticeably well—FWHM spectral width is ~1.5 times narrower compared with the spectrum of master oscillator pulses. Compression provides 265 fs output pulses under this condition as shown in **Figure 13(c)**.

This gain narrowing effect can be suppressed, for example, by making the pump power of Yb:KYW crystals not to be equal [9]. We changed the pump power launched on the N_p-cut and N_g-cut crystals to the ratio of 3:2 [15]. The experimental measurements showed that the spectral width became broader and its shape was modified considerably. In this case, the spectral width was measured to be 11 nm and the pulse length was measured to be 210 fs for assuming sech^2 profile. This method has a drawback that the restriction of pumping power on one crystal results in the restriction of total output power in expense of pulse width. For example the output power dropped 37% in our experimental conditions.

Another way to suppress the spectrum narrowing is to use preliminary spectrum shaping [11] as it was discussed earlier. The example of output spectrum for extra-cavity spectrum shaping by filter Lyot is shown in **Figure 13(b)**. The optical spectrum showed a characteristic "bell" shape with a spectral FWHM bandwidth of 11 nm. Such a bandwidth provides smooth output pulse with width of intensity autocorrelation trace 305 fs that gives the pulse length of τ_{FWHM} = 182 fs for sech^2 pulse profile as shown in **Figure 13(d)**. This pulse length is close to the pulse length of 160 fs defined by aberrations in the stretcher-compressor module. To measure the ultrashort pulse width, we used a PulseCheck autocorrelator (APE GmbH). The inset of **Figure 13(d)** shows that there is no noticeable peak beyond the range of 1.5 ps.

Inserting a spectrum shaper inside the cavity of regenerative amplifier, we obtained approximately the same spectral width but less output power of about 20%. It is connected with accumulated effect of intra-cavity losses by Lyot filter inside the cavity. Thus combination of Lyot filter outside the cavity as a spectral shaper and identical pump power for two slabs in the dual-slab regenerative amplifier provides optimal condition of output power and pulse length.

Beam quality M^2 of output beam was below 1.2 at output power <12 W that allows the beam focusing to small spot size of 5–10 µm. High average output power, with more than tens of µJ, and beam quality are important for industrial microprocessing applications.

4. Microhole drilling for processing drawing dies using ultrafast Yb:KGW laser

Laser micromachining techniques are currently used due to the broad applications across the manufacturing sectors. Among the major applications, laser microhole drilling and cutting have much attention. For microhole drilling, the conventional fabrication method, lithography, which requires advanced facilities and numerous multiple steps, is limited in material type and geometry. Currently used drilling and cutting with nanosecond (ns) or longer pulsed laser are always accompanied with contamination to the surrounding material, melt zone, and recast layer. Although the geometrical precision could be improved by using ns laser techniques, the quality and precision achievable are still limited due to the subsequently uncontrolled deposition of the melt. Due to the high energy input and thermally induced stress, drilling and cutting using picosecond pulsed laser still have disadvantages, e.g. cracks and heat-affected zone in the surrounding area. Not only the accuracy but also the reliability of the process is affected. For ultrafast lasers, energy deposition occurs on a timescale that is short compared to atomic relaxation processes. Ultrafast lasers suppress thermal diffusion and thus reduce heat-affected zone.

When processing transparent materials with ultrafast laser, the high intensity of focal volume induces multiphoton or tunnel ionization, and then subsequent electron heating or avalanche ionization, which gives rise to the efficient absorption of light. This phenomenon is observed by the nonlinear nature of the ultrafast laser interaction with the transparent materials. This generates the unique capability of transparent material processing using ultrafast laser.

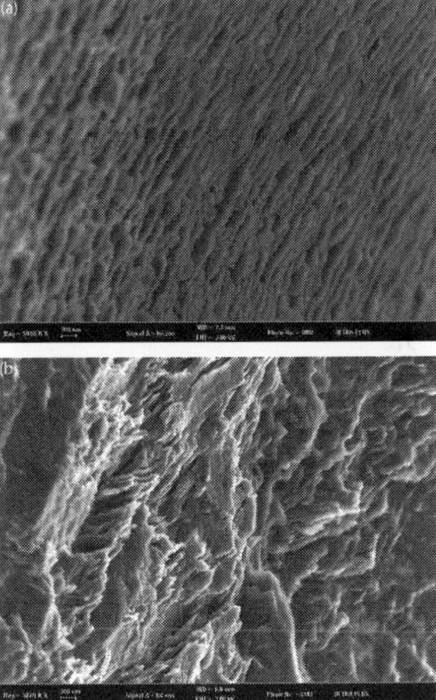

Figure 14. Scanning electron microscope (SEM) views of the structure on bearing land in drawing dies (a) using femtosecond (fs) laser and (b) using nanosecond (ns) laser.

Its promising application is drawing dies-hole drilling. Wire drawing is a deformation and metalworking process used to reduce the cross-section of a wire by pulling the wire through a drawing die. For drawing very fine wire, a single crystal diamond die is used. The drawing die consists of three zones: cone-shape entrance, bearing land, and back relief. The die bearing determines the size of the wire. As demand for microwire increases across the manufacturing sectors, large scale machines are currently used to produce wire of microlevel, but it is not economical in an industry. Micromachining with ultrafast laser, which made the hole size small, has been reported. Smooth surfaces, also, are generally preferred for precision machining. **Figure 14** shows the SEM view of the bearing surface of drilled hole in dies. The comparison with fs and ns laser drilling results shows the advantage of fs laser drilling. **Figure 14(a)** shows a ripple for which spacing is generally <200 nm. The orientations of the ripple structures are parallel to each other. Similar ripple structure has been observed in various materials for fs laser drilling. Uneven and rough structures are shown in **Figure 14(b)**. It is clear that the material removal during the dies-hole drilling is accomplished by the formation of melt. Compared with two methods, micromachining with ultrafast laser creates much cleaner and smoother hole.

Ultrafast laser micromachining is an emerging technology for high-precision and cold-ablation material processing. For its advantages and potential uses, suitable ultrafast laser and laser operating parameters such as wavelength, repetition rate, average power, pulse duration, spot size, beam quality, and sample moving speed must be selected to achieve desired high-quality micromachining. In the near future, ultrafast laser micromachining will be used in various sectors including sub-micron material processing, surface structuring, photonics devices, biomedical devices, microfluidics, displays, and solar applications.

5. Conclusion

In conclusion, a room-temperature, diode-pumped, dual-crystal Yb:KGW laser operating as a Q-switched oscillator or regenerative amplifier has been developed where the gain bandwidth was extended by using N_g-cut + N_p-cut or N_g-cut + N_g-cut configuration. It was shown that fine-tuning the mode sizes in the crystals in the resonator with high pump power is important to obtain the maximal output power since thermal effects change the operation point in the stability zone and mode matching conditions. It was demonstrated simply by shifting the position of the end mirror along optical axis in the resonator. Optimization of the laser resonator increased the output power from 18 to 24 W in case of Q-switched oscillator and from 17 to 21 W in case of regenerative amplifier. Such optimization of laser resonator improves not only output power but also stability of laser operation, especially for N_g-cut + N_g-cut crystal configuration, that is manifest in reduction of output power fluctuations.

The use of this regenerative amplifier enabled to create compact, high average power, high brightness, diode-pumped femtosecond Yb:KGW laser system. This laser, which utilized a CPA MOPA laser scheme, consisted of master oscillator, regenerative amplifier, and stretcher-compressor module. It was capable of delivering 15 W of average output power with a pulse duration down to 182 fs high in a nearly diffraction limited output beam ($M^2{\sim}1.2$) at pulse repetition rates of 50–500 kHz.

This level of output power and quality of a laser beam are practically the same as the output power of Yb:KGW/Yb:KYW thin-disk lasers with medium level of output power [6, 9].

However laser heads based on volume laser media are considerably easier and cheaper compared with those based on thin-disk configuration. Subsequent development of multi-crystal laser approach will be able to increase output power on the level of high-power thin-disk lasers.

The achieved level of radiation parameters allows to successfully use this femtosecond laser for ultrafast micromachining in various applications including sub-micron material processing, surface structuring, creating photonics devices, biomedical devices, microfluidics, displays, and solar applications.

Acknowledgements

Parts of this chapter are reproduced from authors' previous publications [15, 25, 26].

Author details

Guang-Hoon Kim[1*], Juhee Yang[1], Byunghak Lee[1], Bosu Jeong[1], Sergey Chizhov[1], Elena Sall[1], Vladimir Yashin[2] and Uk Kang[1]

*Address all correspondence to: ghkim@keri.re.kr

1 Advanced Medical Device Research Division, Korea Electrotechnology Research Institute, Ansan-si, Republic of Korea

2 State Optical Institute, St. Petersburg, Russia

References

[1] Dausinger F., Lichtner F., and Lubatschowski H. (2004) Femtosecond Technology for Technical and Medical Applications. Berlin: Springer.

[2] Diels J.-C. and Rudolph W. (2006) Ultrashort Laser-Pulse Phenomena: Fundamentals, Techniques, and Applications on Femtosecond Time Scale. Boston: Academic Press.

[3] Fermann M. E., Galvanauskas A., and Sucha G. (2003) Ultrafast Lasers: Technology and Applications. New York: Marcel Dekker.

[4] Kuleshov N. V., Lagatsky A. A., Podlipensky A. V., Mikhailov V. P., and Huber G. (1997) Pulsed laser operation of Yb-doped $KY(WO_4)_2$ and $KGd(WO_4)_2$. Opt. Lett. 22: 1317–1320.

[5] Liu H., Nees J., Mourou G., Biswal S., Spuehler G.J., Keller U., and Kuleshov N. V. (2002) Yb:KGd(WO$_4$)$_2$ chirped-pulse regenerative amplifiers. Opt. Commun. 203: 315–321.

[6] Nickel D., Stolzenburg C., Giesen A., and Butze F. (2004) Ultrafast thin-disk Yb:KY(WO$_4$)$_2$ regenerative amplifier with a 200-kHz repetition rate. Opt. Lett. 29: 2764.

[7] Barty C., Korn G., Raksi F., Rose-Petruck C., Squier J., Tian A., Wilson K., Yakovlev V., and Yamakawa K. (1996) Regenerative pulse shaping and amplification of ultrabroadband optical pulses. Opt. Lett. 21: 219–221.

[8] Raybaut P., Balembois F., Druon F., and GeorgesP. (2005) Numerical and experimental study of gain narrowing in ytterbium-based regenerative amplifiers. IEEE J. Quantum Electron. 41: 415–426.

[9] Buenting U., Sayinc H., Wandt D., Morgner U., and Kracht D. (2009) Regenerative thin disk amplifier with combined gain spectra producing 500 mJ sub 200 fs pulses. Opt. Express 17: 8046–8050.

[10] Strickland D., and Mourou G. (1985) Compression of amplified chirped optical pulses. Opt. Commun. 56: 219–221.

[11] Kim G. H., Kang U., Heo D., Yashin V. E., Kulik A. V., Sall' E. G., and Chizhov S. A. (2010) A compact femtosecond generator based on an Yb:KYW crystal with direct laser-diode pumping. J. Opt. Technol. 77: 225–229.

[12] Software package for LASer Cavity Analysis and Design (LAS-CAD). 1996–2015 LAS-CAD GmbH, Germany, www.las-cad.com.

[13] Brunner F., et al. (2000) Diode-pumped femtosecond Yb:KGd(WO$_4$)$_2$ laser with 1.1-W average power Opt. Lett. 25: 1119.

[14] Buettner A., Buenting U., Wandt D., Neumann J., and Kracht D. (2010) Ultrafast double-slab regenerative amplifier with combined gain spectra and intracavity dispersion compensation. Opt. Express 18(21): 21973–21980.

[15] Kim G. H. Yang J., Chizhov S. A., Sall E. G., Kulik A. V., Yashin V. E., Lee D. S., and Kang U. (2012) High average-power ultrafast CPA Yb:KYW laser system with dual-slab amplifier Opt. Express 20: 3434–3442.

[16] Koechner W. (2006) Solid-state Laser Engineering. New York: Springer.

[17] International Standard ISO 11670:2003: Lasers and Laser-related Equipment – Test Methods for Laser Beam Parameters – Beam Positional Stability. This standard was last reviewed in 2015.

[18] Kim G. H., Yang J., Lee B., Sall E. G., Chizhov S. A., Yashin V. E., and Kang U. (2015) Investigation of diode-pump absorption efficiency and thermo-optical effects in a high-power Yb:KGW laser. Quantum Electron. 45: 211–215.

[19] Biswal S., O'Connor S. P., Bowman S. R. (2005) Thermo-optical parameters measured in ytterbium-doped potassium gadolinium tungstate. Appl. Opt. 44: 3093–3097.

[20] Chenais S., Balembois F., Druon F., Lucas-Leclin G., and Georges P. (2004) Thermal lensing in diode-pumped ytterbium lasers – part I: theoretical analysis and wavefront measurements. IEEE J. Quantum Electron. 40: 1217–1234.

[21] Kim G. H., Yang J., Chizhov S. A., Sall E. G., Kulik A. V., Yashin V. E., Kang U. (2013) High brightness Q-switched oscillator and regenerative amplifier on dual-crystal Yb:KGW laser. Laser Phys. Lett. 10: 125004.

[22] Kim G. H., Yang J., Chizhov S. A., Sall E. G., Kulik A. V., Yashin V. E., Kang U. (2014) High-power directly diode-pumped femtosecond Yb:KGW lasers with with optimized parameters. Proc. of SPIE, 8959: 8959B-1–89591B-8.

[23] Kim G. H., Yang J. H., Lee D. S., Yashin V. E., Kulik A. V., Sall E. G., Chizhov S. A., Kang U. (2012) High power and high efficiency lasers on crystals Yb:KYW with end pumping operating in CW and pulsed regimes. Quantum Electron. 42: 292–297.

[24] Kim G. H., Yang J. H., Yashin V. E., Kulik A. V., Sall E. G., Chizhov S. A., Kang U. (2013) Power limitation and pulse distortions in Yb:KGW laser system with chirped pulse amplification. Quantum Electron. 43: 725–730.

[25] Kim G. H., Yang J., LeeD. S., KulikA. V., Sall'E. G., ChizhovS. A., YashinV. E., and KangU. (2012) Directly diode-pumped femtosecond laser based on an Yb:KYW crystal. Proc. SPIE 8247: 82471C.

[26] Kim G. H., Yang J., Sall E., Chizhov S., Kulik A., Lee D. S., Kang U., Yashin V. (2012) Development of compact femtosecond Yb: KYW oscillators: simulation and experiment. J. Korean Phys. Soc. 61: 365–370.

Chapter 3

Yb:YAG-Pumped, Few-Cycle Optical Parametric Amplifiers

Hanieh Fattahi

Additional information is available at the end of the chapter

http://dx.doi.org/10.5772/64438

Abstract

In this chapter, the principle, design, and characteristics of high-efficiency, short-pulse-pumped, few-cycle optical parametric chirped-pulse amplification (OPCPA) systems are reviewed. To this end, the feasibility of two techniques to increase the conversion efficiency of few-cycle OPCPA systems is demonstrated and discussed. The techniques result in 2.5 mJ, 7.5 W pulses and correspond to a pump-to-signal conversion efficiency of 30%. The broadband amplified spectrum supports 5.7 fs. Finally, the feasibility of extending the amplified spectrum to a near-single-cycle regime by using the combination of different crystals and phase matching is shown.

Keywords: optical parametric chirped-pulse amplification, Yb:YAG thin-disk laser, attosecond pulse, femtosecond laser

1. Introduction

The new generation of femtosecond technology based on short-pulse-pumped optical parametric chirped-pulse amplification (OPCPA) [1] holds promise for scaling the peak power and average power of few-cycle pulses simultaneously. This progress would benefit a number of fields, notably attosecond science [2, 3] by allowing to scale attosecond pulse generation at higher photon energy and higher flux.

In OPCPA, the amplified bandwidth is not limited by the energy level structure of a laser medium or gain narrowing [4], as is the case in laser amplifiers [5]. Therefore, OPCPA appears to be the method of choice for the production of ultrashort pulses (down to the few-cycle regime) at high peak and average power. Employing short pulses (several-ps to sub-ps) to

pump an OPCPA allows higher peak intensity in the nonlinear medium as the damage threshold intensity of materials scales with the inverse square root of the pulse duration [6]. The high pump intensity makes it possible to achieve the required gain in a shorter crystal, which leads to greater amplification bandwidth.

Further advantage with a short crystal is that the effect of transverse walk-off is reduced, the temporal contrast can be enhanced, and stretchers and compressors can be simpler. However, the crystal length does not decrease as rapidly as the pulse duration, so the temporal walk-off relative to the pulse duration increases for short pulses. A simple analytical analysis shows that the optimum pump-pulse duration to achieve a high conversion efficiency and a broadband gain is around 1 ps [7].

Nevertheless, all these advantages of short-pulse-pumped OPCPA remain useless without an efficient, reliable, and powerful pump source. Such pump lasers are required to deliver high-energy near-1-ps pulses with near-diffraction-limited beam quality at repetition rates in the kHz to MHz range.

Heretofore, due to the lack of suitable pump lasers, the few-cycle OPCPA delivered either high-energy pulses at a low repetition rate [8, 9] or low-energy pulses at a high repetition rate [10]. Nowadays, Yb-doped lasers in the thin-disk, fiber or slab geometries [1, 11–15] are capable of delivering high-energy, high average power pulses with ps-pulse duration. Among these laser technologies, the recent advances in Yb:YAG thin-disk lasers have started to fulfill the criteria for suitable pump sources for OPCPA systems and hold promise to change the current state of the art of OPCPA systems to few-cycle pulses with higher energy and average power [1, 16].

This chapter is devoted to the recent progress in Yb:YAG-pumped, few-cycle OPCPA systems. In Section 1, a brief overview on the fundamentals of OPCPA is presented. In Sections 2 and 3, novel techniques for increasing the conversion efficiency are discussed. In Section 4, a technique for extension of the amplification bandwidth is discussed.

In a medium with second-order nonlinearity, a high-energy photon (called pump) can decay to two newly generated photons with lower frequencies (called seed and idler). In the presence of initial seed photons, the decay of pump photons is stimulated and consequently more photons at the seed frequency are generated. The seed photons after amplification are named signal and the process is called optical parametric amplification (OPA). The frequency of the generated signal and idler photons is defined by the conservation of energy. However, the amplification bandwidth can be increased by fulfilling conservation of momentum between pump, signal, and idler pulses, which can be tuned by the type, thickness, and temperature of the nonlinear medium and also the geometry of the three interacting beams.

To obtain a strong pump-to-signal and idler energy conversion, the spatial and temporal overlap between seed and pump pulses in the nonlinear medium should be maximized. The optimum temporal overlap between the pump and seed pulses can be ensured by temporal stretching of the seed pulses to the temporal window of pump pulses. This technique is the combination of chirped-pulse amplification (CPA) [17] and OPA, hence called optical parametric chirped-pulse amplification.

In addition to the above-mentioned parameters, the conversion efficiency in OPA or OPCPA systems also depends on the thickness of the nonlinear medium, peak intensity of the pump pulses, and the initial seed energy. Conversion efficiency scales up by increasing the thickness of the nonlinear medium as long as the phase-matching condition (conservation of momentum) between the three interacting beams is satisfied. At higher pump peak intensity to induced nonlinear polarization in the nonlinear medium is stronger and therefore larger amplification is achieved. Moreover, by increasing the seed-to-pump energy ratio, the conversion efficiency at lower amplification gain can be achieved.

The optimization of the pump-to-signal conversion efficiency of multicycle OPCPA systems has been the subject of several studies [18–22]. In the next two sections, the feasibility and realizability of two techniques to increase the conversion efficiency of few-cycle OPCPA systems are discussed.

2. Recycling the pump energy

In optical parametric amplification, the behavior of the gain over the length of the nonlinear medium can be divided into three main regions (**Figure 1(a)**). In the beginning, the energy of the amplified signal has an exponential growth due to the generation of the idler field that enhances the amplification process (region A). In region B, the gain drops gradually as the pump energy is reduced, and the growth of the signal power becomes approximately linear. When the pump beam has been locally depleted at some point in time and space, back conversion sets in and further reduces the gain. In region C, back conversion dominates, and the signal power drops. In the case of pulses with Gaussian spatiotemporal profile, the depletion mainly occurs at the center of the pulse, where the intensity is highest. Therefore, the back conversion already starts before the complete depletion of the pump. Because back conversion depends on both signal and idler beams, it can be reduced by removing the idler between the stages of a multistage OPCPA.

To explore this option, three different designs (as shown in **Figure 1(b)**) are simulated and compared using the SISYFOS code [23]. In all designs, the amplification takes place in a type-I BBO crystal, where the angle between the Optical axis and the signal is 22°, and the noncollinear angle between the pump and the signal in the tangential phase-matching geometry is 2.7°. The pump have a Gaussian beam and pulse shape, and the seed have the Gaussian beam shape and a super-Gaussian spectrum of order 4, ranging from 600 to 1100 nm and linearly chirped to 1.1 ps pulse duration. Higher-order nonlinear effects and parasitic processes were not taken into account.

Figure 1(c) compares the three simulated OPCPA systems. The first configuration (Design 1 in **Figure 1(b)**) consists of a single OPCPA stage using a 2 mm thick BBO crystal, with 7 mJ of pump energy at a peak intensity of 80 GW/cm^2, which results in a conversion efficiency of 14%.

The second design (Design 2 in **Figure 1(b)**) has two stages, and the idler beam is removed between the two stages. This reduces back conversion in the second stage and allows operation in a regime with stronger pump depletion. Temporal and spatial overlap of the beams could be readjusted between the stages, and a further advantage with the two-stage design is that the phase-matching of the crystals can be tuned slightly differently to optimize the total bandwidth. The crystal lengths for the two stages are 1.2 and 0.7 mm, respectively. The second stage is pumped by the residual pump energy from the first stage resulting in a conversion efficiency of 34%.

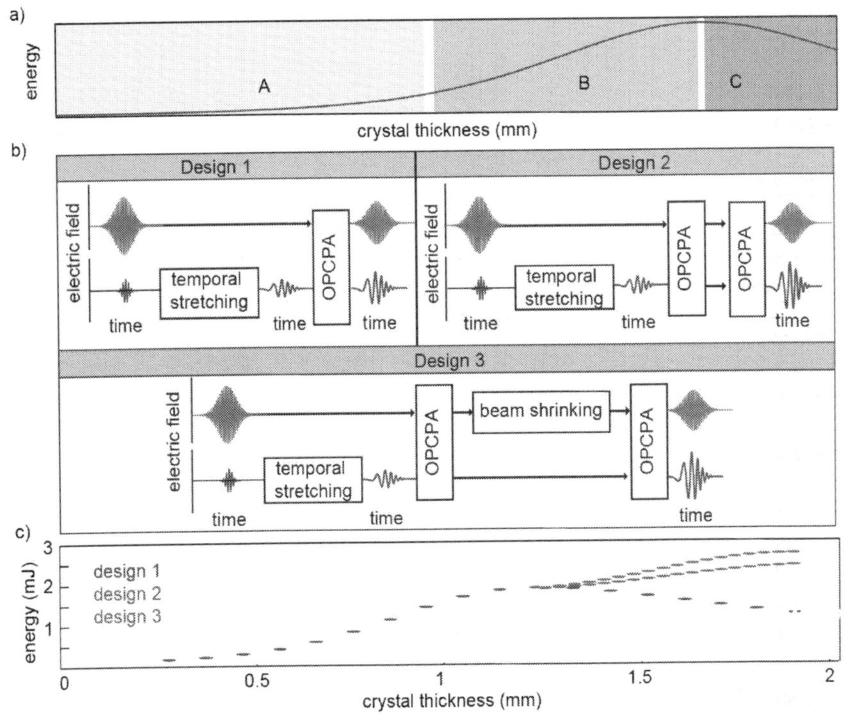

Figure 1. (a) Qualitative behavior of OPCPA amplified energy over the length of the nonlinear medium. (b) Three OPCPA designs are discussed in the main text. Design 1 consists of one OPCPA stage. Designs 2 and 3 consist of two OPCPA stages, where the residual pump energy after the first OPCPA stage is reused in the second stage. In Design 3, the pump after the first amplification stage is resized to increase the pump peak intensity. (c) Calculated amplified signal energy over the crystal length for the three different designs.

The third configuration (Design 3 in **Figure 1(b)**) is a two-stage OPCPA system similar to the second design except that the pump-beam size between the two stages is reduced to compensate the reduction in the pump intensity after the first stage of amplification and therefore the efficiency in the third design reaches 39%.

All three designs are capable of supporting the ultrabroad amplification bandwidth necessary for a few-cycle pulse durations. The detailed parameters of the simulations are shown in

Table 1. In what follows, the experimental realization of the third design is demonstrated and discussed. The third design is chosen as it supports the highest conversion efficiency in the above-mentioned study.

	Design 1	Design 2		Design 3	
	1st stage	1st stage	2nd stage	1st stage	2nd stage
L (mm)	2	1.2	0.7	1.2	0.7
E amp (mJ)	1	1.8	2.4	1.8	2.7
ϕ_p (mm)	2.5	2.5	–	2.5	2
ϕ_s (mm)	2.5	2.5	–	2.5	D^*_{amp}
Efficiency (%)	14	26	14	26	21
Overall efficiency (%)	14	34		39	

D^*_{amp}: the diameter of the amplified beam.

Table 1. Parameters used in simulations: L_c, crystal thickness; $E_{s\,amp}$, amplified signal energy; ϕ_p, pump-beam diameter at full width at half maximum; ϕ_s, seed beam diameter at FWHM.

2.1. System description

2.1.1. Front end

The experimental OPCPA setup (as shown in **Figure 2**) consists of a Ti:Sa-based oscillator and amplifier followed by a broadband nonlinear seed generation scheme, a pump laser, a temporal jitter compensation system, three OPCPA stages, and a chirped-mirror compressor [24]. The Yb:YAG regenerative amplifier [25], optically synchronized with the OPCPA seed [26], delivers 16 mJ, 1.6 ps pulses at full width at half maximum (FWHM) at 3 kHz repetition rate and its frequency doubled output is used for pumping the OPCPA. However, due to the long optical beam-path difference between seed and pump pulses, timing fluctuations occur due to air turbulence, mechanical vibrations of optical components, temperature drifts, and the finite stability of the front end, which need to be compensated by an active stabilization system. The timing jitter in our system is reduced to a level of 24 fs (root mean square) by using an active stabilization system based on spectrally resolved cross-correlation between the stretched seed and the pump pulse [27].

The broadband OPCPA seed was generated by using a small portion of the output of the Ti:Sa multipass amplifier (Femtolasers GmbH), providing a spectral bandwidth of 60 nm (FWHM) centered at 790 nm. These pulses, containing 30 µJ of energy, focused on a 15 cm long hollow core fiber (HCF) with an inner diameter of 120 µm filled with 4.5 bar of krypton. The pressure of krypton in the HCF and the group delay dispersion (GDD) of the input pulse were optimized to obtain the maximum spectral broadening, covering a spectral range from 500 to 1050 nm.

With this combination of parameters, an overall throughput of 10 µJ pulse energy in a near-diffraction-limited output beam containing the broad spectrum was achieved (**Figure 3(a)**).

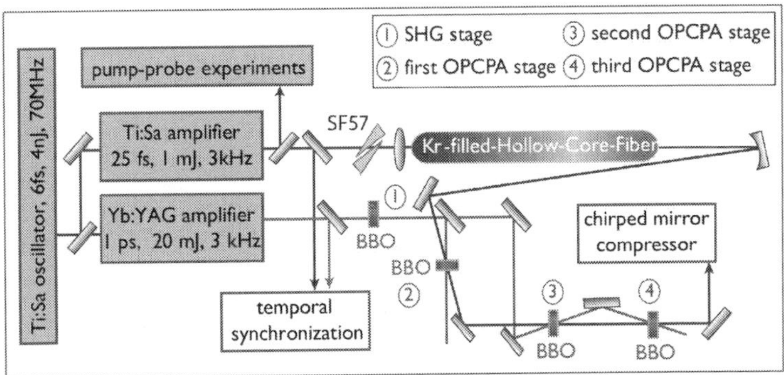

Figure 2. Block diagram of the OPCPA system. The OPCPA broadband seed pulses are generated from a Ti:Sa regenerative amplifier. The output of a Yb:YAG thin-disk amplifier, after frequency doubling, is used to pump the three OPCPA stages. Finally, a chirped-mirror compressor is used for pulse compression of the broadband amplified pulses.

In order to measure the amount of introduced material dispersion, which needs to be compensated after the final OPCPA stage, the broadband seed was sent through the entire beam path without any pump. Pulse dispersion was caused over merely 5.5 mm beam path in BBO, 4 mm path length in SF57 glass, and over 10 m propagation in air. The stretched seed pulses were characterized by a multishot, second harmonic generation (SHG)-XFROG device incorporating a 20 µm thick BBO crystal cut at 29° as the nonlinear medium, while a fraction of the multipass amplifier's output provided the reference beam. From these measurements, a second-order spectral phase of 1433 fs^2 evaluated at 850 nm was retrieved. This is in excellent agreement with the GDD introduced by the above components, evaluated as 1403 fs^2. The pulse duration of the seed pulses assuming a Gaussian fit for the retrieved time structure is 1.1 ps (FWHM), which ensures a sufficiently good temporal matching between the seed and pump pulses in the OPA stages.

For frequency doubling of the pump laser, a BBO crystal was used. Its high nonlinearity allowed the use of a relatively short crystal keeping the accumulated B-integral in the system negligible. Using a BBO crystal of 1.5 mm length, a conversion efficiency as high as 70% with a good beam quality was obtained. The high SHG efficiency confirms the excellent beam quality and clean output pulses of the regenerative amplifier. However, in order to definitely avoid problems related to the B-integral in the OPCPA stages, a 1 mm BBO crystal for the SHG was chosen, which resulted in a conversion efficiency of 57%.

2.1.2. OPCPA stages and pulse compression

The OPCPA setup consists of three stages. We added an OPCPA-based preamplifier stage in the experiment in order to boost the seed energy before the two power amplifier stages. This

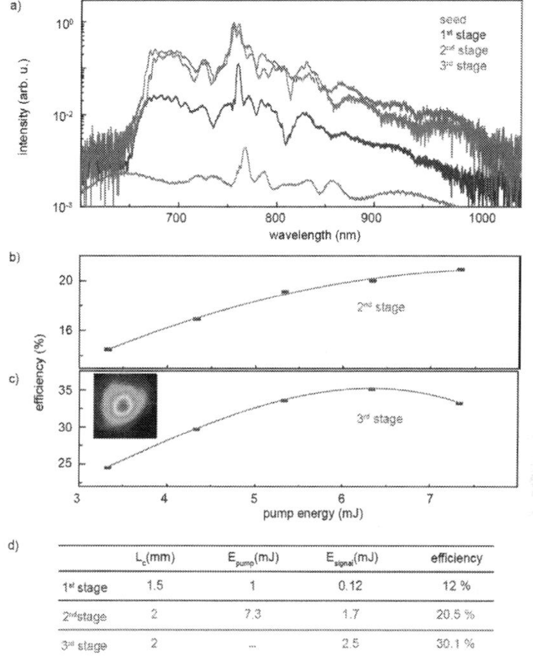

	L_c(mm)	E_{pump}(mJ)	E_{signal}(mJ)	efficiency
1st stage	1.5	1	0.12	12 %
2nd stage	2	7.3	1.7	20.5 %
3rd stage	2	...	2.5	30.1 %

Figure 3. (a) The seed spectrum and the amplified spectra of three OPCPA stages, normalized to the output energy of each stage. (b) and (c) Conversion efficiencies after the second and third stages, respectively. The conversion efficiency is defined as the net increase in signal energy divided by the input pump energy of stage two. (d) The detailed parameters of each OPCPA stage. The total efficiency after each stage is defined as the net increase in signal energy after the stage divided by the total pump energy of the OPCPA chain. Inset: amplified beam profile after the third stage [7].

stage is necessary to drive the following stages into saturation [7]. Therefore, 1 mJ of the frequency doubled output of the thin-disk amplifier was used to pump a 1.5 mm BBO crystal, and an amplified energy of 120 µJ and a 350 nm broad spectrum were achieved in the first stage (**Figure 3(a)**). Here, the OPCPA crystal length was chosen to minimize the superfluorescence at the third stage [28].

The following two stages were designed for reaching the highest possible pump-to-signal conversion efficiency by controlling the idler energy and recycling the pump energy. Up to 7.3 mJ of the pump energy with the peak intensity of 80 GW/cm^2 at 515 nm was used for the second OPCPA stage that employed a 2 mm thick BBO crystal. In this stage, an amplified pulse energy of 1.77 mJ was obtained. The thickness of the crystal at this stage is adjusted to stop amplification slightly below the saturation while preserving a good residual pump-beam quality.

Subsequently, the size of the remaining pump beam was reduced to increase the peak intensity to 80 GW/cm^2 in the third amplification stage. Here, by employing a 2 mm thick BBO crystal, the amplified energy reached 2.5 mJ. In the last two OPCPA stages, an optical-to-optical conversion efficiency of >32% was achieved (**Figure 3(b)** and **(c)**), which to the best of our knowledge is the highest reported conversion efficiency for few-cycle OPCPA systems [9, 10, 29, 30]. No measurable superfluorescence background was observed when blocking the signal beam in front of the first stage.

The simulated boost efficiency in our design study is in good agreement with the experimental results. Quantitative comparison shows, however, that higher conversion efficiencies were yielded for a shorter crystal in the simulation than in the experiment. We relate the deviation from the theoretical prediction to a slight ellipticity in our pump beam, caused by the compressor of the Yb:YAG amplifier, which limited the effective interaction area between pump and signal beams.

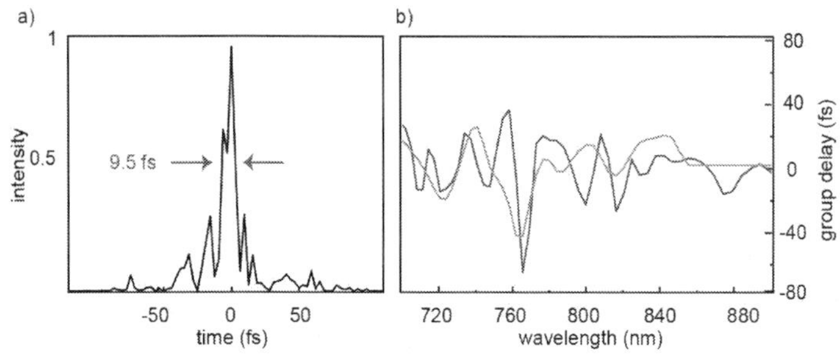

Figure 4. (a) Retrieved temporal intensity of the compressed pulses after 12 reflections in a double-angle chirped-mirror compressor measured by SH-FROG. The pulse is compressed to 9.5 fs and holds Fourier transform limit of 5.7 fs. (c) The calculated GD of retrieved spectral phase for the pumped (blue curve) and unpumped (green curve) OPCPA chains.

The 350 nm broad amplified signal measured with the Si-based spectrometer supports a transform-limited pulse duration of 5.7 fs. Preliminary compression, by using 12 reflections on double-angle chirped mirrors with −30 fs^2 GDD per reflection, resulted in a pulse duration of 9.5 fs (FWHM). The compressor had a total throughput of 80%. The retrieved temporal intensity profile and retrieved residual group delay (GD) of the pulses are shown in **Figure 4(a)** and **(b)**. Our analytical study shows that the pulse can be compressed to 7 fs by adding the GD of a 0.5 mm thick fused silica to the measured GD of the pulse. However, to investigate the origin of the fine oscillation in the retrieved GD, a frequency resolved optical gating (FROG) measurement of the whole OPCPA chain was performed, but this time without any pumping. The comparison between two cases in **Figure 4(b)** shows that oscillations were enhanced by amplification but did not originate from the OPCPA phase [31]. The peak of the GD is at 760 nm, which coincides with the wavelength of Ti:Sa amplifier's pulses and the peak in the spectral intensity of the seed pulses after the HCF. Therefore, it can be concluded that the measured residual higher-order chirp is due to the self-phase modulation in the HCF, OPCPA phase, and the residual oscillations in group delay dispersion of the double-angle chirped-mirror compressor [32, 33]. The higher-order dispersion and the satellite pulses can be compensated by using specially designed chirped mirrors for this system along with the implementation of spectral smoothing techniques, such as cross-polarized wave generation [34] after the HCF.

The demonstrated highly efficient compact OPCPA system delivers broadband pulses with 2.5 mJ energy supporting a two-cycle pulse at a repetition rate of 3 kHz. Our simple OPCPA design shows that, by extraction of the idler energy and optimization of the pump peak intensity, a higher conversion efficiency can be achieved. The output of the system ensures to be compressible to its two-cycle transform limit by using specially designed chirped mirrors. The system also has the capability to operate with a stabilized carrier envelope phase (CEP) by stabilizing the Ti:Sa oscillator. These features make the reported OPCPA system a suitable driver for high harmonic generation (HHG) [35].

In the next section, an alternative method to achieve high conversion efficiency as well as uniform amplified spectrum is discussed.

3. Controlling the deposition of pump energy

In this section, the realizability of a second novel technique that allows the simultaneous increase in the spectral bandwidth and optical conversion efficiency of OPCPA systems is discussed. This approach is based on a patent application by Deng and Krausz [36].

In the conventional OPCPA systems, similar to the one described in Section 2, the seed-pulse duration is designed to be a fraction of pump-pulse duration in order to maximize the energy conversion. Here as the seeds are strongly chirped, the temporal intensity profile of the pump pulses has to be nearly constant to ensure uniform amplification for the entire seed spectrum. Therefore, for pump pulses with Gaussian temporal profile, the seed pulses have to be considerably shorter than the pump pulses. Consequently, the pump energy is not consumed efficiently and the relative seed-to-pump pulse duration ratio will be a compromise between the amplification bandwidth and the conversion efficiency.

These deficiencies can be overcome if OPCPA seed pulses are linearly stretched to several times longer than the pump pulses. Subsequently, different fraction of seed pulses can be temporally overlapped with pump pulses and are amplified in different OPCPA stages.

This technique enables the controlled deposition of pump energy in the subsequent temporal/spectral locations along the chirped seed pulse [36]. Furthermore, by controlling the amplification gain in each stage, the spatiotemporal profile of the pump pulses can be shaped into a flat-top pulse. By tuning the phase-matching angle of the crystal to the central wavelength of the seed pulse, the ultimate amplified spectrum can be shaped and a broader amplified bandwidth is gained. In addition, by reusing the pump energy after each amplification stage, the total conversion efficiency is increased.

3.1. Theoretical analysis

Figure 5 shows simulation results for three different OPCPA designs. The simulation's input parameters are similar to the ones presented in Section 2.

In the first design, the residual pump energy after the first amplification stage is used to pump the second stage and ultimately the residual pump energy after the second stage is used to pump the third stage.

Due to the Gaussian shape of the pump in time and space, the energy extraction takes place primarily in the middle of the pump. Therefore, the wings are mostly left unaffected with a signature of energy back conversion at the center, due to the fact that this part of the pump possesses the highest peak intensity. In this design, the OPCPA pump-to-signal conversion efficiency is increased to 43% compared to the OPCPA system demonstrated in Section 2 (**Figure 5(b)** and **(c)**).

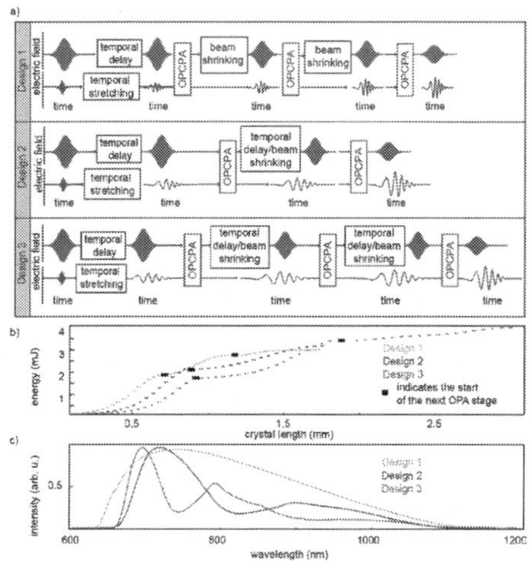

Figure 5. (a) Three designs are discussed in the main text. Design 1 consists of three OPCPA stages, where the residual pump energy after each amplification stage is reused in a subsequent OPCPA stage. Design 2 consists of two OPCPA stages. Here, the seed pulses are temporally stretched to twice the pump-pulse duration, and the residual pump energy after the first OPCPA stage is reused in the second stage. In Design 3, the seed pulses are temporally stretched to triple of the pump-pulse duration, and the residual of the pump energy is reused after each amplification stage. (b) Calculated amplified signal energy over the crystal length and their corresponding spectra (c) for the three different designs [7].

In the second design, the seed pulses are stretched temporally to twice the pump-pulse duration. The blue part of the seed spectrum is amplified in the first OPCPA stage by adjusting the temporal overlap between pump and seed pulses. Subsequently, the pump pulses after the first stage are reused to amplify the red part of the spectrum at the second stage. In this design, the pump-to-signal conversion efficiency reaches 45% indicating the good pump-energy extraction while the beam quality of the amplified signal is maintained (**Figure 5(b)** and **(c)**).

The 45% conversion efficiency achievable from the second design is not drastically different from the 38.6% efficiency achievable from the case where the seed and pump pulses have the same pulse duration (as discussed in Section 2) as the pump pulses after the first amplification stage in both cases maintain a good spatiotemporal profile.

The gain and the shape of the amplified spectrum can be further optimized by adjusting the phase-matching angles of the crystal in each stage to tune the amplification for the selected part of the spectrum, which is not investigated in this study.

In the third design, the seed pulses are stretched three times the pump pulses and amplified in three subsequent OPCPA stages, while the residual pump energy of the preceding stage is used to pump the subsequent stage. The bluest frequencies of the seed spectrum are amplified in the first stage, while the reddest frequency components are amplified in the third OPCPA stage. The pump-to-signal conversion efficiency in this design reaches 57%, which is a noticeably higher value compared to the other designs (**Figure 5(b)** and **(c)**). The spectral narrowing for Designs 2 and 3 is caused by a suboptimal stretching factor of the input signal and phase-matching angle of the crystal. The optimizations of these parameters are cumbersome in simulation but straightforward in an experimental setup.

3.2. Experimental setup

The seed pulses of the OPCPA system described in Section 2 are stretched after the first amplification stage by using an 8 mm thick SF57 plate at Brewster's angle. The blue frequencies of the seed spectrum were amplified to 4 W by adjusting the temporal delay between the seed and pump pulses and adjusting the phase-matching angles of the BBO crystal (blue curve in **Figure 6(a)**). In the next OPCPA stage, the amplification is moved to the second half of the seed spectrum gaining 6 W of the total amplification (**Figure 6(a)**, pink curve).

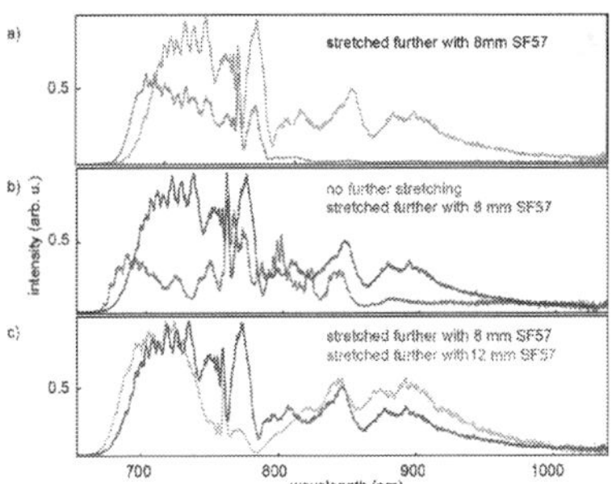

Figure 6. (a) The experimental demonstration of Design 2. The seed pulses are heavily stretched by using an 8 mm thick SF57 plate. The higher frequencies in the spectrum are amplified first (blue curve). The residual of the pump energy is used to amplify the lower frequency components of the spectrum (pink curve). (b) The gray curve shows the amplified spectrum in the similar OPCPA system without any spectral stacking (gray curve) compared to the amplified spectrum shown in (a) (blue curve). (c) Further stretching of the seed pulses by using a 12 mm thick SF57 plate results in appearing of a hole in the amplified spectrum (purple curve) [7].

The same setup, after removing the bulk stretcher, resulted in 6.2 W of amplification after the third stage. As shown in **Figure 6(b)**, the amplified spectrum of the stacked OPCPA is more uniform due to the even amplification gain. The red wing of the spectrum carries more energy in the stacked OPCPA compared to the amplified spectrum achieved from the system discussed in Section 2. Moreover, the spectral spike at 780 nm is heavily suppressed. **Figure 6(c)** shows amplified spectrum for a similar system but with a larger stretching factor. Here, after stretching the seed pulses using a 12 mm thick SF57 plate, 5 W of the average power was obtained. It can be seen that the amplified spectrum contains a hole, leaving the 8 mm thick SF57 plate, the optimum thickness for temporal stretching of the seed pulses.

Temporal stretching of the signal pulses to twice the pump-pulse duration, demonstrated in this section experimentally, did not show further increase in the OPCPA conversion efficiency compared to the scheme realized in Section 2. This similarity in the conversion efficiency is due to the fact that the spatiotemporal quality of the residual pump pulses after one amplification stage is preserved for both cases.

However, it is shown analytically that the further temporal stretching of seed pulses results in the increase in the conversion efficiency, as the spatiotemporal quality of the remaining pump pulses after two amplification stages is preserved just for the case of heavily chirped input seed.

4. Gain bandwidth engineering

The bandwidth of the amplified spectra, as discussed in the previous sections, can be extended further by using different crystals or a crystal with different phase-matching angles. The combination of BBO and LiB_3O_5 (LBO) crystals can be used to extend the amplified spectrum to longer frequencies, as the BBO crystal does not support amplification for spectral components above 1.1 µm. The amplified spectrum in both crystals supports near-single-cycle pulses, which is not unobtainable with solely either of them.

To this end, the broadened seed spectrum generated in the HCF subsequently focused on a 4 mm $Y_3Al_5O_{12}$ (YAG) crystal to extend the spectrum to 1400 nm and is amplified in an OPCPA chain similar to the system described in Section 2. Combinations of LBO and BBO crystals at different OPCPA stages are used to increase the amplification bandwidth. The first OPCPA stage was optimized to amplify a broad spectral range from 750 to 1400 nm up to 50 µJ energy in a 2 mm LBO crystal. In the second stage, a 2 mm BBO crystal was employed. The amplified spectrum measured in this stage, using an Si-based spectrometer, spans from 670 to 1100 nm and contained 1.1 mJ energy. Finally at the third stage, 1.8 mJ energy was obtained in a 3 mm LBO crystal.

The amplified spectra at each OPCPA stage, normalized to their energy, are shown in **Figure 7**. The amplified spectrum obtained after the third stage supports 4.3 fs transform-limited pulses (FWHM). The preliminary pulse compression was performed by using a set of chirped-mirror compressor designed for spectral wavelength of 700–1300 nm. **Figure 7(b)** shows the pulse compression to 9 fs measured with an SH-FROG containing a 10 µm BBO crystal. The retrieved spectrum from the FROG measurement is in a good agreement with a spectrum measured after the third OPCPA stage. Pulse compression to its Fourier transform limit would require a specially designed chirped-mirror compressor for compensating the higher-order chirp.

The conversion efficiency of the system can be optimized further by using a longer crystal in the last OPCPA stage without relinquishing the amplified spectral bandwidth.

As shown in this section, the utilization of different well-selected nonlinear crystals extends the OPCPA gain bandwidth substantially. The realized three-stage OPCPA system, using one BBO and two LBO crystals, delivers 1.8 mJ pulses with a Fourier transform limit of 4.5 fs. The system supports shorter pulse duration than an all-LBO three-stage OPCPA system with 5.3 fs (FWHM) pulses. The reported extension of the amplified spectral bandwidth is crucial for experiments that rely on high-energy, single-cycle pulses.

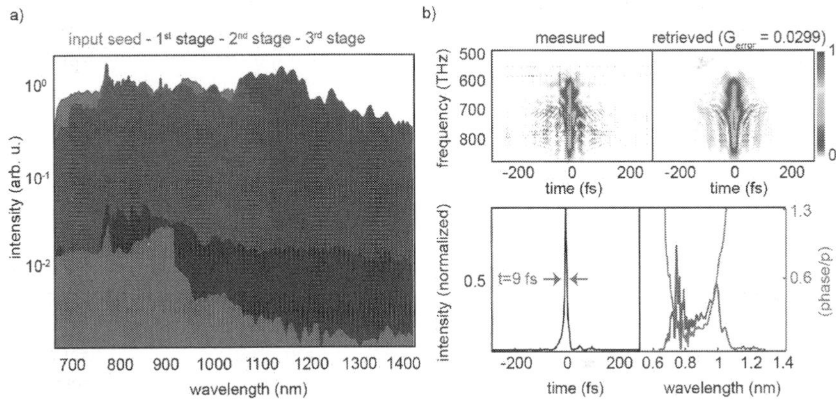

Figure 7. (a) Amplified spectra in a three-stage OPCPA system. A 2 mm LBO crystal is used to amplify the spectral components from 750 to 1400 nm in the first OPCPA stage. In the second stage, the spectral components from 680 to 1100 nm were amplified in a 2 mm BBO crystal and finally in the last stage a 3 mm LBO crystal is used to boost the amplification to 1.8 mJ. (b) Measured and retrieved SH-FROG traces (top) and the retrieved spectrum and temporal profile of the pulses (bottom) of the OPCPA system [7].

5. Summary

In this chapter, three few-cycle OPCPA systems operating at the near-infrared spectral range and pumped by the second harmonic generation of a Yb:YAG thin-disk amplifier were reviewed. The feasibility of increasing the conversion efficiency of the system by reusing the pump energy after each amplification stage, in the subsequent OPCPA stages, was demonstrated. It was shown that by controlled deposition of pump energy in different parts of the seed spectrum, high conversion efficiency along with a smooth amplified spectrum can be achieved. Furthermore, the feasibility of TW-level monocycle OPCPA systems was studied by using different crystals in different amplification stages. In addition to the presented systems, different harmonics of the Yb:YAG amplifier can be used to pump few-cycle pulses in visible or mid-infrared spectral range [16].

6. Outlook

The capabilities of the current high harmonic generation sources, based on CPA Ti:Sa technology, are limited to energies around a few hundred eV and to pulse durations of several tens of attoseconds. This limitation originates from a deficiency of the current laser technology that can either provide pulses with ultrahigh (petawatt) peak powers at relatively low repetition rates or moderate peak power (gigawatt) pulses at kHz repetition rates. Scaling attosecond pulses to high repetition rates and photon energies as high as several keV demands few-cycle laser systems with high peak and average power, which is beyond the performance of the current laser technology.

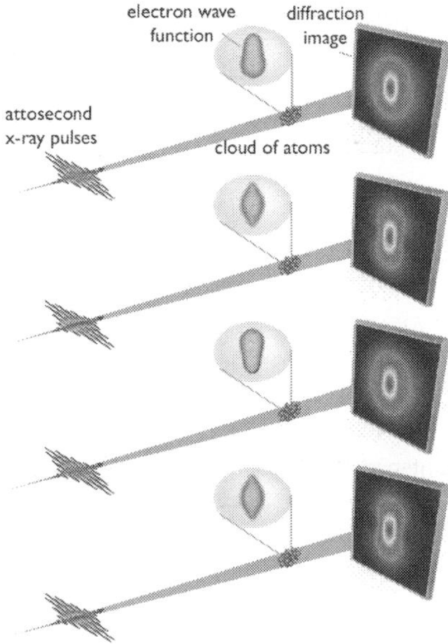

Figure 8. Attosecond X-ray diffraction: as a coherently induced charge oscillation takes place in an atom or molecule, an incident X-ray pulse takes a diffraction snapshot of the electron distribution at the time of interaction; changing the time delay between the source of the excitation and the attosecond pulse allows for the temporal evolution of the charge density to be directly measured in time and space. The figure represents the simulated dynamic of hydrogen atoms when they are exposed to 100-as, X-ray pulses and are excited into the 1S-2P coherent superposition state. As shown, the electron dynamic can be reconstructed by means of attosecond X-ray diffraction spectroscopy [3].

OPCPAs are scalable in terms of peak and average power and directly benefit from the availability of turn-key, industrial-grade ps-pump lasers. For more than two decades, powerful, cost-effective ps pump sources have been unavailable. Nowadays, diode-pumped ytterbium-doped lasers in the thin-disk geometry are able to deliver 1 ps scale pulses at kilowatt-scale average power, in combination with terawatt-scale peak powers. By merging these two existing technologies, and comprising OPCPAs driven by ytterbium-based pump lasers, the new generation of femtosecond technology will combine terawatt-scale peak powers with kilowatt-scale average powers in ultrashort optical pulse generation opening new path in the generation of isolated attosecond pulses with higher flux and photon energies.

The increased photon flux will greatly expand the applicability of attosecond spectroscopy to scrutinizing phenomena where current-generation sources delivered signal near or below the noise level, whereas shorter wavelengths provide direct access to electronic motions on increasingly shorter length scales from nanostructures toward atomic dimensions.

The availability of attosecond X-ray pulses could lead to resolve the spatiotemporal motion of electrons in their ultimate resolution, picometer-attosecond resolution, via attosecond X-ray diffraction spectroscopy (**Figure 8**).

Acknowledgements

I wish to thank Prof. Ferenc Krausz for fruitful discussions about the material of this chapter.

Author details

Hanieh Fattahi

Address all correspondence to: hanieh.fattahi@mpq.mpg.de

Max-Planck Institut für Quantenoptik, Garching, Germany and Department für Physik, Ludwig-Maximilians-Universität München, Garching, Germany

References

[1] H. Fattahi, H. G. Barros, M. Gorjan, T. Nubbemeyer, B. Alsaif, C. Y. Teisset, M. Schultze, S. Prinz, M. Haefner, M. Ueffing, A. Alismail, L. Vámos, A. Schwarz, O. Pronin, J. Brons, X. T. Geng, G. Arisholm, M. Ciappina, V. S. Yakovlev, D. E. Kim, A. M. Azzeer, N. Karpowicz, D. Sutter, Z. Major, T. Metzger, and F. Krausz, "Third-generation femtosecond technology," Optica 1, 45–63 (2014).

[2] M. Hentschel, R. Kienberger, C. Spielmann, G. A. Reider, N. Milosevic, T. Brabec, P. Corkum, U. Heinzmann, M. Drescher, and F. Krausz, "Attosecond metrology," Nature 414, 509–513 (2001).

[3] F. Krausz and M. Ivanov, "Attosecond physics," Rev. Mod. Phys. 81, 163–234 (2009).

[4] A. Dubietis, G. Jonušauskas, and A. Piskarskas, "Powerful femtosecond pulse generation by chirped and stretched pulse parametric amplification in BBO crystal," Opt. Commun. 88, 437–440 (1992).

[5] C. Le Blanc, P. Curley, and F. Salin, "Gain-narrowing and gain-shifting of ultra-short pulses in Ti: sapphire amplifiers," Opt. Commun. 131, 391–398 (1996).

[6] M. Lenzner, J. Krüger, S. Sartania, Z. Cheng, C. Spielmann, G. Mourou, W. Kautek, and F. Krausz, "Femtosecond optical breakdown in dielectrics," Phys. Rev. Lett. 80, 4076–4079 (1998).

[7] H. Fattahi, "Third-generation femtosecond technology," Springer Theses (Springer International Publishing, Cham, 2015).

[8] D. Herrmann, L. Veisz, R. Tautz, F. Tavella, K. Schmid, V. Pervak, and F. Krausz, "Generation of sub-three-cycle, 16 TW light pulses by using noncollinear optical parametric chirped-pulse amplification," Opt. Lett. 34, 2459–2461 (2009).

[9] N. Ishii, L. Turi, V. S. Yakovlev, T. Fuji, F. Krausz, A. Baltuska, R. Butkus, G. Veitas, V. Smilgevicius, R. Danielius, and A. Piskarskas, "Multimillijoule chirped parametric amplification of few-cycle pulses," Opt. Lett. 30, 567–569 (2005).

[10] M. Schultze, T. Binhammer, G. Palmer, M. Emons, T. Lang, and U. Morgner, "Multi-µJ, CEP-stabilized, two-cycle pulses from an OPCPA system with up to 500 kHz repetition rate," Opt. Express 18, 27291–27297 (2010).

[11] F. Röser, T. Eidam, J. Rothhardt, O. Schmidt, D. N. Schimpf, J. Limpert, and A. Tünnermann, "Millijoule pulse energy high repetition rate femtosecond fiber chirped-pulse amplification system," Opt. Lett. 32, 3495–3497 (2007).

[12] P. Russbueldt, T. Mans, G. Rotarius, J. Weitenberg, H. D. Hoffmann, and R. Poprawe, "400W Yb:YAG Innoslab fs-amplifier," Opt. Express 17, 12230–12245 (2009).

[13] L. E. Zapata, H. Lin, A.-L. Calendron, H. Cankaya, M. Hemmer, F. Reichert, W. R. Huang, E. Granados, K.-H. Hong, and F. X. Kärtner, "Cryogenic Yb:YAG composite-thin-disk for high energy and average power amplifiers," Opt. Lett. 40, 2610–2613 (2015).

[14] O. H. Heckl, J. Kleinbauer, D. Bauer, S. Weiler, T. Metzger, and D. H. Sutter, "Ultrafast Thin-Disk Lasers," 93–115 (2016) DOI: 10.1007/978-3-319-17659-8_5.

[15] H. Fattahi, A. Alismail, H. Wang, J. Brons, O. Pronin, T. Buberl, L. Vámos, G. Arisholm, A. M. Azzeer, and F. Krausz, "High-power, 1-ps, all-Yb:YAG thin-disk regenerative amplifier," Opt. Lett. 41, 1126–1129 (2016).

[16] Y. Deng, A. Schwarz, H. Fattahi, M. Ueffing, X. Gu, M. Ossiander, T. Metzger, V. Pervak, H. Ishizuki, T. Taira, T. Kobayashi, G. Marcus, F. Krausz, R. Kienberger, and N. Karpowicz, "Carrier-envelope-phase-stable, 1.2 mJ, 1.5 cycle laser pulses at 2.1 µm," Opt. Lett. 37, 4973–4975 (2012).

[17] D. Strickland and G. Mourou, "Compression of amplified chirped optical pulses," Opt. Commun. 55, 447–449 (1985).

[18] M. Guardalben, J. Keegan, L. Waxer, V. Bagnoud, I. Begishev, J. Puth, and J. Zuegel, "Design of a highly stable, high-conversion-efficiency, optical parametric chirped-pulse amplification system with good beam quality." Opt. Express 11, 2511–2524 (2003).

[19] L. J. Waxer, V. Bagnoud, I. A. Begishev, M. J. Guardalben, J. Puth, and J. D. Zuegel, "High-conversion-efficiency optical parametric chirped-pulse amplification system using spatiotemporally shaped pump pulses," Opt. Lett. 28, 1245–1247 (2003).

[20] L. Yu, X. Liang, L. Xu, W. Li, C. Peng, Z. Hu, C. Wang, X. Lu, Y. Chu, Z. Gan, X. Liu, Y. Liu, X. Wang, H. Lu, D. Yin, Y. Leng, R. Li, and Z. Xu, "Optimization for high-energy and high-efficiency broadband optical parametric chirped-pulse amplification in LBO near 800 nm," Opt. Lett. 40, 3412–3415 (2015).

[21] P. Wnuk, Y. Stepanenko, and C. Radzewicz, "Multi-terawatt chirped pulse optical parametric amplifier with a time-shear power amplification stage." Opt. Express 17, 15264–15273 (2009).

[22] J. Moses, C. Manzoni, S.-W. Huang, G. Cerullo, and F. X. Kaertner, "Temporal optimization of ultrabroadband high-energy OPCPA," Opt. Express 17, 5540–5555 (2009).

[23] G. Arisholm, "General numerical methods for simulating second-order nonlinear interactions in birefringent media," J. Opt. Soc. Am. B 14, 2543–2549 (1997).

[24] R. Szipocs, K. Ferencz, C. Spielmann, and F. Krausz, "Chirped multilayer coatings for broadband dispersion control in femtosecond lasers." Opt. Lett. 19, 201–203 (1994).

[25] T. Metzger, "High-repetition-rate picosecond pump laser based on a Yb:YAG disk amplifier for optical parametric amplification," PhD thesis (2009).

[26] C. Teisset, N. Ishii, T. Fuji, T. Metzger, S. Köhler, R. Holzwarth, A. Baltuska, A. Zheltikov, and F. Krausz, "Soliton-based pump-seed synchronization for few-cycle OPCPA." Opt. Express 13, 6550–6557 (2005).

[27] A. Schwarz, M. Ueffing, Y. Deng, X. Gu, H. Fattahi, T. Metzger, M. Ossiander, F. Krausz, and R. Kienberger, "Active stabilization for optically synchronized optical parametric chirped pulse amplification," Opt. Express 20, 5557–5565 (2012).

[28] S. Adachi, N. Ishii, T. Kanai, A. Kosuge, J. Itatani, Y. Kobayashi, D. Yoshitomi, K. Torizuka, and S. Watanabe, "5-fs, multi-mJ, CEP-locked parametric chirped-pulse amplifier pumped by a 450-nm source at 1 kHz," Opt. Express 16, 14341–14352 (2008).

[29] M. Puppin, Y. Deng, O. Prochnow, J. Ahrens, T. Binhammer, U. Morgner, M. Krenz, M. Wolf, and R. Ernstorfer, "500 kHz OPCPA delivering tunable sub-20 fs pulses with

15 W average power based on an all-ytterbium laser," Opt. Express 23, 1491–1497 (2015).

[30] J. Rothhardt, S. Demmler, S. Hädrich, J. Limpert, and A. Tünnermann, "Octave-spanning OPCPA system delivering CEP-stable few-cycle pulses and 22 W of average power at 1 MHz repetition rate," Opt. Express 20, 10870–10878 (2012).

[31] F. Tavella, Y. Nomura, L. Veisz, V. Pervak, A. Marcinkevius, and F. Krausz, "Dispersion management for a sub-10-fs, 10 TW optical parametric chirped-pulse amplifier," Opt. Lett. 32, 2227–2229 (2007).

[32] V. Pervak, I. Ahmad, M. K. Trubetskov, A. V. Tikhonravov, and F. Krausz, "Double-angle multilayer mirrors with smooth dispersion characteristics," Opt. Express 17, 7943–7495 (2009).

[33] D. Franz, H. Fattahi, V. Pervak, M. Trubetskov, E. Fedulova, N. Karpowicz, Z. Major, and F. Krausz, "Investigation of temporal compression of few-cycle pulses from an ultrabroadband, multi-mJ optical parametric amplifier," in "Conf. Lasers Electro-Optics—Int. Quantum Electron. Conf." (2013), p. CFIE_P_3.

[34] T. Buberl, A. Alismail, H. Wang, N. Karpowicz, and H. Fattahi, "Self-compressed, spectral broadening of a Yb:YAG thin-disk amplifier," Opt. Express 24, 10286–10294 (2016).

[35] W. Schweinberger, A. Sommer, E. Bothschafter, J. Li, F. Krausz, R. Kienberger, and M. Schultze, "Waveform-controlled near-single-cycle milli-joule laser pulses generate sub-10 nm extreme ultraviolet continua," Opt. Lett. 37, 3573–3755 (2012).

[36] Y. Deng and F. Krausz, "Method and device for optical parametric chirped pulse amplification," WO/2013/020671 PCT/EP2012/003283, 14.02.2013, 01.08.2012 (2013).

Laser-produced Soft X-Ray Sources

Chapter 4

Brilliance Improvement of a Laser-Produced Soft X-Ray Plasma

Tobias Mey

Additional information is available at the end of the chapter

http://dx.doi.org/10.5772/64149

Abstract

The brilliance of a laser-produced soft X-ray source is enhanced for gaseous target concepts. In contrast to solid or liquid target materials, these sources are clean and versatile but provide a comparably low conversion efficiency of laser energy into EUV and soft X-ray radiation. The basic idea is to induce supersonic effects in the gas jet, leading to a local increase of the particle density, and thus, to a larger number of possible emitters. Typically, the target gas is expanded into a vacuum environment and the density rapidly drops in all directions. In the present approach, a low pressure helium atmosphere is used to generate shock waves in the supersonic nozzle flow. Passing through these structures, the target gas is recompressed, and the particle density is raised. By focusing the laser beam into such regions, a higher number of gas atoms can be ionized resulting in a brighter and smaller plasma.

Keywords: soft X-ray, laser-produced plasma, supersonic jet, barrel shock

1. Introduction

Photons of the soft X-ray spectral region ($\approx$ 0.1–10 nm) have very small absorption lengths in all kinds of material due to the strong interaction with matter [1]. This fact together with the short wavelength qualifies this radiation as a tool for structuring and the analysis with nanometer resolution. An important application is the next-generation lithography that further reduces the achievable feature size in computer chip production [2, 3]. Surface analysis becomes extremely precise by means of reflectometry and scatterometry [4–6] and also the binding state of molecules can be studied by spectral investigations [7–9]. Microscopy with radiation at wavelengths in the water window (λ = 2.3–4.4 nm) allows highly resolved direct

imaging of samples in aqueous environments [10–12]. Mostly, these applications are realized at large-scale facilities, such as synchrotron sources or free-electron lasers. However, the demand for beam time is always too large to be satisfied by these institutions, and thus, people endeavor to transfer experiments to their laboratories. This constitutes the need of compact beam sources as can be realized by the principle of laser-produced plasmas (LPP) that is the subject of this chapter.

In order to classify and compare the radiation of different soft X-ray beam sources, the brilliance Br is a commonly used quantity that is the number of photons within a narrow spectral range $\Delta\lambda/\lambda$ emitted into a solid angle Ω from an area A within the time scale τ (typically the wavelength range $\Delta\lambda$ is defined to be 0.1% of the central wavelength λ) [1]:

$$Br = \frac{N_{ph}}{\tau \cdot A \cdot \Omega \cdot \Delta\lambda / \lambda}. \tag{1}$$

The value of Br is given in the unit $1/(\text{s·mm}^2\text{·mrad}^2 \cdot 0.1\%\text{BW})$ with 0.1%BW indicating the bandwidth of 0.1%. A distinction is made between the peak brilliance, where τ denotes the pulse duration, and the average brilliance, where τ is the inverse of the repetition rate.

In comparison to synchrotrons and free-electron lasers, the brilliance of laser-produced plasma sources is several orders of magnitude lower. However, there are strategies to increase their brilliance which involve, e.g., higher power densities of the generating laser pulse. On the other hand, the density of the target material has a strong impact on the achievable number of soft X-ray photons too, whereas basically the source brilliance scales with the density. Thus, the brightest plasmas can be achieved with solids. Respective target materials are deposited on rotating cylinders [13] or quickly moving tapes [14], which provide repetition rates of up to 1 kHz. Prominent elements are gold or tin for the production of radiation at a wavelength of 13.5 nm, which is applied in EUV lithography [15]. Furthermore, there are sources employing cold gases in a solid phase, such as an argon filament that emits soft X-rays in the wavelength range 2–5 nm [16]. Achievable plasma sizes with solid targets are comparably small and on the order of several tens of μm (full-widths at half-maximum, FWHM).

A plasma of similar brilliance and extent is obtained with liquid targets, e.g., xenon [17], methanol [18], or tin [15]. A fluid jet [19] provides high target densities but might lead to size and brightness fluctuations. Going one step further to individual microscopic droplets [20], the advantage is the mass limitation such that the entire target material is converted into a highly ionized plasma state, supporting source stability. However, the drawback of solid and liquid target concepts is the inevitable production of fast particles and ions with kinetic energies of up to several hundred keV [21], which severely damage optics in the beam path. There are mitigation strategies to slow down the debris material such as repeller fields [22] or localized gas jet shields [23], but still the collector optics has a limited lifetime [2]. Contrarily, gaseous targets are almost free from debris [24]. Short gas pulses with durations of μs to ms are expanded from a pressure of several 10 bar into vacuum by a piezomechanical or electromagnetic nozzle, resulting in a supersonic jet. Different target gases feature individual spectra of the resulting radiation, ranging from emitters with characteristic spectral lines (low atomic number, e.g., nitrogen) to broadband emitters (high atomic number, e.g., xenon) [25]. However, here, the conversion efficiency from laser energy into soft X-ray energy is comparably low due to the low density of the target material. Furthermore, achievable plasma sizes of several 100 μm are large. For metrology or scientific applications, though, these sources are very attractive due to their high cleanliness and versatility [8, 26].

In this study, a brilliance enhancement of laser-produced plasmas is demonstrated for gaseous jet targets, making use of supersonic effects. First, the theoretical background is provided to describe the physical properties of laser-produced plasmas as well as the gas dynamics of the related jet target. Experimental techniques are introduced that are employed to characterize the plasma and the gas jet. The effect of supersonic shocks within the target gas is investigated in both ways, theoretically and experimentally, revealing a significant brilliance enhancement for plasmas generated in respective shock regions. This chapter is based on a previous publication by Mey et al. [25] and has been revised and extended partly.

2. Physical properties of a plasma

Initially, the laser beam that irradiates the target material creates ions by multiphoton absorption, tunneling, or field ionization [27]. The resulting free electrons are accelerated by the strong electric field leading to inverse bremsstrahlung and avalanche ionization. A hot dense plasma state is generated. In competition to the heating processes, deionization takes place in terms of diffusion and recombination [27]. Depending on the electron temperature, a continuum of electromagnetic radiation is produced due to bremsstrahlung and recombination of free electrons with ions. Additionally, bound–bound transitions within the ions contribute narrow lines to the emission spectrum. A corresponding scheme is depicted in **Figure 1**.

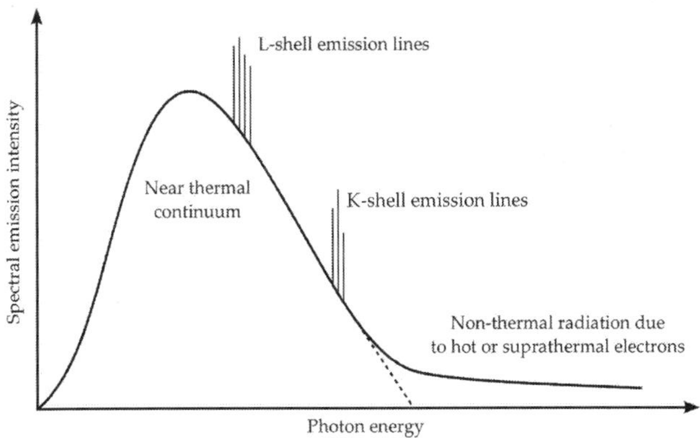

Figure 1. Scheme of the emission spectrum of a hot dense plasma.

The thermodynamics within a hot dense plasma can be approximated by the idealized state of a thermal plasma that is characterized by a single electron temperature T and a corresponding Maxwell velocity distribution. Within that simplification, the plasma may be treated as a blackbody that emits radiation with a continuous spectrum. The assumption that photons are emitted carrying discrete quanta of energy, with energy proportional to frequency, leads to the spectral energy density [1]

$$Br = 3.146 \times 10^{11} \left(\frac{k_B T}{eV} \right)^3 \frac{\left(\frac{\hbar \omega}{k_B T} \right)^3}{e^{\frac{\hbar \omega}{k_B T}} - 1} \frac{Photons}{mm^2 \, mrad^2 s \, (0.1\%BW)} \tag{2}$$

here, given in units of the brilliance with Planck's constant $\hbar$ and the photon frequency ω. Typical electron temperatures for gas targets irradiated by nanosecond laser pulses are 20–200 eV [28, 29]. The corresponding spectral maxima of the Planck distribution are found at the photon wavelengths 2.2–22.0 nm with peak brilliances of 3.6×10^{15} to 3.6×10^{18} 1/(s mm² mrad²). In fact, a laser-produced plasma is far away from thermodynamic equilibrium and a thermal plasma rather is an upper limit for the spectral power density. However, mostly a two-temperature model is already sufficient to adequately describe the continuum radiation by a hot dense plasma, which is then called near-thermal plasma [1]. In addition to the thermal electrons, a suprathermal component is introduced which is raised by nonlinear interactions such as resonant absorption. When these electrons undergo bremsstrahlung or recombination, they give rise to a high photon energy tail in the emission spectrum as indicated in **Figure 1**. Line radiation is emitted when electrons change their energy state within an ion from an outer to an inner electron shell. The resulting photon energy corresponds to the transition energy of the electron as described by Moseley's law, which is an extension of the Rydberg formula [30]

$$\frac{1}{\lambda} = \frac{R_\infty}{1 + m_e / m_{nuc}} (Z_{at} - S_{sh})^2 \left(\frac{1}{n_1^2} - \frac{1}{n_2^2} \right) \tag{3}$$

with the Rydberg constant R_∞, the nuclear mass m_{nuc} and the atomic number Z_{at}. The constant S_{sh} describes the shielding due to electrons between the core and the considered electron. Furthermore, n_1 and n_2 are the principal quantum numbers of the initial and final states of the electron. In plasmas of species with low atomic numbers like nitrogen ($Z_{at} = 7$), comparatively few free electrons are produced and the emitted radiation is dominated by single spectral lines. In contrast, elements with high atomic numbers such as xenon ($Z_{at} = 54$) yield much more free electrons, resulting in a spectrum of numerous closely packed lines and a significant thermal contribution.

Another important plasma parameter is the electron plasma frequency [1]

$$\omega_p = \left(\frac{e^2 n_e}{\epsilon_0 m_e} \right)^{1/2} \tag{4}$$

at which the free electrons tend to oscillate (where e is the electron charge, n_e is the electron density, m_e is the electron mass, and ϵ_0 is the vacuum permittivity). As a consequence, an incident electromagnetic wave can propagate in the plasma only if its frequency ω is greater than ω_p and it is totally reflected if $\omega = \omega_p$. This yields a critical electron density [1]

$$n_c = \frac{\epsilon_0 m_e \omega^2}{e^2} \tag{5}$$

which is $n_c = 1 \times 10^{23}$ cm^{-3} for a common Nd:YAG laser beam with a wavelength of 1064 nm. Thus, when the plasma reaches the critical electron density, it cannot further be heated to pose a limit especially on solid and liquid target concepts. In order to mitigate that limitation, a less intense prepulse can be used to heat the target material and decrease its density precedent to the main pulse [31].

3. Gas dynamics of jet targets

Supersonic gas jets employed as targets inherently exhibit strong density gradients. Here, the basics of supersonic nozzle flows and related shock phenomena are described theoretically, mainly based on [32, 33]. As a result, density estimations of the target gas are provided corresponding to the experimental situation at a laser plasma source.

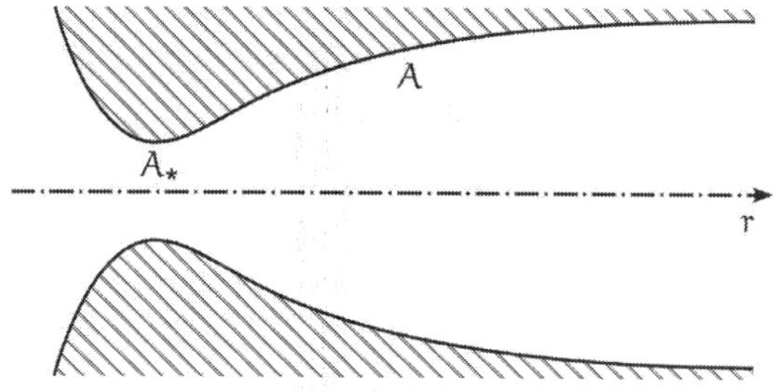

Figure 2. Sketch of a de Laval nozzle. A denotes the local cross-sectional area with the minimum value A_* at its throat position.

Let us first consider the example of a compressible fluid that expands through a convergent-divergent nozzle, a so-called de Laval nozzle as shown schematically in **Figure 2**. In gas dynamics, the basic equations to describe that problem are the conservation laws of mass and energy, formulated for compressible and isentropic flows. It can be shown that these relations lead to the well-known area relation between the local cross-sectional area A, the throat area A_*, and the local Mach number M [33]

$$\frac{A}{A_*} = \frac{1}{M}\left[\frac{2}{\kappa+1}\left(1+\frac{\kappa-1}{2}M^2\right)\right]^{\frac{\kappa+1}{2(\kappa-1)}}. \tag{6}$$

Here, $\kappa = c_p/c_v$ is the ratio of specific heats (c_p at constant pressure and c_v at constant volume) and the Mach number M is defined as the ratio between the local flow velocity and the local speed of sound. In the present example of a convergent-divergent nozzle, a gas is accelerated in the convergent part according to the continuity equation. If the critical Mach number $M_* = 1$ is reached at the throat, this results in supersonic velocities $M > 1$ in the divergent part, and the thermal energy of the gas is efficiently converted into directed kinetic energy. Concurrently, the gas density decreases according to the relation [32]

$$\frac{\rho}{\rho_0} = \left(1+\frac{\kappa-1}{2}M^2\right)^{-\frac{1}{\kappa-1}} \tag{7}$$

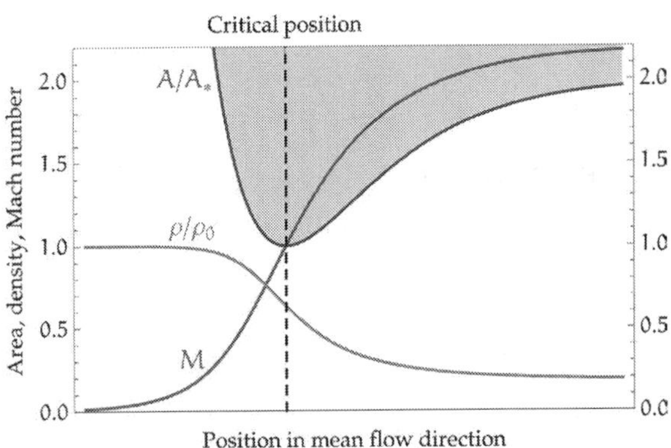

Figure 3. State functions of a flow in a de Laval nozzle: density ρ in terms of its stagnation value ρ_0, Mach number M and the local cross-sectional area A reaching A_* at its throat position. A diatomic gas with $\kappa = 7/5$ is assumed.

The shape of the cross-sectional area A/A_* of a typical de Laval nozzle is shown in **Figure 3** together with the resulting distribution of density ρ/ρ_0 (ρ_0 stagnation density) and Mach number M under the assumption of a diatomic gas with $\kappa = 7/5$. Utilizing a supersonic gas jet as a target for laser-produced plasmas requires large particle densities for high conversions efficiencies of laser energy into soft X-ray energy. Thus, a compromise needs to be found between a directed, but rarefied flow at high Mach numbers and divergent but denser flow at low Mach numbers. This can be achieved by adapting the nozzle geometry [34].

Within this work, shock waves, as they can be observed in supersonic flows, are employed to further optimize the particle density in a jet target. Within very short distances on the order of the mean-free path of the molecules, this phenomenon leads to an increase in density, pressure, and temperature while the Mach number decreases. Based on the conservation laws of mass, momentum, and energy, it is possible to derive equations that relate the initial values of those properties with the conditions right behind a shock wave. Here, it is sufficient to consider the change of the initial density ρ and the Mach number M in the case of a normal shock relative to the flow direction. After passing through the shock structure, these properties are denoted as $\hat{\rho}$ and $\hat{M}$, as indicated in **Figure 4**. The corresponding shock relations read [32]

$$\frac{\hat{\rho}}{\rho} = \frac{(\kappa+1)M^2}{2+(\kappa-1)M^2} \tag{8}$$

$$\hat{M} = 1 - \frac{M^2-1}{1+\frac{2\kappa}{\kappa+1}(M^2-1)}. \tag{9}$$

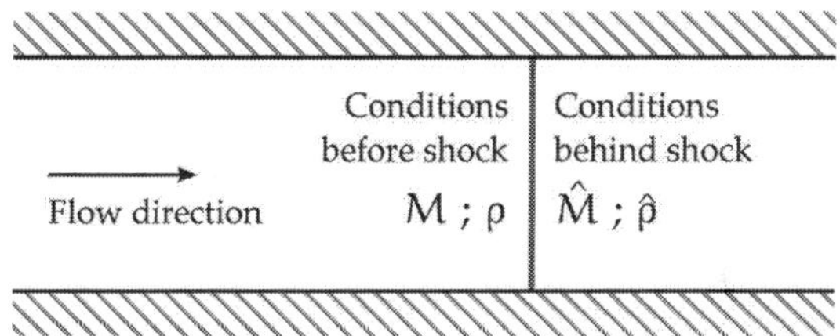

Figure 4. Normal shock structure in a supersonic flow. Gas passing through the shock experiences a decrease from the initial Mach number M to $\hat{M}$ and an increase in density from ρ to $\hat{\rho}$.

Basically, high Mach numbers lead to a strong compression of the fluid when passing through a shock. However, relation (8) defines an upper limit for the density ratio that can be achieved in connection with a shock wave. This limit is approached if $M \rightarrow \infty$, and for diatomic gases it is $\frac{\hat{\rho}}{\rho} \rightarrow 6$ ($\kappa = 7/5$). At the same time, the Mach number behind the shock decreases to $\hat{M} \rightarrow 1/7$. Shock waves appear, e.g., when obstacles perturb a supersonic flow or vice versa, or when objects travel with Mach numbers $M > 1$ through a gas at rest. In the case of a supersonic jet that expands from a stagnation pressure p_0 into an atmosphere with a sufficiently large background pressure p_b, shock waves can also be observed. At a certain distance to the nozzle exit, the collision between the jet particles and the surrounding gas particles leads to a shock structure, which is called barrel shock (see **Figure 5**). With respect to that situation, Muntz et al. introduced the rarefaction parameter [35]

$$\zeta = d_* \frac{\sqrt{p_0 \cdot p_b}}{T_0} \tag{10}$$

where d_* is the throat diameter of the nozzle and T_0 denotes the stagnation temperature. This parameter describes the interaction between jet and background particles, i.e., how strong the expansion flow is influenced by the surrounding gas. Muntz et al. propose a differentiation of the occurring flow into three regimes [35]:

- Scattering regime $\zeta \leq \zeta_s$

    Molecules of the background gas interact with the freely expanding jet by diffusion only, no distinct shock waves evolve.

- Transition regime $\zeta_s < \zeta < \zeta_c$

    Thick lateral shock waves develop and confine an undisturbed core of the jet that is surrounded by a mixing zone of jet and background particles.

- Continuum regime $\zeta_c \leq \zeta$

    The fully evolved barrel shock structure is present, as shown in **Figure 5**. The inner barrel shock waves and the Mach disk spatially delimit the influence of the background gas.

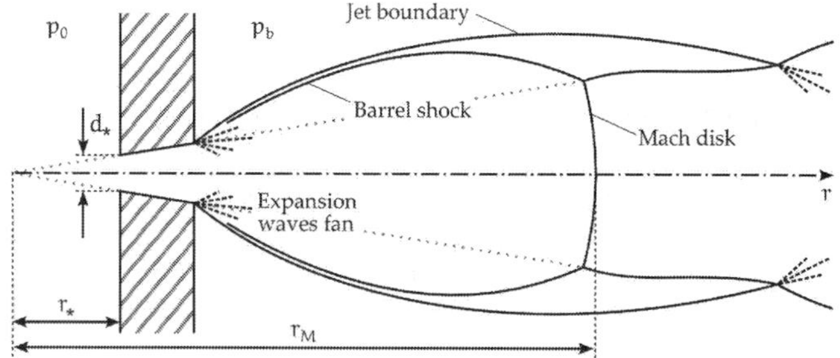

Figure 5. Typical structure of a barrel shock as apparent at supersonic jets in the presence of a background gas. Here, a fluid is expanded from a high pressure p_0 through the conically diverging nozzle into an ambient atmosphere of relatively low pressure p_b. The depicted shock system represents the continuum regime. Adapted from [36].

In the continuum regime, the extent of the shock structure scales with the nozzle pressure ratio p_0/p_b. In particular, within the range $15 < p_0/p_b < 17{,}000$, the distance $l_M = r_M - r_*$ between the nozzle throat and the Mach disk is given by [37]

$$l_M = 0.67 \cdot d_e \sqrt{\frac{p_0}{p_b}} \tag{11}$$

where d_e is the exit diameter of the orifice. It should be noted that this relation has been derived for nozzles with a constant diameter, i.e., for a nondivergent geometry.

In the following, estimations are made for a gas jet with barrel shock structures as it is under experimental investigation in this work too. Nitrogen expands from a pressure of $p_0 = 11$ bar into a helium atmosphere with a pressure of $p_b = 170$ mbar through a conically diverging nozzle (thickness $l_n = 1$ mm, throat diameter $d_* = 300$ μm, and exit diameter $d_e = 500$ μm). At rest, both gases are at room temperature $T_0 = 293$ K. In a simplification, a source flow is assumed corresponding to the dotted cone in **Figure 5** with its apex in a distance of $r_* = 1.5$ mm to the nozzle's throat. According to Eq. (11), the Mach disk appears 2.7 mm behind the nozzle throat, i.e., $r_M = 2.7$ mm. The dimensionless area of the assumed source flow is expressed in terms of the distance r to the virtual source as $A/A_* = (r/r_*)^2$. Solving Eqs. (6) and (7) results in the state functions ρ/ρ_0 and M along the symmetry axis of the nozzle from throat position to the Mach disk, i.e., in the range 1.5 mm $< r < 4.2$ mm. The conditions directly behind the Mach disk are determined by the shock relations (8) and (9). For $r > 4.2$ mm, the flow is assumed to be incompressible (ρ = const.) since the Mach number has decreased sufficiently below $M = 1$. Thus, subsequent behavior of M is approximated by the continuity equation $M(r) = \hat{M} \cdot (r/r_M)^2$.

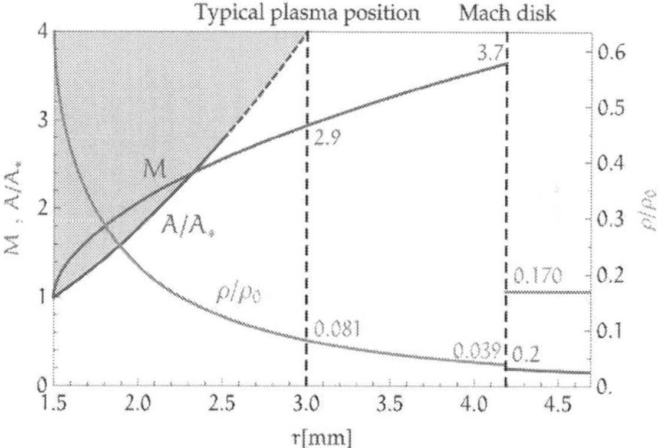

Figure 6. State functions along the symmetry axis of a barrel shock: density ρ in terms of its stagnation value ρ_0, and Mach number M. The solid blue line for A/A_* represents the cross-sectional area of the orifice, whereas the dashed blue line indicates the subsequent conical source flow.

The corresponding state functions $\rho(r)/\rho_0$ and $M(r)$ are depicted in **Figure 6** with respect to the distance r to the virtual source. Usually, the nozzle is operated with a background pressure on the order of $p_b \leq 10^{-4}$ mbar and the plasma is generated in a distance of 500 μm to the nozzle exit. The conditions at the usual plasma position, before and behind the Mach disk, are given in the diagram. It is revealed that due to the shock a two times higher density is achieved in a larger distance to the nozzle as compared to the typical plasma position. In practice, the density increase is even higher since plasma production takes place a few 100 μm besides the symmetry axis of the nozzle. Here, without ambient gas the jet is even more rarefied and with ambient gas the shock structure is present.

4. The laser-produced soft X-ray source

The setup of a standard soft X-ray source based on gas targets is used [8]. It basically consists of a piezoelectrically operated Proch–Trickl gas valve [38] mounted on a vacuum chamber, and a driving Nd:YAG laser that emits radiation at the fundamental wavelength of 1064 nm with a pulse energy of 800 mJ and a pulse duration of 6 ns (InnoLas SpitLight 600), see **Figure 7**. The intensity profile of the unfocused laser beam, measured by a CCD camera (Lumenera Lu160M), reveals a beam diameter of 5.9 mm (determined through $1/e^2$ decay), corresponding to a mean power density of 4.9×10^8 W/cm^2. Plasma production takes place as soon as a critical power density of $\approx 10^{12}$ W/cm^2 is reached in the focused beam at a sufficiently large particle density [39]. This initiates the first ionization of the target gas followed by avalanche ionization, creating large numbers of free electrons.

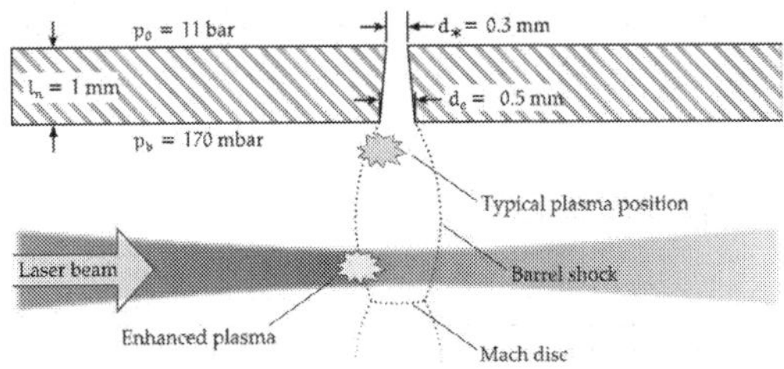

Figure 7. Pinhole camera images of the plasma at a stagnation pressure of p_0 = 11 bar for various background pressures p_b as given below the individual figure. The average of 30 single shots is shown.

The target gas is expanded through a divergent nozzle of conical shape. Over a length of l_n = 1 mm, its diameter increases from the throat diameter d_* = 0.3 mm to the exit diameter d_e = 0.5 mm. The nozzle is opened for a period of 1 ms, generating an underexpanded supersonic jet that expands from stagnation pressure p_0 = 11 bar into vacuum, i.e., the background pressure p_b is as low as 10^{-4} mbar. The laser is focused into the gas as soon as the jet flow is steady. The position where the plasma is produced is located in a distance of 500 µm, i.e., one diameter d_e behind the nozzle exit (see the typical plasma position indicated in **Figure 8**). Although the density is highest at the nozzle exit, the plasma should not be generated closer to the nozzle because of growing degradation effects. By employing different target species, various spectra can be obtained in the EUV and soft X-ray range. Noble gases with high atomic numbers such as xenon, argon, or krypton are broadband emitters, while oxygen or nitrogen each produces several narrow lines. The corresponding spectra can be found in **Figure 9**, produced by a system comparable to that described above and captured with a soft X-ray spectrometer, which is described in detail in [8]. Here, nitrogen is used in combination with a titanium filter, resulting in a monochromatic emittance at λ = 2.88 nm in the water window, corresponding to the transition $1s^2 - 1S2p$ of the valence electron of the N^{5+} ion [40].

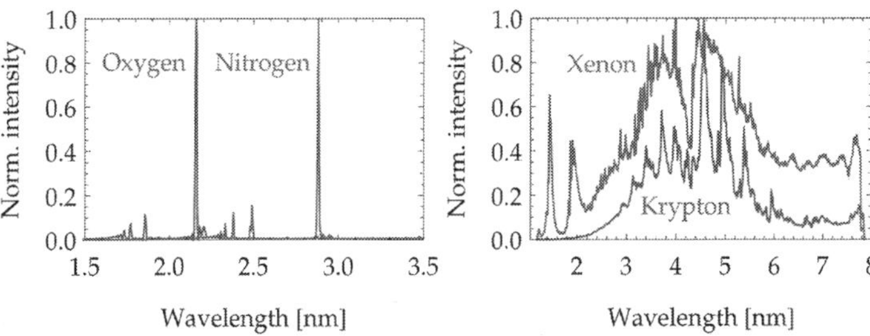

Figure 8. Photograph of experimental setup for plasma generation. The laser beam is focused by a lens into the vacuum chamber and generates the plasma right below the gas nozzle, which appears here in bluish color in the center of the chamber.

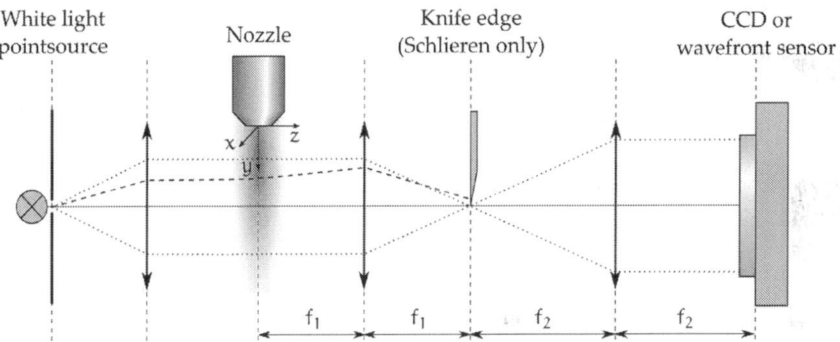

Figure 9. Principle of plasma generation employing jet targets: typically the plasma is generated close to the nozzle under vacuum conditions. Applying a background pressure p_b induces the barrel shock structure, which enhances plasma generation due to a local density increase.

In the approach pursued in this work, the background pressure p_b is increased to several tens of mbar in order to generate a barrel shock in the supersonic jet. For this purpose, helium is utilized as a background gas due to its high transmissivity of photons generated by the plasma. In addition, the optical path length of the resulting soft X-rays through helium is minimized by differential pumping. Another advantage of using helium as a surrounding gas is its large first ionization energy (24.6 eV) compared to that of nitrogen (14.5 eV) [41]. Thus, the critical power density to drive ionization by the incident laser beam is higher for helium, which ensures that only the target species nitrogen is ionized. Right behind the shock system generated in the jet, the particle density increases. In this manner, regions involving high densities of the target gas are obtained at comparably large distances from the nozzle. Thus, the plasma can be generated further away from the nozzle exit, and degradation effects are minimized.

5. Gas jet and soft X-ray diagnostics

In order to characterize the supersonic gas jet and the evolving shock structure, two different methods are employed: the Schlieren technique for highly resolved qualitative imaging of density gradients and wavefront measurements with a Hartmann–Shack sensor in order to quantify the density distribution, but at a lower resolution. Both methods are described in detail in Sections 5.1 and 5.2.

The plasma is imaged by a pinhole camera and the number of the resulting soft X-ray photons is determined with a calibrated photodiode. A description of these tools follows in Section 5.3.

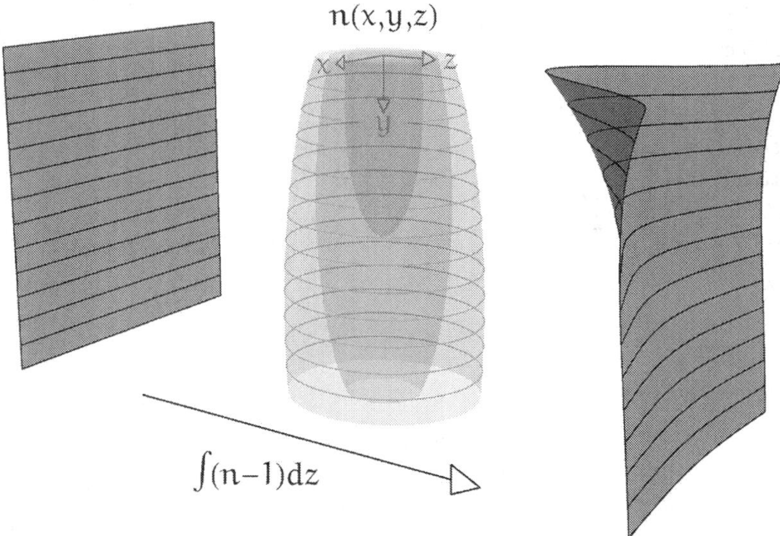

Figure 10. Characteristic emission spectra of various target gases, captured with a soft X-ray spectrometer.

5.1. Schlieren imaging

Schlieren imaging is a common technique in fluid dynamics that enables the qualitative measurement of density gradients [42]. The experimental setup is schematically shown in **Figure 10**. A pinhole with a diameter of 100 µm is illuminated by white light, and a focusing lens collimates the resulting beam, which then travels in the z-direction through the gas distribution of the jet target. The xy-plane at $z = 0$ is imaged by a 4f setup to a CCD camera

(Lumenera Lu160M) and captured with an exposure time of 50 μs. The camera is synchronized with the gas jet at a repetition rate of 10 Hz. Here, imaging lenses with focal lengths of $f_1 = 160$ mm and $f_2 = 300$ mm are used. A knife edge is moved close to the focal spot in between the two lenses, eliminating half of the spatial frequencies in the Fourier plane. The orientation of the blade determines which component of the density gradient will become visible. For example, as depicted in **Figure 10**, a knife edge aligned with the x-axis generates contrast proportional to the gradient of the refractive index $\partial n/\partial y$ corresponding to the density gradient $\partial\rho/\partial y$. Note, however, that in the Schlieren images shown below, the knife edge is aligned with the y-axis, so that density gradients within the jet are visualized in radial direction, thus emphasizing the barrel shock.

5.2. Wavefront monitoring

A Hartmann-Shack wavefront sensor [43, 44] is used to obtain quantitative information on the density distribution in the supersonic gas jet [34]. The experimental setup is mostly the same as that depicted in **Figure 10** for Schlieren imaging. However, the knife edge is removed and the CCD camera is replaced by the wavefront sensor. An initially plane wavefront of a test beam that travels through the target gas is deformed due to the spatial variation of the refractive index $n(x, y, z)$. The sensor splits the test beam into many subbeams by an array of microlenses, each producing a spot on a CCD camera (Lumenera Lu160M). The position of the spots contains the information of the local wavefront gradient. Thus, the deformation of the wavefront can be recovered. The spatial resolution Δx of a measured wavefront is equal to the pitch of the microlens array of 150 μm divided by the magnification factor $f_2/f_1 = 1.88$ of the 4f setup, yielding $\Delta x = 80$ μm.

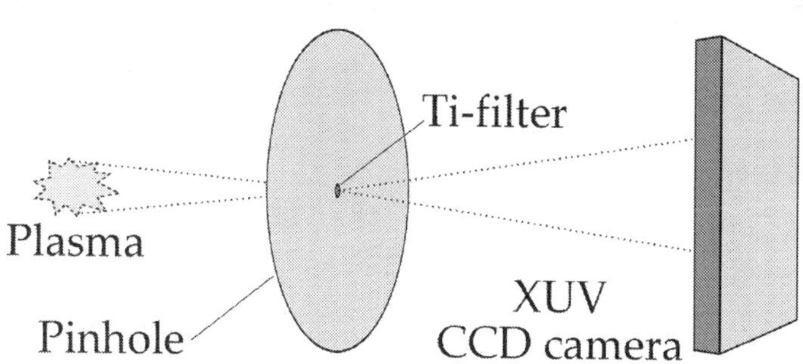

Figure 11. Experimental setup for Schlieren and wavefront measurements. The dotted lines represent the path of unrefracted light. The dashed line indicates a light ray that is refracted by varying distribution of gas density below the nozzle, thus hitting the knife edge and darkening the image. In order to monitor wavefront deformations, the CCD camera is replaced by a Hartmann–Shack sensor and the knife edge is removed.

The particle density distribution $N(x, y)$ in the nozzle plane $z = 0$ is recovered from a measured shape $w(x, y)$ of a deformed wavefront as follows. The test beam integrates $n(x, y, z)$ over the propagation direction z of the light beam, resulting in a difference $w(x, y)$ in the optical path, as illustrated in **Figure 11**. Now, it is assumed that in a plane corresponding to a constant $y = y_0$, $n(x, y_0, z)$ is approximated by a rotationally symmetric Gaussian shape with a maximum value $n_0(y_0) = n(0, y_0, 0)$. Then, the deformation of the wavefront reads

$$w(x, y) = \int \left[n(x, y, z) - 1 \right] dz$$

$$= \int \left[n_0(y) - 1 \right] \cdot \exp\left(-\frac{x^2}{2\sigma(y)^2} \right) \cdot \exp\left(-\frac{z^2}{2\sigma(y)^2} \right) dz$$

$$= \left[n_0(y) - 1 \right] \cdot \sqrt{2\pi}\, \sigma(y) \cdot \exp\left(-\frac{x^2}{2\sigma(y)^2} \right).$$

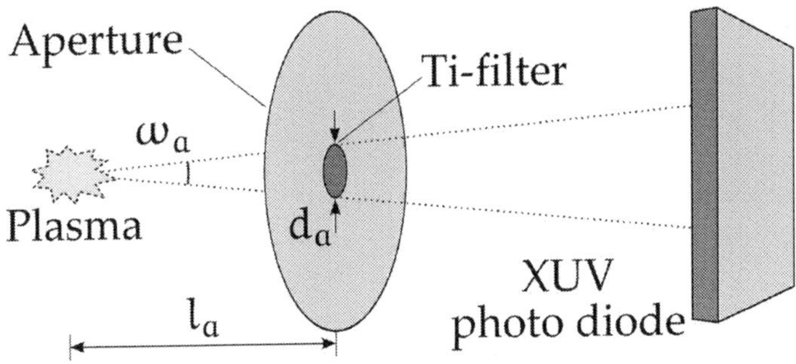

Figure 12. Wavefront deformation induced by the gas jet. The distribution of the optical density $n(x, y, z)$ increases the optical path length, resulting in the indicated wavefront deformation.

The standard deviation $\sigma(y)$ of $n(x, y, z)$ is determined from the shape of the measured deformation of the wavefront $w(x, y)$ by a Gaussian fit. The distribution of the refractive index in the plane $z = 0$ containing the jet axis is recovered by

$$n(x, y, 0) - 1 = \frac{w(x, y)}{\sqrt{2\pi}\sigma(y)}. \tag{12}$$

Conversion of the refractive index $n(x, y, 0)$ into a particle density N is done by using the Lorentz-Lorenz formula [45]

$$\frac{n^2 - 1}{n^2 + 2} = \frac{4}{3}\pi \alpha N \tag{13}$$

where α, the polarizability of the considered gas particles, is derived using the values n = 1.0002974 and N = 2.69 × 10^{19} cm^{-3} for nitrogen under normal conditions [41] (at a temperature of 273.15 K and a pressure of 1013.25 mbar). In these calculations, the surrounding helium atmosphere is neglected because of its low refractive index that amounts to only a few percent as compared to that of the nitrogen jet.

5.3. Plasma characterization

Qualitatively, the plasma is characterized by a pinhole camera system as sketched in **Figure 12**. It consists of a phosphor-coated CCD camera (Lumenera Lu160M with three layers of phosphor P43 with a grain size of ≈ 1 µm) in combination with a titanium-filtered pinhole (100 µm diameter, Ti-layer 200 nm thick). This way, the intensity distribution of radiation at the wavelength λ = 2.88 nm is captured. Here, the luminescent area A is approximated by an ellipsoidal shape with the semiaxes a and b. Then, $A = \pi\, a\, b$, where a and b are defined as the full-widths at half-maximum of the intensity in the x- and y-directions. The uniformity of the plasma is characterized by its eccentricity $\varepsilon = \sqrt{a^2 - b^2}\big/a$. Examples of the intensity images are shown in **Figure 17** in combination with the corresponding Schlieren images of the gas jet for the case of both, gas issuing into vacuum and gas issuing into a background gas and thus forming a barrel shock.

Quantitatively, the peak brilliance Br of the plasma is derived by

$$Br = \frac{N_{ph}}{\tau \Omega A}\tag{14}$$

with the pulse duration τ, the solid angle Ω, and the source area A. The number of photons N_{ph} with a wavelength of λ = 2.88 nm is determined by a calibrated XUV photodiode (International Radiation Detectors, AXUV100), which is equipped with a titanium filter (thickness 200 nm, applied to a nickel mesh with a transmissivity of 0.89). As shown in **Figure 13**, only these photons reach the detector that propagate within a cone confined by a circular aperture with diameter d_a in a distance l_a to the plasma. Thus, the solid angle is defined by the corresponding opening angle $\omega_a = 2 \tan d_a/(2l_a)$ of the cone [46]

$$\Omega = 4\pi \sin^2 \sin^2 \frac{\omega_a}{4}.\tag{15}$$

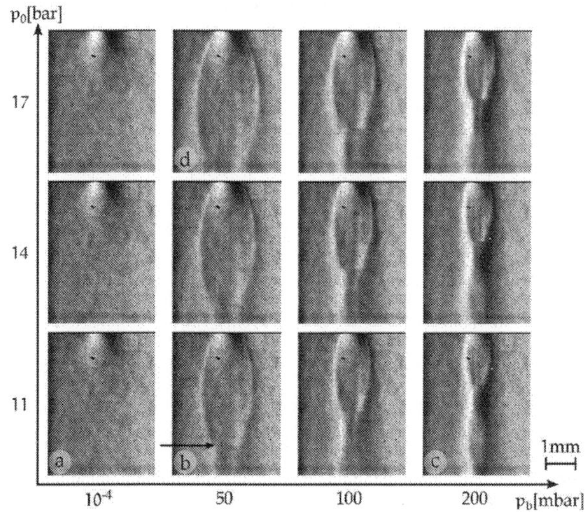

Figure 13. Principle of plasma characterization by pinhole camera.

In good approximation, the lifetime of the plasma is assumed to be $\tau = 6$ ns, which equals the duration of the exciting laser pulse. Finally, the luminescent area A is determined with a pinhole camera as described above.

6. Experimental results

First, the gas jet and the effect of a background pressure on the resulting flow structure are investigated using the techniques described in the previous section. The derived results are compared to theoretical relations discussed in Section 3. Subsequently, the effect of the barrel shock on the plasma generation is explored and the brilliance improvement of the soft X-ray source is quantified.

6.1. Characterization of the target gas jet

Depending on the stagnation and background pressure, the gas jet may form various shapes, which are discussed in the following. In previous studies of laser-produced soft X-ray sources, the nozzle was operated in the range $p_0 = 11$–17 bar at a background pressure of $p_b = 10^{-4}$ mbar, i.e., practically without any background gas. In this case, the emerging flow is in the scattering regime and does not show any discontinuities. Independently of p_0, the density distribution has a maximum value at the nozzle exit and rapidly falls off in all directions. The corresponding Schlieren images are taken with the knife edge aligned with the y-axis and can be found in **Figure 14** for the pressure range $p_0 = 11$–17 bar.

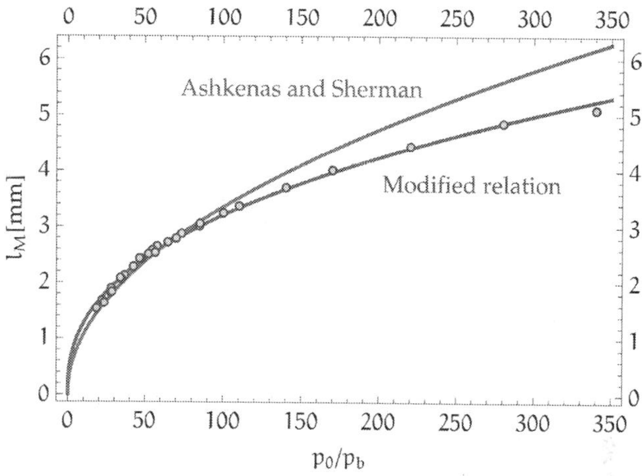

Figure 14. Principle of plasma characterization by diode measurement.

With rising background pressure, particle collisions increasingly affect the gas jet and retard its free expansion. At a certain distance from the nozzle, this results in a shock that is directly connected to a local decrease of the Mach number M. At the same time, the local particle density increases. This becomes evident in regions of the Schlieren images that show strong changes in intensity, implying high-density gradients. As can be seen, e.g., in **Figure 14(b)**, the shape of the resulting shock structure resembles a barrel, why it is referred to as a barrel shock. In the downstream direction, the barrel shock is terminated by the Mach disk, which is indicated in the Schlieren image by an arrow. In the present Schlieren pictures, the Mach disk is reproduced only weakly because the knife edge was aligned perpendicular to the disk and only density gradients parallel to the disk were detected.

Increasing p_b, as from **Figure 14(b)–(c)**, results in a confinement of the gas flow toward the nozzle axis—the lateral shocks approach each other and the Mach disk moves upstream. In contrast to this, increasing p_0 has the opposite effect, i.e., the radius of the barrel shock and the width σ of the density distribution increase and the Mach disk moves downstream, see **Figure 14(b)–(d)**. These two opposite effects allow generation of the same shock structure at different combinations of the pressures, provided that the ratio p_0/p_b stays constant.

From the Schlieren images as shown in **Figure 14**, the distance l_M between the Mach disk and the nozzle throat is derived for pressure ratios in the range $18 \leq p_0/p_b \leq 340$. In **Figure 15**, the resulting data set is compared to the empirical relation (11)

$$l_M = 0.67 \cdot d_e \left(\frac{p_0}{b_b} \right)^{1/2}$$

(16)

which has been derived by Ashkenas and Sherman [37] for a nondivergent nozzle. Apparently, the experimental results deviate from the depicted curve, especially for large pressure ratios. Most likely, this can be attributed to a different nozzle geometry as in the present situation. Here, a divergent orifice initially guides the supersonic expansion of the gas before it expands freely into the helium atmosphere. For that case, a relation of the form

$$l_M = d_e \left(\frac{p_0}{p_b} \right)^a \tag{17}$$

reveals good agreement with the measured shock distances as it is evident in **Figure 15**. By a least-squares fit routine, the exponent is derived to $a = 0.4034$.

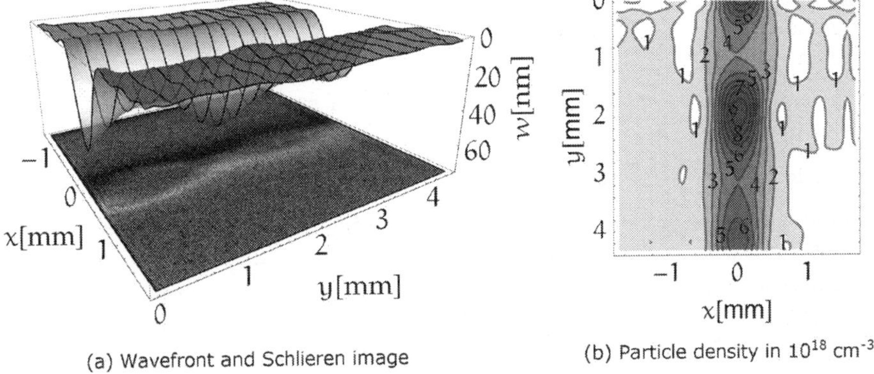

(a) Wavefront and Schlieren image

(b) Particle density in 10^{18} cm^{-3}

Figure 15. Schlieren images indicating the supersonic flow structure of an N_2 jet as a function of stagnation and background pressure (flow direction: top → bottom). (a) Scattering regime, no internal structures evolve; (b) continuum regime with barrel shock structure, the Mach disk is indicated by the arrow; (b) → (c) shock structure contracts for increasing background pressure; (b) → (d) shock structure inflates for increasing stagnation pressure.

In **Figure 16**, wavefront and Schlieren measurements are compared with each other for a stagnation pressure of $p_0 = 11$ bar and a background pressure of $p_b = 170$ mbar. The results of both techniques are well consistent. The particle density $N(x, y)$ shows the mean gas distribution inside the jet. In the downstream direction, along the nozzle axis, N first decreases to $N_{min} = 4.0 \times 10^{18}$ cm^{-3} and then increases again up to a maximum value of $N_{max} = 9.8 \times 10^{18}$ cm^{-3}. Subsequently, the wave-like behavior of the particle density is repeated at lower density values. The observed maxima coincide approximately with the positions where the lateral shocks interfere, forming a Mach disk.

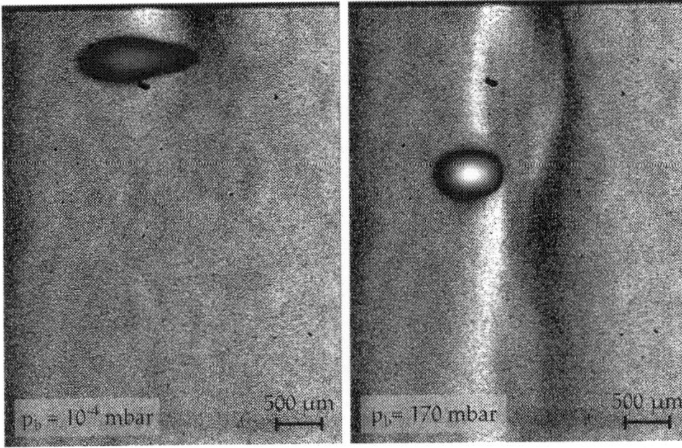

Figure 16. Distance between Mach disk and nozzle exit for various pressure ratios. The points are derived from the Schlieren images, the violet curve represents the empirical relation (11) from Ashkenas and Sherman [37] and the blue curve represents the modified relation given in Eq. (17) with $a = 0.4034$.

Employing the relations of gas dynamics introduced in Section 3, a rough theoretical estimate of the particle density ahead and behind the Mach disk is now compared to the results obtained with the wavefront sensor. Corresponding to the experimental situation, the density distribution along the symmetry axis of the gas jet is shown in terms of its stagnation value ρ_0 as shown in **Figure 6**. There, a normal shock induces a density increase from $\rho_{min} = 0.039\,\rho_0$ to $\rho_{max} = 0.170\,\rho_0$ in a distance of $l_M = 2.7$ mm to the nozzle throat. For the current pressure ratio of $p_0/p_b = 64.7$, this coincides with the position of the Mach disk.

In order to derive absolute density values, the stagnation density ρ_0 of the nitrogen jet is required. Following the ideal gas law, $p_0 = \rho_0\, R_{sp}\, T_0$ results in $\rho_0 = 12.65$ kg/m^3 with the specific gas constant for nitrogen $R_{sp,N_2} = 296.8$ J/(kg · K) [41], the temperature $T_0 = 293$ K, and pressure $p_0 = 11$ bar inside the vessel. Finally, particle densities ahead and behind the Mach disk are evaluated with the molecular mass of nitrogen $m_{N_2} = 4.653 \times 10^{-26}$ kg [41]. A comparison between the theoretical estimation and the measured values is given in **Table 1**.

	Theoretical estimate	Measurement
N_{min} (10^{18} cm^{-3})	10.6	4.0
N_{max} (10^{18} cm^{-3})	46.1	9.8

Table 1. Particle density ahead (N_{min}) and after (N_{max}) the barrel shock, given on the symmetry axis of the jet. Comparison between theoretical estimate and measurement.

The estimated values are of the same order of magnitude but larger than the experimental results. This discrepancy can be attributed to the spatial resolution of the wavefront sensor that is not able to resolve the high density value right behind a shock. Furthermore, the estimate provides an upper limit of the particle density since in a simplification a conical source flow has been assumed. In fact, the stream lines of the flow are bended in lateral direction, stronger than the cone geometry presumes. Consequently, this results in a higher rarefaction of the gas and the typical bulbous barrel shock. This explains why values of the estimated particle densities, both of the maximum and the minimum, are higher than the corresponding measured values.

6.2. Characterization of the plasma enhancement

The effect of an increase in target gas density on the plasma generation is illustrated in **Figure 17** for a stagnation pressure of p_0 = 11 bar. Taking advantage of the barrel shock, obviously the brightness of the plasma is raised, whereas its size has decreased in the direction of the incident laser beam. Due to the increased target density, there are more emitters of soft X-ray radiation in the same volume. Besides, the absorption of laser energy is raised. Thus, the power density of the beam decreases more rapidly below its critical value and no further atoms are ionized. This confines the size of the plasma in the beam direction and explains its smaller size. Another mechanism causing the reduced size might be plasma defocusing [47]. Due to an increased plasma density, a stronger defocusing effect can be expected, limiting the ionization region. During the experiments, it turned out that generation of a plasma right below the Mach disk, where the density is expected to be at a maximum, is not the optimal position. It was found that even brighter and smaller plasmas occur when the laser is focused onto the edge of the jet at a location slightly above the Mach disk and after the barrel shock (see **Figure 17**). This behavior may be caused by reabsorption of soft X-rays by the surrounding nitrogen particles.

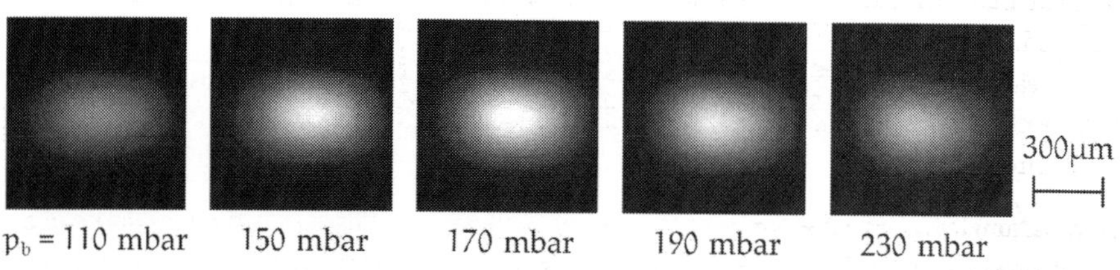

p_b = 110 mbar   150 mbar   170 mbar   190 mbar   230 mbar

Figure 17. (a) Combination of quantitative wavefront and qualitative Schlieren image of the N_2 jet expanding from p_0 = 11 bar into an He atmosphere with p_b = 170 mbar. (b) Density distribution $N(x, y)$ of the N_2 jet in the plane z = 0, which results from the wavefront as described in Section 5.2.

The barrel shock is enclosed by a thin supersonic compressed layer, which becomes thicker at the Mach disk [36], leading to increased reabsorption. In order to study the brilliance improvement depending on the location of plasma generation with respect to shock structures in the jet, the latter were varied by changing the background pressure at a constant stagnation pressure (p_0 = 11 bar). By lowering p_b, the radius of the barrel shock is increased; conversely, with increasing p_b, the radius of the barrel shock decreases. Thus, with the location of the focus of the laser beam fixed, its relative location with respect to high-density regions behind the shock is changed. In **Figure 18**, intensity distributions of the plasma are shown for various background pressures p_b. In this case, the location of plasma generation is kept constant. An optimum value is found at $p_b = 170$ (see also **Figure 17**).

Figure 18. Pinhole camera images of the plasma superimposed on the Schlieren images of gas jet at $p_0 = 11$ bar. Left: under vacuum conditions $p_b = 10^{-4}$ mbar. Right: with ambient He atmosphere at $p_b = 170$ mbar. Both plasma images are an average of 30 single shots.

Unexpectedly, increasing both p_0 and p_b while preserving the pressure ratio p_0/p_b, does not lead to a considerable further increase in the brilliance of the source. Approaching high-pressure values ($p_0 = 17$ bar), quite the reverse happens: the plasma appears even darker. It can be assumed that, in fact, more soft X-ray photons are generated since the target density is increased. However, the density of the background gas is increased as well, which leads to higher reabsorption of the generated photons. The latter effect seems to dominate the former. It is expected that further efforts in differential pumping can shorten the path length of the soft X-rays through the outer helium gas so that the brilliance of the source can further be increased.

Now, parameters characterizing the plasma in the optimal case are compared with those of a plasma produced near the nozzle exit with a jet in the scattering regime. In both cases, the same stagnation pressure of $p_0 = 11$ bar is considered. Regarding the shape of the resulting plasma, which is represented by its luminescent area, it can be seen that the radiating area is reduced by a factor of 0.71 to $A = 0.063$ mm^2, and its eccentricity decreases slightly from $\varepsilon = 0.91$ to $\varepsilon = 0.80$ when a barrel shock is present. This results in a better brilliance and improves the coherence properties due to a smaller source size and a more uniform shape. The number of photons emitted per pulse and solid angle from the nitrogen plasma at a wavelength of $\lambda = 2.88$ nm is raised by a factor of 7.1 to a value of 1.2×10^{13} sr^{-1}. Based on these values, the peak brilliance can be computed. One finds an improvement by a factor of 10 to a value of $Br = 3.15 \times 10^{16}$ photons/(mm^2 mrad^2 s). This clearly demonstrates the advantage of utilizing the density increase across a barrel shock system. An overview of the characteristic parameters of the plasma is given in **Table 2**.

	Without barrel shock	With barrel shock	Factor
Radiating area (mm²)	0.088	0.063	0.71
Eccentricity [1]	0.91	0.80	0.88
Photons/(solid angle · pulse) (sr⁻¹)	1.66×10^{12}	1.18×10^{13}	7.10
Peak brilliance (mm⁻² mrad⁻² s⁻¹)	3.15×10^{15}	3.15×10^{16}	10.0

Table 2. Comparison of plasma emission characteristics at $\lambda = 2.88$ nm obtained with a nitrogen jet issuing into vacuum (no barrel shock) and into a background gas (with barrel shock).

7. Conclusion

Laser-produced plasmas based on gas targets serve as versatile and nearly debris-free soft X-ray sources at a table-top size. In this chapter, a method has been shown by which the brilliance of gas targets can be improved. To this end, a background pressure is applied to the gas jet that leads to a strong recompression of the target particles. For the example of an underexpanded supersonic nitrogen jet, the resulting barrel shock has been qualitatively visualized by the Schlieren photography. The corresponding density distribution was obtained by a quantitative Hartmann-Shack measurement. Measured values of the shock location and particle densities are of the same order of magnitude as those of a first estimate that was partly based on correlations. The size of the resulting plasma is reduced by a factor of 0.71 and its shape becomes more uniform, thus improving the coherence properties of the source. At the same time, the number of photons per solid angle at $\lambda = 2.88$ nm is raised by a factor of 7.1. In this manner, the brilliance of the source is increased by a factor of 10.0 to $Br = 3.15 \times 10^{16}$ photons/(mm² mrad² s). Even greater increases may be obtained by using hydrogen as the background gas since H_2 shows a 13 times lower absorption of the generated X-rays compared to He [48]. However, for safety reasons H_2 has not been employed here. A further increase in the plasma's brilliance is to be expected with increasing the stagnation and background pressure. An essential condition for achieving this is an improvement in the differential pumping system in order to lower the reabsorption of the soft X-rays by the background gas.

Acknowledgements

The author kindly acknowledges support by Deutsche Forschungsgemeinschaft within Sonderforschungsbereich 755 "Nanoscale photonic imaging." Furthermore, the author acknowledges the permission to reproduce material [25], which has been previously published by IOP Publishing & Deutsche Physikalische Gesellschaft under CC BY-NC-SA license.

Author details

Tobias Mey

Address all correspondence to: tobias.mey@llg-ev.de

Laser-Laboratorium Göttingen e.V., Göttingen, Germany

References

[1] Attwood D. Soft X-Rays and Extreme Ultraviolet Radiation. Cambridge: Cambridge University Press; 1999.

[2] Bakshi V, editor. EUV Lithography. SPIE Press; Bellingham, Washington, USA. 2008. 702 pp.

[3] Wu B, Kumar A. Extreme ultraviolet lithography: a review. Journal of Vacuum Science & Technology B. 2007; 25: 1743–1761. DOI: 10.1116/1.2794048

[4] Scholze F et al. New PTB beamlines for high-accuracy EUV reflectometry at BESSY II. Soft X-Ray and EUV Imaging Systems. .Proc. SPIE 4146. 2000; 72. DOI: 10.1117/12.406678

[5] Scholze F, Laubis C. Use of EUV scatterometry for the characterization of line profiles and line roughness on photomasks. In: Mask and Lithography Conference (EMLC), 24th European Conference; 2008.

[6] Banyay M, Juschkin L. Table-top reflectometer in the extreme ultraviolet for surface sensitive analysis. Applied Physics Letters. 2009; 94(6): 063507. DOI: 10.1063/1.3079394

[7] Chen JG. NEXAFS investigations of transition metal oxides, nitrides, carbides, sulfides and other interstitial compounds. Surface Science Reports. 1997; 30: 1–152. DOI: 10.1016/s0167-5729(97)00011-3

[8] Peth C, Barkusky F, Mann K. Near-edge X-ray absorption fine structure measurements using a laboratory-scale XUV source. Journal of Physics D. 2008; 41: 105202. DOI: 10.1088/0022-3727/41/10/105202

[9] Stöhr J. NEXAFS Spectroscopy (Springer Series in Surface Sciences). Springer; Berlin Heidelberg, Germany. 2003.

[10] Da Silva LB et al. X-ray laser microscopy of rat sperm nuclei. Science. 1992; 258: 269–271. DOI: 10.1126/science.1411525

[11] Berglund M et al. Compact water-window transmission X-ray microscopy. Journal of Microscopy. 2000; 197: 268–273. DOI: 10.1046/j.1365-2818.2000.00675.x

[12] Hertz HM et al. Laboratory X-ray microscopy for high-resolution imaging of environmental colloid structure. Chemical Geology. 2012; 329: 26–31. DOI: 10.1016/j.chemgeo.2011.07.012

[13] Barkusky F et al. Direct photoetching of polymers using radiation of high energy density from a table-top extreme ultraviolet plasma source. Journal of Applied Physics. 2009; 105: 014906. DOI: 10.1063/1.3054565

[14] Haney SJ et al. Prototype high-speed tape target transport for a laser plasma soft-x-ray projection lithography source. Applied Optics. 1993; 32: 6934–6937. DOI: 10.1364/ao.32.006934

[15] Wagner C, Harned N. EUV lithography: lithography gets extreme. Nature Photonics. 2010; 4: 24–26. DOI: 10.1038/nphoton.2009.251

[16] Peth C et al. XUV laser-plasma source based on solid Ar filament. Review of Scientific Instruments. 2007; 78: 103509. DOI: 10.1063/1.2801882

[17] Hansson BAM. Characterization of a liquid-xenon-jet laser-plasma extreme-ultraviolet source. Review of Scientific Instruments. 2004; 75: 2122–2129. DOI: 10.1063/1.1755441

[18] Vogt U et al. High-resolution spatial characterization of laser produced plasmas at soft x-ray wavelengths. Applied Physics B. 2004; 78: 53–58. DOI: 10.1007/s00340-003-1338-9

[19] Jansson PAC et al. Liquid-tin-jet laser-plasma extreme ultraviolet generation. Applied Physics Letters. 2004; 84: 2256–2258. DOI: 10.1063/1.1690874

[20] Richardson M et al. High conversion efficiency mass-limited Sn-based laser plasma source for extreme ultraviolet lithography. Journal of Vacuum Science & Technology B. 2004; 22: 785–790. DOI: 10.1116/1.1667511

[21] Wieland M et al. EUV and fast ion emission from cryogenic liquid jet target laser-generated plasma. Applied Physics B. 2001; 72: 591–597. DOI: 10.1007/s003400100542

[22] Bakshi V, editor. EUV Sources for Lithography. SPIE Press; Bellingham, Washington, USA. 2006.

[23] Rymell L, Hertz HM. Debris elimination in a droplet-target laser-plasma soft x-ray source. Review of Scientific Instruments. 1995; 66: 4916–4920. DOI: 10.1063/1.1146174

[24] Rakowski R et al. Characterization and optimization of the laser-produced plasma EUV source at 13.5 nm based on a double-stream Xe/He gas puff target. Applied Physics B. 2010; 101: 773–789. DOI: 10.1007/s00340-010-4327-9

[25] Mey T et al. Brilliance improvement of laser-produced soft x-ray plasma by a barrel shock. New Journal of Physics. 2012; 14: 073045. DOI: 10.1088/1367-2630/14/7/073045

[26] Fiedorowicz H et al. Compact laser plasma EUV source based on a gas puff target for metrology applications. Journal of Alloys and Compounds. 2005; 401: 99–103. DOI: 10.1016/j.jallcom.2005.02.069

[27] Morgan CG. Laser-induced breakdown of gases. Reports on Progress in Physics. 1975; 38: 621–665. DOI: 10.1088/0034-4885/38/5/002

[28] Fiedorowicz H et al. Investigation of soft X-ray emission from a gas puff target irradiated with a Nd:YAG laser. Optics Communications. 1999; 163: 103–114. DOI: 10.1016/s0030-4018(99)00100-5

[29] Müller M et al. Emission properties of ns and ps laser-induced soft X-ray sources using pulsed gas jets. Optics Express. 2013; 21: 12831. DOI: 10.1364/oe.21.012831

[30] Mayer-Kuckuk T. Atomphysik. Teubner Verlag; Stuttgart, Germany. 1997.

[31] Komori H et al. EUV radiation characteristics of a CO_2 laser produced Xe plasma. Applied Physics B. 2006; 83: 213–218. DOI: 10.1007/s00340-006-2172-7

[32] Wieghardt K. Theoretische Strömungslehre. Universitätsverlag; Göttingen, Germany 2005.

[33] Spurk J, Aksel N. Strömungslehre. Berlin: Springer; 2010.

[34] Peth C et al. Characterization of gas targets for laser produced extreme ultraviolet plasmas with a Hartmann-Shack sensor. Review of Scientific Instruments. 2004; 75: 3288–3293. DOI: 10.1063/1.1791314

[35] Muntz EP et al. Some characteristics of exhaust plume rarefaction. AIAA Journal. 1970; 8: 1651–1658. DOI: 10.2514/3.49856

[36] Rebrov AK. Free jets in vacuum technologies. Journal of Vacuum Science & Technology A. 2001; 19: 1679–1987. DOI: 10.1116/1.1382649

[37] Ashkenas H, Sherman FS. The structure and utilization of supersonic free jets in low density wind tunnels. In: Proceedings of the Fourth Symposium on Rarefied Gas Dynamics; 1966. pp. 84–105.

[38] Proch D, Trickl T. A high-intensity multi-purpose piezoelectric pulsed molecular beam source. Review of Scientific Instruments. 1989; 60: 713. DOI: 10.1063/1.1141006

[39] Phuoc TX. Laser spark ignition: experimental determination of laser-induced breakdown thresholds of combustion gases. Optics Communications. 2000; 175: 419–423. DOI: 10.1016/s0030-4018(00)00488-0

[40] Moore CE. Selected tables of atomic spectra, atomic energy levels and multiplet tables — N IV, N V, N VI, N VII. National Standard Reference Data Series, NSRDS-NBS3 (Sect. 4). Washington, D.C. 1971.

[41] Haynes WM, editor. CRC Handbook of Chemistry and Physics, 94th ed. CRC Press; Boca Raton, Florida 2013.

[42] Settles G. Schlieren & Shadowgraph Techniques. Springer; Berlin Heidelberg, Germany 2006.

[43] Hartmann J. Bemerkungen über den Bau und die Justierung von Spektrographen. Zeitschrift für Instrumentenkunde. 1900; 20: 47.

[44] Platt BC, Shack R. History and principles of Shack-Hartmann wavefront sensing. Journal of Refractive Surgery. 2001; 17: 573–577.

[45] Born M, Wolf E. Principles of Optics: Electromagnetic Theory of Propagation, Interference and Diffraction of Light. Cambridge University Press; Cambridge, UK. 1999.

[46] Bronstein IN et al. Taschenbuch der Mathematik. Europa-Lehrmittel; Haan-Gruiten, Germany.2013.

[47] Auguste T et al. Defocusing effects of a picosecond terawatt laser pulse in an underdense plasma. Optics Communications. 1992; 89: 145–148. DOI: 10.1016/0030-4018(92)90148-k

[48] Henke BL et al. X-ray interactions: photoabsorption, scattering, transmission, and reflection at E = 50-30,000 eV, Z = 1-92. Atomic Data and Nuclear Data Tables 1993; 54: 181–342. DOI: 10.1006/adnd.1993.1013

Chapter 5

High-Brightness Solid-State Lasers for Compact Short-Wavelength Sources

Akira Endo

Additional information is available at the end of the chapter

http://dx.doi.org/10.5772/64147

Abstract

Various types of compact short-wavelength sources are emerging in the region from EUV to hard X-ray and further to gamma ray. These high-energy photons are usually accessible in a large-scale facility such as SR or FEL, and the compactness of these new technologies provides new possibilities for broader applications in dedicated laboratories or factories. Laser-produced plasma is used for soft X-ray laser and high average power EUV sources for lithography. Laser Compton short-wavelength sources are now entering into practical applications in medical imaging. The performance of these sources critically depends on the laser driver performance. This chapter describes the recent progress of high-brightness, short-pulse solid-state laser technology in close relation to these new compact short-wavelength sources. Pulsed picosecond thin disc laser progress is reviewed with kW average power specifications. Cryogenic laser is reported for the advantage of higher beam quality in large-pulse energy operation.

Keywords: thin disc laser, cryogenic laser, plasma EUV source, laser Compton X-ray source

1. Introduction

A rapid progress is recently observed in the field of compact extreme-ultraviolet (EUV) and X-ray sources with high brightness and small footprint enough to be installed in laboratories in educational and research institutions, manufacturing facilities, hospitals, and other suitable sites [1]. This may advance scientific and technical disciplines in practical applications by complementing large-scale synchrotron radiation and free-electron laser sources. Applications span a wide range from biomedical, semiconductor, fundamental and applied

research, environmental engineering to industrial nondestructive testing. Component technology progress is one of the key factors in these advancements of the compact EUV through hard-X-ray sources. These key elements are instrumentation, optics, detectors, data management and processing, and one of the most significant factors is the progress of high average power, short-pulse solid-state lasers.

Semiconductor industry has been struggling in the past two decades to establish a technological system of extreme-ultraviolet lithography as the ultimate scheme, and the establishment of reliable, high average power (>100 W) 13.5 nm source has been always the most critical challenge. The basic architecture is now realized as the LPP (laser-produced plasma) EUV source, in which the conditioning of the mist target from a liquid tin droplet is essential for higher conversion efficiency and perfect recovery of the injected tin atoms [2]. The mist formation is performed by a diverging shock wave inside the microdroplet, which is driven by an impulse generated by an irradiation of picosecond solid-state laser pulse of mJ level pulse energy. The system repetition rate is typically 100 kHz, and the laser average power is more than 100W. The size of the droplet is 10 μm in diameter, and the required laser beam quality and stability must meet the requirements.

Lasing was reported in the EUV spectrum region by efficient excitation of dense plasma columns at 100 Hz repetition rate using a tailored pump pulse profile of a 1 J picosecond cryogenic Yb:YAG laser [3]. The average power of the 1 J picosecond laser is 100 W. The tabletop soft-X-ray laser average power is 0.1 mW at λ = 13.9 nm and 20 μW at λ = 11.9 nm from transitions of Ni-like Ag and Ni-like Sn, respectively. Lasing on several other transitions with wavelengths between 10.9 and 14.7 nm was also reported. The efficient X-ray laser operation was realized by an optimized pump pulse design as a nanosecond prepulse followed by two picosecond pulses to create higher density plasma of Ni-like ions of higher temperature for higher gain in longer time and in larger space. The high average power of these compact soft X-ray lasers promises to enable various applications requiring high photon flux with coherence.

Laser Compton X-ray source has been established as a compact, high-brightness short-wavelength source. The basic principle is similar to an undulator emission, and a high-intensity laser field is used as the modulating electromagnetic field. The laser Compton X-ray source is demonstrated as a compact short-wavelength imaging approach combined with the phase contrast method of biosamples. Single-shot imaging is critical for many practical applications, and the required specification depends on the usable laser pulse with some threshold parameters because all other component technologies are well matured. The optimization of the laser Compton hard X-ray source by single-shot base is already studied in detail [4, 5]. Experimental results well agreed with theoretical predictions. Highest peak brightness is obtained in the configuration of counterpropagating laser pulse and electron beam bunch, in the minimum focusing area before nonlinear threshold [6, 7]. A single-shot phase contrast bioimaging was demonstrated in the hard X-ray region [8]. The employed laser was a picosecond CO_2 laser of 3 J pulse energy [9], but the laser system was not an easy and compact one for further broad applications in various laboratories and hospitals. The Extreme Light Infrastructure–Nuclear Physics (ELI–NP) facility will have a brilliant γ-beam of 10^4 photons/s/eV, ≤0.5% bandwidth, with $E\gamma$ < 19.5 MeV, which is obtained by the laser Compton method from an intense electron beam (Ee > 700 MeV) produced by a warm linac [10]. The main purpose is to provide an opportunity for the production of radioisotopes for medical research. The repetition rate is 100 Hz with a 1 J, picosecond Yb:YAG laser. A standard laser Compton X ray source is under construction as the STAR project at the University of Calabria (Italy) to generate monochromatic tunable, ps-long, polarized X-ray beams, ranging from 20 to 140 keV. The X-rays will be devoted to experiments of material science, cultural heritage, advanced radiological imaging with microtomography capabilities [11]. An S-band RF gun produces electron bunches at 100 Hz, boosted up to 60 MeV by a 3 m long S-band cavity. It is critical to use a high-brightness linac of low emittance and high pointing stability to focus higher charge bunch to a smaller spot size down to 10 µm. The allowed spatial stability is a few µm. The research and development of the X-ray generation laser is the key technology for higher and stable X-ray generation. The Yb:YAG laser is ideal for a compact, high pulse energy picosecond pulse and should be synchronized to the RF system in less than picosecond time jitter.

Compact short-wavelength sources are emerging due to the progress of extreme ultraviolet lithography (EUVL) in semiconductor industry. The EUVL has been intensively developed in the field of various component technologies, for example, Mo/Si high reflectivity mirror at 13.5 nm wavelength, new types of resist of higher sensitivity at this wavelength, and plasma-based 100 W class stable EUV sources. Further increase of average power is expected for large-scale manufacturing to kW level and shorter wavelength to 6.7 nm where a higher reflectivity mirror seems available. The necessity to evaluate an alternative approach is recently proposed based on high repetition rate free electron laser (FEL), to avoid a risk of the source power limit by the plasma-based technology. The possibility is indicated to realize a high repetition rate (superconducting) FEL to generate a multiple kW 13.5 nm light. It is important to note that the present FEL pulses are characterized typically as 0.1 mJ pulse energy, 100 fs pulse duration, and 1 mm beam diameter, and generated in the SASE mode. The beam fluence is higher than the ablation threshold of typical resists, and the beam has a higher spatial coherence, which leads to speckle patterns. The beam is composed of many short spikes with high peak intensities [12]. Seeding an FEL with an external coherent source has been studied together with SASE operation to increase the brightness and pointing/energy stability compared to SASE mode. An efficient seeding method was established by using UV wavelength laser in which the seed laser modulates the electron beam into coherent bunching at the harmonics of the seed laser wavelength. The bunching is intensified in another undulator for coherent FEL action, and the method is named as high-gain higher harmonic generation (HGHG). A successful demonstration is reported from FERMI as a double stage-seeded FEL with a fresh bunch injection technique [13]. The fresh bunch scheme was demonstrated as the FEL radiation produced by one HGHG stage acts as an external seed for a second HGHG stage. A 10 Hz demonstration was reported in the EUV wavelength region. The development of higher repetition rate FEL requires new optical laser developments to meet the needs of laser-induced FEL seeding. Conventional copper accelerating cavities operate up to tens to hundreds of hertz, but superconducting (SC) cavities, allow a much higher repetition rate of up to few megahertz. FLASH at DESY has a maximum repetition rate of 1 MHz within a burst structure (electron bunch train) of 800 µs at 10 Hz. Future linear accelerator designs plan an SC linear accelerator capable of a continuous repetition rate of up to 1 MHz. This presents major challenges for the design and operation of laser-seeded FELs in both burst and continuous mode. At lower repetition rates, conventional Ti:Sapphire lasers are currently used for laser-induced FEL seeding at, for example, FERMI FEL-1. The future requirements of a tunable, high repetition rate laser with sufficient pulse energy can be met with optical parametric chirped-pulsed amplification (OPCPA). A tunable OPCPA is demonstrated at 112 W in burst mode. The center wavelength is located in the wavelength region of 720–900 nm. The repetition rate is 100 kHz and the pulse energy is 1.12 mJ with 30 fs pulse duration. The OPCPA pumping laser power limits the scalability of the OPCPA output, and it was demonstrated for a 6.7–13.7 kW (burst mode) thin-disk OPCPA-pump amplifier, increasing the possible OPCPA output power to many hundreds of watts. Furthermore, the third and fourth harmonic generation experiments are performed for the FEL seeding purpose [14].

Recent solid-state laser progress is closely related to the demands in the field of laser microablation in industry. Fiber laser is advancing in the high repetition rate, short-pulse operation mode in the subpicosecond pulse length. Significant progress has been made on the scaling of the performance of subpicosecond fiber laser systems in the past decade. The current limitation exists in the achievable peak power and average power of a linear amplifier. The maximum of the available average power in a single fiber laser is determined by the mode instabilities. Several hundred watts is the typical maximum power, depending on the properties of the fiber and other system parameters. The pulse energy is ultimately limited by the extractable energy of the fiber, nonlinear pulse distortions, and damage issues. Four coherently combined fiber amplifiers were reported as a single CPA system [15]. The average power was 530 W and combined pulse energy was 1.3 mJ. It is expected to realize higher system parameters from a beam combined fiber laser, especially in higher average power in pulsed mode. The beam quality was excellent and the beam combination efficiency was as high as 93%. It is expected that with the coherent combination concept and further progress in fiber laser technology, average powers in the range of 1 kW and pulse energy of 10 mJ are realistic parameters in the future. A 10 J, 10 kHz femtosecond laser system is under conceptual design by a coherent combination of 10,000 fibers as the extension of the coherent combining scheme for high repetition rate PW laser [16].

Another promising laser is the InnoSlab laser, which is a thin slab laser cooled from both surface and is reported as a Yb:YAG InnoSlab amplifier with femtosecond pulses of <3 mJ pulse energy with a repetition rate of 100 kHz. The chirped pulse amplification is essential to achieve high average power generation in the power amplifier stage. The laser system is consisted of a 10 mW seed laser with a pulse repetition rate of 100 kHz to MHz, and a preamplifier stage, and a high power InnoSlab amplifier which is followed by a grating pulse compressor. This laser system is ideal for OPCPA pumping and micromaterial processing [17]. The highest average power picosecond laser was reported from a thin-disk multipass laser amplifier, delivering 1.4 kW with pulses of 4.7 mJ pulse energy and duration of 8 ps at 300 kHz repetition rate [18]. The beam quality factor was better than $M2 = 1.4$. The experiments showed that the thin-disk multipass amplifier can scale pulse energy and average output power independently in the repetition rates between 300 and 800 kHz. Frequency doubling by means of an LBO crystal led to 820 W of average power at a wavelength of 515 nm with 1170 W of incident IR power which corresponded to a conversion efficiency of 70% and an SHG pulse energy of 2.7 mJ. By sum-frequency generation between the beams at 1030 and 515 nm in a second LBO crystal, an average UV power of 234 W (780 µJ of pulse energy) was generated at the wavelength of 343 nm with a conversion efficiency of 32%. The output powers in the green and UV spectral region are limited by thermal effects and the apertures of the crystals employed. Future work may try to use shorter seed pulses as well as to increase the output power by implementing a higher number of passes in the amplifier and the pump module and by increasing the pump power. For the higher harmonic generation, crystals with larger apertures and an improved temperature control is critical to further improve the performance.

Carbon fiber-reinforced plastic (CFRP) is the most promising light material in aircraft or similar machines. CFRP was processed with the kW picosecond laser with 8 ps pulses and an average output power of up to 1.1 kW at a pulse repetition rate of 300 kHz with a maximum pulse energy of 3.7 mJ. Heat accumulation influences are studied for the processing quality in high average power operation [19]. The pulse overlapping and repetitive scans are studied for the heat accumulation effect in the report. The study indicates an estimation of optimized feed rates and maximum scan speeds. The kW picosecond thin disc laser demonstrated its applicability in the cutting application of a 2 mm CFRP with a high cutting speed of 0.9 m/min and smaller thermal damage less than 20 µm. These lasers, such as fiber, InnoSlab, and thin disc, have been proving solutions for high beam quality, short-pulse generation in the high average power regime in the past two decades. An alternative approach was reported by a cryogenically cooled Yb:YAG by demonstration to have significant potential for efficient near-diffraction-limited high average power lasers [20]. A single-pass amplifier was reported with 250 W output power, 54% optical-optical efficiency, M2= 1.1 and a power oscillator with 300 W output power demonstrated 64% optical-optical efficiency, and M2= 1.2. In each case, the laser systems were based on end-pumped laser rod gain modules cryogenically cooled in liquid nitrogen cryostats. The single-pass amplifier is a simple way, compared to fiber or thin disc, to boost the power of a laser oscillator. The output power in the experiments was limited only by the incident pump power. The cryogenically cooled, bulk Yb:YAG four-pass amplifier was operated at 100 kHz repetition rate [21]. The amplified optical pulses were 2.5 mJ pulse energy with <20 ps pulse length before compression and the spectrum for 3.6ps in transform limited duration. The measured power stability was less than 0.5% in half an hour full power operation. A flat-top spatial profile was measured with near-diffraction-limited beam divergence. This compact amplifier is ideal for pumping of OPCPA. This chapter describes recent progress of high average power, picosecond thin-disc laser from the research and development of the HiLASE project during 2012–2015. HiLASE R&D laser center is a technological infrastructure in Dolní Břežany near Prague in the Czech Republic, which was founded in a close connection to the ELI activity. Major effort is to develop lasers for high-tech application, in which the short-wavelength generation is one of the dominant ones. HiLASE focuses on the development of kW-class thin-disk-based picosecond and subpicosecond lasers from mJ to sub-1-J pulse energy. Laser pulses are emitted at repetition rates from 1 to 100 kHz with prospective upgrade up to 1 MHz near fundamental wavelength. In order to cover the broadest application potential of the lasers, it was also initiated high-power harmonic frequency generation and high-power mid-IR picosecond system consisting of an OPG followed by double OPA systems (**Figure 1**).

Figure 1. Building of the HiLASE R&D Centre in Dolní Březany, Czech Republic.

2. High repetition rate picosecond Yb:YAG thin disc-laser in LPP EUV source

Continuous shrinking of the microcircuit is the natural law for lower cost, higher yield, short time to the market in the semiconductor industry. The microlithography has been the central manufacturing technology, and the continuous shrinking of the wavelength is the principal architecture. The proposal of the application of EUV wavelength appeared long before the perspective of the light source itself. The shift of source technology to the ArF excimer is followed by immersion technology and the ArF laser is the long-life light source technology. The EUV lithography is now entering into the mass production phase in the 22 nm node, and the wavelength is 13.5 nm (92.5 eV) supported with Mo/Si high reflectivity mirrors. 13.5 nm wavelength is the first generation of ionizing radiation in the mass production of semiconductor industries. The laser-produced plasma (LPP) EUV source has been established as the basic architecture of the EUV source technology, after one decade of focused research and engineering. The present concern is the stability and cleanness of the source itself and further engineering is continued [22]. The EUV light source is essentially incoherent spherical emission from highly ionized Tin plasma. The source is composed of three parts, namely driving laser, plasma generation/exhaust, and EUV light collector. A large Mo/Si collector mirror has peak reflectivity at 13.5 nm with 2% bandwidth. It is located close to the high-power plasma source and the extension of the lifetime is the most critical engineering concern. It is reported in a recent conference that the power available at the intermediate focus (IF) in the field is 125 W, and a test source is operated in a company laboratory aiming at 250 W [23]. A typical configuration of the LPP EUV source for high volume manufacturing (HVM) is shown in **Figure 2**, where a train of 100 kHz Sn droplet is injected and irradiated by a solid-state laser prepulse (purple), dispersed into a mist bunch, and irradiated by a CO_2 laser main pulse (red). A discharge pumped EUV source is now employed for metrology purpose in less than 100 W level. The typical configuration is shown to the right of the LPP system. A small laser pulse initiates Sn vapor for main discharge from a rotating disc immersed in Sn liquid [24]. It is called as laser-assisted discharge plasma (LDP).

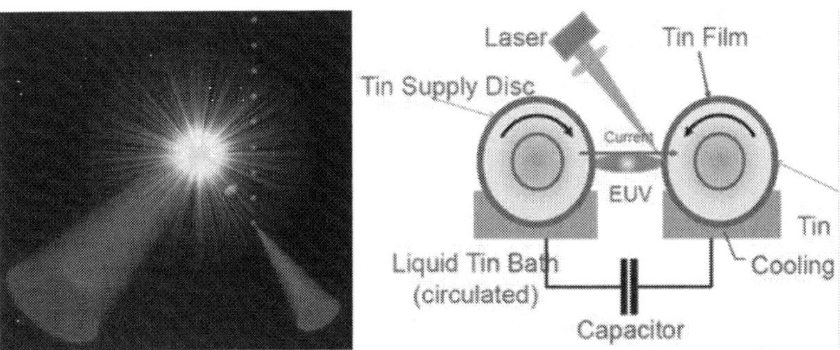

Figure 2. Configuration of double pulse method in LPP (left) and LDP (right) EUV sources.

The initial state of the injected Sn droplet is liquid phase of 10–20 μm diameter in the LPP system, and the direct laser irradiation results in a lower conversion efficiency (CE) and messy split of liquid Sn inside the chamber. The solution is the double-pulse method, as the initial pulse converts the liquid Sn droplet into nanocluster bunch (mist) for better laser absorption and ionization. An experiment demonstrated that the prepulse is much efficient in the case of picosecond pulse length compared to the nanosecond one. **Figure 3** shows the experimental results reported in a conference [25].

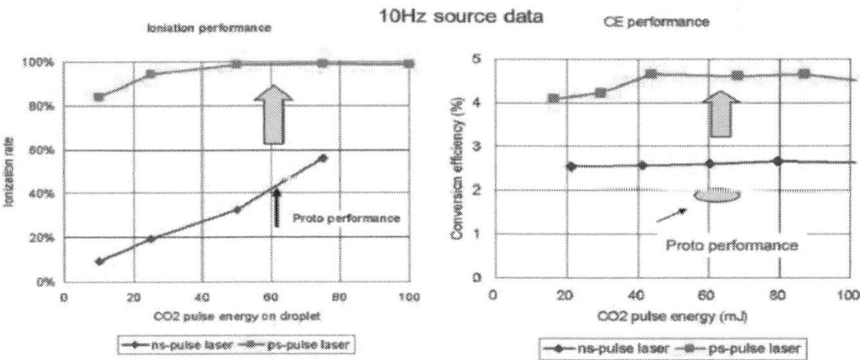

Figure 3. Ionization rate of Sn and CE depending on the pulse length of pre-pulse. Left: Ionization rate. Right: Conversion efficiency (CE).

The picosecond laser has typical parameters as pulse energy more than mJ, pulse length is 10 ps or less, and focusing diameter is a few times larger than the droplet diameter of 10–20 μm. The average power is more than 100 W at the repetition rate of 100 kHz. The laser specification is not easily covered by any commercial products and must be specifically developed. Thin-disc laser is suitable for the required specification among other types of advanced lasers such as fiber or thin slab with its larger beam diameter. HiLASE project was dedicated in a research and development of kW class picosecond thin-disc lasers in the period of 2012–2015. One of the laser beamlines is PERLA (Pearl) C, which is aimed to realize a compact, stable 500 W picosecond thin-disc laser with 100 kHz repetition rate [26]. The research and development of the laser system is briefly described in the following.

Design of the laser comes from the thin-disk laser concept. **Figure 4** shows the configuration of the thin-disc laser module with a parabolic mirror that collimates and images the pump radiation from laser diodes. The parabolic mirror images several times the unabsorbed pumping radiation with a set of roof mirrors. The thermal lensing is limited minimum due to the axial thermal flow from the gain medium to the water cooled heat sink. The nonlinear effects in the solid-state medium (self-phase modulation, B-integral) are controlled at low level in the multiple optical passes in the thin disc. The cooling is efficient due to the small thickness of the disc. The typical discs are characterized by the gain thickness as 100–300 μm and the disc diameter as 8–30 mm. Special optical design is required to compensate the low single-pass amplification gain together with pump absorption. Regenerative amplifier is selected for medium-power amplifier, and multipass amplifier is designed for higher average power or higher pulse energy amplifier. Regenerative amplifiers allow very compact and robust laser systems. High-power regenerative amplifier concept is based on a ring cavity, which is in fact a new approach. High average power and high repetition rate regenerative amplifiers usually suffer from Pockels cell issues. A new kind of large aperture BBO Pockels cell was developed to overcome this obstacle (**Figure 4**). A kW-class regenerative amplifier with a ring cavity is a novel approach in the field of picosecond thin disk lasers.

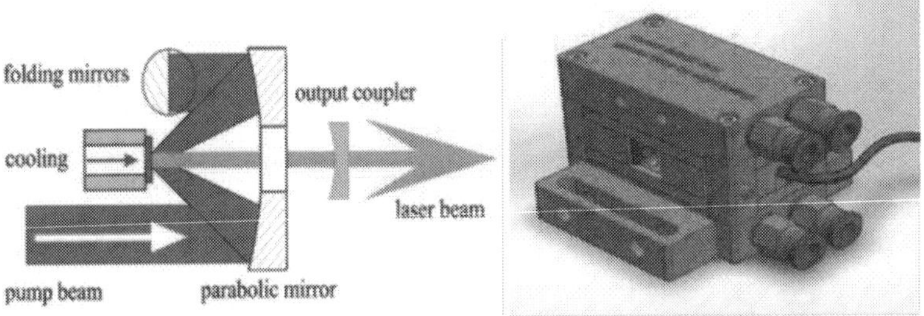

Figure 4. Left: Concept of efficient pumping (blue beam) of thin-disk lasers. Right: in-house developed large-aperture and water-cooled BBO Pockels cell.

Various solid-state materials are applied in thin disc modules, and the Yb:YAG is the most favored one due to high quality in fabrication and picosecond pulse generation. Yb:YAG is studied for more than two decades in its growing, cutting, and polishing, and its thermome-chanical characteristic is well fitted for picosecond and subpicosecond pulse generation. One of the disadvantages of the thin-disc laser is the bonding technology of large diameter thin Yb:YAG disc to the heatsink basement to be robust in high-temperature and high optical fluence environment. Several bonding methods are available to 10 mm diameter and further new techniques are still tested for higher reliability. In the present stage, HiLASE Centre uses two types of bonding methods, namely soldering to a copper-tungsten heatsinks and bonding to a diamond substrate. The diamond substrate is advantageous for its higher thermal conductivity for lower disc temperature under high pumping fluence. Popular pumping source is a laser diode with 940 nm center wavelength where the absorption has a broader bandwidth. Yb:YAG has a narrower but high-peak absorption wavelength at 968.8 nm, which is called as zero-phonon line, and a specific laser diode at this wavelength is used for efficient pumping. The quantum defect decreases from 8.7% with 940 nm pumping to 5.9 % with zero-phonon line pumping. Zero-phonon line pumping is also better in its suppression of nonlinear phonon relaxation in the Yb:YAG medium. The resulting steady-state disc temperature is kept lower compared to 940 nm pumping, and better stability of the amplification and higher output pulse energy is the positive result [27]. Pump diodes should have bandwidth <1 nm. Since the absorption line near 968.8 nm is very narrow, the diodes are stabilized by volume Bragg gratings (**Figure 5**).

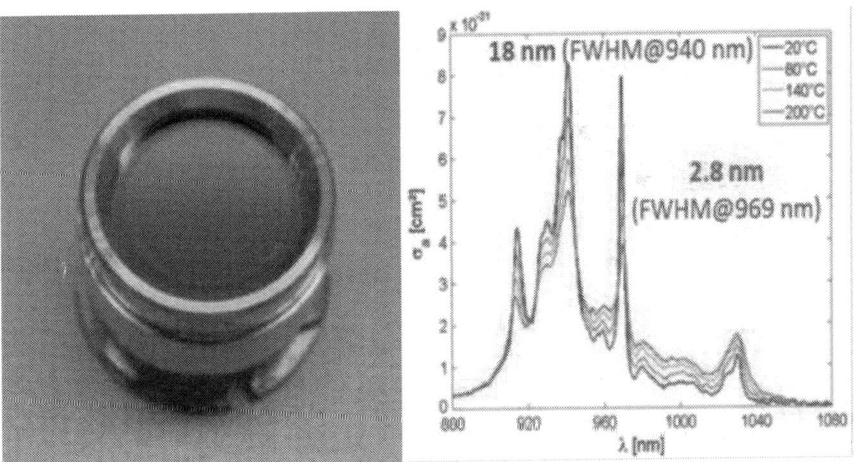

Figure 5. Left: Diamond-heat spreader-bonded thin disc. Right: Absorption cross section of Yb:YAG.

The high repetition rate beamline PERLA C operates at 100 kHz and provides picosecond pulses from 1 to >4 mJ in a compressed pulse. The seeder of the laser system is a commercially available Yb-doped fiber laser from Fianium. The pulse length is 12 ps at 50 MHz repetition rate, and the pulse energy is 6 nJ with 20 nm broad bandwidth. The pulses are stretched to 0.5 ns pulse length by a small Bragg grating. The pulse bandwidth is filtered to 2.2 nm by the bandwidth of the grating, and the bandwidth limited pulse length is less than 2 ps. The seeder pulses are amplified by a semiconductor optical amplifier (SOA) and a single-mode fiber amplifier before injection into a regenerative amplifier. The advantage of the SOA is its electric controllability of the gain time window and used as a pulse picker to reduce the repetition rate from 50 to 1 MHz. The average power is 300 mW before the regenerative amplifier.

The regenerative amplifier is composed of a single Yb:YAG module with a standing wave cavity for 100 W operation (**Figure 6**). The total footprint is compact as 900 × 1200 mm including a pulse compressor. The pump spot size of the thin disc is 2.7 mm in diameter with cavity length of 2 m. A double Pockels Cell system optically switches the input and output pulses. The size of the BBO crystal is 8 × 8 mm^2. The crystal holder is engineered to avoid damages to the BBO by piezo ringing in high repetition rate switching. The maximum available BBO aperture is 12 × 12 mm^2, and the repetition rate is 1 MHz and the voltage is 10 kV. As described earlier, the pumping is by zero-phonon line continuous fiber-coupled laser diodes. The maximum amplified pulses are 1.2 mJ of energy at 100 kHz repetition rate with M2 = 1.3 beam quality

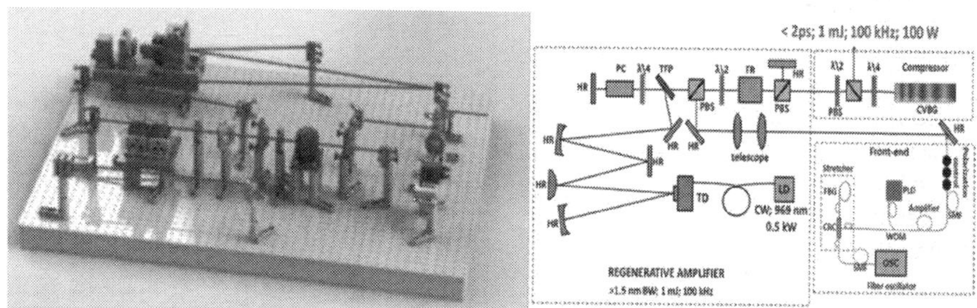

Figure 6. Left: 100 W regenerative amplifier. Right: Optical scheme including a CVBG pulse compressor (HR, highly reflective mirror; TD, thin disk; LD, pump laser diodes; PC, Pockels cell; TFP, thin-film polarizer; FR, Faraday rotator; PBS, polarizing beam splitter; CVBG, chirped volume Bragg grating; λ/2 and λ/4, half- and quarter-wave plates; SMF, single-mode fiber; PLD, pump laser diode; WDM, multiplexer; CRC, circulator; FBG, fiber Bragg grating; OSC, oscillator).

This is critically important for precise irradiation like prepulse in an LPP EUV source. Pulses are compressed by a chirped volume Bragg grating (CVBG) compressor, which is a very robust, compact, and easy to align bulk compressor with 8 × 8^2 mm aperture. The CVBG compressor was tested for long time operation and demonstrated a reliable pulse compression of high average power pulse train with >85 % diffraction efficiency under optimized cooling condition. Compressed pulses measured by intensity autocorrelation (**Figure 7**, right) have temporal width 1.6 ps (sech^2). The pulse-to-pulse energy stability measured over 4 million pulses was better than 1.7%, and the long-term average power stability measured over 1 h was <1.5 % (RMS value). Better housing and active stabilization can even improve the stability.

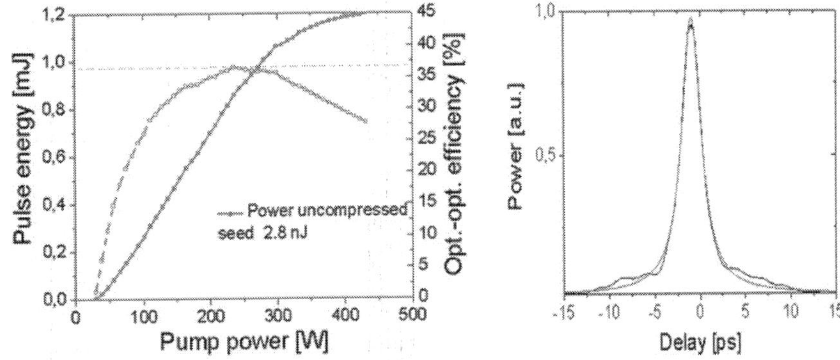

Figure 7. Left: 1.2 mJ of the output pulse energy at 100 kHz repetition before compression has been achieved from the 100 W PERLA C in a nearly diffraction-limited beam. Right: the pulses were compressed to 1.6 ps (FWHM) by a CVBG as shown by the intensity autocorrelation trace.

A higher average power regenerative amplifier was developed with a ring cavity (**Figure 8**). The amplifier is switched by a reliable Pockels cell, which is in-house design for 10×10 mm^2 aperture with effective cooling. The fundamental spatial mode operating cavity is designed for a 5.2 mm pump spot and the cavity contains a single diamond-bonded Yb:YAG thin disc. The disc is zero-phonon line-pumped by VBG-stabilized fiber-coupled diodes. Laser cavity was tested in the CW regime to evaluate the thermal distortion. 550 W output was observed with almost 50% optical-optical efficiency and >4 mJ was achieved in a 100 kHz pulse train with a nearly diffraction-limited output beam (**Figure 8**).

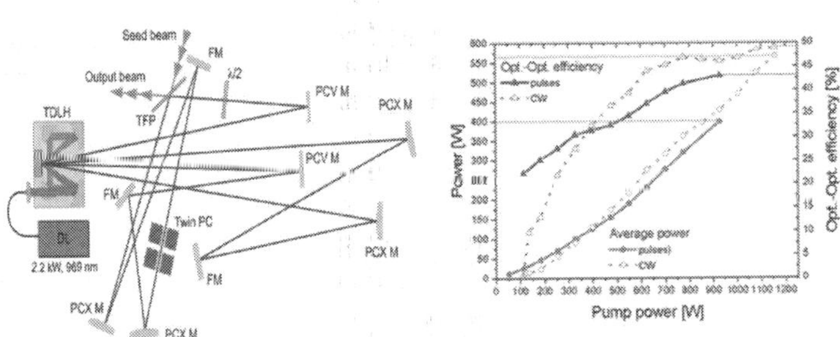

Figure 8. Left: ring cavity of the 500 W PERLA C laser system. FM, folding mirror; PCX M, planoconvex mirror; PCV M, planoconcave mirror; TFP, thin-film polarizer; PC, Pockels cell; TDLH, thin-disc laser head; DL, pump diodes; $\lambda/2$, half-waveplate. Right: performance of the 500 W ring cavity in CW and pulse mode (PERLA C).

3. High average power wavelength conversion of picosecond solid-state lasers

The extreme ultraviolet lithography is now in an introductory phase in semiconductor industry. The EUVL has been developed in the field of various component technologies such as Mo/Si high reflectivity mirror at 13.5 nm wavelength, new types of resist of higher sensitivity at this wavelength, and plasma-based 100 W class stable EUV sources. Further increase in average power is expected for large-scale manufacturing to kW level and shorter wavelength to 6.7 nm where a higher reflectivity mirror seems available. The present source architecture is the laser-produced plasma (LPP) and is recently considered in its practical scaling limitation in average power in the range of kW. Free electron laser has been emerging as the new short-wavelength source in the EUV to X-ray region in the past decade. The present generation is based on lower repetition rate operation for scientific applications, but the next generation is aiming at high repetition rate for high average power. Several research papers are discussing on the possibility of high repetition rate FEL by superconducting RF cavity technology for the generation of more than kW average power at 13.5 nm wavelength [28, 29]. The present FEL is operated in the SASE mode, in which the pulses are generated in undulator and composed of many short-pulse length spikes. The typical pulse parameters are 0.1 mJ pulse energy, 100 fs pulse length, and the beam diameter is 1 mm. The beam fluence is higher than the ablation threshold of a resist [30], and the high spatial coherence results in much higher localized peak fluence on the resist. The interaction mechanism is now in a basic study to overcome these effects compared to the present LPP-generated 100 kHz, mJ EUV pulses with no coherence and longer pulse length as 10 ns.

The scaling of the FEL technology to kW average power level requires the photocathode operation in higher repetition rate in industrial environment together with optical technology to optimize the FEL beam for lithography application and scaling to the 6.7 nm wavelength region.

The first consideration is the industrial operation of photocathode at >MHz repetition rate. The bunch charge is typically 1 nC. Metal photocathode is robust, but the quantum efficiency (QE) is lower for higher charge generation. Several semiconductor cathodes were studied for higher efficiency to reduce the requirement for the driver laser average power in the repetition rate mode. The Advanced Photo-Injector (APEX) experiment in Lawrence Berkley National Laboratory is working to realize a high repetition rate at MHz, high-brightness photocathode. The photocathode is a normal conducting, 187 MHz RF cavity in the CW mode, and designed for short bunches as 1–10 ps of 750 keV energy up to 1 nC charges. Several semiconductor cathode materials are tested for better beam emittance for various operational conditions. CsK_2Sb is irradiated by SHG of Yb fiber pulses, and Cs_2Te is irradiated by 4HG. Both semiconductor cathodes have nearly 1% quantum efficiency. The laser pulse energy is 0.5 µJ with the MHz repetition rate, and the average power is 0.5 W [31]. Cu and Mg photocathodes were studied for use in an RF photocathode. The gun was manufactured by a technique of hot isostatic pressing with diamond polishing and tested under a peak electric field of 57 MV/m. The quantum efficiency of the Cu cathode was 10^{-4}, while Mg cathode achieved a high QE of up to 10^{-3} under 262 nm laser-light illumination. The QE of the Mg cathode under 349 nm laser-light illumination was measured to be 2.2×10^{-5}. The experimental setup and the results of the photocathode QE measurement are shown in **Figure 9** [32].

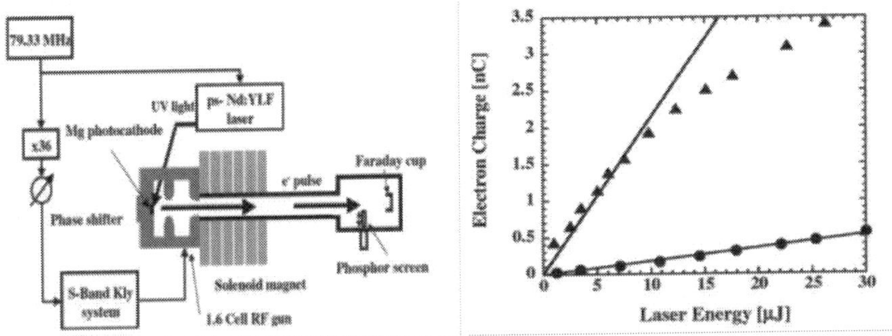

Figure 9. Left: Experimental setup of photocathode QE measurement. Right: Electron charge vs. input laser energy (266 nm) from Mg photocathode, • is for before laser cleaning.

It is concluded that 5 μJ, 266 nm, picosecond pulse is enough for Mg photocathode operation and the required average power at the MHz repetition rate is 5 W. Cu photocathode is proven to be robust material and usable by a 50 W 4HG picosecond laser. The progress of the laser technology is now making the metal photocathode again usable for the emerging requirement for long-life industrial application.

The other consideration is the reduction of the temporal microspikes in the SASE FEL pulses. Coherence is characterized in a report on the FLASH operation at 8.0 nm wavelength [33]. The single FEL femtosecond beam is passed through double pinholes for diffraction pattern, and the measured transverse coherence length is 6.2 ± 0.9 μm in the horizontal and 8.7 ± 1.0 μm in the vertical directions. The mutual coherence function K is given as 0.42, and a measurement of K by a laser plasma source is 3.2×10^{-9}. It is concluded from these measurements that a beam spatial homogenization is required at EUV wavelength by using total reflection. Temporal coherence was also reported by using a split and delay unit. The coherence time of the pulses produced in the same operation conditions of FLASH was measured to be 1.75 fs. The measured coherence time has a value, which corresponds to about 65.5 ± 0.5 wave cycles ($c\tau/\lambda$). It is well known that the SASE FEL pulses are composed of many small spikes and random spectrum due to SASE process. It is reported that the averaged spectrum has a 1.4% bandwidth typically, which is favorable for the Mo/Si EUV multilayer mirror at 13.5 nm (bandwidth 2%). It is necessary to smooth the temporal spikes to avoid random EUV flux change in the resist absorption process. The requirement is similar to most FEL applications, and we must consider efficient seed technology for MHz repetition rate operation.

It is desirable to increase the brightness and pointing/energy stability compared to SASE mode. An efficient seeding method was established by using a UV wavelength laser, in which the seed laser modulates the electron beam into coherent bunching at the harmonics of the seed laser wavelength. The bunching is intensified in another undulator for coherent FEL action, and the method is called as high-gain higher harmonic generation (HGHG). FERMI is the leading institute in this specific technology, and it is reported on the double-stage-seeded FEL with the fresh bunch injection technique [13]. The main limitation for the direct extension of the HGHG to shorter wavelength is the required small electron beam energy spread and higher average power seed laser source. The fresh bunch scheme is the solution for this problem, in which the FEL radiation is initially produced in an earlier stage undulator and used as the seeder for shorter wavelength generation (**Table 1**).

	Fermi FEL-2	EUV FEL
Seed wavelength (nm)	260	324
1nd FEL (nm)	32	40.5
2nd FEL (nm)	10.8	13.5

Table 1. Comparison of wavelengths for HGHG operation in FERMI FEL-2 and EUV FEL.

The external seed laser was the third harmonic of a Ti:Sapphire laser with a duration of ~180 fs (FWHM) and up to 20 μJ energy per pulse. Its transverse size in the modulator was made larger than the electron beam size to ensure as uniform as possible electron beam energy modulation. Once the same laser energy is required for MHz EUV FEL, 20 W average power is required for 324 nm with 180 fs at MHz repetition rate. There are two approaches to generate such laser pulses, the first is based on the MHz repetition rate Ti:Sapphire laser with 100 μJ level pulses, and the second one is based on OPCPA.

The short-pulse, short-wavelength laser technology is now advancing to realize the specification described here in a compact box, due to the new suitable laser configuration as thin-disc laser and an efficient wavelength conversion method.

An ultrafast thin-disc multipass laser amplifier demonstrated the advantage recently by delivering 1.4 kW of average output power with 4.7 mJ pulse energy and duration of 8 ps at a repetition rate of 300 kHz [18]. The beam quality factor was better than M2 = 1.4. The experiments show that the thin-disc multipass amplifier can scale pulse energy and average output power independently for the investigated repetition rates between 300 and 800 kHz. Frequency doubling by means of an LBO crystal generated 820 W SHG average power at the wavelength of 515 nm with 1170 W of incident IR power, which corresponds to a conversion efficiency of 70% and an SHG pulse energy of 2.7 mJ. By sum-frequency generation between the beams at 1030 and 515 nm in a second LBO crystal, an average UV power of 234 W (780 μJ of pulse energy) was generated at the wavelength of 343 nm THG with a conversion efficiency of 32%.

A wavelength conversion experiment was performed in the HiLASE project to evaluate the high average power generation of picosecond harmonics, namely, SHG (515 nm) and FHG (257.5 nm), in LBO and BBO/CLBO crystals, respectively [34]. The pumping of the crystals was performed by the PERLA C Yb:YAG thin-disc laser operating at 100 kHz and 60 W average power with 4 ps pulse duration. The average output power of 6 W DUV was achieved in CLBO at a spectral bandwidth of 0.2 nm and the FHG/fundamental conversion efficiency was 10%. The basic optical configuration is shown in **Figure 10** together with a photo. The input beam (upper left) is reflected by two motorized mirrors controlled by a beam stabilizer ensuring pointing stability better than 20 μrad (RMS). The following half-wave plate and polarizer is used for energy tuning. The beam is frequency doubled in an LBO at 50°C and 10 mm long, cut for the critically phase-matched generation at $\theta = 90°$ and $\phi = 12.8°$, and antireflection coated for 1030 and 515 nm. The second harmonic beam passes two dichroic mirrors and is injected into an argon-filled box with a BBO or CLBO crystal. To ensure a stable long-term functioning of the crystals the temperature was kept at 150°C. The experimental results are shown in **Figure 11**. It is visible that the 4HG/SHG conversion in the CLBO crystal has 30% higher efficiency than in the BBO crystal. The next step of the experiment is to increase the pumping power to 500 W to confirm the linearity of the conversion efficiency for 50 W FHG output power.

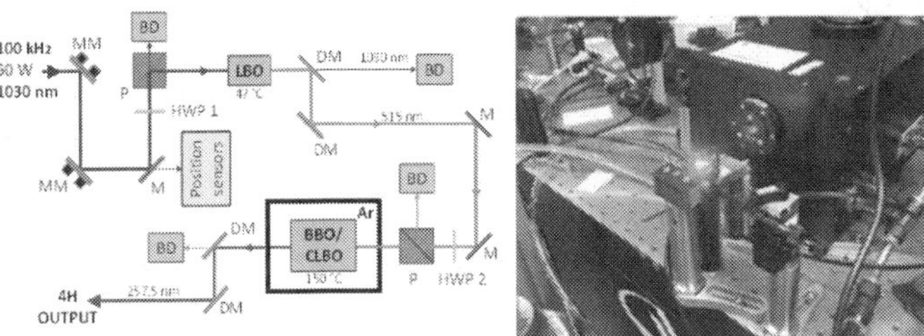

Figure 10. Optical configuration of the SHG and FHG, and SHG light is introduced into a box filled with argon.

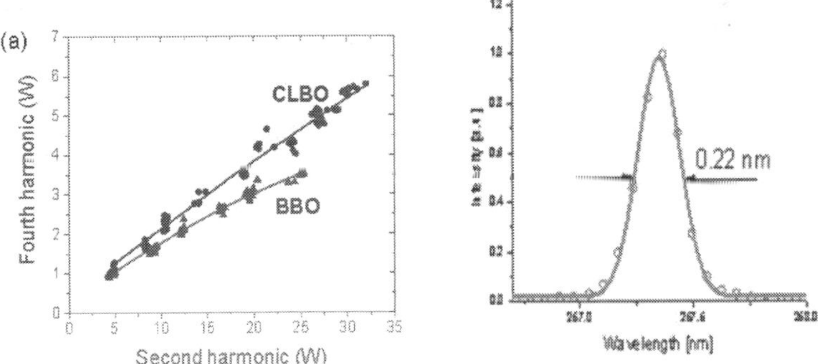

Figure 11. Left: Fourth harmonic output power dependence on the second harmonic in BBO (AR coated) and CLBO (uncoated) crystals. Right: Relevant FHG spectra from CLBO.

A small part of the beam is absorbed in the crystal and converted into heat that leads to temperature gradients in the crystal in the high average power wavelength conversion. This causes partial phase mismatch and reduces the conversion efficiency. It is estimated that the fundamental power absorption at the 60 W input is <20 mW in the 10 mm long LBO crystal. The total absorbed power may be higher due to the fact that a green laser beam has higher absorption than the fundamental beam [35]. The absorption in the antireflective coating increases the temperature, which causes the mismatch more than the bulk absorption [36].

A tunable, 112 W optical parametric chirped-pulse amplifier (OPCPA) was demonstrated for FEL seeding in a burst mode with center frequencies ranging from 720 to 900 nm, pulse energies up to 1.12 mJ, and a pulse duration of 30 fs at a repetition rate of 100 kHz [14]. The results demonstrated the feasibility of 112 W femtosecond OPCPA in a burst mode with a duty cycle of 8×10^{-3}, where no heating effects were observed. It was indicated from the measurements of absorption coefficients of BBO and LBO and calculations, the feasibility of much higher powers up to 1 kW in continuous mode was expected. Absorption causes a spatially and temporally varying temperature distribution in the sample. This leads to local changes of the refractive index and results in the development of a thermal lens. Especially in anisotropic crystals, this has consequences on increased phase mismatch in optical parametric processes with a conversion efficiency decrease. In the case of anisotropic crystal in electro-optical devices such as Pockels cells, the thermally induced depolarization reduces the contrast ratio. Though the absorption of the anisotropic crystal in these applications is usually very low, the related effects can be significant with input powers at the kilowatt level. In order to estimate the influence of thermal effects and taking it into account in the optical system design, the comprehensive knowledge of material absorption at the operation wavelength is unavoidable. An extended photothermal method was demonstrated for the quantitative determination of laser-induced wavefront deformations, which enables the separation of bulk and surface contributions to absorption in the more complex case of optically anisotropic crystalline media [36]. Experimental setup is shown in **Figure 12**. The wavefront deformations of the test beam (light source) are measured and used for absorption evaluations. The results show that the absorption is highest at the AR-coated KTP surface of input side (**Figure 13**, left), while it is higher at the surface of the output side of noncoated KTP (**Figure 13**, right). This photothermal method is usable in the real OPCPA for a better cooling system installation.

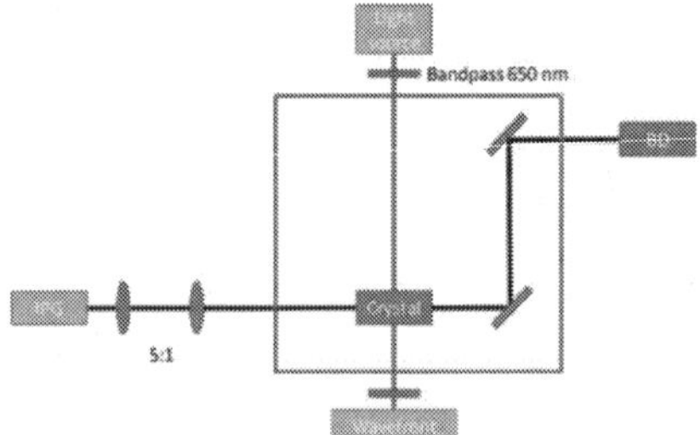

Figure 12. Setup of the photothermal method by crossed beam measurement. Wavefront measurement is performed by a Hartmann-Shack sensor.

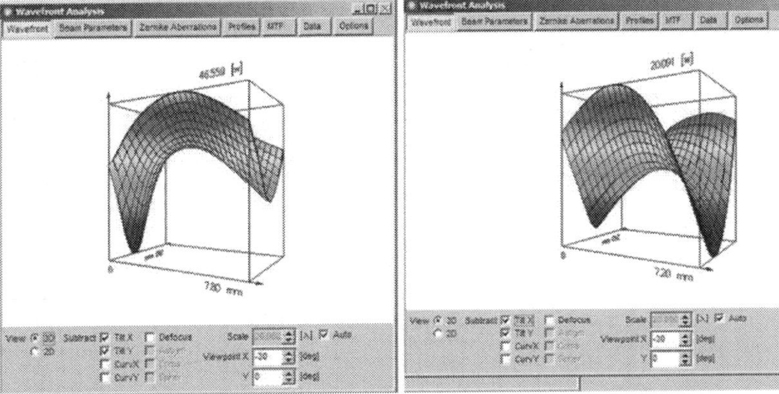

Figure 13. Wavefront deformation measurement results for AR-coated (left) and noncoated KTP samples. Blue indicates the largest wavefront deformation. The heating laser beam comes into the crystal horizontally from the left, while the probe beam passes vertically to the readers.

4. Cryogenic laser technology for high pulse energy picosecond amplifier

The basic principle of the laser Compton short-wavelength source is similar to an undulator emission, and high-intensity laser field is used as the modulating electromagnetic field. Basic principle of the laser Compton X-ray source is well studied, and a single-shot imaging is critical for many practical applications. The required specification is explained as the laser pulse must exceed some threshold parameters. It is known that the highest peak brightness is obtained in the case of counterpropagating laser pulse and electron beam bunch, in the minimum focusing area before nonlinear threshold. **Figure 14** describes the schematic of the laser Compton interaction between the electron beam and the laser.

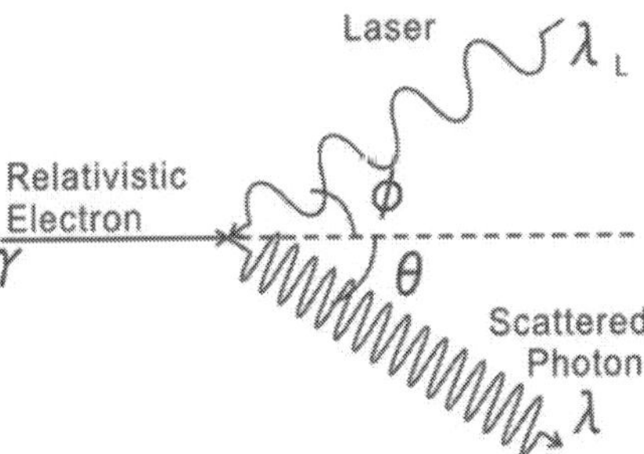

Figure 14. Schematic of the laser Compton scattering process.

The general formula of obtainable X-ray photon flux N_0 is calculated in the counter collision by the following expression:

$$N_0 \propto (\sigma_c \, Ne \, Np) / (4\pi r^2)$$

where σ_c is the Compton cross section (6.7×10^{-25} cm^2), Ne is the total electron number, Np is the total laser photon number, and r is the interaction area radius. It is predicted that an increase in Ne and Np, and the reduction in r results in the increase in the photon flux N_0. The practical limitation of these operations are the instrumental condition of electron beam emittance in higher charge, M2 of laser beam at higher pulse energy, and optimization for reduced focusing diameter r. It is possible to assume these parameters as 1 nC charge with 3 ps pulse duration to be focused down to 10 μm at 38 MeV voltage. Another limitation is the maximum of single-pulse laser intensity to reach the nonlinear threshold of the higher harmonics generation in the X ray region. The nonlinear Compton threshold is characterized by the laser field strength

$$a0 = eE/m\omega LC$$

where parameters E, ωL, and C correspond to the amplitude of the laser electric field, laser frequency, and the speed of light, respectively. The laser field strength is a function of the laser wavelength. The nonlinear threshold a_0 is given around 0.6 which corresponds to 1 J pulse energy in 1 ps pulse duration at 10 μm focusing intensity in the solid-state laser wavelength. The threshold laser energy for a single-shot imaging is similar to this critical laser pulse energy in the expected tight focus condition. The laser technology was not matured to realize such parameters simultaneously in the past, and usual approach was to increase the repetition rate of the event to increase the effective obtainable X-ray photon average flux in the affordable imaging time period such as <millisecond for bioimaging. The first approach is the pulsed laser storage in an optical enhancement cavity for laser Compton X-ray sources [37]. It is descri-bed in the experimental report that "the enhancement factor P inside the optical cavity was 600 (circulating laser power was 42 kW), in which the Finess was more than 2000, and the laser beam waist of 30 μm (2σ) was stably achieved using a 1 μm wavelength Nd:Vanadium mode-locked laser with repetition rate 357 MHz, pulse width 7 ps, and average power 7 W." The second approach is the multipass optical cavity, in which the laser Compton generation focus exists inside the multipass cavity. The minimum focusing diameter is limited due to the requirement of the cavity design. SHG picosecond pulse of 0.2 J pulse energy is circulated 32 times to collide electron bunches [38].

An approach is undertaken by the thin-disc laser technology to generate 1 J, picosecond high beam quality pulses at 100 Hz repetition rate in the Max Born Institute, Berlin, Germany. The development is based on a ring cavity concept combined with chirped pulse amplification (CPA) [39]. The regenerative amplifier produced more than 300 mJ energy when pumped with the maximum available pump power of 1.7 kW. The regenerative amplifier is followed by a large aperture ring amplifier that increases the pulse energy further to 600 mJ. This ring amplifier consists of a Pockels cell and a set of polarizers for the in- and out-coupling, two amplifier heads and a spatial filter in between. The amplifier heads are equipped with 750 µm thick Yb:YAG (7%) discs of 25 mm diameter. Each disc is pumped by 1 ms long pulses of 4 × 1.5 kW. Booster amplifier for 1 J pulse is based on the large aperture ring amplifier design without internal Pockels cell. The amplifier discs were pumped by diode modules that deliver 6 kW peak power out of a 2 mm fiber. Each amplifier is equipped with two of these pump modules, which together provide about twice the pump power compared to the large aperture ring amplifier. The booster amplifier is changed from former multipass configuration to a large aperture ring amplifier. The result is a multiple amplifier stage configuration with many thin-disc laser modules.

Cryogenic laser technology is suitable for the generation of large-pulse energy in a laser configuration of lower stages. A cryogenic thick-disc Yb:YAG laser was reported as 1 J was generated at 100 Hz repetition rate [3]. The picosecond CPA laser was developed for driving high average power soft X-ray lasers. This is one of the greatest breakthroughs in the history of high-energy solid-state laser, and it is described in the report on the configuration and operation as "Seed pulses of 100 mJ energy were produced by the laser frontend and ampli-fied to 1.5 Joules pulse energy by the five-pass power amplifier which consists of two Yb:YAG disks mounted in vacuum on a single cryo-cooling head. The Yb:YAG disks are bonded on all lateral sides with a Cr:YAG cladding to eliminate feedback of spontaneous emission into the active region to prevent amplified spontaneous emission (ASE) losses and transverse parasit-ic lasing. Cryogenic cooling of Yb:YAG to liquid nitrogen temperature increases the heat conductivity and reduces the saturation fluence, allowing for efficient high energy pulse generation at high repetition rates. High capacity cooling was accomplished by flowing cryogenic liquid coolant through the laser head. Each disk was pumped with 1.5 ms dura-tion, 4 kW pulses from a λ = 940 nm laser diode array. At the maximum pump power, 1.5 J laser pulses were obtained. These pulses were compressed by a dielectric grating pair producing 1 J, 5 ps FWHM duration pulses at 100 Hz repetition rate." The repetition rate is recently increased to 500 Hz and the picosecond pulse energy is 1 J, and the resulting aver-age power is 500 W. Temporally pulse-shaped laser pulses were focused into a ~5 mm long, 30 µm FWHM wide line on a solid target using cylindrical optics. The beam quality is indicated by the focusing specification. The resulting plasma was in the Ni-like stage, and strong collisional excitation leads to a large transient population inversion on the $4d1S_0 \rightarrow 4p1P_1$ transition of Ni-like ions at wavelengths ranging from 10.9 to 18.9 nm.

Cryogenic solid-state laser is preferred for power scalability with better beam quality, especially in higher pulse energy mode, and improvement of efficiency at the cost of longer pulse length [40]. Yb:YAG is the most tested material due to its low quantum defect and still broadband absorption in low temperatures. Various thermal optical properties are reported for base materials as YAG (ceramic and single crystal), GGG, GdVO4, and Y_2O_3 on the thermal conductivity, thermal expansion, refractive index, absorption cross section, emission cross section, and fluorescence lifetime in the cryogenic condition.

One of the key features of the cryogenic laser is its better beam quality. A quantitative evaluation is important for a practical laser design for dedicated applications and a measurement was performed on the wavefront distortion caused by the thermal origin in a cryogenic Yb:YAG crystal in the temperature range 250–130 K in nonlasing condition [41]. The wavefront aberration was evaluated by a wavefront sensor. The measurement results showed a significant reduction of the wavefront aberration in lower temperature. The thermal defocus was concluded as originated to the thermal lensing effect together with electronically induced change of the refractive index by the excitation of ion activators (electronic lensing). The dominant reason of the aberration was found as the thermal lensing in the experimental condition as 6.3 kW/cm^2 pumping intensity and pumping repetition rate of 100 Hz. The Strehl ratio was observed to be improved in the lower temperature even the absorbed energy was increased. The experiment showed the advantage of the cryogenic technology in terms of efficiency and beam quality.

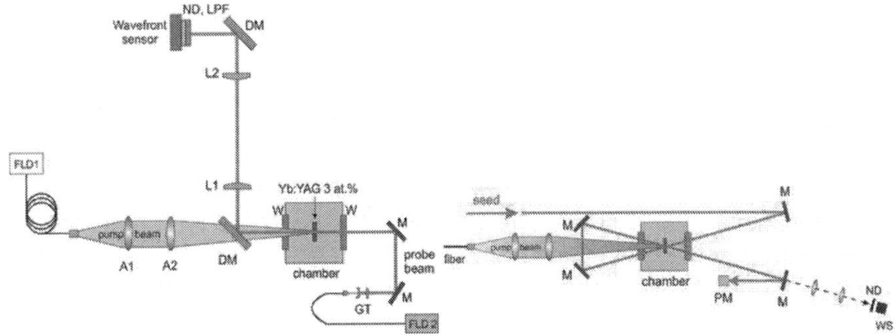

Figure 15. LEFT: Experimental configuration of the aberration measurement. FLD1, fiber-coupled pump diode at 936.6 nm; FLD2, fiber-coupled probe beam laser diode at 1065 nm; GT, Galilean telescope; W, windows; M, turning mirrors; DM, dichroic mirrors; A1, 2, achromatic doublets with focal lengths of 100 and 250 mm, respectively, L1, 2, lenses with a focal length of 250 mm; ND, neutral density filters; LPF, longpass filter with cutoff wavelength at 1050 nm. Right: Multipass amplification configuration. WS, wavefront sensor; PM, power meter.

The experimental setup used for the measurement of the wavefront aberrations in a cryogenically cooled Yb:YAG slab is shown in **Figure 15** (left). The right side figure shows the configuration of multipass amplification from the 100 mJ level input. A Yb:YAG crystal is mounted in a copper holder in a closed-loop pulse tube cryostat (Q Drive). The cooling capacity is 12 W at 100 K. The crystal was supplied from Crytur, Czech Republic, and the specification was thickness 2 mm, diameter 10 mm, and doping concentration 3 at%. A fiber-coupled laser diode (DILAS) pumped the crystal from one side. The peak wavelength was 936.9 nm, and the peak intensity was 6.3 kW/cm^2. Two achromatic doublet lens of focal length 100 and 250 mm imaged the 1 mm core of the fiber with NA 0.22 to the Yb:YAG surface. The resulting pump spot size was 2.5 mm (1/e2) in super-Gaussian intensity distribution. **Table 2(a)** summarizes the absorbed energy per pump pulse at each temperature. The absorbed energy increases by about 19 % if the temperature decreases by 120 K from initial 250 K. The absorbed power and thus generated heat is higher with decreasing temperature, and the aberrations are lower because of higher thermal conductivity, lower dn/dT, and lower expansion coefficient. Theoretical thermal decay time constants were calculated according to the formula $t_T = r_p^2/4\kappa$, where r_p is the radius of the pump, and κ is the thermal diffusion coefficient, which is defined as $k/(\rho cp)$ where k is the thermal conductivity, ρ is the mass density, and cp is the specific heat. The estimated values are shown in **Table 2(b)** for the thermal relaxation time constants for different temperature conditions in 3 at% doped Yb:YAG as an estimation from data for 2 and 4 at% doped crystals. The thermal decay time is around 31 ms at 150 K, which is three times longer than the time interval between pumping at 100 Hz. This value is 93 ms at 250 K.

(a)

Temperature (K)	Absorbed energy (mJ)	Absorbed energy (%)
250	128	41.3
210	137	44.3
170	145	46.8
130	152	49.0

(b)

Temperature (K)	Thermal decay time for 3 at% Yb:YAG crystal (ms)
250	93
200	60
150	31

Table 2. (a) Absorbed energy per pump pulse by the 2 mm thick, 3 at% doped Yb:YAG slab pumped by energy of 310 mJ at a wavelength of 936.6 nm for different temperatures of the cooling finger. (b) Calculated thermal decay time for 3 at% Yb:YAG crystal.

Cryogenic cooling is usually applied in booster amplifiers with more than one pass of the seed beam through the active medium in order to efficiently extract the stored energy. Therefore, it is assumed to evaluate four beam passes, and the measured wavefront with subtracted tilt and defocus was four times multiplied to calculate the real Strehl ratio in multipass amplification. The calculated Strehl ratio was 0.96 at 130 K and decreased to 0.93 at 250 K as shown in **Figure 16**. The practical Strehl ratio to obtain the same pulse energy decreases at higher temperature to obtain the same pulse energy in lower gain. The measurement indicates the linearity of the Strehl ratio to the temperature decrease, and it is expected further increase in the beam quality is possible in lower temperature of about less than 130 K.

In the last part of this chapter, a large aperture cryogenic laser is evaluated. The performance of a gas-cooled multislab laser is recently reported from the DiPOLE project within the Central Laser Facility (CLF RAL STFC), UK. The development is aiming at an efficient high pulse energy diode-pumped solid-state laser (DPSSL) architecture based on cryogenic gas-cooled, multislab ceramic Yb:YAG amplifier technology. A prototype amplifier is delivering up to 10.8 J pulse energy at 1030 nm wavelength with 10 Hz repetition rate. The optical-optical conversion efficiency is 22.5% [42]. The long-term energy stability was observed as 0.85% RMS with 7 J pulse energy for 48 h operation (2 million shots). An extension of the cryogenic technology is now under test in the DiPOLE 100 to confirm the cryogenic concept at 100 J, 10 Hz region (kW average power). The present laser system is built for the HiLASE project and will deliver 100 J temporally shaped ns pulses at 10 Hz with a fully integrated control system. A second system is also under development for the high-energy density (HED) beamline of the European XFEL project.

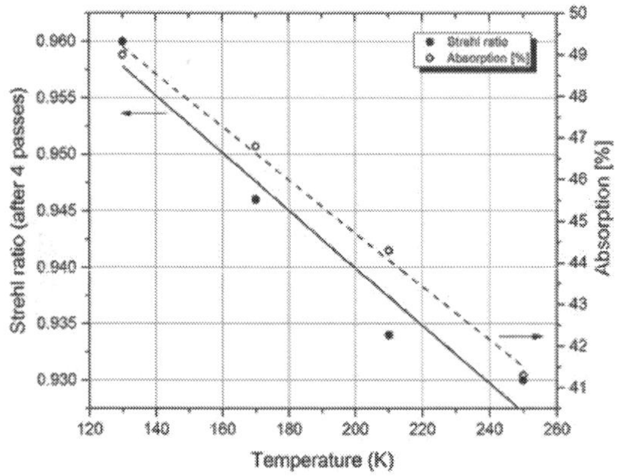

Figure 16. Strehl ratio for the four passes of the probe beam through the Yb:YAG slab and absorbed pump power per single pass as a function of temperature. Lines represent a linear fit with slopes of -6×10^{-4}/K and -0.064%/K, respectively.

The 10 J amplifier architecture is based on the multislab approach. The gain medium is composed of four circular Yb:YAG slabs with two different Yb doping levels as 1.1 and 2.0 at % to confirm a uniform temperature distribution among each slab. The diameter of the circular slab is 45 mm with a 5 mm thickness, and the pump area is square of 23×23 mm^2. The pump beam is supplied from 939 nm diodes in stack with a pumping time duration of 700 µs at 10 Hz. The Yb:YAG circular slab is cladded with a 5 mm wide Cr:YAG absorber with 6 cm^{-1} absorption coefficient. This is effective to prevent amplified spontaneous emission (ASE) and parasitic oscillations. A cryogenic He flow cools the slab and keeps the temperature as 150–170 K. The pressure of the He flow is 10 bar. **Figure 17** shows the optical arrangement for amplification. A seed beam is injected into the amplifier through a dichroic mirror and then image relayed by a spatial filter ($f = 1$ m) to a back reflector and reflected back to the amplifier module. One spatial filter locates on each side of the amplifier head. Each pass is composed by a set of separate mirrors. A deformable mirror is placed in the amplifier after the third pass for the aberration compensation. After seven passes, the beam is extracted from the amplifier with pulse energy increased to 9 J with a size of 21 mm × 21 mm^2.

Numerical modeling of the multislab amplifier is conducted in the HiLASE project to ensure the scaling of the cryogenic technology for further increased parameter region in pulse energy, repetition rate, and better beam quality. Comsol Multiphysics software was chosen to model the thermal and stress effects in the amplifiers [43]. The sources of heat were calculated in the ASE code [44]. The axial surfaces of the slabs are assumed to be cooled only by flowing helium gas at 160 K. The slab was assumed to have no thermal contact with its 2 cm thick Invar holder; and all heat is removed by convection through the faces. From the temperature and stress maps of the slab, the optical path difference (OPD) and birefringence depolarization losses were calculated for a single slab according to a prior approach. The gradual decrease of cooling efficiency in the direction of gas flow, caused by He heating, results in the loss of left-right symmetry of the temperature, stress, depolarization, and OPD maps.

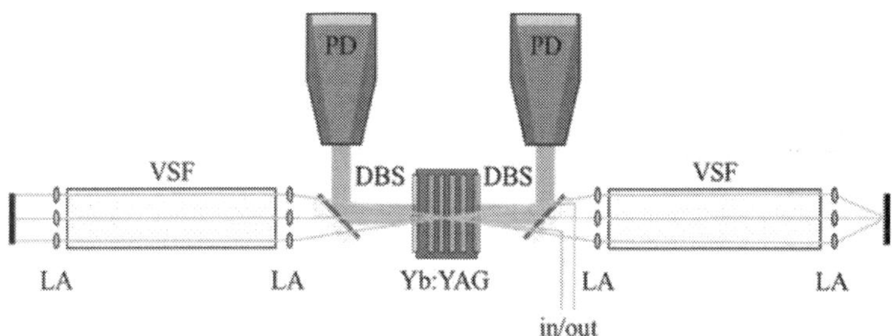

Figure 17. Schematic of the 10 J cryogenic multislab amplifier. It consists of Yb:YAG ceramic slabs in the laser head, dichroic beam splitters (DBSs), lens arrays (LAs), vacuum spatial filters (VSFs), and homogenized pump diode laser modules (PDs).

It is planned to increase the repetition rate of the 10 J amplifier to 100 Hz by keeping the basic performance. Once the operation is as expected, the cryogenic laser offers a technological stage for a single module to generate picosecond multipulses of a few joules of energies, for a single-shot imaging Compton source.

5. Conclusion

Recent progress of thin-disc lasers is promising to realize a high-brightness pumping source of laser plasma or laser Compton short-wavelength sources. Further progress is possible by an advanced cryogenic technology with its higher thermal conductivity. These laser progresses are contributing in the practical applications in short-wavelength imaging and material processing. Picosecond thin-disc laser technology is now in the stage of 1 kW level with >100 kHz repetition rate. Further research and developments are aiming at 1 kW with kHz repetition rate (1 J, ps, kHz), pulse length reduction into subpicosecond region with MHz repetition rate (mJ, fs, MHz), and increase in the average power to 10 kW region. These challenges require further improvements of the achieved technology bases and evaluation of new schemes.

Cryogenic technology is now offering an option for these challenges in the solid-state laser technology.

Acknowledgements

The author appreciates kind collaboration of many researchers in HiLASE Centre in Czech Republic, Waseda University, and Gigaphoton Inc., in Japan. Dr. M. Smrz is especially acknowledged for his excellent work in the 100 kHz thin-disc picosecond laser, Dr. H. Turcicova and Dr. O.Novak for their high average power wavelength conversion, and Dr. P. Sikocinski for his aberration measurement in cryogenic Yb:YAG. Dr. Sakaue is highly appreciated for his experimental and theoretical work in the Laser Compton and FEL seeding research and Dr. T. Yanagida for his pioneering work of tin droplet atomization by a picosecond laser. Experimental work of this chapter is cofinanced by the state budget of the Czech Republic (project HiLASE: Superlasers for real world: LO1602). This work is also supported by the Czech Science Foundation (GACR) under project GA16-12960S.

Author details

Akira Endo

Address all correspondence to: endo@fzu.cz

1 Research Institute for Science and Engineering, Waseda University, Tokyo, Japan

2 HiLASE Centre, Institute of Physics AS CR, Dolní Břežany, Czech Republic

References

[1] Assoufid, L. and Naulleau, P. (2016) Topical Meeting, Compact (EUV & X-ray) Light Sources, OSA High-Brightness Sources and Light-Driven Interactions Congress, 20–22 March 2016, Hilton Long Beach, Long Beach, California, USA

[2] Mizoguchi, H. Nakarai, H. Abe, T. Nowak, K.M. Kawasuji, Y. Tanaka, H. Watanabe, Y. Hori, T. Kodama, T. Shiraishi, Y. Yanagida, T. Soumagne, G. Yamada, T. Yamazaki, T. Okazaki, S. and Saitou, T. (2015) "Performance of one hundred watt HVM LPP-EUV source," Proceedings of SPIE 9422-11

[3] Reagan, B.A. Berrill, M. Wernsing, K.A. Baumgarten, C. Woolston, M. and Rocca, J.J. (2014) "High-average-power, 100-Hz-repetition-rate, tabletop soft-X-ray lasers at sub-15-nm wavelengths," Phys. Rev. A 89, 053820

[4] John, R.W (1998) "Brilliance of X rays and gamma rays produced by Compton back scattering of laser light from high energy-electrons," Laser Particle Beams 16, 115–127

[5] Endo, A. Yang, J. Okada, Y. Yanagida, T. Yorozu, M. and Sakai, F. (2001) "Characterization of the monochromatic laser Compton X-ray beam with picosecond and femtosecond pulse widths," Proceedings SPIE 4502, pp. 100–108

[6] Babzien, M. Ben-Zvi, I. Kusche, K. Pavlishin, I.V. Pogorelsky, I.V. Siddons, D.P. and Yakimenko, V. (2006) "Observation of the second harmonic in Thomson scattering from relativistic electrons," Phys. Rev. Lett. 96, 054802

[7] Kumita, T. Kamiya, Y. Babzien, M. Ben-Zvi, I. Kusche, K. Pavlishin, I.V. Pogorelsky, I.V. Siddons, D.P. Yakimenko, V. Hirose, T. Omori, T. Urakawa, J. Yokoya, K. Cline, D. and Zhou, F (2008) "Observation of the nonlinear effect in relativistic Thomson scattering of electron and laser beams," Laser Phys. 16, 267–271

[8] Oliva, P. Carpinelli, M. Golosio, B. Delogu, P. Endrizzi, M. Park, J. Pogorelsky, I. Yakimenko, V. Williams, O. and Rosenzweig, J (2010) "Quantitative evaluation of single-shot inline phase contrast imaging using an inverse Compton x-ray source," Appl. Phys. Lett. 97, 134104

[9] Pogorelsky, I.V. Babzien, M. Pavlishin, I. Stolyarov, P. Yakimenko, V. Shkolnikov, P. Pukhov, A. Zhidkov, A. and Platonenko, V.T. (2006) "Terwatt CO2 laser; a new tool for strong field research," Proceedings of SPIE, 6261, 18

[10] Ur.C.A. Balabanski, D. Cata-Danil, G. Gales, S. Morjan, I. Tesileanu, O. Ursescu, D. Ursu, I. and Zamfir, N.V. (2015) "The ELI–NP facility for nuclear physics," Nucl. Instrum. Method B 355, 198–202

[11] Bacci, A. Palmer, D. Serafini, L. Torri, V. Petrillo, V. Tomassini, P. Puppin, E. Alesini, D. Anania, M. Bellaveglia, M.P. Bisesto, F. Pirro, G.Di. Esposito, A. Ferrario, M. Gallo, A. Gatti, G. Ghigo, A. Spataro, B. Vaccarezza, C. VillaF. Cianchi, A. Agostino, R.G. Borgese, G. Ghedini, M. Martire, F. Pace, C. Levato, T. Dauria, G. Fabris, A. and Marazzi, M. (2014) "The STAR Project," Proceedings of IPAC2014, WEPRO115A

[12] Endo, A. Sakaue, K. Washio and M. Mizoguchi, H. (2014) "Optimization of high average power FEL for EUV lithography application," Proceedings of FEL2014, FRA04

[13] Allaria, E. Castronovo, D. Cinquegrana, P. Craievich, P. Dal Forno, M. Danailov, M.B. D'Auria, G. Demidovich, A. De Ninno, G. Di Mitri, S. Diviacco, B. Fawley, W.M. Ferianis, M. Ferrari, E. Froehlich, L. Gaio, G. Gauthier, D. Giannessi, L. Ivanov, R. Mahieu, B. Mahne, N. Nikolov, I. Parmigiani, F. Penco, G Raimondi, L. Scafuri, C. Serpico, C. Sigalotti, P. Spampinati, S. Spezzani, C. Svandrlik, M. Svetina, C. Trovo, M. Veronese, M. Zangrando, D. and Zangrando, M. (2013) "Two-stage seeded soft-X-ray free-electron laser," Nature Photon. 7, 913–918

[14] Höppner, H. Hage, A. Tanikawa, T. Schulz, M. Riedel, R.Teubner, U. Prandolini, MJ. Faatz, B. and Tavella, F. (2015) "An optical parametric chirped-pulse amplifier for seeding high repetition rate free-electron lasers," New J. Phys. 17, 053020

[15] Klenke, A. Breitkopf, Kienel, S.M. Gottscha, T. Eidam, T. Hädrich, S. Rothhardt, J. Limpert, J. and Tünnermann, A. (2013) "530 W, 1.3 mJ, four-channel coherently combined femtosecond fiber chirped-pulse amplification system," Opt. Lett. 38, 2283–2285

[16] Brocklesby, W.S. Nilsson, J. Schreiber, T. Limpert, J. Brignon, A. Bourderionnet, J. Lombard, L. Michau, V. Hanna, M. Zaouter, Y. Tajima, T. and Mourou, G. (2014) "ICAN as a new laser paradigm for high energy, high average power femtosecond pulses," Eur. Phys. J. Special Topics 223, 1189–1195

[17] Mans, T.R. Graf, R. Dolkemeyer, J. Schnitzler, C (2014) "Femtosecond Innoslab amplifier with 300 W average power and pulse energies in the mJ regime," Proceedings of SPIE 8959-43

[18] Negel, J.P. Loescher, A. Voss, A. Bauer, D. Sutter, D. Killi, A. Ahmed, M.A. and Graf.T, (2015) "Ultrafast thin-disk multipass laser amplifier delivering 1.4 kW (4.7 mJ, 1030 nm) average power converted to 820 W at 515 nm and 234 W at 343 nm," Opt. Exp. 23, 21064

[19] Freitag, C. Wiedenmann, M. Negel, J.P. Loescher, A. Onuseit, V. Weber, R. Ahmed, M.A. Thomas Graf, T. (2015)" High-quality processing of CFRP with a 1.1-kW picosecond laser," Appl. Phys. A 119, 1237–1243

[20] Ripin, D.J. Ochoa, J.R. Aggarwal, R.L. Fan, T.Y. (2005) "300-W Cryogenically Cooled Yb:YAG Laser," IEEE J. Quantum Electron, QE-41, 1274–1277

[21] Zapata, L.E. Reichert, F. Hemmer, M. Kaertner, F.X. (2016) "250 W average power, 100 kHz repetition rate cryogenic Yb:YAG amplifier for OPCPA pumping," Opt. Lett. 41, 492–495

[22] Endo, A. Lithography, Chapter 9 (2010) "CO_2 laser produced Tin plasma light source as the solution for EUV lithography," edited by Michael Wang, InTech, Janeza Trdine 9, 51000 Rijeka, Croatia

[23] Pirati, A. Peeters, R. Smith, D.A. Lok, S. Noordenburg, M. Es, R. Verhoeven, E. Meijer, H. Minnaert, A. Horst, W. Meiling, H. Mallmann, J. Wagner, C. Stoeldraijer, J. Fisser, G. Levasier, L. Finders, J. Zoldesi, C. Stamm, U. Boom, H. Brandt, D.C. Brown, D.J. and Fomenkov, I.V. (2016) "EUV lithography performance for manufacturing: status and outlook," Proceedings of SPIE 9776-10

[24] Teramoto, Y. Santos, B. Mertens, G. Kops, R. Kops, M. Wezyk, A. Bergmann, K. Yabuta, H. Nagano, A. Ashizawa, N. Shirai, T. Nakamura, K. and Kasama, K. (2016) "High-radiance LDP source: Clean, reliable, and stable EUV source for mask inspection," Proceedings of SPIE 9776-22

[25] Mizoguchi, H. Saitou, T. Yamazaki, T. Okazaki, S. Nakarai, H. Abe, T. Kodama, T. Yanagida, T. Hori, T. Nowak, K.M. Kawasuji, Y. Tanaka, H. Shiraishi, Y. Watanabe, Y.

Yamada, T. and Soumagne, G. (2014) "Sub-hundred Watt operation demonstration of HVM LPP-EUV source," Proceedings of SPIE 9048-22

[26] Smrž, M. Miura, T. Chyla, M. Muzik, J. Nagisetty, S.S. Novák, O. Turcicova, H. Linnemann, Huynh, J. Severová, P. Sikocinski, P. Endo, A. and Mocek, T. (2016) "Progress in kW-class picosecond thin-disk lasers development at the HiLASE," Proceedings of SPIE 9726-43

[27] Smrž, M. Miura, T. Chyla, M. Nagisetty, S. Novák, O. Endo, A. and Mocek, T. (2014) "Suppression of nonlinear phonon relaxation in Yb:YAG thin disk via zero phonon line pumping," Opt. Lett. 39, 4919–4922

[28] Schneidmiller, E.A. , Vogel, V.F. Weise, H. and Yurkov, M.V. (2011) "A kilowatt-scale free electron laser driven by L-band superconducting linear accelerator operating in a burst mode," International Workshop on EUV and Soft X-ray Sources, November 7–9, 2011, Dublin, Ireland

[29] Hosler, E.R. Wood, O.R. Barletta, W.A. Mangat, P.J., Preil, M.E. (2015) "Considerations for a free-electron laser-based extreme-ultraviolet lithography program," Proceedings of SPIE, 9422-12

[30] Chalupský, J. Juha, L. et.al. (2007) "Characteristics of focused soft X-ray free-electron laser beam determined by ablation of organic molecular solids," Opt. Exp. 15, 6036

[31] Filippetto D. Byrd, J. Chin, M. Cork, C. Santis, S.De. Feng, J. Norum, W.E. Doolittle, L. Papadopoulos, C. Portmann, G. Quintas, D.G. Sannibale, F. Stuart, M. Wells, R. and Zolotorev, M. (2011) "Low energy beam diagnostic for APEX, the LBNL VHF photo-injector," Proceedings of 2011 Particle Accelerator Conference, WEP222

[32] Nakajyo, T. Yang, J. Sakai, F. and Aoki, Y. (2003) "Quantum efficiencies of Mg photocathode under illumination with 3rd and 4th harmonics Nd: LiYF4 laser light in RF gun," Jpn. J. Appl. Phys. 42 1470–1474

[33] Singer, A. Sorgenfrei, F. Mancuso, A.P. Gerasimova, N. Yefanov, O.M. Gulden, J. Gorniak, T. Senkbeil, T. Sakdinawat, A. Liu, Y. Attwood, D. Dziarzhytski, S. Mai, D.D. Treusch, R. Weckert, E. Salditt, T. Rosenhahn, A. Wurth, W. and Vartanyants, A. (2012) "Spatial and temporal coherence properties of single free electron laser pulses," Opt. Exp. 20, 17480

[34] Turcicova, H. Nowak, O. Smrz, M. Miura, T. Endo, A. and Mochek, T. "Deep ultraviolet (257.5 nm and 206 nm) picosecond pulses produced a high-power 100 kHz solid-state thin-disk laser," Proceedings of SPIE 9893-1

[35] Riedel, R. Rothhardt, J. Beil.K, Gronloh, B. Klenke, A. Höppner, H. Schulz, M., Teubner, Kränkel, U.C. J.Limpert, J. Tünnermann, A. Prandolini, M.J. F. and Tavella, F. (2014) "Thermal properties of borate crystals for high power optical parametric chirped-pulse amplification," Opt. Exp. 22, 17607.

[36] Stubenvoll, M. Schäfer, B. Mann, K. and Novak, O. (2016) "Photothermal method for absorption measurements in anisotropic crystals," Rev. Sci. Instrum. 87, 023904, Proceedings of SPIE 9726-43

[37] Sakaue, K. Araki, S. Fukuda, M. Higashi, Y. Honda, Y. Sasao, N. Shimizu, H. Taniguchi, T. Urakawa, J. and Washio, M. (2011) "Development of a laser pulse storage technique in an optical super-cavity for a compact X-ray source based on laser-Compton scattering," Nucl. Instru. Meth. A 637, S107–S111

[38] Courjaud, A. Tropheme, B. Falcoz, F. Eric, P. Mottay, E.P. and Riboulet, G. (2015) "High power lasers for gamma source," Proceedings of SPIE 9342-23

[39] Jung, R. Tümmler, J. Nubbemeyer, T. and Will, I. (2015) "Two- Channel thin-disk laser for high pulse energy," Advanced Solid State Lasers Conference© OSA 2015, AW3A.7.

[40] Fan, T.Y. Ripin, D.J. Aggarwal, R.L. Ochoa, J.R. Chann, B. Tilleman, M. and Spitzberg, J. (2007) "Cryogenic Yb3+-Doped Solid-State Lasers," IEEE J. Quantum Electron. QE-13, 448–458

[41] Sikocinski, P. Novak, O. Smrz, M. Pilar, J. Jambunathan, V. Jelínková, H. Endo, A. Lucianetti, A. and Mocek, T. (2016) "Time-resolved measurement of thermally induced aberrations in a cryogenically cooled Yb:YAG slab with a wavefront sensor," Appl.Phys.B. April 2016, 122:73

[42] Mason, P.D. Banerjeea, S. Ertela, K. Phillipsa, P.J. Thomas, J. Butchera, T.J. Smitha, J.M. Vidoa, M.D. Tomlinsona, S. Oleg Chekhlova, O. Shaikha, W. Blakea, S. Holligana, P. Divoky, M. Pilar, J Hernandez-Gomeza, C. Justin, R. Greenhalgha, S. and Colliera, J.L. (2015) "DiPOLE100: A 100 J, 10 Hz DPSSL using cryogenic gas cooled Yb:YAG multi slab amplifier technology," Proceedings of SPIE 9513 02-1

[43] Slezak, O. Lucianetti, A. Divoky, M. Sawicka, M. and Mocek, T. (2013) "Optimization of wavefront distortions and thermal-stress induced birefringence in a cryogenically-cooled multislab laser amplifier," IEEE J. Quantum Electron, QE 49, 960–966

[44] Sawicka, M. Divoky, M. Novak, J. Lucianetti, A. Rus, B. and Mocek, T. (2012) "Modeling of amplified spontaneous emission, heat deposition, and energy extraction in cryogenically cooled multislab Yb3+:YAG laser amplifier for the HiLASE Project," J. Opt. Soc. Am. B 29, 1270–1276

Large-scale High-power Laser Systems

Chapter 6

Multiterawatt Hybrid (Solid/Gas) Femtosecond Systems in the Visible

Leonid D. Mikheev and Valery F. Losev

Additional information is available at the end of the chapter

http://dx.doi.org/10.5772/63972

Abstract

A novel hybrid (solid/gas) approach to the development of femtosecond high-intensity laser systems operating in the visible is presented in this chapter. Behind this approach is a combination of a solid-state front end relying on widespread and highly developed techniques for femtosecond pulse generation in the near infrared with a photochemically driven boosting amplifier operating in the visible spectral range. Historical background of developing photochemically pumped gas lasers on broad bandwidth electronic transitions in molecules and physical principles of their operation are briefly summarized as well. The architecture and the design issues of the hybrid femtosecond systems relying on the amplification of the second harmonic of Ti:sapphire front ends in the photodissociation XeF(C-A) power-boosting amplifiers driven by the VUV radiation from electron-beam-to-VUV-flash converters are described, as well as breakthrough results of proof-of-principle experiments demonstrating a high potential of the hybrid approach. Wavelength scaling of laser-matter interaction is shortly discussed to demonstrate advantages of shorter driver wavelengths for some applications with main emphasis placed on recombination-pumped soft X-ray lasers.

Keywords: hybrid femtosecond systems, visible spectral range, photochemically driven laser media, laser-matter interaction

1. Introduction

Significant progress in the development of all-solid-state femtosecond laser systems relying on chirped-pulse amplification (CPA) technique has resulted in reaching petawatt (PW) peak powers [1] and focused intensities as high as 10^{22} W/cm^2 [2] that provides great opportunities for experimental studies in the area of the extreme high-field science. The all-solid-state

laser systems providing pulses shorter than 100 fs are based on the Ti:sapphire and optical parametric chirped-pulse amplification (OPCPA) technologies.

Presented in this chapter are milestones and main results obtained in the course of the realization of a novel hybrid (solid/gas) approach to the development of femtosecond systems that, unlike the all-solid-state systems operating in the near-infrared (NIR) region, allow for producing super-intense optical pulses in the blue-green spectral range. This approach aims to marry robust solid-state laser technologies highly developed for femtosecond pulse generation and amplification in the NIR spectral range with advantages of photochemically driven gaseous gain media of the visible range.

Most extensive development of the photochemical method of pumping gaseous active media dates back to the 1960s–1990s of the last century (see [3–5] and references cited therein). Being applied to optical excitation of gas lasers on electronic molecular transitions by radiation from such unconventional pump sources as high-temperature electrical discharges and strong shock waves in gas, This method resulted in emerging a new class of gaseous active media for lasers emitting in the spectral range extending from the NIR to UV with a high output energy increasing in proportion to the active volume and pump energy. Among a variety of molecules lasing upon optical excitation, there are three broadband active media (XeF(C-A), Kr_2F, and Xe_2Cl), which offer a number of characteristics extremely attractive for the amplification of femtosecond optical pulses up to ultrahigh peak powers. The gaseous nature and visible spectrum of emission of these media promise important virtues of the hybrid systems. First of all, the gaseous nature of these media is characterized by low nonlinear refractive index allowing amplification of optical pulses with much higher intensities as compared with solid media. Secondly, the visible spectrum of emission requires nonlinear frequency upconversion of a seed pulse generated in the NIR spectral region by a solid-state front end, thereby providing efficient temporal cleaning of the ultrashort optical pulse and the high temporal contrast ratio of an output pulse, which is of primary importance for a number of high-field experiments.

The main motivation for the development of hybrid systems in the visible is favorable drive frequency scaling of laser-matter interaction in a number of high-field applications. Of overriding importance is the dramatic improvement of the recombination soft X-ray laser excitation in an optical field ionized (OFI) plasma with drive frequency.

2. Photochemical lasers

The photochemical method of gaseous active media pumping by radiation from broadband optical sources originates from the development of the high-power photodissociation laser in metastable iodine atoms ($\lambda = 1.315$ μm) (for example, see [6–8]), resulted in variety of remarkable results: 1 MJ of output energy in a single beam [9], 2 and 30 kJ in a short laser pulse obtained, respectively, with flash lamp [10] and surface discharge [11] optical pumping. The breakthrough results obtained in the course of the iodine photodissociation laser development stimulated extensive studies of the potentialities of the gaseous active media optical pumping and the search for new active media for high-power optically driven lasers in many laboratories around the world.

It is important to stress the major impact of untraditional pumping sources on the photo-chemical method development, such as high-current open discharges initiated with exploding wires or sliding sparks, as well as strong shock waves driven by detonation of chemical explosives [4]. These pumping sources, initiated directly in laser working mixtures, had no shell separating them from an active medium that removes limitations on energy deposition into the pumping sources and makes it possible to utilize the radiation in any spectral range, including the UV and VUV, where the most intense absorption bands of the overwhelming majority of molecules are located. As compared with ordinary flash lamps, these pumping sources have much higher brightness temperature reaching 30–35 kK.

The main emphasis in these studies was placed on molecular transitions that, unlike atoms, do not require complete population inversion of the electronic states participating in the laser transition and enable a fairly simple depletion of the lower laser levels due to vibrational relaxation and/or dissociation of the lower state. The application of such optical sources has led to emerging new class of gas lasers emitting in the spectral range from the NIR to UV regions due to the development of a variety of photochemical excitation techniques relying on the photolysis and direct optical excitation of molecular gases, as well as secondary photo-chemical reactions (see [3–5] and references cited therein).

One of the most remarkable results obtained in the course of these studies is the optical excitation of lasing on broadband bound-free $Kr_2F(4\ ^2\Gamma\text{-}1,2\ ^2\Gamma)$, $Xe_2Cl(4\ ^2\Gamma\text{-}1,2\ ^2\Gamma)$, and XeF(C-A) transitions in the visible. These active media are very sensitive to the internal losses because they have much lower small-signal gain as compared with excimers emitting in the UV spectral range on the B-X transition due to the large width of their luminescence spectra and long radiative lifetime of the excited states. For this reason, e-beam or fast discharge pumping of these active media turned out to be ineffective since electron excitation technique is based on plasmochemical reactions involving ionized and highly excited atoms and molecules that are characterized by strong absorption in the visible [12]. For example, laser action from electron-beam-excited Kr_2F and Xe_2Cl active media was observed only in the afterglow of the pump pulse when the transient absorption is significantly reduced [13]. The photochemical techni-que, relying on reactions of neutrals excited to low-lying energetic states, free of the short-comings associated with transient absorption. This makes optical pumping to be superior to electron excitation in the efficiency of producing laser action on the aforementioned transitions.

Due to extremely broad gain bandwidth supporting pulses of shorter than 10 fs, these active media are of practical interest for amplifying femtosecond optical pulses to ultrahigh peak powers [4, 5, 14, 15]. The main spectroscopic characteristics of the transitions are listed in the **Table 1**. One of most important parameters is the saturation fluence accounting for the maximum energy extraction per surface unit of an amplifier output aperture. This parameter, ranging from 0.05 J/cm^2 for XeF(C-A) to 0.2 J/cm^2 for Kr_2F due to rather long radiative lifetime of the upper laser states, promises peak power, I_{out}, of up to ~10 TW per square cm of an output aperture in a 25 fs pulse. At the same time, the gas active media are easily scalable to large volumes with wide aperture (several hundred cm^2) to reach PW level of output peak power.

Moreover, the combined use of two media with spectrally shifted emission bands (Kr_2F with XeF(C-A) or Xe_2Cl) in an amplifier chain makes promising for the amplification of even shorter optical pulses due to twofold broadening a gain bandwidth (**Figure 1**) and spectrally inhomogeneous gain saturation.

Transition	XeF(C-A)	$Kr_2F(4^2\Gamma\text{-}1,2^2\Gamma)$	$Xe_2Cl(4^2\Gamma\text{-}1,2^2\Gamma)$
λ_{max}, nm	480	420	510
$\Delta\lambda$, nm	70	80	100
τ_{FT}, fs	5	3	4
τ_{sp}, ns	100	180	250
σ_{st}, cm^2	10^{-17}	2.3×10^{-18}	2.8×10^{-18}
ε_{sat}, J/cm^2	0.05	0.2	0.15
I_{out} ($\tau = 25$ fs), TW/cm^2	2	8	6

λ_{max} is the wavelength of the gain maximum.
$\Delta\lambda$ is the gain bandwidth.
τ_{FT} is the transform-limited pulse width (for Gaussian profile).
τ_{sp} is the spontaneous lifetime.
σ_{st} is the stimulated emission cross-section.
ε_{sat} is the saturation fluence.
I_{out} is the estimated maximum peak power density related to a laser amplifier output aperture.

Table 1. Characteristics of broadband active media.

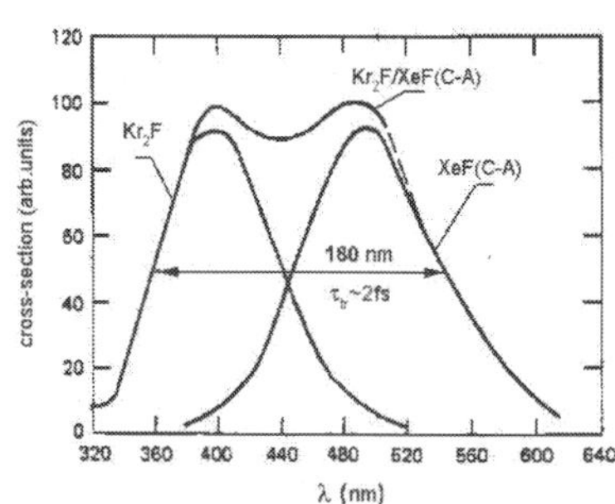

Figure 1. Emission spectrum for the mixed Kr_2F/XeF(C-A) system.

An important advantage of gaseous active media resides in the much lower value of the nonlinear index of refraction as compared with solid-state materials, which allows amplification of optical pulses at much higher intensities than in condensed gain media. For the first time, hybrid architecture employing a gaseous active medium was realized in the system comprising a dye femtosecond oscillator and e-beam or fast discharge-driven boost amplifier on the UV B-X transitions of ArF, XeCl, KrF, and XeF rare-gas-halide excimer molecules. The highest peak power obtained in the hybrid systems of this type reached ∼1 TW in a 310 fs pulse [16]. However, gain bandwidth on the B-X transition does not exceed 2 nm in full width at half-maximum (FWHM) corresponding to the spectrally limited pulse width of ∼50 fs. With spectral gain narrowing in amplifiers, it is difficult to count on the possibility of producing pulses shorter than 0.1 ps at the TW level of peak power. Moreover, due to the short spontaneous lifetime of the B state and rather narrow spectral bandwidth, these transitions exhibit as low as 1 mJ/cm² saturation fluence, which corresponds to the output intensity of 10^{-2} TW/cm² in a 0.1 ps pulse, requiring too large output aperture to produce multiterawatt output peak power.

Compared with the B-X transition, the broadband photochemical media are characterized by more than an order of magnitude larger gain bandwidth and saturation fluence allowing for several TW/cm² to be obtained in hybrid systems comprising these active media. In addition, unlike to the electron-impact excitation, the optical pumping, which is practically free of transient absorption within the laser transition spectral band, makes the entire transition bandwidth to be accessible for the femtosecond pulse amplification. These active media are briefly reviewed in the subsequent sections.

2.1. XeF(C-A) active medium

A schematic energy diagram of the upper and lower XeF laser levels is shown in **Figure 2**. Behind the optical pumping XeF active medium is the photolysis of XeF$_2$ vapor in the spectral range of <204 nm to produce XeF excimers mainly in the B state [17–19]. The C state, lying lower than the B state, is populated due to collisional relaxation of the latter in the presence of a buffer gas. Depending on composition of working mixture, laser action is observed on the B-X (353 nm) or C-A (480 nm) transition [20]. Broad gain bandwidth on the C-A transition (Δλ =70 nm [21]) is accounted for by the repulsive nature of the A state. On the other hand, repulsive character of the lower A state provides its virtually instantaneous depopulation and ensures population inversion independently on the presence of buffer gas.

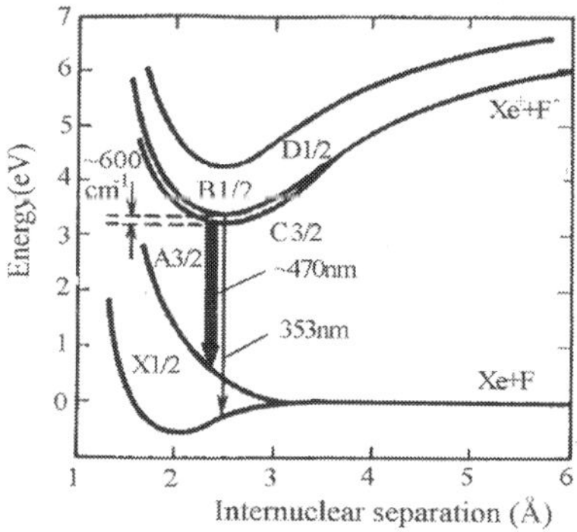

Figure 2. The potential curves of low-lying energetic states of the XeF excimer molecule.

Lasing in the photochemically driven XeF active medium was obtained for the first time in 1977 on the *B-X* transition with an exploding wire as a pump source [22]. Later on, in a series of papers, laser action on XeF(B-X) and XeF(C-A) was reported upon optical pumping by broadband VUV radiation from exploding wires [20, 23], surface discharges [24–27], formed-ferrite flash [28, 29], and strong shock waves [30], as well as by the Xe_2 spontaneous emission at 172 nm excited by an electron beam [31–33]. These studies have led to a number of significant achievements. Among them are high output energies attaining as much as 1 kJ and 170 J in the UV region with the strong shock wave [30] and surface discharge optical pumping [24], respectively, as well as 120 J and 10 J obtained in the visible under optical pumping by radiation from surface discharges operating in the single shot [24] and 1 Hz repetitive rate [27] modes respectively.

Studies of the XeF(C-A) laser showed that, besides minor transient absorption discussed above, the optical pumping XeF(C-A) active medium has an additional advantage over the electron-impact excitation, which consists in the much weaker competition of the B-X transition. The point is that upon optical excitation, relative populations of the closely lying B and C states are determined by the thermodynamic equilibrium at a buffer-gas temperature that is close to the room temperature, while, upon e-beam and fast discharge pumping, the main role in the energy exchange between these states is played by secondary electrons, which have a temperature of ~1 eV characteristic of this pumping. In the optically driven active medium, the electron concentration is negligibly low, thereby providing an efficient laser action on the C-A transition.

From the viewpoint of femtosecond pulse amplification, the pump technique relying on the e-beam-driven spontaneous emission of Xe_2 excimers is of particular concern because, along with surface discharge optical pumping, it paved a new way for the development of the hybrid (solid/gas) femtosecond systems in the blue-green region. With the use of this technique to pump the XeF(C-A) laser, as high as 6 J of output energy was reported in [33].

Finally, it is essential to note that one of the most important conclusions gained from the optically driven XeF(C-A) laser studies is that at a proper composition of the active medium its broad amplification band is not practically modified by transient absorption making the entire transition bandwidth to be accessible for the femtosecond pulse amplification. A drawback of the XeF(C-A) active medium is that it must be replaced in a laser chamber after each shot because of the photodecomposition of the parent XeF_2 molecules.

2.2. Xe_2Cl active medium

Among the broad bandwidth photochemically driven active media listed in **Table 1**, Xe_2Cl is of special interest because initial working mixture is not consumed throughout pump flash allowing for operation in the repetitive mode without replacing the working gas mixture after each shot. Moreover, as compared with the XeF(C-A) transition, this active medium has a much longer radiative lifetime of the upper state, which was measured to be 245 ns [34].

This active medium, as well as Kr_2F discussed below, is currently studied to a much lesser extent as compared with XeF. First observation of lasing in the Xe_2Cl triatomic excimer was reported in 1980 upon e-beam pumping [35]. This was followed in 1985 by successful operation of the Xe_2Cl laser optically pumped by the VUV radiation at 137 nm from the open discharge initiated by an exploding wire [36]. Behind the laser action is the reaction of optically excited molecular chlorine with Xe to form XeCl(B) excimers that then recombine with xenon into Xe_2Cl^* emitting in the blue-green region with a fluorescence quantum yield experimentally measured to be ~75%, with respect to the absorbed pumping photons [37].

However, the application of this technique for the femtosecond pulse amplification is of little practical significance since, due to the narrow Cl_2 absorption bandwidth at 137 nm, an efficiency of the Xe_2Cl pumping by the thermal pump source is expected to be low. Moreover, to take advantage of repetitive pulse mode of operation, it requires the development of a powerful large-area pumping source radiating in the short-wave part of the VUV spectral range in a repetitive pulse regime.

On the other hand, as was shown in [38], more promising is the excitation of the active medium in mixtures of Xe and Cl_2 vapor due to the photoassociation process

$$Xe + Cl + h\nu \rightarrow XeCl(B,C)$$

at the wavelength of a XeCl laser (308 nm) followed by the three-body XeCl(B,C) recombination with xenon to form $Xe_2Cl(4\ ^2\Gamma)$. Production of Cl atoms is provided by the same pump pulse of the XeCl laser via photodissociation of Cl_2. In the mixtures of Xe and Cl_2 at pressures of 1–2 bar and 1–2 Torr, respectively, the quantum efficiency for the energy transfer from XeCl(B,C) to $Xe_2Cl(4\ ^2\Gamma)$ is close to 100% due to the extremely high rate constant (1.3×10^{-30} cm^6 s^{-1} [39]) of the XeCl(B,C) recombination with xenon. The main loss process in the pump mechanism considered here is the Cl_2 photodissociation to accumulate a sufficient number of Cl atoms and thereby ensure a high enough absorption of pump photons in the longitudinal geometry of excitation. Nevertheless, the overall efficiency for the conversion of the pump energy into the energy stored in the Xe_2Cl active medium is estimated to be as high as 5% at the pump intensity of 5 MW/cm² in a 100 ns pump pulse. According to Ref. [38], this mechanism of pumping can be of great practical importance, since it enables to produce a small-signal gain in the Xe_2Cl active medium, which is expected to be even higher than that obtained in the XeF(C-A) amplifier to be discussed later. Moreover, molecular chlorine is not consumed upon optical excitation because chlorine atoms, generated in the Cl_2 photodissociation, recombine back to the molecular state at a time scale of 1 ms. This makes it far easier to operate a Xe_2Cl femtosecond amplifier in the pulse repetition regime.

For the sake of completeness, it should be noted that the gain on the Xe_2Cl laser transition excited due to photoassociation at 308 nm was observed for the first time in chlorine-doped solid [40] and liquid [41] xenon. However, unlike gaseous active media, realization of the femtosecond pulse amplifier based on this technique is a serious technological problem.

2.3. Kr_2F active medium

Emission band of Kr_2F at 420 nm is spectrally shifted relative to emissions of XeF(C-A) at 480 nm and Xe_2Cl at 490 nm that enables twofold broadening of the amplification band (**Figure 1**) with the use of two different active media in an amplifier chain.

The Kr_2F excitation mechanism relies on the KrF_2 photodissociation absorption in the VUV spectral range around 164 nm [42] to produce KrF(B) excimers. The utilization of KrF(B) in secondary processes is different, depending on the composition and pressure of the working mixture. For example, in low-concentration Xe admixtures, exchange processes take place, resulting in XeF(B) formation with a yield close to 100% and laser action at 353 nm upon pumping by radiation from exploding wire [42]. Being mixed with Kr at a pressure of ~1 bar, KrF(B) forms $Kr_2F(4\,{}^2\Gamma)$ excimers in three-body recombination collisions [43].

It was found that the laser action in Kr_2F^* also occurs if, instead of Kr, nitrogen is admixed to the working mixture. This observation was attributed to the formation of KrN_2F^* four-atomic excimers that produce Kr_2F^* in exchange reactions with Kr atoms generated upon photochemical decomposition of KrF_2 vapor by VUV pump radiation [43]. Note that, despite a complex Kr_2F^* formation mechanism, which involves three stages of chemical transformations in mixtures with nitrogen (at one of the stages, products of photochemical processes react with each other), the Kr_2F^* yield is rather high providing ~70% of KrF(B) molecules to be transformed into Kr_2F^* excimers [43]. Lasing in Kr_2F at 450 nm was observed upon optical pumping of KrF_2:N_2 = 1:1500 Torr and KrF_2:CF_4:Kr = 1:300:1200 Torr gas mixtures by the VUV radiation from an open discharge initiated by an exploding wire [44].

The most detailed results of experimental investigations of above considered broadband active media and a complete bibliography on the works can be found in Refs. [3, 20, 45].

3. Hybrid systems with XeF(C-A) amplifiers

Among the gaseous active media with broad amplification band, XeF(C-A) is the most widely studied. That is why this transition was the first to study the hybrid approach with the use of the broadband excimer molecules. Studies of the femtosecond pulse amplification on the XeF(C-A) transition were pioneered by demonstration of 1 TW peak power due to the direct amplification of 250 fs seed pulses from a mode-locked dye laser in the e-beam-driven active medium [15]. However, application of electron-beam excitation to the XeF(C-A) active medium faces serious difficulties associated with the above mentioned transient absorption and competition of B-X and C-A transitions due to strong mixing of closely lying B and C states of the XeF excimer by electrons. As it was discussed above, transient absorption results in the afterglow gain formation. Strong B-C states mixing causes C-state depopulation via ASE depletion of the B-state which has two orders of magnitude higher stimulated emission cross section for the B-X transition compared with that for the C-A transition. Using five-component mixture comprising F_2, NF_3, Xe, Kr, and Ar has circumvented this problem due to formation of Kr_2F excimers strongly absorbing emission at the wavelength of the B-X transition (353 nm). As discussed above, optical pumping of this active media is free of both of these shortcomings.

3.1. Hybrid systems based on surface discharge-driven XeF(C–A) amplifiers

Beginning with the proposal of femtosecond pulse amplification in the optically driven XeF(C-A) active medium [14], experimental studies in the field of the hybrid (solid/gas) technology started with the development of the XeF(C-A) amplifier optically driven by the VUV radiation from large-area multichannel surface discharges (see [5] and references cited therein). For this purpose, two versions of the photochemically driven XeF(C-A) amplifier based on the surface discharge as an optical pump source have been built at the P.N. Lebedev Physical Institute (LPI, Moscow, Russia) and at the Lasers Plasmas and Photonic Processes (LP3) Laboratory (Marseille, France) (**Figure 3**). They differ from each other by an output aperture (3 × 11 and 5 × 18 cm², respectively), discharge initiation technique, and pump energy. Pumping sources are based on the multichannel surface discharges initiated along the side walls of rectangular half a meter long dielectric chambers filled with a mixture of XeF_2 vapor, argon, and nitrogen at 1 atm. The pumping scheme, in which two planar sources pump an active medium placed between them, provides spatially homogenous excitation of the medium. Moreover, rectangular aperture of the designed amplifiers offers a simple approach to the development of the multipass optical scheme for energy extraction from XeF(C-A) active medium characterized by rather low values of small-signal gain upon optical pumping. In a multipass scheme of the "wedge-trap configuration," a seed pulse runs between two tilted intracell mirrors allowing for up to 45 double passes through the active medium to be realized.

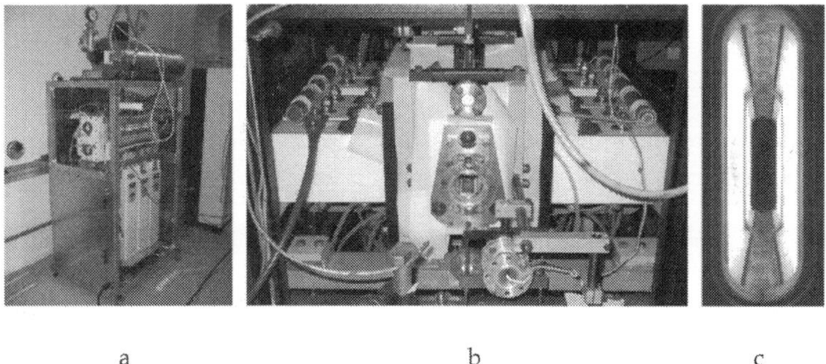

abc

Figure 3. Photographs of the XeF(C-A) amplifiers built at (a) LP3 and (b) LPI. (c) Inside view of the amplifier cell with multichannel surface discharges fired along its side walls.

Operating performances of the XeF(C–A) amplifier, which were measured at the LP3 Laboratory with the use of a hybrid Ti:sapphire/optical-parametric-amplifier front end system, demonstrated a total multipass gain factor of 10^2, corresponding to a small-signal gain of 2×10^{-3} cm^{-1}, with spectrally and spatially homogeneous amplification.

More details and a complete bibliography on the operating characteristics of the laser amplifier and pump sources relying on the multichannel surface discharge are summarized in [5].

3.2. Hybrid systems based on XeF(C-A) amplifiers pumped by the VUV radiation of e-beam converter

The promising results obtained in the course of the studies of the surface discharge pumped XeF(C-A) amplifiers motivated the development of an alternative pump technique based on the conversion of e-beam energy to the VUV radiation, which is expected to be more practical from the viewpoint of the XeF(C-A) amplifier scaling. The main principle of laser action upon pumping by the e-beam excited xenon emission was introduced for the first time in [31] to pump a XeF(B-X) laser operating in the UV region. Later on, this technique was applied to pump a multijoule XeF(C-A) laser in the visible [33].

To study this approach, two hybrid fs systems, THL-30 with a design peak power of ~10 TW and THL-100 designed for 50–100 TW peak power, have been built at the LPI and Institute of High-Current Electronics (IHCE, Tomsk, Russia), respectively. Both of these systems comprise Ti:sapphire front ends (Avesta Project Ltd), frequency doublers, prism pair stretcher, and power-boosting XeF(C-A) amplifiers driven by e-beam-to-VUV-flash converters made at the IHCE.

3.2.1. THL-30 hybrid system

Photos of the front end and XeF(C-A) amplifier incorporated into the THL-30 hybrid system are presented in **Figure 4**. **Figure 5** shows the cross sectional schematic diagram of the XeF(C-A) amplifier. The e-beam converter (2) of cylindrical form is filled with pure xenon at a pressure of 3 bars. Emission of Xe_2^* is excited by four 120 cm long × 15 cm wide radially converging beams of electrons accelerated up to 450 keV in the vacuum diode (1) and injected into the converter through 40 μm Ti foils. The laser cell (3) consists of a 12 × 12 × 128 cm square cross-sectional tube with arrays of 10 rectangular CaF_2 windows (12 × 12 cm) sealed on its side walls. The cell containing XeF_2/N_2 mixture at 0.25–1 bar is housed into the e-beam converter along its axis. In this configuration, the laser cell is thus immersed into the xenon. The distance of 7.5 cm between the Ti foil and CaF_2 windows is chosen to assure that the electrons issued from the e-beams are stopped in the converter at the xenon operating pressure of 3 bars. The e-beams excite the xenon over a 250 ns pulse to produce Xe_2^* fluorescence at 172 nm with a ~30% average fluorescence efficiency [33]. The Xe_2^* radiation is transmitted through arrays of CaF_2 windows into the laser cell to photodissociate XeF_2. The e-beam energy deposited into xenon

Figure 4. THL-30: Photographs of the Ti:sapphire front end (left) and XeF(C-A) amplifier (right).

is measured to be 2.5–3 kJ, which yields a small signal gain of $(1.5–2.5) \times 10^{-3}$ cm^{-1} in the laser cell [46, 47].

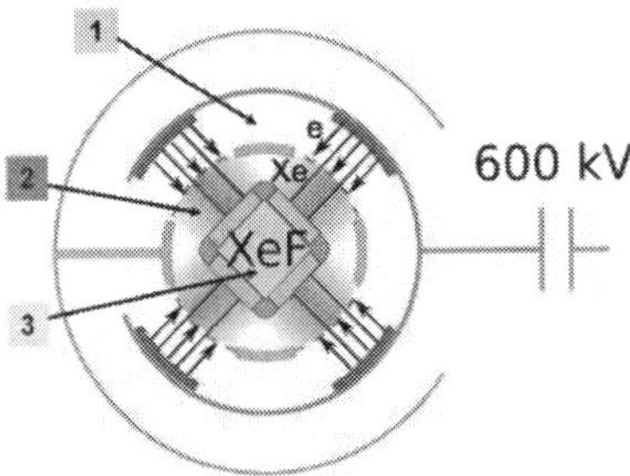

Figure 5. Cross-sectional schematic diagram of the XeF(C-A) amplifier: (1) vacuum diode, (2) e-beam converter, and (3) photolytic laser cell.

Seed pulses with 50 fs duration and 5 mJ energy at 475 nm are produced in the solid-state front end consisting of a Ti:sapphire oscillator operating at 950 nm, regenerative and multipass amplifiers, and KDP frequency doubler spectrally matching the front end to the boosting XeF(C-A) amplifier. Nonlinear frequency upconversion also allows for temporal cleaning of seed pulses injected into the final gas amplifier.

Before the final amplification, the seed pulses are negatively chirped to 1 ps with the use of a prism-pair arrangement. Besides avoiding nonlinear pulse distortion in the amplifier, seed pulse stretching is required to exceed the rotational reorientation time of the XeF(C) molecules, which is estimated to be about 0.8 ps [48, 49]. If the seed pulse is linearly polarized and shorter than the above value, saturation of only a portion of all excited XeF molecules is possible (because of random molecular orientation), thus limiting energy extraction from the amplifier. Down-chirped pulses can be then recompressed due to the positive group velocity dispersion in bulk glass and/or chirped mirrors.

A multipass optical scheme for energy extraction from the active medium has a 3D "wedge-trap" configuration formed by two pairs of tilted intracell mirrors providing the displacement of a beam in two mutually orthogonal directions.

In test experiments, an output energy of 0.25 J has been extracted from the amplifier seeded with a 4 mJ pulse [46, 47], indicating that 5 TW peak power can be obtained in THL-30. Presently, this system is mainly used for the development of key technologies purposed for the implementation in the THL-100 system to be discussed below in more detail.

3.2.2. THL-100 hybrid system

Architecture of THL-100 laser is substantially similar to that described above for the THL-30 hybrid system. Its optical scheme is shown in **Figure 6**.

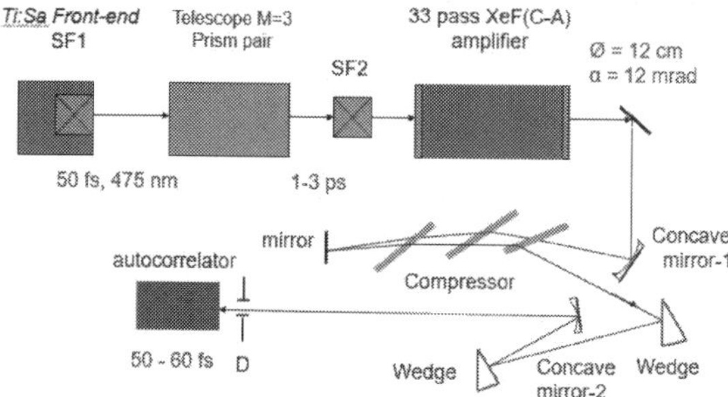

Figure 6. Optical scheme of THL-100 laser system. Ti:Sa front end—Start-480M; SF1 and SF2—vacuum spatial filters; compressor—4-cm-thick fused silica plates at Brewster angle; D—260 μm diaphragm; concave mirror-1—F = 7.5 m; concave mirror-2—F = 12 m.

3.2.2.1. Front end

The front end (Start-480M manufactured by "Avesta Project Ltd") shown in **Figure 7a** consists of a Ti:sapphire master oscillator pumped by a CW pump laser (Verdy-8) at a wavelength of 532 nm, grating stretcher, regenerative and two multipass amplifiers pumped by repetitively pulsed lasers (SOLAR Laser Systems) at a wavelength of 532 nm, spatial filter after the final amplifier, grating compressor, and generator of the second harmonic. The output energy of 20 mJ is produced in a 50 fs pulse at the wavelength of the second harmonic (~475 nm) [50]. The front end operates in the single pulse mode and with a repetition rate of 10 Hz. The output beam of 2.5 cm diameter is directed to the prism pair arrangement (**Figure 7b**) with negative group velocity dispersion, which allows a seed pulse to be stretched. It consists of a mirror telescope with a magnification M = 3, two fused silica prisms allowing a 75-mm-diameter beam to pass, and the mirrors for beam transportation between the prisms in the forward and backward directions. The maximum distance that can be realized between the prisms is 9.6 m, corresponding to the 2.4 ps pulse duration. After the prism pair, the laser beam is directed into the XeF(C-A) amplifier.

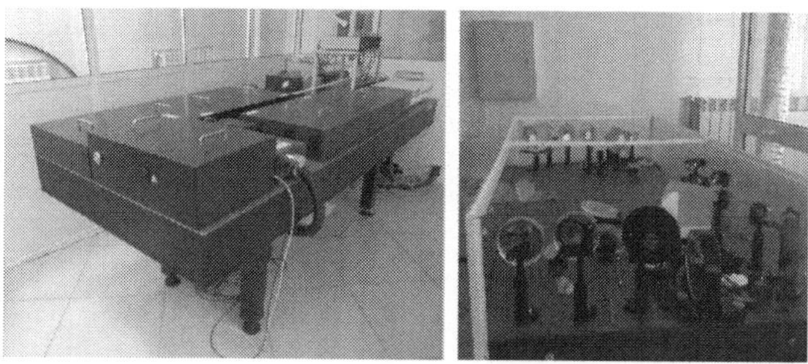

Figure 7. (a) Femtosecond Ti:sapphire front end. (b) Prism pair arrangement.

3.2.2.2. Photodissociation XeF(C-A) amplifier

Figure 8 shows a photograph of the XeF(C-A) amplifier. Its principle of operation and the design is similar to the above described XeF(C-A) amplifier of THL-30 laser system. However, it has its differences: the e-beam converter is driven by six electron beams instead of four and its pump energy is four times higher. The amplifier includes two high-voltage pulse generators operating at $U_0 = 90$ kV or 95 kV charging voltage, a vacuum electron diode, gas chamber with a foil support structure, which filled with xenon, and laser cell with two mirror units for multipass amplification of the laser beam. The design and specifications of the XeF(C-A) amplifier are described in detail in [47, 51–54].

Figure 8. General view of the XeF(C-A) amplifier.

An electron accelerator generates six 100 cm long × 15 cm wide electron beams with a maximum energy of 550 keV (at $U_0 = 95$ kV) at the total diode current of 250 kA in a 150 ns (FWHM) pulse. Bunches are injected into the chamber filled with xenon at a pressure of 3 bar. The electron beam energy is converted to the VUV radiation of Xe_2^* excimers at a wavelength of 172 nm with an efficiency of 30–40%. Inside the gas chamber along its axis is the hexagon laser cell (**Figure 9a**), on the side faces of which there are a total of 54 windows made of CaF_2. The windows with size of 12 × 12 cm and 2 cm thick are set in grooves on the rubber gasket (Viton) and sealed by means of clamping flanges. The windows are arranged opposite grates with foil, through which the electron beam is injected into the gas chamber. This provides the best geometric coupling of the cell with laser pump source. The output aperture of the laser cell has a diameter of 24 cm. Both ends of the laser cell are sealed with fused silica windows with a diameter of 30 cm. Inside the cell, there are two mirror units that provide multiple passage of the seed pulse through the active region. Each unit has 16 mirrors of different diameters arranged along its perimeter. The mirror reflection coefficient is 99.5–99.7%. The laser cell is filled with a gas mixture consisting of 0.25–0.5 bar high purity nitrogen and 0.1–0.4 Torr XeF_2 vapor. **Figure 9b** shows an internal view of the laser chamber with the mirror unit at the rear end. To maximize the efficiency of conversion of the electron beam energy into the VUV radiation, the high purity (99.9997%) xenon is used and the e-beam converter gas chamber is evacuated to a pressure of 10^{-4} Torr. As xenon purity decreases due to exposure of the electron beams to the structural elements of gas chamber, its purification is carried out by a filtering device "Sircal MP-2000."

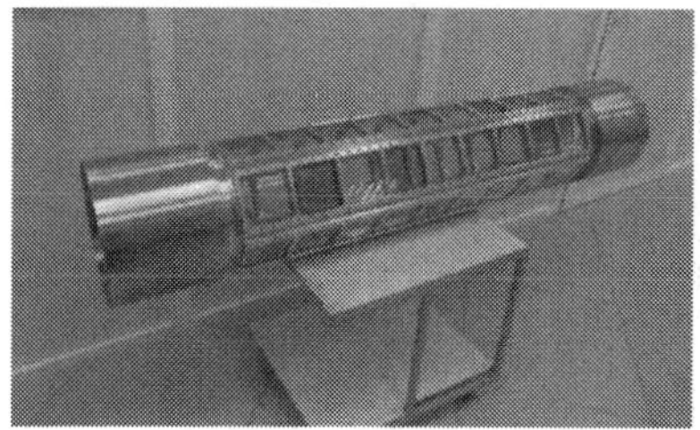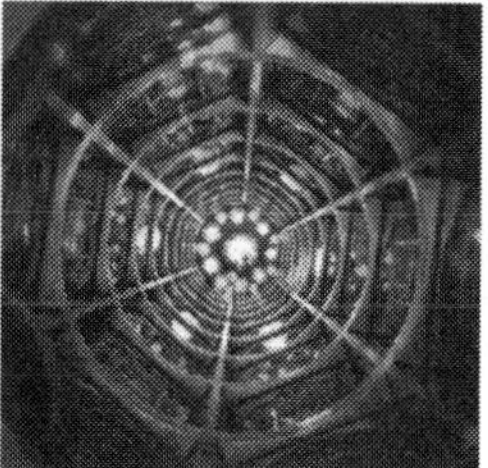

Figure 9. (a) Laser cell of the XeF(C-A) amplifier. (b) Inside view of the laser cell with the mirror unit.

3.2.2.3. Experimental techniques

The active medium of the XeF(C-A) amplifier is created under the action of the VUV radiation at a wavelength of 172 nm. XeF excimer molecules are formed in the B-state via XeF_2 photo-dissociation. The upper state of the XeF(C-A) laser transition is formed as a result of the relaxation of the XeF(B) molecules in collisions with the molecules of the N_2 buffer gas. Amplification of the seed pulses was carried out in a multipass optical scheme (33 passes). The laser beam entering into the XeF(C-A) amplifier was made slightly divergent, so that during amplification, it was steadily increased in diameter from 20 mm (inlet) to 62 mm (the penulti-mate mirror), making a double round-trip along the inside perimeter of the laser cell. The penultimate convex mirror directed the beam to a flat mirror 100 mm in diameter, located on the optical axis. After reflection from this mirror, the beam propagated along the optical axis of the cell reaching 120 mm in diameter at the amplifier output.

Small-signal gain distribution over the laser chamber cross section was measured at four passes through the active medium of the XeF(C-A) amplifier with the help of the Sapphire-488 CW semiconductor laser (488 nm). In addition, the total small-signal gain, G, was measured in the 33 pass amplification scheme. For measuring the power of amplified spontaneous emission (ASE) of the XeF(C-A) amplifier, the output radiation without a seed pulse was focused by concave mirror with a focal length of 22 m on an aperture with diameter 1.1 mm, behind which a filter and calibrated photodiode were located.

Seed pulses of 50 fs duration from the front end were pre-lengthened to 1–2.4 ps in the prism pair with negative group velocity dispersion. After amplification, the down-chirped laser pulses were recompressed in a double pass of a collimated beam with a diameter of 20 cm through three 4-cm-thick fused silica plates at the Brewster angle. Energy losses in the compressor did not exceed 2%. To measure the pulse duration, the central part of the beam (with the help of two fused silica wedges and a spherical mirror with diameter of 90 mm and focal length of 12 m) was assigned to the autocorrelator ASF-20–480 through an aperture with a diameter of 260 μm.

3.2.2.4. Experimental results

The gain characteristics of the active medium (viz. value and spatial distribution of the small-signal gain) depending on the initial composition of the working mixture were examined using 4-passes amplification of the probe laser. The optimal composition of the gas mixture (0.25 Torr XeF_2 and 190 Torr nitrogen) corresponding to the maximum value of the unsaturated total gain ($G = (5–6) \times 10^3$) was found as a result of a compromise between the uniformity of the radial distribution of the small-signal gain and its maximum lying in the peripheral region of the laser cell, where the main part of the beam trajectories is located. It should be noted that the heterogeneity of the small-signal gain distribution over the chamber cross section had almost no effect on the uniformity of the beam intensity, as the diameters of the laser beam and laser cell differ by almost an order of magnitude.

In the test mode, when the maximal pump energy (at $U_0 = 95$ kV) of the VUV radiation was 240 J, the energy of output radiation attained 1 J at amplification of a 1 ps seed pulse with an energy of 1.8 mJ. At the level of 0.7 J output energy, the pulse duration after the bulk fused silica compressor was measured with and without activation of the XeF(C-A) amplifier. In both cases, the pulse width was measured to lie in the range of 50–60 fs (**Figure 10a**). This indicates that a peak power of 14 TW was reached at the output of the laser system [47].

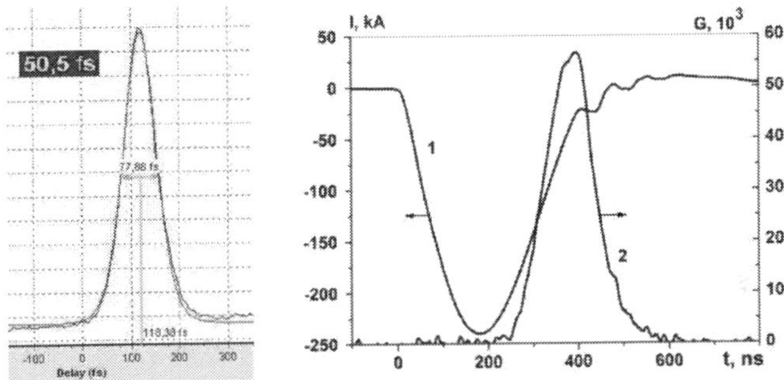

Figure 10. (a) Autocorrelation function of the output pulse with an energy of 0.7 J (in the Sech² approximation). (b) The time behavior of the diode current (1) and total gain (2). $U_0 = 95$ kV.

The ASE power from the XeF(C-A) amplifier, which was measured in the angle of 0.5 mrad with the seed pulse blocked, turned out to be as low as 0.7 W. At the 14 TW peak power obtained in the above experiments, this ASE power corresponds to the temporal contrast ratio of 2×10^{13}. Since the temporal contrast at the 10^{10} level is now routine to obtain in Ti:sapphire systems, and taking into account the nonlinear frequency doubling required to spectrally match the Ti:sapphire front end to the XeF(C-A) boosting amplifier, the temporal contrast of the hybrid (solid/gas) systems seems to be determined only by the ASE of the XeF(C-A) amplifier and is expected to reach $10^{12}–10^{13}$ at a peak power of about 100 TW.

For further improvements of the THL-100 operating performances, the XeF(C-A) amplifier pump system has been upgraded [51, 52] to increase the maximum VUV pump energy up to

300 J at $U_0 = 95$ kV. As a result, the total small-signal gain at 33 passes was enhanced up to (5–6) $\times 10^4$. **Figure 10b** shows the temporal behavior of the diode current (1) and total small-signal gain (2) at 33 passes through the active medium. **Figure 11** shows the dependence of the small-signal gain in arbitrary units vs the partial pressures of nitrogen and XeF_2 vapor.

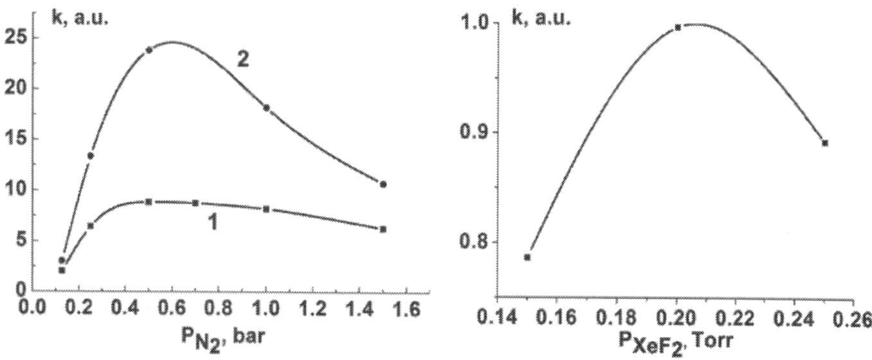

Figure 11. (a) Dependence of the small-signal gain vs the nitrogen pressure, $p(XeF_2) = 0.25$ Torr, $1-U_0 = 90$ kV, $2-U_0 = 95$ kV. (b) Dependence of the small-signal gain vs the XeF_2 pressure, $p(N_2) = 0.5$ bar, $U_0 = 90$ kV.

Amplification of a chirped pulse in the XeF(C-A) amplifier was carried out with the use of the laser mixture containing 0.2 Torr XeF_2 and 0.5 bar nitrogen at $U_0 = 95$ kV. An output energy of 2.5 J was reached when a 2.4 ps seed pulse with an energy of about 1 mJ in a super-Gaussian beam was injected into the XeF(C-A) amplifier [51]. An autograph of the output laser beam on a photographic paper sheet is shown in **Figure 12**. The energy obtained in this experiment promises a peak power as high as ~50 TW to be attained in the visible after pulse recompression to the initial duration of 50 fs.

Figure 12. Imprint of the output laser beam with 2.5 J energy.

4. Wavelength scaling of laser-matter interaction

The visible spectrum of the hybrid femtosecond systems based on the XeF(C-A) amplifiers allows for the λ-scaling laser-matter interaction to be studied. Just to illustrate the role of laser-driver wavelength in laser-matter interaction we shall mention briefly some of applications in which the shorter laser wavelength provides advantages due to favorable wavelength scaling with special emphasis placed on the possible realization of a recombination soft X-ray laser operating in the "water window". The examples considered below do not cover all the applications in which the shorter laser wavelength may be favorable as compared with near IR solid-state femtosecond systems. Nevertheless, they show that the extension of laser wavelength to the visible region allows for wider experimental conditions to be realized providing different laser-matter interaction parameters and better understanding of the underlying physics.

4.1. Laser wake-field acceleration

Laser-driven plasma wake-field acceleration (LWFA) capable of producing high-energy electron beams is widely studied both theoretically and experimentally. In plasma-based acceleration, an intense laser beam drives large amplitude plasma waves via the ponderomotive force. The plasma wave can support very high longitudinal electric fields trapping and accelerating electrons. LWFA experiments have demonstrated acceleration gradients >100 GV m^{-1} enabling electrons to be accelerated well beyond GeV energy on a distance of about 1 cm using a 100 TW-class laser [55].

The electron energy gain ΔE is proportional to the acceleration length, L_{acc}, and longitudinal electric field, E_z, averaged over the acceleration length: $\Delta E = eE_z L_{acc}$. Among the factors limiting the effective acceleration length, laser diffraction, electron dephasing, and pump depletion are the most important. In experiments, the limiting role of the first factor is usually mitigated due to relativistic self-guiding or by using a preformed plasma channel. Electron dephasing originates from the difference of electron and plasma-wave propagation velocities. As a result, highly relativistic electrons, accelerated up to a velocity approaching the speed of light, outrun the accelerating phase region of the plasma wave propagating with a phase velocity, v_p, that is close to the laser pulse group velocity v_g and less than the electron velocity. Electrons are accelerated until their phase slips by one-half the plasma-wave period. In the most promising and efficient high-intensity limit corresponding to a nonlinear wake-field acceleration, referred to the blow-out, bubble, or cavitation regime, the radial ponderomotive force expels all the plasma electrons outward to create a electron density structure resembling a spherical ion cavity behind the laser pulse. Coulomb forces pull the electrons back to the axis in about a plasma period at the rear of the cavity to be trapped and accelerated by the wake-field until they reach the cavity center where they diphase. The acceleration length strongly depends on the plasma-wave phase velocity, which is close to the laser pulse group velocity, v_g, obeying the plasma dispersion low: $v_g = d\omega/dk \approx c(1-\omega_p^2/\omega_0^2)^{1/2}$ with ω_0 and ω_p being the laser and plasma frequencies, respectively. Due to the plasma dispersion, shorter wavelength laser pulses propagating with higher group velocity provide longer dephasing length, L_d. According to the estimations made in [51] with pump depletion taken into account, the dephasing length, given by $L_d \approx 4ca_0^{1/2}\,(\omega_0^2/3\omega_p^3)$ with $a_0 = eA/m_e c^2$ being the relativistically normalized laser amplitude, scales as ω_0^2 showing that shorter laser wavelengths are highly beneficial from the viewpoint of an increase in the acceleration length.

The detailed consideration [56] based on the phenomenological 3D theory for LWFA in the blowout regime, valid at laser power, P, exceeding the critical power $P_c = 17(\omega_0^2/\omega_p^2)$ [GW] for relativistic self-focusing, predicts the electron energy gain

$$\Delta E\left[\text{GeV}\right] \;=\; 1.7\left(P\left[\text{TW}\right]/100\right)^{1/3}\left(10^{18}/n_p\left[\text{cm}^{-3}\right]\right)^{2/3}(0.8/\lambda_0[\mu m])^{4/3}$$

where n_p is the plasma density, and λ_0 stands for the laser wavelength. This indicates that the λ-scaling of LWFA could be of great practical interest. Practically, the same λ-scaling has been obtained in Ref. [57]. However, it should be noted that the gain in energy is achieved at the expense of reducing the number of accelerated electrons, which is proportional to the laser wavelength [56, 57].

4.2. High-order harmonic generation

High-order harmonic generation (HHG) is nowadays widely used to generate spatially and temporally coherent short-wavelength radiation when an intense optical field interacts with a gas or solid target. HHG can provide a single burst or train of attosecond pulses, which allow for ultrafast dynamics of electrons in atoms, molecules, or even solids to be explored [58]. (For a detailed review of experimental and theoretical developments in HHG, see, e.g., [58, 59].)

According to the generally accepted semiclassical three-step model of HHG in gases, the highest possible photon energy (cutoff energy, E_{cutoff}) in the high harmonic spectrum that can be generated from a single atom or ion is predicted by the universal law $E_{cutoff} = I_p + 3.17 U_p$ [60]. Here, I_p is the ionization potential and $U_p = 9.33 \times 10^{-14}\,I_0\lambda_0^2$ is the ponderomotive energy, which is the cycle-averaged kinetic energy of an electron in the laser electric field of intensity I_0 and wavelength λ_0. The λ_0^2 dependence of U_p implies that the use of long excitation wavelengths should result in extending the harmonic cutoff energy further into the X-ray region. On the other hand, conversion efficiency of HHG in gases strongly depends on laser wavelength. The λ-scaling at constant laser intensity has revealed the dependences of HHG efficiency to be between λ_0^{-5} and λ_0^{-6}, which have been obtained experimentally and from numerical simulations [61–63]. General scaling analysis of HHG efficiency as a function of drive laser parameters and material properties is given in [64], which predicts the scaling of the HHG efficiency with the driving wavelength to be λ_0^{-5} at the cutoff and λ_0^{-6} at the plateau region for fixed harmonic wavelength. The severe wavelength dependence of the HHG efficiency is associated with the single-atom dipole response and phase matching. Shorter driver wavelengths are advantageous for both of these factors, if the final objective is not to produce as high-energetic photons as possible. The experimental results obtained in Ref. [65] for different noble gases confirm this wavelength scaling and show two orders of magnitude higher HHG intensity in the energy range of 20–70 eV with 400 nm pulses as compared with 800 nm laser driver.

However, HHG in gases has fundamental physical restrictions arising from the limitation on the laser intensity since plasma generation, caused by strong ionization of the gaseous medium at intensities above 10^{16} W/cm^2, results in phase mismatch, thereby suppressing harmonic generation [66]. This limitation is not present in the case of HHG from solid. According to the oscillating plasma mirror model [67], the laser field produces a relativistic oscillation of the overcritical plasma surface with the laser frequency, inducing incident pulse modulation that gives rise to the high-order harmonics in the spectrum of the reflected emission due to a transient Doppler frequency upshift. (For more detailed analysis of the basic generation mechanisms lying behind HHG from solids, see, e.g., [59, 68, 69].)

Using particle-in-cell (PIC) simulation to accurately model the HHG with a plasma target, Teubner and Gibbon [59] have obtained the laser-to-harmonic conversion efficiency, η_H, in the laser intensity range $I_0 = 10^{17}$–10^{19} W/cm^2. Summarized by an empirical relation for high harmonic orders ($N = \lambda_0/\lambda_H \gg 1$, where λ_H stands for harmonic wavelength), the results of the numerical simulation take the form [54]:

$$\eta_H \approx 9'10^{-5}(I_0\lambda_0{}^2 / 10^{18}\,W\,cm^{-2}\mu m^2)^2 \left(N/10\right)^{-\alpha}$$

with α depending on the laser intensity and ranging from $\alpha = 6$ at $I_0 = 10^{17}$ W/cm^2 to $\alpha = 3.5$ at $I_0 = 10^{19}$ W/cm^2 at $\lambda_0 = 1$ μm. On the basis of this empirical scaling, the authors came to the conclusion that shorter wavelength lasers are highly beneficial from the viewpoints both of extending to shorter-wavelength harmonics and harmonic efficiency enhancement. The wavelength scaling of the same form but with $\alpha = 5$ independent of laser intensity was argued with the use of 1D-PIC code in the earlier paper of Gibbon [70]. The analysis made in [59] of a variety of experimental results obtained with different laser wavelengths confirms the prediction of the strong harmonic yield increase with laser frequency. This makes powerful blue-green hybrid systems to be very promising as the drivers for generation of soft X-ray harmonics well within the water-window spectral range.

4.3. Soft X-ray lasers

Development of coherent X-ray sources is motivated by a variety of their applications in science and technology. X-ray lasers offer new capabilities in understanding the nanoscale structure of complex materials, including biological systems, and X-ray matter under extreme conditions. One of the greatest challenges is the high-resolution 3D holographic microscopy of a wide range of biological objects in the living state. For this purpose, coherent ultrafast X-ray sources of high power are required. One of the most important milestones for the high contrast X-ray imaging of living biological structures in a natural aqueous environment is the "water window" lying between the K absorption edges of carbon (4.37 nm) and oxygen (2.33 nm), where carbon is highly opaque, while water is largely transparent. A primary challenge of X-ray exposure is the realization of "diffraction-before-destruction" approach allowing for diffraction patterns to be obtained on time scales shorter than the onset of radiation damage of samples. This approach has been successfully demonstrated in studies of biological samples with the use of X-ray free electron lasers (XFEL) producing femtosecond pulses of high intensity (For example, see [71]).

There has been remarkable progress in the development of XFELs that hold the great promise for user experiments ranging from atomic physics to biological structure determination. Despite the fact that these lasers are powerful tools in studies of matter structure and physics of light-matter interaction, they have limited accessibility because of their high cost and large-scale. This makes the current search for alternative X-ray sources of laboratory scale to be of great importance.

Presently, there are two main approaches to the development of compact ultrashort-pulsed sources of coherent soft X-ray: above considered HHG by gas and solid targets, as well as the generation of coherent X-ray radiation in the laser plasma. The first one is characterized by low-intensity soft X-ray radiation, insufficient for the realization of holographic imaging methods. The laser plasma enables generation of lasing in the soft X-ray region with beam performances close to those of XFELs [72].

Actually, only collisional and recombination schemes of active media excitation to produce soft X-ray in a laser plasma are of practical interest. The first of them was realized in a laser plasma with high electron temperature, providing a population inversion on transitions between excited states of ions, which typically lie in the range 10–50 nm [72]. The most promising way to extend the spectral range of the X-ray lasers deeper into the X-ray region, including the "water window," lies in the further developing recombination scheme of excitation of transitions to the ground state of recombining fully stripped ions.

The first observation of the amplification on the transition to the ground state dates back to 1983 [73] when hydrogen-like lithium ions were excited in the laser plasma produced due to optical field ionization (OFI) by the UV radiation from a subpicosecond KrF laser. Later that year, this observation was confirmed in different experimental conditions [74–76]. The OFI approach to excitation of recombination soft X-ray lasers is particularly attractive since it produces fully stripped ions on a time scale of one period of the incident laser electric field and enables formation of cold electrons with low residual energy (for a linearly polarized laser pulse) providing favorable conditions for high-rate three-body recombination. Residual energy is proportional to the square of a pump laser wavelength and can be reduced by using a short-wavelength driver pulse.

An electron removed from an atom due to OFI interacts with the plane polarized laser field and acquires quiver energy of the coherent electron oscillation in the field and energy of electron drift along the laser field direction [77, 78]. For ultrashort pulses, the quiver energy is returned to the wave, and it does not contribute to residual energy. Most of the electrons are ionized within a narrow interval near the crest of the oscillating electric field because of the exponential dependence of the ionization rate on the electric field amplitude. Classically, the average drift energy, ε, of an electron depends on the phase mismatch, $\Delta\phi$, between the phase at which the electron is freed and the crest of the electromagnetic wave:

$$\varepsilon = 2\varepsilon_q \sin^2 \Delta\phi$$

where ε_q is the quiver energy ($\varepsilon_q = e^2 E_0^2/4m_e\omega^2$ with E_0 and ω being the peak amplitude and angular frequency of the laser electric field $E = E_0\sin\omega t$, respectively). Thus, the residual energy of electrons produced by OFI can be much lower than the electron quiver energy, and, secondly, shorter wavelength ionizing lasers are beneficial to achieve gain in the recombination scheme.

To demonstrate advantages of short-wavelength pumping, λ-scaling of the recombination excitation efficiency for the transitions to ground state was experimentally and numerically studied by different groups [76, 78, 79]. Numerical simulations of the small-signal gain on the $4s_{1/2}$–$3p_{3/2}$ transition at 23.2 nm in Ar^{7+} show that 400 nm pump laser radiation allows an increase in the small-signal gain on the $4s_{1/2}$–$3p_{3/2}$ transition at 23.2 nm in Ar^{7+} by more than an order of magnitude as compared with 800 nm pumping [80].

Simulation of the recombination gain formation on the $2 \rightarrow 1$ transition at 3.4 nm in H-like CVI pumped with a 400 nm pump laser has been performed in [81]. It was shown that the recombination gain as high as 180 cm^{-1} can be achieved on this transition using the driving pulse duration of 20 fs with peak intensity of 8×10^{18} W/cm^2 and 10 μm diameter focal spot. The key factor playing important role in the recombination mechanism of pumping is the non-Maxwellian nature of the distribution function after OFI [79, 81], which is strongly peaked near the zero electron energy. Ultrashort pumping time (<100 fs) is required to minimize heating and Maxwellization of electron energy distribution at the time scale of three-body recombination.

Less encouraging results have been obtained in Ref. [82], indicating that there are a number of issues, which have to be investigated experimentally for better understanding of the physical processes lying behind the optical production of recombination plasmas. This requires the development of ultrashort (20–50 fs) multiterawatt lasers in the UV or visible range as pumping sources.

5. Conclusions

Development of the photochemical method for exciting active media has resulted in the emerging of the new class of gas lasers in the spectral range extending from the NIR to UV regions. The most remarkable achievements of these studies belong to the visible range where no alternatives are available so far for the excitation of broadband gaseous active media (XeF(C-A), Xe_2Cl, and Kr_2F) that would not be strongly modified by transient absorption. This pave the way for the development of hybrid (solid/gas) laser systems towards petawatt peak power in the blue-green spectral region due to their broad amplification bandwidths, able to support as short as 10 fs pulses, and their relatively high saturation fluences (0.05–0.2 J cm^{-2}), promising as high as 10 TW peak power to be obtained from square cm of an output aperture.

To demonstrate the high potential of the hybrid approach relying on the optically driven broadband active media in the visible, two femtosecond hybrid systems are now under development with the aim of conducting proof-of-principle experiments: THL-30 at LPI, designed for about 5 TW of output peak power, and THL-100 at IHCE, designed to be ten times more powerful. Behind these systems is the amplification of the second harmonic of Ti:sapphire front ends in the power-boosting XeF(C-A) amplifiers driven by the e-beam-to-VUV flash converters. In the pilot experiments performed in the THL-100 system, peak power of 14 TW has been attained in the 50 fs pulse at the output energy of 0.7 J. After upgrading pumping source, an energy output has been enhanced up to 2.5 J in the 2.4 ps pulse before its recompression promising a peak power of 50 TW to be obtained. Besides spectral matching between a solid-state frond-end and gas XeF(C-A) amplifier, the nonlinear frequency upconversion results in efficient temporal cleaning of the ultrashort optical pulse, thereby providing a high contrast ratio for the output blue-green pulses produced by a hybrid laser chain. This was confirmed by the results of ASE measurements in the XeF(C-A) amplifier of the THL-100 system, which argue that a contrast ratio of 10^{12}–10^{13} is feasible in the blue-green hybrid femtosecond systems with a peak power of about 100 TW.

By the example of LWFA, HHG, and recombination soft X-ray lasers, it was shown that, in some cases, application of shorter wavelength lasers (as compared to Ti:sapphire lasers operating in the NIR) for laser-matter interaction may be advantageous and extends the frontiers of experimental ability to provide deeper insight into the physical mechanisms of the laser-matter interaction. One of the greatest challenges is the development of recombination-pumped soft X-ray lasers that have potential to extend SXRL spectral range towards "water window" and beyond.

Actually, the above-discussed blue-green hybrid concept can be considered as an alternative to the direct nonlinear upconversion of intense NIR laser radiation to the visible with the use of second harmonic generation (SHG) technique. However, to the best of our knowledge, the highest peak power reached so far in the visible with SHG does not exceed 4 TW, producing the peak intensity in a focal spot diameter of about 3 μm as low as 3×10^{18} W/cm^2 because of poor beam quality [83]. Achieving higher parameters in Ti:sapphire laser systems with SHG meets serious technical problems arising from a variety of nonlinear effects in crystals at high intensities leading to a significant spatiotemporal degradation of beam quality [84]. Moreover, a broad spectrum of femtosecond pulses and strong nonlinear wave front distortion require application of very thin (0.5–1 mm) nonlinear crystals of large diameter (>10 cm). The technology of such crystals manufacture is not yet available. Nevertheless, a large ongoing effort is presently devoted to overcome these difficulties in SHG and to reach hundreds of TW at wavelength of the second harmonic [85, 86]. The hybrid (solid/gas) laser technology is free of these problems because peak powers of 0.1–1 TW are required for a seed pulse generated by the solid-state front end in order to extract most of the energy stored in the final gaseous amplifier.

At the same time, it is necessary to say that the hybrid systems relying on the photochemically driven boosting amplifiers are inferior to the all-solid-state systems from the view point of a pulse-repetition rate reaching 1 kHz at moderate output peak powers. The hybrid systems operating in the visible could be of interest for the use in low repetition rate experiments, which require an output peak power of tens and hundreds of TW. In the case of the Xe_2Cl active medium, repetition rates up to 10 Hz seems to be attainable with proper engineering.

Acknowledgements

The work was supported by the Russian Foundation for Basic Research (Grant No. 15-19-10021).

Author details

Leonid D. Mikheev[1,2*] and Valery F. Losev[3,4]

*Address all correspondence to: mikheev@sci.lebedev.ru

1 P.N. Lebedev Physical Institute, Russian Academy of Sciences, Moscow, Russia

2 National Research Nuclear University MEPhI (Moscow Engineering Physics Institute), Moscow, Russia

3 Institute of High Current Electronics, Siberian Branch, Russian Academy of Sciences, Tomsk, Russia

4 National Research Tomsk Polytechnic University, Tomsk, Russia

References

[1] Danson C, Hillier D, Hopps N, Neely D. Petawatt class lasers worldwide. High Power Laser Sci. Eng. 2015;3:e3. doi:10.1017/hpl.2014.52

[2] Bahk S-W, Rousseau P, Planchon TA, Chvykov V, Kalintchenko G, Maksimchuk A, Mourou GA, Yanovsky V. Characterization of focal field formed by a large numerical aperture paraboloidal mirror and generation of ultra-high intensity (10^{22} W/cm^2). Appl. Phys. B. 2005;80:823–832. doi:10.1007/s00340-005-1803-8

[3] Zuev VS, Mikheev LD. Photochemical Lasers. Chur: Harwood Acad. Publ; 1991. 103 p.

[4] Mikheev LD. Photochemical lasers on electronic molecular transitions. Quantum Electron. 2002;32:1122–1132. doi:10.1070/QE2002v032n12ABEH002355

[5] Mikheev LD, Tcheremiskine VI, Uteza OP, Sentis ML. Photochemical gas lasers and hybrid (solid/gas) blue-green femtosecond systems. Progr. Quantum Electron. 2012;36:98–142. doi:10.1016/j.pquantelec.2012.03.004

[6] Basov NG, Zuev VS. Short-Pulse Iodine Laser. Il Nuovo Cimento. 1976;31B:129–151.

[7] Brederlow G, Fill E, Witte KJ. The High-Power Iodine Laser. Berlin Heidelberg: Springer-Verlag GmbH; 1983. 183 p. doi:10.1007/978-3-540-39491-4

[8] Zuev VS, Ka1ulin VA. Scientific foundations of powerful photodissociation lasers (history of research in the 1960s at the Division of Quantum Radiophysics of the P N Lebedev Physics Institute). Quantum Electron. 1997;27:1073–1080. doi:10.1070/QE1997v027n12ABEH001112

[9] Zarubin PV. Academician Basov, high-power lasers and the antimissile defence problem. Quantum Electron. 2002;32:1048–1064. doi:10.1070/QE2002v032n12ABEH002348

[10] Brederlow G, Brodmann R, Eidmann K, Nippus M, Petsch R, Witkowski S, Volk R, Witte KJ. Performance of the Asterix III high power iodine laser. IEEE J. Quantum Electron. 1980;QE 16:122–125. doi:10.1109/JQE.1980.1070441

[11] Kirillov GA, Murugov VM, Punin VT, Shemyakin VI. High-power laser system Iskra-V. Laser Part. Beams. 1990;8:827–831. doi:10.1017/S0263034600009198

[12] Tittel FK, Marowsky G, Wilson WL, Smiling M. Electron beam pumped broad-band diatomic and triatomic excimer lasers. IEEE J. Quantum Electron. 1981;QE 17:2268–2281. doi:10.1109/JQE.1981.1070705

[13] Tang KY, Lorents DC, Huestis DL. Gain measurements on the triatomic excimer Xe_2Cl. Appl. Phys. Lett. 1980;36:347–349. doi:10.1063/1.91498

[14] Mikheev LD. Possibility of amplification of a femtosecond pulse up to the energy 1 kJ. Laser Part. Beams. 1992;10:473–478. doi:10.1017/S0263034600006716

[15] Hofmann T, Sharp T, Dane B, Wisoff P, Wilson W, Tittel F, Szabó G. Characterisation of an ultrahigh peak power XeF(C-A) excimer laser system. IEEE J. Quantum Electron. 1992;QE 28:1366–1375. doi:10.1109/3.135279

[16] Simon J. Ultrashort light pulses. Rev. Sci. Instrum. 1989;60:3597–3624. doi:10.1063/1.1140516

[17] Tellinghuisen P, Tellinghuisen J, Coxon JA, Velazco JE, Setser DW. Spectroscopic studies of diatomic noble gas halides. IV. Vibrational and rotational constants for the X, B, and D states of XeF. J. Chem. Phys. 1978;68:5187–5198. doi:10.1063/1.435583

[18] Bibinov NK, Vinogradov IP, Mikheev LD, Stavrovskii DB. Determination of spectral dependence of the absolute quantum yield of XeF(B,C,D) excimer formation upon photolysis of XeF_2. Sov. J. Quantum Electron. 1981;11:1178–1181. doi:10.1070/QE1981v011n09ABEH008227

[19] Kono M, Shobatake K. Photodissociative excitation processes of XeF_2 in the vacuum ultraviolet region 105–180 nm. J. Chem. Phys. 1995;102:5966–5978. doi:10.1063/1.469331

[20] Mikheev LD, Stavrovskii DB, Zuev VS. Photodissociation XeF laser operating in the visible and UV regions. J. Rus. Las. Res. 1995;16:427–475. doi:10.1007/BF02581226

[21] Brashears HC, Setser DW. Transfer and quenching rate constants for XeF(B) and XeF(C) state in low vibrational levels. J. Chem. Phys. 1982;76:4932–4946. doi:10.1063/1.442839

[22] Basov NG, Zuev VS, Mikheev LD, Stavrovskii DB, Yalovoi VI. Stimulated emission due to the $B(1/2)–X^2\Sigma^+$ transition in the XeF molecule formed by photodissociation of XeF_2. Sov. J. Quantum Electron. 1977;7:1401. doi:10.1070/QE1977v007n11ABEH008201

[23] Basov NG, Zuev VS, Kanaev AV, Mikheev LD, Stavrovskii DB. Laser action due to the bound-free C(3/2)–A(3/2) transition in the XeF molecule formed by photodissociation of XeF_2. Sov. J. Quantum Electron. 1979;9:629. doi:10.1070/QE1979v009n05ABEH009058

[24] Zuev VS, Kashnikov GN, Mamaev SB. XeF laser with optical pumping by surface discharges. Quantum Electron. 1992;22:973–979. doi:10.1070/QE1992v022n11ABEH003645

[25] Knecht BA, Fraser RD, Wheeler DJ, Zietkiewich CJ, Mikheev LD, Zuev VS, Eden JG. Compact XeF (C → A) and iodine laser optically pumped by a surface discharge. Opt. Lett. 1995;20:1011–1913. doi:10.1364/OL.20.001011

[26] Knecht BA, Fraser RD, Wheeler DJ, Zietkiewich CJ, Mikheev LD, Zuev VS, Eden JG. Optical pumping of the XeF(C-A) and iodine 1.315-μm lasers by a compact surface discharge system. Opt. Eng. 2003;42:3612–3621. doi:10.1117/1.1751133

[27] Yu L, Liu J, Ma L, Yi A, Huang C, An X, Li H, Chen G. 10 J energy-level optically pumped XeF(C–A) laser with repetition mode. Optics Lett. 2007;32:1087–1089. doi:10.1364/OL.32.001087

[28] Gross RWF, Schneider LE, Amimoto ST. XeF laser pumped by high-power sliding discharges. Appl. Phys. Lett. 1988;53:2365–2367. doi:10.1063/1.100231

[29] Sentis ML, Tcheremiskine VI, Delaporte PC, Mikheev LD, Zuev VS XeF(C-A) laser pumped by formed-ferrite open discharge radiation. Appl. Phys. Lett. 1997;70:1198–1200. doi:10.1063/1.118529

[30] Anisimov SV, Ermilov YuA, Kashnikov GN, Kazanskii VM, Mikheev LD, Nesterov RO, Stavrowskii DB, Theremiskine VI, Zemskov EM, Zuev VS. Spectrally selective time resolved actinometry of VUV radiation of a moving gas-dynamic discontinuity. Laser Phys. 1994;4:416–418.

[31] Eden JG. XeF(B → X) laser optically excited by incoherent Xe_2^* (172-nm) radiation. Opt. Lett. 1978;3:94–96. doi:10.1364/ol.3.000094

[32] Bischel WK, Nakano HH, Eckstrom DJ, Hill RM, Huestis DL, Lorents DC. A new blue-green laser in XeF. Appl Phys. Lett. 1979;34:565–567. doi:10.1063/1.90868

[33] Eckstrom DJ, Walker HC, Jr. Multijoule performance of the photolytically pumped XeF (C-A) laser. IEEE J. Quantum Electron. 1982;QE 18:176–181. doi:10.1109/JQE.1982.1071517

[34] McCown AW, Eden JG. Ultraviolet photoassociative production of XeCl(B,C) molecules in Xe/Cl$_2$ gas mixtures: Radiative lifetime of Xe$_2$Cl(4 $^2\Gamma$). J. Chem. Phys. 1984;81:2933–2938. doi:10.1063/1.448042

[35] Tittel FK, Wilson WL, Stickel RE, Marowsky G, Ernst WE. A triatomic Xe2Cl excimer laser in the visible. Appl. Phys. Lett. 1980;36:405–407. doi:10.1063/1.91533

[36] Basov NG, Zuev VS, Kanaev AV, Mikheev LD. Stimulated emission from an optically pumped Xe$_2$Cl laser. Sov. J. Quantum Electron. 1985;15:1289–1290. doi:10.1070/QE1985v015n09ABEH007742

[37] Zuev VS, Kanaev AV, Mikheev LD. Determination of the absolute quantum efficiency of the luminescence of Xe$_2$Cl* in Cl$_2$–Xe mixtures. Sov. J. Quantum Electron. 1987;17:884–885. doi:10.1070/QE1987v017n07ABEH009473

[38] Mikheev LD. Evaluating the prospects of exciting the Xe$_2$Cl active medium by laser radiation for amplifying femtosecond pulses. Quantum Electron. 2005;35:984–986. doi:10.1070/QE2005v035n11ABEH013027

[39] Quiñones E, Yu YC, Setser DW, Lo G. Decay kinetics of XeCl(B,C) in Xe and in mixtures of Xe with Kr, Ar, Ne, and He. J. Chem. Phys. 1990;93:333–344. doi:10.1063/1.459605

[40] Okada F, Apkarian VA. Electronic relaxation of Xe$_2$Cl in gaseous and supercritical fluid xenon. J. Chem. Phys. 1991;94:133–144. doi:10.1063/1.460387

[41] Wiedeman L, Fajardo ME, Apkarian VA. Cooperative photoproduction of Xe$_2^+$Cl$^-$ in liquid Cl$_2$/Xe solutions: Stimulated emission and gain measurements. Chem. Phys. Lett. 1987;134:55–59. doi:10.1016/0009-2614(87)80013-1

[42] Zuev VS, Isaev IF, Kanaev AV, Mikheev LD, Stavrovskii DB, Shchepetov NG. Lasing as a result of a B–X transition in the excimer XeF formed as a result of photodissociation of KrF$_2$ in mixtures with Xe. Sov. J. Quantum Electron. 1981;11:221–222. doi:10.1070/QE1981v011n02ABEH005884

[43] Zuev VS, Kanaev AV, Mikheev LD, Stavrovskii DB. Investigation of luminescence in the 420 nm range as a result of photolysis of KrF$_2$ in mixtures with Ar, Kr, and N$_2$. Sov. J. Quantum Electron. 1981;11:1330–1335. doi:10.1070/QE1981v011n10ABEH008468

[44] Basov NG, Zuev VS, Kanaev AV, Mikheev LD, Stavrovskii DB. Stimulated emission from the triatomic excimer Kr$_2$F subjected to optical pumping. Sov. J. Quantum Electron. 1980;10:1561–1562. doi:10.1070/QE1980v010n12ABEH010282

[45] Mikheev LD. Use of photoprocesses with charge transfer to excite active laser media. J. Sov. Las. Res. 1990;11:288–304. doi:10.1007/BF01120629

[46] Aristov AI, Grudtsin YaV, Zubarev IG, Ivanov NG, Konyashchenko AV, Krokhin ON, Losev VF, Mavritskii AO, Mamaev SB, Mesyats GA, Mikheev LD, Panchenko YuN, Rastvortseva AA, Ratakhin NA, Sentis M, Starodub AN, Tenyakov SYu, Uteza O, Tcheremiskine VI, Yalovoy VI. Hybrid femtosecond laser system based on photochemical XeF(C-A) amplifier with an aperture of 12 cm. Optica Atmosfery i Okeana J. 2009;22:1024–1029 (in Russian).

[47] Alekseev SV, Aristov AI, Grudtsyn YaV, Ivanov NG, Koval'chuk BM, Losev VF, Mamaev SB, Mesyats GA, Mikheev LD, Panchenko YuN, Polivin AV, Stepanov SG, Ratakhin NA, Yalovoi VI, Yastremskii AG. Visible-range hybrid femtosecond systems based on a XeF(C–A) amplifier: state of the art and prospects. Quantum Electron. 2013;43:190–200. doi:10.1070/QE2013v043n03ABEH015096.

[48] Hofmann T, Sharp TE, Dane CB, Wisoff P, Wilson WL, Jr, Tittel FK, Szabó G. Characterisation of an Ultrahigh Peak Power XeF(C-A) Excimer Laser System. IEEE J. Quantum Electron. 1992;QE 28:1366–1375. doi:10.1109/3.135279

[49] Sharp TE, Hofmann Th, Dane CB, Wilson WL, Tittel FK, Wisoff PJ, Szabó G. Ultrashort-laser-pulse amplification in a XeF(C-A) excimer amplifier. Opt. Lett. 1990;15:1461–1463. doi:10.1364/OL.15.001461

[50] Alekseev SV, Ivanov MV, Ivanov NG, Losev VF, Mesyats GA, Mikheev LD, Panchenko YN, Ratakhin NA. Modernization of THL-100 hybrid femtosecond laser system. Izvestiya Vysshikh Uchebnykh Zavedenii, Seriya Fizika. 2014;57(12/2):101–105 (in Russian).

[51] Alekseev SV, Ivanov MV, Ivanov NG, Losev VF, Mesyats GA, Panchenko YN, Ratakhin NA. Parameters of the THL-100 hybrid femtosecond laser system after modernization. Russ. Phys. J. 2015;58:1087–1092. doi:10.1007/s11182-015-0616-4

[52] Abdullin EN, Ivanov NG, Losev VF, Morozov AV. Production of a large cross-section electron beam in electron diode with rod reverse current conductors. Laser and Particle Beams. 2013;31:697–702. doi:10.1017/S026303461300075X

[53] Alekseev SV, Ivanov NG, Kovalchuk BM, Losev VF, Mesyats GA, Mikheev LD, Panchenko YN, Ratakhin NA., Yastremsky AG. Terawatt laser hybrid THL-100 system based on the photodissociation XeF(C-A) amplifier. Optica Atmosfery i Okeana J. 2012;25:221–225 (in Russian).

[54] Alekseev SV, Aristov AI, Ivanov NG, Kovalchuk BM, Losev VF, Mesyats GA, Mikheev LD, Panchenko YuN, Ratakhin NA. Multi-terawatt femtosecond hybrid system based on a photodissosiation XeF(C-A) amplifier in the visible range. Quantum Electron. 2012;42:377–378. doi:10.1070/QE2012v042n05ABEH014902

[55] Leemans WP, Gonsalves AJ, Mao H-S, Nakamura K, Benedetti C, Schroeder CB, Tóth CS, Daniels J, Mittelberger DE, Bulanov SS, Vay J-L, Geddes CGR, Esarey E. Multi-GeV

Electron beams from capillary-discharge-guided subpetawatt laser pulses in the self-trapping regime. Phys. Rev. Lett. 2014;113:245002. doi:10.1103/PhysRevLett.113.245002

[56] Lu W, Tzoufras M, Joshi C, Tsung FS, Mori WB, Vieira J, Fonseca RA, Silva LO. Generating multi-GeV electron bunches using single stage laser wakefield acceleration in a 3D nonlinear regime. Phys. Rev. ST AB 2007;10:061301. doi:10.1103/PhysRevSTAB. 10.061301

[57] Gordienko S, Pukhov A. Scalings for ultra relativistic laser plasmas and quasimonoe-nergetic electrons. Phys. Plasmas 2005;12:043109. doi:10.1063/1.1884126

[58] Eden JG. High-order harmonic generation and other intense optical field–matter interactions: review of recent experimental and theoretical advances. Prog. Quantum Electron. 2004;28:197–246. doi:10.1016/j.pquantelec.2004.06.002

[59] Teubner U, Gibbon P. High-order harmonics from laser-irradiated plasma surfaces. Rev. Mod. Phys. 2009;81:445–479. doi:10.1103/RevModPhys.81.445

[60] Krause JL, Schafer KJ, Kulander KC. High-order harmonic generation from atoms and ions in the high intensity regime. Phys. Rev. Lett. 1992;68:3535–3538. doi:10.1103/ PhysRevLett.68.3535

[61] Yakovlev VS, Ivanov M, Krausz F. Enhanced phase-matching for generation of soft X-ray harmonics and attosecond pulses in atomic gases. Opt. Express 2007;15:15351–15364. doi:10.1364/OE.15.015351

[62] Tate J, Auguste T, Muller HG, Salières P, Agostini P, DiMauro LF. Scaling of wave-packet dynamics in an intense midinfrared field. Phys. Rev. Lett. 2007;98:013901. doi: 10.1103/PhysRevLett.98.013901

[63] Colosimo P, Doumy G, Blaga CI, Wheeler J, Hauri C, Catoire F, Tate J, Chirla R, March AM, Paulus GG, Muller HG, Agostini P, DiMauro LF. Scaling strong-field interactions towards the classical limit. Nat. Phys. 2008;4:386–389. doi:10.1038/nphys914

[64] Falcão-Filho EL, Gkortsas VM, Gordon A, Kärtner FX. Analytic scaling analysis of high harmonic generation conversion efficiency. Opt. Express 2009;17:11217–11229. doi: 10.1364/OE.17.011217

[65] Falcão-Filho EL, Lai C-J, Gkortsas V-M, Huang S-W, Chen L-J, Hong K-H, Kärtner FX. Scaling of high harmonic generation efficiencies with 400-nm and 800-nm driver pulses. In: Proceedings of Conference on Lasers and Electro-Optics and Quantum Electronics and Laser Science Conference (CLEO/QELS-2010); 16–21 May 2010; San Jose, California United States: IEEE; paper JThI4. doi:10.1364/CLEO.2010.JThI4

[66] Chen M-C, Arpin P, Popmintchev T, Gerrity M, Zhang B, Seaberg M, Popmintchev D, Murnane MM, Kapteyn HC. Bright, coherent, ultrafast soft X-ray harmonics spanning the water window from a tabletop light source. Phys. Rev. Lett. 2010;105:173901. doi: 10.1103/PHYSREVLETT.105.173901

[67] Lichters R, Meyer-ter-Vehn J, Pukhov A. Short-pulse laser harmonics from oscillating plasma surfaces driven at relativistic intensity. Phys. Plasmas 1996;3:3425–3437. doi: 10.1063/1.871619

[68] Thaury C, Quéré F. High-order harmonic and attosecond pulse generation on plasma mirrors: basic mechanisms. J. Phys. B: At. Mol. Opt. Phys. 2010;43:213001. doi: 4075/43/21/213001

[69] Baeva T, Gordienko S, Pukhov A. Theory of high-order harmonic generation in relativistic laser interaction with overdense plasma. Phys. Rev. E. 2006;74:046404. doi: 10.1103/PhysRevE.74.046404

[70] Gibbon P. Harmonic generation by femtosecond laser-solid interaction: a coherent "Water-Window" light source?. Phys. Rev. Lett. 1996;76:50–53. 10.1103/PhysRevLett. 76.50

[71] Redecke L, et al. Natively inhibited *Trypanosoma brucei* cathepsin B structure determined by using an X-ray laser. Science. 2013;339:227–230. doi:10.1126/science.1229663

[72] Suckewer S, Jaeglé P. X-ray laser: past, present, and future. Laser Phys. Lett. 2009;6:411–436. doi:10.1002/lapl.200910023

[73] Nagata Y, Midorikawa K, Kubodera S, Obara M, Tashiro H, Toyoda K. Soft-x-ray amplification of the Lyman-alpha transition by optical-field-induced ionization. Phys. Rev. Lett. 1993;71:3774–3777. doi:10.1103/PhysRevLett.71.3774

[74] Nagata Y, Midorikawa K, Kubodera S, Obara M, Tashiro H, Toyoda K, Kato Y. Production of an extremely cold plasma by optical-field-induced ionization. Phys. Rev. A. 1995;51:1415–1419. doi:10.1103/PhysRevA.51.1415

[75] Korobkin DV, Nam CH, Suckewer S, Goltsov A. Demonstration of soft X-ray lasing to ground state in Li III. Phys. Rev. Lett. 1996;77:5206–5209. doi:10.1103/PhysRevLett. 77.5206

[76] Goltsov A, Morozov A, Suckewer S, Elton R, Feldman U, Krushelnick K, Jones T, Moore C, Seely J, Sprangle P, Ting A, Zigler A. Is efficiency of gain generation in LiIII 13.5-nm laser with 0.25-μm subpicosecond pulses the same as with 1 μm?. IEEE J. Sel. Top. Quantum Electron. 1999;5:14531459. doi:10.1109/2944.814984

[77] Corkum PB, Burnett NH, Brunel F. Above-Threshold Ionization in the Long-Wavelength Limit. Phys. Rev. Lett. 1989;62:1259–1262. doi:10.1103/PhysRevLett.62.1259

[78] Penetrante BM, Bardsley JN. Residual energy in plasmas produced by intense subpicosecond lasers. Phys. Rev. A 1991;43:3100–3113. doi:10.1103/PhysRevA.43.3100

[79] Ditmire T. Simulations of heating and electron energy distributions in optical field ionized plasmas. Phys. Rev. E. 1996;54:6735–6740. doi:10.1103/PhysRevE.54.6735

[80] Spence DJ, Hooker SM. Simulations of recombination lasing in Ar^{7+} driven by optical field ionization in a capillary discharge waveguide. Opt. Commun. 2005;249:501–513. doi:10.1016/j.optcom.2005.01.031

[81] Avitzour Y, Suckewer S. Feasibility of achieving gain in transition to the ground state of C VI at 3.4 nm. J. Opt. Soc. Am. B 2007;24:819–828. doi:10.1364/JOSAB.24.000819

[82] Pert GJ. Scaling relations for the design of a recombination laser using tunneling ionization. J. Phys. B: At. Mol. Opt. Phys. 2009;42:225401. doi:10.1088/0953-4075/42/22/225401

[83] Toth R, Kieffer JC, Fourmaux S, Ozaki T, Krol A. In-line phase-contrast imaging with a laser-based hard x-ray source. Rev. Sci. Instrum. 2005;76:083701. doi:10.1063/1.1989407

[84] Begishev IA, Kalashnikov M, Karpov V, Nickles P, Schönnagel H. Limitation of second-harmonic generation of femtosecond Ti:sapphire laser pulses. J. Opt. Soc. Am. B. 2004;21:318–322. doi:10.1364/JOSAB.21.000318

[85] Mironov S, Lozhkarev V, Ginzburg V, Khazanov E. High-efficiency second-harmonic generation of superintense ultrashort laser pulses. Appl. Opt. 2009;48:2051–2057. doi:10.1364/AO.48.002051

[86] Ginzburg VN, Lozhkarev VV, Mironov SYu, Potemkin AK, Khazanov EA. Influence of small-scale self-focusing on second harmonic generation in an intense laser field. Quantum Electron. 2010;40:503–508. doi:10.1070/QE2010v040n06ABEH014126

Chapter 7

Nuclear-Induced Plasmas of Gas Mixtures and Nuclear-Pumped Lasers

Mendykhan U. Khasenov

Additional information is available at the end of the chapter

http://dx.doi.org/10.5772/63823

Abstract

We briefly describe the basic processes of formation and relaxation of nuclear-induced plasmas of gas mixtures, especially the processes of inverse population creation in nuclear-pumped lasers (NPL). A review of the work to create and research nuclear-pumped lasers is in progress: on transitions of atoms and atomic ions and on molecular transitions. An increased focus is on the gas media, which we also study on WWR-K nuclear reactor and DC-60 ion accelerator. The studies on emission of heteronuclear ionic molecules of inert gases are also reviewed.

Keywords: laser, ionizing pumping, mechanism of population, recombination, direct excitation

1. Introduction

Direct conversion of nuclear energy into light energy is of great interest as it provides for application of compact and energy-intensive nuclear energy sources to create high-power generators of coherent and incoherent optical radiation. Nuclear energy pumping into active laser medium was first proposed with the appearance of first lasers [1, 2]. At present, the research on nuclear-pumped lasers (NPL) has progressed to the stage where design and engineering developments of continuous and pulse nuclear laser equipment for various purposes have become possible, that is integrated units based on nuclear engineering and physics, quantum electronics, physics of low-temperature plasma, optics, gas dynamics, and other areas of science and technology [3].

Nuclear-pumped lasers have potential in a wide range of applications, especially in cases requiring high-power and compact lasers to be placed on autonomous remote facilities. The most promising areas of nuclear-pumped laser application are as follows: laser thermonuclear fusion, long-distance transmission of radiant energy and information, rocket laser engine, laser isotope separation and photochemistry, stratospheric ozone layer recovery, and space junk removal. Considerable interest in this area research is also associated with significant difference between the mechanisms of level population during nuclear pumping and population processes in conventional gas-discharge lasers. Application of nuclear energy for active laser medium pumping can be considered not as the way to create high-power laser, but as the way to obtain energy from nuclear reactor. This necessitates consideration of fundamentally new equipment—a reactor laser designed to spatially combine nuclear laser active medium and nuclear reactor core. This approach opens up opportunity to generate qualitatively new energy.

Attempts to achieve laser action during pumping of condensed media with nuclear radiation did not yield positive results. The main obstacle on the road of creating condensed media NPLs is their radiation damage: radiation defects of crystal lattice in solid-state laser, radiolysis, and gas bulb generation on the tracks of nuclear particles in liquid lasers. Presently known gas NPLs [3] radiate in spectral range 391–5600 nm in about 50 atomic transitions of Xe, Ar, Kr, Ne, C, N, Cl, O, I, Hg; Cd^+, Zn^+, Hg^+ ions, CO molecules, and N_2^+ molecular ion.

2. Methods and sources of gas excitation by nuclear reaction products

NPLs include active media that are excited directly using nuclear radiation, or with the use of intermediate nuclear-optical converters. There are three basic sources of nuclear radiation, which can be used to pump NPLs or convert nuclear energy into light energy on transitions of atoms and molecules:

1. nuclear explosions,

2. radioactive isotopes,

3. neutron radiation of nuclear reactors.

Using the γ-ray radiation from nuclear explosion as the pump source was apparently done for the first time in VNIIEF (All-Union Scientific Research Institute of Experimental Physics) in 1971 [3]. Xenon emitted as Xe_2^* excimer molecules was used as an active media. Experiments on xenon excimer laser pumping ($\lambda \sim 170$ nm) were done in a testing area in Nevada in 1973 [4]. Experiments to develop NPLs using nuclear explosive devices were carried out up until 1987 when underground test ban was introduced.

Optical radiation of gases excited by radioactive nucleus decay products (^{210}Po, ^{238}Pu, ^{239}Pu, ^{241}Am, etc.) was studied before to create gas scintillators [5, 6]. Radioactive isotopes usage for laser pumping [7] or pre-ionization in electric discharge lasers [8, 9] is limited by lower power density deposited in gas (up to 0.6 W/cm^3). The main volume of works on NPL active media search and study of their parameters was performed on stationary and pulsed nuclear reactors. Nuclear reactors are the source of neutron and γ-radiation. Neutrons were used to pump lasers, as in this case the energy input to laser medium is several times higher than due to γ-radiation. Direct pumping of active media is usually carried out not by neutron radiation but by nuclear reaction products with thermal neutrons (**Table 1**).

Reaction (energy of reaction, MeV)	Natural composition of isotopes	Kinetic energy of reaction products, MeV	Cross-section of reaction for thermal neutrons, barns
$^{3}He(n,p)T$, (0,76)	$^{4}He(100\%)+^{3}He(1,4°10^{-4}\%)$	$p-0{,}57$; $T-0{,}19$	5400
$^{10}B(n,\alpha)^{7}Li$, (2,35)	^{11}B (80,1)% $+^{10}B$ (19,9%)	$^{4}He-1{,}5$; $^{7}Li-0{,}85$	3837
$^{6}Li(n,T)^{4}He$ (4,7)	^{6}Li (7,5%)$+^{7}Li$ (92,5%)	$T-2{,}7$; $^{4}He-2{,}0$	945
$^{235}U(n,ff)FF$ (167)	$^{238}U(99{,}28\%)+^{235}U(0{,}72\%)$	Fragments: light-99; heavy-68	582

Table 1. Nuclear reactions [10] for NPL pumping.

When nuclear reactors are used as neutron sources, two basic types of laser-medium excitation are utilized:

1. a gaseous isotope or compound thereof (^{3}He, $^{235}UF_{6}$, $^{10}BF_{3}$) is a component part of the laser medium,

2. internal surface of the gas-filled laser cell is coated with a thin layer of isotope (^{10}B, ^{6}Li, ^{235}U) or compound thereof ($^{235}UO_{2}$, $^{235}U_{3}O_{8}$).

In the case of volumetric source of pumping using ^{3}He, the non-uniformity of pumping comes from the absorption of slow neutrons in ^{3}He and from the reduction of energy contribution in area near the wall, owing to removal of reaction products to the walls of the cell, in case of $^{235}UF_{6}$—fission fragments energy loss on the walls of the cell. Results of computation of the total energy deposition and spatial distribution of the deposited energy depending on ^{3}He pressure and diameter of cylindrical cell are given in [11, 12], while [13] shows the results of computation for $^{235}UF_{6}$-He at different pressure of mixture and content. In Ref. [14], authors show summary of results of experimental and theoretical studies dedicated to definition of energy contribution in NPL cells. Three experimental methods were considered: pressure shock method, interferometric method, and string calorimeter method. The cell size and path length of nuclear reaction products in the gas mixture determines spatial non-uniformity at surface pumping source use. Various calculation models for spatial distribution of energy deposition, influence of non-uniformity of uranium-containing layers, and analysis of experimental data on determination of uranium fission fragments energy loss in gas medium are given in [3].

Due to high ^{3}He and ^{10}B neutron-absorption cross-section, loading laser devices using ^{3}He or ^{10}B on the walls can significantly affect reactivity charge of nuclear reactor and even lead it to subcritical state. A laser cell containing ^{235}U also serves as a fuel element of reactor. This has triggered an idea of laser reactor, which must spatially combine active laser medium and nuclear reactor core [2, 15]. Initially considered option included uranium-235 hexafluoride serving as uranium-containing medium, the only uranium compound existing in gas phase at moderate temperatures. However, the use of $^{235}UF_6$ complicated due to the chemical aggressiveness of uranium hexafluoride and products of radiolysis. Laser radiation absorption by UF_6 molecules, high speed of quenching of excited atoms and molecules in collisions with UF_6, electron attachment to molecules of UF_6 also prevents the use of uranium hexafluoride as a component of the laser mixture [16]. At present, the most realistic designs are heterogeneous reactor lasers using thin-film uranium fuel [17, 18]. The core of this reactor laser is a specific quantity of laser cells with uranium layers appropriately placed in a neutron moderator matrix. With appropriate selection of components, the conditions for the reactor-laser operation are provided without utilizing additional fuel (uranium). The number of laser cells may vary from a few 100 to 1000, the total weight of uranium from 5 to 70 kg, and characteristic linear dimensions are 2–5 m [3, 18].

Studies in recent years were set out to explore opportunity to load uranium in active laser mixture in the form of fine dust with a particle size of 100–500 nm substantially lower than the path length of fragments [19]. In this case, it is possible to minimize energy loss in the fuel and substantially improve the efficiency and uniformity of pumping. It is necessary to ensure the transparency of such a mixture at the lasing wavelength. Theoretically, it is possible by arranging the dust particles in the form of periodic structure with a mutual distance comparable to laser wavelength, that is, in the form of dust crystals.

2.1. Basic processes of formation and relaxation of nuclear-induced plasmas of gas mixtures

Currently, direct nuclear pumping is implemented in gas media in which the populating of lasing levels occurs in a low-temperature plasma formed by ionizing radiation, in nuclear-induced plasmas. This section describes the basic processes of formation and relaxation of such plasma, in relation to active media of lasers with nuclear pumping and conversion of nuclear energy into the energy of spontaneous gas emission. The most complete information about the processes in plasma of active media of gas lasers with nuclear pumping is contained in monographs [3, 20].

Initial stage of ionization processes in gas media. Gas medium ionization occurs by various types of radiation: uranium fission fragments and transuranium elements, fast electrons, protons and tritons, lithium nuclei, α-particles, γ-quanta. Ionization in γ-radiation of gas is induced by fast electrons produced in the process of Compton scattering, photoeffect, and effect of electron-positron pair formation. In the initial stage of ionization process, primary ionization during the immediate interaction of charged particles and secondary ionization in interaction of media atoms with electrons formed as a result of primary ionization.

The process of ionization of an atom may be viewed as a binary collision of oncoming charged particle and one of the electrons of the atom's shell [21]. Due to the large difference in masses of heavy charged particles and the electron, only a comparatively small percentage of fragment energy can be transferred to the orbital electron. The spectrum of electrons produced by ionization of heavy particles is softer compared with the spectrum produced by ionization of gas by fast electrons [22, 23]. The average energy of electrons formed in neon as a result of ionization by fission fragments is 40 eV and fast electrons 150 eV [22]. In the case of fission fragments, the secondary electron may provide additional one or two acts of ionization on average, while in the case of fast electrons, it is from 5 to 10 [23].

However, differences in the effects on gas media by different types of ionizing particles are not substantial, because the ultimate result is a combined effect of primary and secondary ionization. It follows from calculations in that the electron energy distribution and energy formation of electron-ion pair in the gas does not depend on the type of charged particles [24, 25]. The same conclusion can be drawn from the luminescent properties of plasma and gas NPLs output parameters, which do not depend on the type of charged particles, but depend on the power and duration of pumping [18].

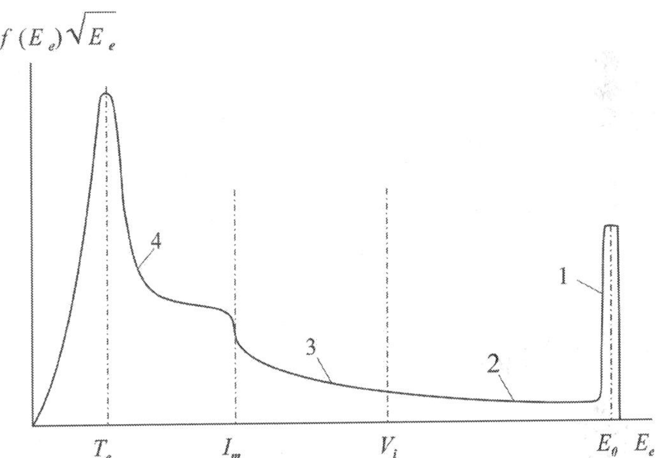

Figure 1. Electron energy distribution in the ionized gas. 1—primary electrons of source; 2—electrons of ionization cascade; 3—electrons in the inelastic excitation region; 4—thermal and subthreshold electrons.

Usually, the calculation and analysis of parameters of nuclear-induced plasmas, development of laser action and NPLs radiation, ionization of gas medium are assumed to be homogeneous. One of the features of nuclear-induced plasma is associated with the formation of tracks when passing through dense gas of heavy charged particles [26, 27]. Depending on parameters of gas media, transverse dimensions of the tracks are 1–10 μm, and the track lifetime or the time to establish uniform ionization through diffusion is 0.1–1 μs [28]. Non-uniformity of ionization associated with track structure of plasma will be most noticeable in the following cases:

- at ionization by fission fragments and other heavy ions,

- in dense gases with high atomic weight,

- at a very low degree of ionization, when there is no overlapping of tracks.

Fluctuations of plasma component concentrations induced by the track structure may have some effect on NPL characteristics excited by fission fragments, if population of upper levels is due to the fast charge process, for example, in laser on a mixture of He-Cd [28]. The influence of plasma track structure on recombination processes will be insignificant, as the track lifetime is much less than the characteristic time of recombination processes [29, 30]. Tracks' overlapping occurs at high pumping power densities; therefore, the track structure of plasma disappears. Estimates show that the overlap of tracks in atmospheric pressure helium occurs at excitation power densities ~2 W/cm^3 [29].

Ionization of gas at the initial stage is carried out directly by charged particles and secondary electrons. A picture of the electron energy distribution in the gas is shown in **Figure 1**, where f_e is the energy distribution function of the electrons, and E_e is the electron energy.

The entire electron energy range can be divided into three regions [24]:

1.   The region of ionization cascade $V_i < E_e < E_0$ (E_0 is the particle initial energy, V_i is the ionization potential of the gas atoms and molecules), in which the electrons energy is sufficient for ionization of gas particles.

2.   In elastic excitation region $I_m < E_e < V_i$ (I_m is the minimal threshold of electron or vibrational excitations), in which the energy of the electrons is reduced, primarily due to excitation of electron and vibrational states of the gas particles.

3.   In the subthreshold region ($E_e < I_m$), electrons lose energy in small "portions" due to elastic collisions with gas particles, thus creating electron thermalization. In subthreshold region, the electrons are involved in the processes that are important in population kinetics and NPL levels' deactivation. These processes include the following: the electron-ion recombination, quenching of excited states, processes of attachment to electronegative gas, excitation, and ionization in collisions with gas particles in excited states.

At $E_e > I_m$, the function of electron energy distribution differs greatly from Maxwell distribution. Electrons in this region do not participate in recombination processes and this region supplies electrons in the subthreshold region. The electrons of inelastic region excitation and ionization cascade possibly play a major role in population of 3p levels of neon [30, 31].

2.2. Kinetics of plasma processes at nuclear pumping of gas mixtures

In quantum system, the gain (absorption) factor of the medium is described by [32]:

$$\alpha = \sigma(N_2 - N_1 \frac{g_2}{g_1})$$

(1)

where indexes (1, 2) refer to upper level 2 and lower 1, N—level population, g—statistical weight of levels. The cross section of stimulated transition:

$$\sigma = \frac{\lambda^2}{2\pi}\frac{A}{\Delta\omega} \tag{2}$$

where λ—transition wavelength, $\Delta\omega$—line width, A—transition probability. The amplifying medium ($\alpha > 0$) requires maintaining population inversion: Population of the upper level must exceed population of the lower level (adjusted to degeneracy multiplicity). Formation of population inversion requires selectivity of population of the upper or lower level. The inversion can be provided not only by the predominant population of upper laser level, but also through selective cleaning of the lower level.

The active media of gas NPLs are often the double mixtures A-B (A—a buffer gas with a high ionization and excitation potential, B—a gas with a lower ionization potential, and lasing occurs in its transitions) or triple A-B-C. In triple mixtures, the third C gas usually plays the role of deactivator of lower level and is not involved in upper level population, but can quench it to some extent. Therefore, we consider the kinetics of processes in plasma in the example a two-component mixture.

The first stage includes ionization and excitation of the buffer gas atoms A (formation of A^+ ions and excited atoms A^*), in some cases, direct excitation of active gas B [31, 33]. The main channels of energy transfer from A^+ and A^* to particles B are as follows:

1.   Charge exchange processes

$$A^+ + B \rightarrow (B^+)^* + A \tag{3}$$

$$A_2^+ + B \rightarrow (B^+)^* + 2A \tag{4}$$

2.   Penning process (if A^* energy is higher than B ionization potential)

$$A^* + B \rightarrow (B^+)^* + A + e \tag{5}$$

3.   Excitation transfer

$$A^* + B \rightarrow B^* + A \tag{6}$$

The main type of ions in high-pressure plasma is molecular ions A_2^+, B_2^+, $(AB)^+$ which are formed in triple processes:

$$A^+ + A + M \rightarrow A_2^+ + M \tag{7}$$

$$B^+ + B + M \rightarrow B_2^+ + M \tag{8}$$

$$A^+ + B + M \rightarrow (AB)^+ + M \qquad (9)$$

where M—third particle (A or B). Plasma neutralization occurs as a result of recombination processes, which, depending on specific conditions, may prevail or have dissociative recombination.

$$A_2^+ + e \rightarrow A^* + A \qquad (10)$$

$$B_2^+ + e \rightarrow B^* + B \qquad (11)$$

$$(AB)^+ + e \rightarrow B^* + A \qquad (12)$$

triple or shock-radiative recombination

$$A^+(B^+) + 2e \rightarrow A^*(B^*) + e \qquad (13)$$

$$A^+(B^+) + e + M \rightarrow A^*(B^*) + M \qquad (14)$$

Population of laser levels occurs during processes (3–5) for B^+ atomic ions, (6, 11–14) for B neutral particles, as well as in cascade transitions from B^{**} high levels. It was previously considered [34, 35] that the processes of dissociative recombination of molecular ions are predominantly populated by p-states of atoms, but in recent years this conclusion was questioned [3].

In [36] based on radioluminescence intensity dependence from mercury vapor pressure (10^{-3}–10^{-7} Torr) in ^{3}He-Hg mixture was made a conclusion that in the process of Hg^+ three-body recombination mostly populated d-states of mercury atoms. Thus, D-levels may also be populated in the processes of dissociative recombination of molecular ions (at a higher density of mercury atoms).

In low-pressure gas discharge laser, the lower laser level is usually deactivated in optical transitions to lower levels, and in lasers with nuclear pumping of atmospheric pressure, the deactivation occurs in collisions with media atoms or plasma electrons, and in Penning reaction with the particles of additional gas. In excimer lasers, where at photon emission the excimer molecule passes in the lower dissociated or weakly coupled state, the lower level deactivation is feasible.

Characteristics of laser radiation at pumping by hard ionizer depend on the power and duration of energy input into active medium, but do not depend on the type of ionizer [18]. This means that kinetics of processes in active media of lasers excited by an electron or ion beam and for nuclear-pumped lasers will be identical [37, 38]. Calculation of plasma parameters and laser characteristics implies for kinetic models, representing the balance of rates of formation and decay of individual components in plasma. Kinetic equations are supplemented by the equations of electron energy balance. In some kinetic models, the number of plasma chemical reactions reaches several hundred (see, for example, [29, 39]). Typically, the relevant description of plasma and laser parameters' calculation suffices it to include 10–15 basic reactions. In this regard, it is sometimes advisable to use the so-called small models for calculation, which includes only the basic plasma processes, examples of such models [40, 41]. It should be noted that in many cases the basic level population process is either not defined (e.g., lasers with Xe, Kr, Ar IR transitions), or under discussion (e.g., laser on the 3p–3s transitions of neon [30, 31]). In other cases, relevant calculation is hindered by a large uncertainty in the values of (or even the order of magnitude) processes rate constants [42], the uncertainty in coefficient of light absorption by active medium particles [43].

2.3. Design and development of experimental methods for nuclear-induced plasma research

Pulsed nuclear reactors were used as a source of neutron radiation for NPLs research [3, 27, 44–47]: in Russia—VIR-1, VIR-2, TIBR-1M, BR-1, BIGR (VNIIEF), EBR-L (VNIITF, All-Russian Scientific Research Institute of Technical Physics), BARS-6 (FEI, Institute of Physics and Power Engineering), IIN-3 (Kurchatov Institute of Atomic Energy); in USA—TRIGA Mark-II (University of Illinois), SPR-III (Sandia Labs), APRF (NASA), Godiva-IV (Los Alamos National Laboratory). Relatively recent reports were issued about experimental NPL investigations in China on the CFBR-II [48]. First, NPL works in China were carried out on a stationary INPC nuclear reactor [49].

Thermal neutron flux density at stationary nuclear reactor reaches 10^{13} to 10^{14} n/cm^{2}s, and gas mixtures' pumping power does not exceed a few W/cm^3. Therefore, research on stationary reactors (IRT-2000 in Moscow Engineering Physics Institute (MEPhI), our works on WWR-K reactor) was mainly associated with the study of the spectral characteristics of plasma [50, 51], as well as the development of lasers with non-self-maintained discharge (WWR-K reactor) [52].

WWR-K reactor (**Figure 2**) is a heterogeneous unit of water-cooled type, operating on thermal neutrons. Desalinated water serves as moderator, reflector, and coolant. Uranium is used as reactor fuel enriched by uranium-235 isotope to 36%. WWR-K reactor is a powerful source of neutrons and gamma rays. The maximum thermal neutron flux in the central channel of the reactor reaches $2°10^{14}$ n/cm^{2}s. The reactor core is placed in an aluminum tank filled with water and is designed as hexagonal lattice containing fuel elements, control and protection system channels and experimental channels. The water temperature is kept constant and does not exceed 40°C. The core has a shape similar to cylindrical, with diameter of 645 mm and height of 600 mm. Central vertical experimental channels with diameter of 96 mm, which was used for nuclear-pumped laser works, pass through the core center. Biological protection of staff from the reactor radiation in horizontal plane is provided by layers of water of 850 mm wide, cast iron—210 mm and limonite concrete—2250 mm. Biological protection in vertical direction is formed of 3700 mm of water layer and removable cast iron lids of 800 mm wide.

Figure 2. General view of WWR-K reactor.

The upper part of reactor tank includes rotating cast-iron lid, through which experimental channels are loaded. The left side includes the base made of plates on which vacuum pumps for pumping mixtures from laser devices under the reactor lid were placed.

Three designs of reactor core laser systems [53] were developed and tested on WWR-K reactor. One design was intended for testing mixtures of xenon laser pumped by uranium fission fragments, the second for lasers on inert gas mixtures excited by the products of ^{3}He(n,p)^{3}H reactions and the third to run the laser on transitions of mercury triplets [54]. **Figure 3** shows the design intended for lasers on mixtures of inert gases excited by the products of ^{3}He(n,p)T reactions. The laser cell is designed as electropolished pipe of 36 mm in diameter with flanges for mirrors on the edges. The distance between the mirrors is 2.1 m, and the mirrors are on a quartz substrate with a dielectric multilayer coating. Mirrors are distanced for 0.5 m (non-transmitting) and 1.0 m (half-transmitting) from the reactor core. Laser channels cutting after a 6-month settling showed that KU-1 quartz substrate has sufficiently high radiation resistance. De-gassing and gas puffing were also conducted through waveguide pipe Ø36 mm. The laser light from this tube was extracted through the window of LiF or CaF_2 and analyzed by registration system. IR radiation was simultaneously recorded by calorimeter located above the exit window and reflected on the LiF plate portion through the matrix of five PD-7G photodiodes. Radiation within the visible range was recorded by photodiodes matrix and the system for measuring the luminescence spectra based on the SPM-2 monochromator and FEU-106 photomultiplier operating in photon counting mode.

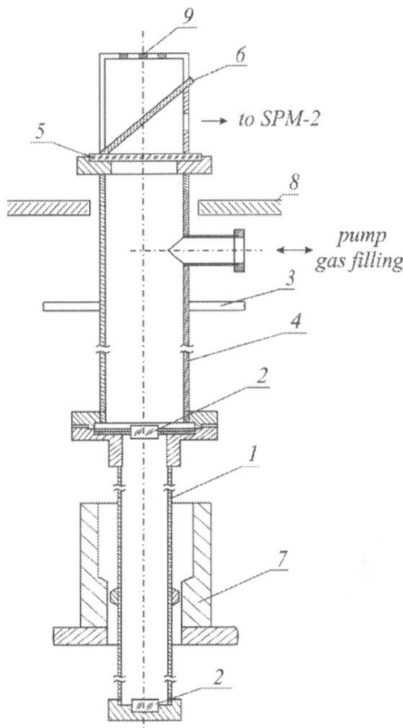

Figure 3. Scheme of laser cell of inert gases excited by ^{3}He(n,p)T reaction products. 1—laser pipe, 2—mirrors, 3—aluminum lid of reactor, 4—waveguide pipe, 5—window, 6—LiF plate, 7—propellant, 8—cast iron lid of reactor, 9—photodiodes matrix.

Although we were not able to create a continuous laser with a direct nuclear pumping, the use of continuous ionizing radiation sources: stationary nuclear reactors and radioactive sources provide for detailed study of plasma properties of gas mixtures, and stationary state of pumping simplifies the analysis of processes kinetics in plasma [53].

Currently, the research is being conducted on DC-60 heavy ion accelerator [55]. The main parameters of the accelerated ion beam: ion type—from lithium to xenon, and ion energy—from 0.5 to 1.75 MeV/nucleon. The intensity of ion beam is from 10^{12} to 10^{14} particles/s depending on type and energy of ions. Ion impulse duration is several nanoseconds and repetition rate of ions is 1.84–4.22 MHz. Mainly, the ions of argon were used as a source of ionization and excitation. The accelerated ion beam passes from evacuated transportation channel through 3-mm hole in the flange to irradiation chamber (**Figure 4**). The hole in the flange is closed by membrane of 2-μm titanium foil or 2.5-μm-thick havar foil. The gas pressure in the cell is measured by capacitance diaphragm gauge mounted at the top of the chamber. The ions passed through the foil ionize and excite the gas mixture in irradiation chamber (**Figure 5**). Argon ion energy after separation foil is about 50 MeV. The emerging light radiation passes through quartz window and condenser and focused on optical fiber. The beam falls on a compact spectrometer through the fiber, and the recorded spectrum is displayed on a computer.

Figure 4. View of the experimental setup on the DC-60 accelerator.

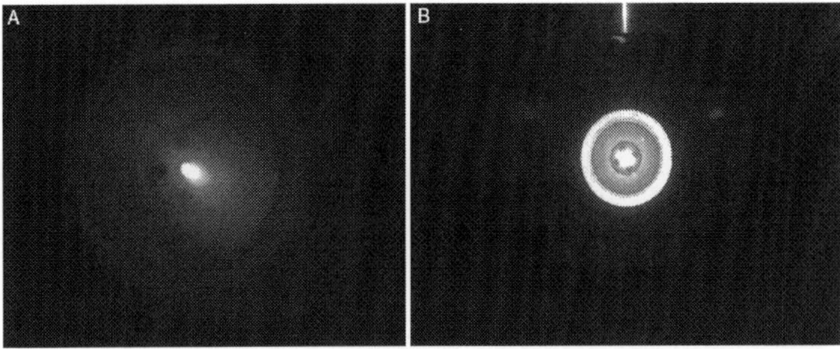

Figure 5. 400 Torr helium (a) and 600 Torr neon (b) luminescence induced by argon ions.

Beam from the narrow ion-excited region is reflected on the separating flange (a, b) and on the top of the chamber (b).

Studies on luminescence spectra of gaseous media were also conducted with the use of radioactive isotopes [36, 56–58] and pulsed nuclear reactors [59–61].

3. NPL active media on transitions of atoms and atomic ions

3.1. IR lasers operating on transitions of Xe, Kr, and Ar

Nuclear-pumped lasers operating on IR transitions of Xe, Kr, and Ar were investigated in detail and have maximum output parameters for NPL. The research by VNIIEF in 1972 during the first experiments was performed with the VIR-2 reactor; the output power of xenon laser with an optimal pressure and composition of the He-Xe was 25 W with efficiency ~0.5%, but the results were not published that time [62]. In 1974, obtained laser action on He-Xe composition (λ = 3.51 μm) with excitation by uranium fission fragments [63]. It was one of the first publications on the achievement of lasing under direct nuclear pumping [63, 64].

Most of the lines with laser action refer to nd-(n + 1)p transitions of Xe, Kr, and Ar atoms (n = 5, 4, 3 for Xe, Kr, Ar, respectively). Xenon laser (λ = 1.7–3.5 μm) has received the most studies, as it has the highest output parameters. He, Ar, Kr, and compounds thereof served as buffer gases. The first NPL xenon laser has the maximum achieved energy parameters:

- output power 1.3 MW and energy 526 J per pulse with duration 400 ms in He-Ar-Xe compounds, with a wavelength 2.03 μm [65],

- 5.6% efficiency in Ar–Xe mixture [66], 2.5% in He-Ar-Xe [67] with λ = 1.73 μm and ~3% with 2.03 μm [65]. The differences in efficiency values may be related to differences in the evaluation of energy deposited in gas [3].

Xenon laser also has the lowest lasing threshold: $1.5°10^{12}$ n/cm²s in Ar-Xe mixture (λ = 2.03 μm) [7, 68]. These advantages, as well as the absence of degradation of the laser mixture as a result of radiation and chemical reactions, join the lasers operating on IR transitions of Xe, Kr, Ar the ranks of most promising in terms of reactor-laser creation [18].

Upper location of laser levels suggests a weak temperature dependence of its output parameters. However, in the case of lasers operating on IR transitions of inert gases, the output power is halved at the temperature of the mixture 350–550 K [3, 69]. The reasons for this are still not fully understood and are the matter of discussions. The most probable reasons were considered:

- reduction in the rate of formation of heteronuclear $ArXe^+$ ions and mixing laser levels by electrons [69],

- destruction of $ArXe^+$ ions by buffer gas atoms [70],

- collisional quenching and mixing laser levels by buffer gas atoms [71],

- active medium contamination by impurities as a result of their desorption from laser cell walls as long as it is warming [72].

Although lasers operating on IR transitions of inert gases are studied for over 40 years and considered to be the most promising, there is still no clarity with mechanism of upper nd-levels population [3, 73]. The processes of deactivation of lower (n + 1)p-levels can be considered well established; it is a collisional quenching in collisions with atoms of active medium, and with electrons at high-power pumping. The main problem in determining nd-levels population mechanism is associated with complexities of IR radiation registration within the reactor experiment.

The main presently discussed mechanisms of levels population (B—Xe, Ar, Kr; A—buffer gas atom) [73] are as follows:

1. Shock-radiative recombination: $B^+ + e + e\ (M) \rightarrow B(nd) + e(M)$, M— third particle.

2. Electron-ion recombination: $B_2^+ + e \rightarrow B(nd) + e$.

3. Recombination of heteronuclear ionic molecules: $AB^+ + e \rightarrow B(nd) + A$.

4. Transfer of excitation in inelastic collisions: $Ar^* + Xe \rightarrow Xe(5d) + Ar$.

5. Step excitation: $Xe(6s, 6s') + e \rightarrow Xe(5d) + e$.

Widely held hypothesis implies for levels population in processes (3) of dissociative recombination of heteronuclear ionic molecules with electrons [29, 74]. In works [3, 73, 75], the main channel of population is considered to be the process (2) of electron-ion recombination of B_2^+ ions. In recombination mechanism of Xe levels' population, the lasing failure with addition of 5 Torr of uranium hexafluoride to Ar-Xe mixture [16] can be explained not only by quenching xenon levels by UF_6 molecules, and also by electron attachment to UF_6, by recombination of xenon ions with negative ions in the mixture.

According to [33, 76], dissociative electron-ion recombination of molecular ions of inert gases cannot be the main process of nd-levels population of inert gas atoms. It is expected to populate levels by direct excitation of secondary electrons from the ground state of atom, as well as transfer of excitation from buffer gas atom [33].

3.2. Visible-range lasers operating on Ne atom transitions

Atomic neon laser created in 1961 by Javan A. and others is the first laser with active gas medium. Therefore, the first proposals [1] and first experiments [2] on NPL creation were associated with well-known transitions of He-Ne laser with a wavelength of 632.8 nm and 1.15, 3.39 μm. In 1980, it was reported [77] on laser action of $5s'[1/2]_1^0-3p'[3/2]_2$ neon atom transition ($\lambda = 632.8$ nm, **Figure 6**) after excitation of ^{3}He–Ne mixture by products of nuclear reaction ^{3}He(n,p)T in stationary nuclear reactor. At the same time, laser efficiency was approximately 0.03% and the lasing threshold was reached at a very low thermal neutron flux $F = 2°10^{11}$ n/cm²s. High gain in a mixture of ^{3}He-Ne ($1.7°10^{-2}$ cm^{-1} at $F = 2°10^{12}$ n/cm²s) was also measured in work of Chinese authors [49]. The results of these studies are questionable and subject to discussion [78, 79]. Simple estimates [78] show that in these conditions [77] population cannot be obtained even in the extreme case, when all power deposited in gas is fully transferred to the upper laser level, and the level is not quenched in collisions with atoms. In our experiments, the neutron flux density was gradually changed from 10^{11} to 10^{14} n/cm²s; however, the lasing threshold in ^{3}He:Ne = 5:1 mixture was not reached [53, 79]. In addition, the luminescence spectrum of He-Ne mixtures at ionized radiation pumping has no 632.8 nm line (**Figure 7**).

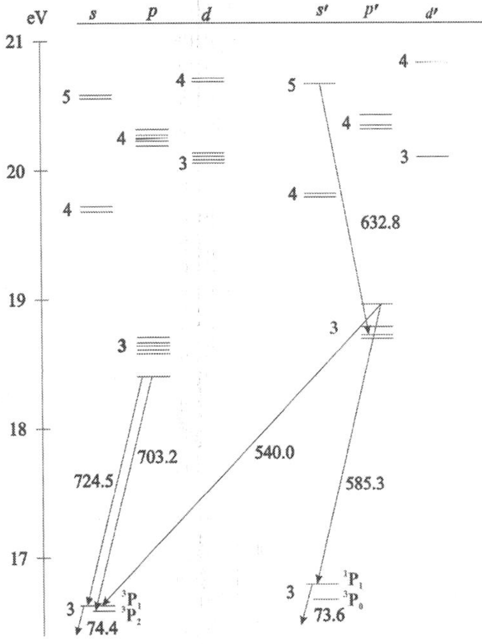

Figure 6. Scheme of laser transitions in neon. The wavelengths of laser and resonant transitions are indicated in nm.

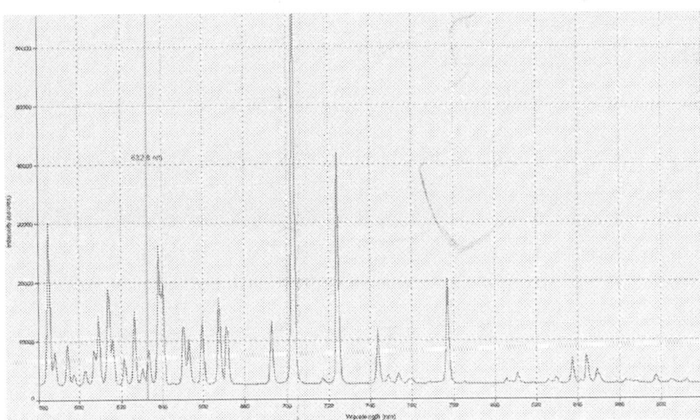

Figure 7. Emission spectrum of neon at a pressure of 605 Torr under ion beam excitation in the 570–900 nm region. The vertical green line indicates the wavelength of 632.8 nm.

3p levels of neon atoms are efficiently populating during excitation by the products of nuclear reactions of neon and its mixtures [57, 80] (see **Figure 7**). To create laser operating on 3p–3s transitions of neon, it is necessary to solve the problem of lower s-levels deactivation. These levels are metastable or resonantly coupled to the main level and have trapped radiation at neon pressures important for nuclear pumping radiation. Rapid depletion of lower laser levels at a relatively low concentration of quenching particles can be achieved in the processes occurring with Coulomb cross-sections. In Ref. [20], authors show the possibility of deactivation of excited states in the Penning process. The required selectivity of deactivation of the upper and lower levels can be achieved using the lower levels states as resonantly coupled to the main, for which the ionization cross-section of quenching by additives is particularly high [81].

Quasi-continuous lasing in allowed neon atom transition $3p'[1/2]_0 - 3s'[1/2]_1^0$ ($\lambda = 585.3$ nm) was first observed in the afterglow of discharge [82], and then at pumping by powerful electron beam of Ne-H$_2$, Ne-Ar, Ne-Ar-He mixtures [83]. In [84] by reducing the concentration of neon and quenching the lower level of additives by an order compared to [83], and pumping power by electron beam ~100 kW/cm^3 with He:Ne:Ar = 96:3:1 mixtures at a pressure of 3 atm at the same transition obtained laser efficiency of 1–2%. The laser efficiency with ionizing pumping at $\lambda = 585$ nm in the opinion of other authors cannot exceed 0.5% [85]. The electron beam pumping provided for laser action in a row of 3p–3s neon transitions in the spectrum red region [86] (see **Figure 6**).

Progress in creating efficient lasers of visible range operating on neon at the electron beam pumping power 10–100 W/cm^3 stimulated work on nuclear-pumped lasers operating on 3p–3s transitions of neon atom. Laser action operating on 3p–3s neon transitions at pumping by uranium fission fragments was obtained in 1985 at VNIIEF on VIR-2 reactor. These results were published in 1990 [3, 87]. In 1985, we have also conducted experiments on WWR-K stationary nuclear reactor for excitation of triple mixtures of argon or krypton, neon, and ^{3}He, which has led to a negative result [53, 79].

Obtained within the experiment efficiency values for NPL operating on neon (to 0.1%) are much lower than at pumping by electron beam [84, 85]. Threshold values of thermal neutron flux density are significantly higher than 10^{14} n/cm^{2}s [87–93]. Report [90] on NPL lasing in ^{3}He-Ne-H$_2$ mixture with a lasing threshold of 10^{14} n/cm^{2}s is doubtful, as hydrogen at a pressure of 0.57 bars must be considerably quenching the upper laser level. Unlike lasers operating on IR transitions of Xe, Kr, Ar, the lasing mechanism in lasers on 3p–3s transitions of neon was considered reliably established [39, 87, 94]. Already the first works on neon lasers pumped by an electron beam or nuclear radiation contain roughly the same ideas about the lasing mechanism. The lower 3s-states are deactivated in the processes of:

-   Penning for metastable states $3s[3/2]_2^0$, $3s'[1/2]_0^0$ [20],

-   release of an electron for resonance states $3s[3/2]_1^0$, $3s'[1/2]_1^0$ [81].

The main process of 3p levels population is considered to be dissociative recombination of Ne$_2^+$ ions with electrons; most studies assumed that all of these levels are populated in recombination of molecular ions in the main vibrational state. In [80, 84] to explain changes in the efficiency of various levels pumping with the neon pressure, was made a conclusion on population of $3p'[1/2]_0$-level in processes of recombination of vibrationally excited levels of

Ne_2^+. Population efficiency of level $3p'[1/2]_0$ increases in triple mixtures with high helium content, which is explained in [80] by the formation of vibrationally excited levels of Ne_2^+ in the following processes:

$$Ne^+ + 2He \rightarrow HeNe^+ + He \tag{15}$$

$$HeNe^+ + Ne \rightarrow Ne_2^+(v>0) + He \tag{16}$$

Another mechanism for neon levels' population when excited by a hard ionizer is proposed in [79]. Based on the study of spectral-temporal characteristics of pure neon and He-Ne mixtures pumped by heavy charged particles, the conclusion was made that the population of neon levels occurs in direct excitation by nuclear particles and secondary delta electrons, and in He-Ne mixtures also in the process of excitation transfer from metastable helium atoms (He^m):

$$He^m + Ne + He \rightarrow Ne(3p) + 2He \tag{17}$$

However, this work did not receive recognition: In the review article [18], it was not mentioned, and in monograph [3], it was questioned with reference to [95]. During the study of luminescence of He-Ne mixtures with quenching additives [30], we have obtained results confirming the main conclusion in [79]—the principal and clearly dominant channels of neon 3p levels population in pumping by a hard ionizer are the processes unrelated to dissociative recombination of molecular ions of neon.

3.3. Metal vapor lasers

3.3.1. Mercury vapor lasers

HgII laser. The first report on the lasing of ion NPL operating on mercury vapor appeared in 1970 [96]. This work presents results obtained at ^3He(350 Torr)+Hg(3 Torr) mixture pumping on IIN pulsed reactor with neutrons flux up to $5°10^{16}$ n/cm²s. According to authors, the registered lights power ~10 mW was associated with laser action of mercury ion transition $7^2P_{3/2}$–$7^2S_{1/2}$ (λ = 615.0 nm). Further on, these results were questioned, and in ^4He(600 Torr) +Hg (2.5–10 mTorr) mixture pumping by $^{10}B(n,\alpha)^7Li$ reaction products has proved lasing with a wavelength of 615 nm [97]. The conditions [97] had no lasing at partial pressure of mercury vapor ~3 Torr.

Pumping mechanism at the transition of $7^2P_{3/2}$–$7^2S_{1/2}$ mercury ion at ionizing pumping was considered in [53, 98, 99]. Discrepancy between the results of [96, 97] is possibly due to the difference in mechanisms of upper laser level population at low density and high density of mercury vapor [53, 98]. Calculations [99] indicate a low energy laser characteristics at λ = 615 nm (0.04% maximum efficiency) even with optimal parameters.

HgI laser. The emission spectra of mixtures of ^{3}He + Hg and ^{3}He + Hg + Kr in excitation by products of nuclear reaction ^{3}He(n,p)T studied in [100]. It was concluded that the excitation of low-lying HgI levels by electron impact is predominant in high-pressure plasma:

$$Hg(6^1S_0)+e \rightarrow Hg(6^3P_1)+e \tag{18}$$

$$Hg(6^3P_{0,1,2})+e \rightarrow Hg(nx)+e \tag{19}$$

Continuous laser λ = 546.1 nm with optical pumping [101] uses similar scheme of excitation (electron replaced with photon): a beam with a wavelength of 253.7 nm, corresponding the mercury resonance line, excites 6^3P_1 mercury level and beam λ = 435.8 nm transmits excitation to 7^3S_1-level. However, an attempt to ensure deactivation of 6^3P_2 state by N_2 molecules according to the scheme used in optically pumped laser at ionizing radiation pumping was not successful [102].

We have proposed another scheme of creating inverse population in laser on mercury triplet lines [54]. Our works [36, 98] have shown that population of 7^3S_1-level of mercury atom occurs in the process of dissociative recombination of molecular ions and not in direct or stepwise excitation by electrons. It is proposed to use H_2 to destruct the lower level at the transition of 7^3S_1–6^3P_2; H_2, D_2—on the transition of 7^3S_1–6^3P_1. As the pumping is carried out through ion channel, xenon should be used as buffer gas and charge exchange from xenon to hydrogen is slow. The use of krypton is less justified due to low value of rate constant of Kr_2^+ ion recharge on mercury atoms. High selectivity of dissociative recombination of Hg_2^+ is largely driven by relatively low temperatures involved in recombination of electrons. Therefore, by increasing the pumping power to the level required for laser operation, it is useful to use helium to cool the secondary electrons. Thus, the optimal gas mixture of laser operating on mercury triplet must be four-component He-Xe-Hg-H_2 [54].

Quasi-continuous lasing at 7^3S_1–6^3P_2 transition of mercury atoms using this scheme was obtained on pulsed nuclear reactor EBR-L in VNIITF [103]. Kinetic model of He-Xe-Hg-H_2 nuclear-pumped laser based on VNIITF experiments was developed in [104]. Externally similar scheme is implemented in excitation of mercury mixtures with inert gases by electron beam [105]. This work uses a mixture of He+Ne+Ar as a buffer gas at a total pressure of 2300 Torr. With reference to the paper later than [36, 98], recombination of Hg_2^+ as the main population channel of 7^3S_1 has been specified. An attempt to use H_2 for 6^3P_2 level population was unsuccessful, and at hydrogen pressure of 20 Torr laser action failed. What was also interesting in this study was the absence of molecular additives quenching the lower level. Apparently, the de-excitation of lower laser level took place in formation of excimer molecules (HgR)*:

$$Hg(6^3P_2)+R+M \rightarrow (HgR)*+M \tag{20}$$

where R$-$Ne or Ar.

3.3.2. *Cadmium and zinc vapor lasers*

The first nuclear pumping of visible range laser operating on cadmium vapor was carried out by MEPhI researches on BARS pulse reactor [106]. In ^{3}He-^{116}Cd mixture was obtained laser action on $4f^2F^0_{5/2,7/2}$–$5d^2D_{3/2,5/2}$ transitions of ion Cd$^+$ (λ = 533.7 and 537.8 nm), and later on transition from λ = 441.6 nm [107]. The first successful experiments on laser pumping on metal vapor (mixture of He-^{116}Cd, λ = 441.6; 533.7 and 537.8 nm) by uranium fission fragments were carried out in 1982 by researches of VNIIEF and VNIITF on EBR-L reactor [3]. Maximum parameters of Cd$^+$-laser with nuclear pumping is also obtained in this reactor: 1000 W at an efficiency of 0.4% on the blue line and 470 W on the green lines [44].

At present, the basic processes of population of upper laser levels of Cd$^+$ are considered to be established:

$4f^2F^0_{5/2,7/2}$ upper levels of laser transitions from λ = 533.7 to 537.8 nm are populated due to charge exchange processes

$$He^+ + Cd \rightarrow (Cd^+)^* + He \tag{21}$$

forming higher-lying levels of 6f, 6g, 8d, 9s [108] and subsequent cascade transitions in 4f state.

$5s^{22}D_{3/2,5/2}$ upper levels for transitions with λ = 441.6 and 325 nm are populated in Penning and charge exchange processes

$$He_2^+ + Cd \rightarrow (Cd^+)^* + He \tag{22}$$

$$He^*(2S) + Cd \rightarrow (Cd^+)^* + He \tag{23}$$

Laser action at 325 nm was achieved only when pumped by an electron beam, although the threshold pumping power was only 10 W/cm^3 [109].

Deactivation of lower levels can occur as a result of radiative transitions, quenching by electrons, conversion processes of atomic ions into molecular ions

$$(Cd^+)^* + Cd + He \rightarrow Cd_2^+ + He \tag{24}$$

and for the levels lying above $6^2S_{1/2}$, in Penning process on its own atom

$$(Cd^+)^* + Cd \rightarrow Cd^+ + Cd^+ + e \tag{25}$$

Lasing mechanisms of NPLs on ion transitions of cadmium and zinc are very similar. Laser action by pumping ^{3}He-Zn mixture observed on transition of $4s^{22}D_{5/2}$–$4p^2P_{3/2}$ (λ = 747.9 nm) [110], 60 W output power obtained during pumping by uranium fission fragments at this transition [44].

Information about cadmium-vapor atomic lasers with ionizing pumping is scarce. During experiments was registered laser action (1–2 W) with a threshold density of 10^{16} n/cm^{2}s neutron flux on the lines of 1.648 and 1.433 μm at pumping He-Cd mixture by uranium fission fragments [44]. When pumping He-Cd mixture by an electron beam was obtained laser action on the line 361 nm of cadmium atom [111]. Kinetic model [112] included processes involving excited cadmium atoms and attempted to calculate some laser characteristics for 1.648 μm line.

Luminescence of ^{3}He-Cd and ^{3}He-Xe-Cd mixtures in the radiation region of stationary nuclear reactor investigated in [113]. Measured value of rate constant of Xe_2^+ charge exchange on cadmium atoms is small (~10^{-13} cm^{3}s^{-1}), in contrast to the constant of charge exchange on mercury atoms. A sufficiently high density of cadmium vapor (~3×10^{18} cm^{-3}) was established at a temperature of about 700°C, and such density of cadmium requires consideration of quenching 6^3S_1 state by its own atoms. Perhaps krypton ions charge exchange on Cd will be faster. In addition, cadmium atoms in krypton can be ionized in the Penning processes.

4. NPL active media on molecular transitions

4.1. Lasers operating on first negative system of nitrogen and carbon monoxide

N_2^+. When pumping active medium of a high pressure by electron beam with moderate (~3 A/cm^2) current density was obtained quasi-continuous lasing mode on the first negative system of nitrogen (λ = 391.4 and 427.8 nm) with an efficiency ~1% [114, 115]. In this collision lasers on $B^2\Sigma^+_{u,v\,=\,0} \rightarrow X^2\Sigma^+_{g,v''\,=\,0,1}$ transitions of nitrogen ion, deactivation of lower level was carried out at low hydrogen concentrations (~0.1%) in He-N_2-H_2 mixture in the process with proton transfer:

$$N_2^+\left(X\right)+H_2 \rightarrow N_2H^+ +H \tag{26}$$

In 1996 was obtained laser action at λ = 391.4 nm with excitation of He-N_2-H_2 mixture by uranium fission fragments on EBR-L [116]. This was the first NPL emitting in the UV spectral region: Laser power was ~10 W, efficiency ~0.01%. To date, it remains to be the most short-wavelength nuclear-pumped laser. On BARS-6 reactor were conducted studies of facility with active core length of 250 cm and volume of 4 liters with the average over the length of the laser element-specific energy deposition to 300 W/cm^3 [117]. The lasing threshold on B-X nitrogen ion transitions with λ = 391.4 and 427.8 nm was achieved with low for molecular laser power density of energy deposition 50–60 W/cm^3. It was determined that active medium had non-resonant losses ~5×10^{-5} cm^{-1}. These authors explain the laser efficiency (~0.1–0.2%) was significantly lower than expected. In [118] developed a multicomponent spatially homogeneous model of kinetic processes of NPL active medium on He-N_2-H_2 mixture. According to the model, the maximum laser efficiency (0.5–0.8%) was achieved at pumping power of 0.5–3 kW/cm^3, for specific pumping power 90–130 W/cm^3 the maximum instantaneous efficiency is 0.1%.

The [119] searched mixture compounds that improve NPL efficiency on 1$^-$ — nitrogen system, including study of deuterium and neon impact on the laser parameters. Replacement of hydrogen additive deactivating lower level by deuterium provided no improvements to laser efficiency. The use of neon as the buffer gas at partial pressure of 30 Torr resulted in efficiency decrease for 30–40% for both wavelengths. It has previously been shown [120] that helium replacement with neon impairs laser parameters for B-X transitions of N_2^+. At pumping by an electron beam, lasing energy of mixture with addition of 60 Torr of neon was 1.5 times less than for three-component mixture of He-N_2-H_2 with a total pressure of 6 atm.

In [121], the constants of quenching processes rates of N_2^+ (B) with nitrogen and helium, as well as two- and three-particle charge exchange processes of He_2^+ to H_2, D_2, Kr, CO, were determined on the luminescence in 0–0 transition of the first negative system of nitrogen in excitation by $\propto$–particles of polonium-210.

CO$^+$. Gain on the first negative system of carbon monoxide by powerful electron beam pumping was obtained in 1975, the same way as in N_2^+-laser, by Waller et al. [122]. Quasi-continuous lasing mode can be implemented with addition of H_2, D_2 or Kr to He-CO [123]. In [123, 124] by spectra of radioluminescence of the gas mixtures with carbon monoxide was determined constants of quenching rate of CO$^+$(B) with helium, neon, CO molecules, evaluated the upper limit of constant of quenching rate with hydrogen, deuterium and krypton.

Kinetic model of nuclear-pumped laser on $B^2\Sigma^+_{u,v'} \rightarrow X^2\Sigma^+_{g,v''}$ transition of carbon monoxide ion was developed in [125]. Unfortunately, the only information given on the results of calculations of medium gain, efficiency, but there is no information on considered processes, values of constants for processes rate. This provides for the opportunity to discuss the work. In addition, specified calculated wavelength transition for which the calculations were made, 210 nm; the closest wavelength — 211.2 nm — corresponds to 1–0 transition [126]. That is, laser action, according to [125], should not take place from the ground vibrational level of CO$^+$(B), as the wavelength of 0–0 transition — 219.0 nm. It should be noted that the gain in [122] was observed at 0-2 transition (242 nm) and Baldet-Johnson's system (391 and 425 nm, B$\rightarrow$A transitions) from the ground vibrational level of CO$^+$(B).

4.2. Excimer lasers operating on halides of inert gases

Excimer lasers operating on halides of inert gases have been studied for a long time [127]. At present, they are the most powerful lasers that emit in UV region of spectrum. Optimal operation of excimer lasers corresponds to pumping powers of several megawatts per cm^3 and a pressure of several atmospheres. Such pumping powers are achieved by electron beams or space discharge. Radiation of nuclear explosions [128] and ion beams [129] also has been used to pump these lasers.

Excimer lasers can operate in a quasi-continuous mode, as in the photon emission, an excimer molecules pass into the lower dissociating or weakly bound state. Much of the research on creation of NPL operating on the inert-gas halides is associated with XeF-laser (λ = 351 and 353 nm). Experiments on SPR-III reactor have shown that in ^{3}He-Xe-NF$_3$ mixture the gain on the 351 nm band is about $7°10^{-3}$ cm^{-1} at pumping power q≈5 kW/cm^3 [130], and at pumping by uranium fission fragments of Ne(Ar)-Xe-NF$_3$ mixture, registered gain of ~$2°10^{-3}$ см$^{-1}$ at q≈2 kW/cm^3 [131]. Experiments in pulsed nuclear reactors with neutron flux up to 10^{17} n/cm^{2}s aimed at obtaining lasing at XeF [131] and XeF, KrF [59] did not yield positive result. The work [132] contains theoretically investigated characteristics of laser operating on Ne-Xe-NF$_3$ mixture at 1 atm with nuclear pumping duration 0.1–1 ms at half-height in the near-threshold lasing region. The lasing threshold is 400–500 W/cm^3, and the maximum calculated efficiency of ~1% is achieved at pumping power of 1.5–5 kW/cm^3. Predicting the possibility to create XeF-nuclear pumped laser is mainly limited by the uncertainty of absorption coefficient in active medium, which greatly affects calculation results.

The luminescence efficiency in chlorine-containing gas mixtures with xenon (to 15% [133], ~11% [134]) is about three times higher than the efficiency of emission at the transition of XeF. The maximum luminescence efficiency can be achieved at low chloride content (<0.7 Torr [134], 0.05 Torr [135]). In [133] reported about the lasing on XeCl molecule (λ = 308 nm) by pumping Ar-Xe-HCl(CCl$_4$) mixture by uranium fission fragments in experiments on EBR-L reactor. According to the authors, the narrowing in emission spectrum in 308 nm band, the height of film blackening, 1.3–1.5 times less than at other wavelengths, indicates the presence of laser radiation. However, these data are not sufficient for this conclusion [3].

High radiation resistance of ^{3}He-Xe-CCl$_4$ mixture was noted in [134, 136]. The densities of electrons and negative ions in the plasma of gas mixture of ^{3}He-Xe-CCl$_4$ and ^{3}He-Xe-NF$_3$ irradiated with thermal neutron flux 10^{11}–10^{14} n/cm^{2}s into the active core of stationary nuclear reactor, measured in [137].

4.3. Lasers operating on vibrational transitions of carbon monoxide and carbon dioxide

CO. Molecular nuclear-pumped laser operating on carbon monoxide was one of the first NPLs, the creation of which was reported in the press [64]; the laser action was observed in vibrational-rotational transitions of the CO molecule with λ = 5.1–5.6 µm. Potential of medium on carbon monoxide was determined by the fact that, unlike lasers on electronic transitions, active medium of CO-laser does not require high pumping selectivity. It is important to get energy into the broad band of vibrational levels of the ground electronic state of CO molecule. Furthermore due to the autonomy of vibrational subsystem and anharmonicity of molecular vibrations, this energy at relatively low translational temperature redistributing in the process of exchange of vibrational molecules to provide full or partial inversion on vibrational levels of CO [138]. This property associated with the anharmonicity of the vibrational levels, as well as the cascade mechanism of lasing in CO, allows directing significant energy share of nuclear reaction products to the laser level.

Experiments have not yet confirmed this finding. In [64], carbon monoxide pumping at a pressure of 0.13 atm and temperature of 77 K was carried out by uranium fission fragments, and the radiation power was 2–6 W with an efficiency of 0.1–0.3%. In further work, the authors [64] achieved lasing power of about 100 W by using a multiple-pass resonator with an active length of 120 cm. Lasing on vibrational transitions of CO was also obtained in [139] at excitation of ^{3}He-CO by nuclear reaction products of ^{3}He(n,p)T. Laser radiation power at the mixture pressure of 3 atm exceeded 200 W from an active volume of 300 cm^3, and the lasing threshold was reached at F = 3×10^{16} n/cm^{2}s.

Excitation of vibrational levels of CO can occur due to molecular collisions with plasma electrons. In [140] based on calculation of electrons spectrum formed in a molecular gas under the action of ionizing radiation, it was shown that the efficiency of CO nuclear-pumped laser cannot exceed 0.5%. Further work [141] showed the main mechanism of molecular forma- tion in the form of dissociative recombination of cluster ions with formation of electronically excited molecules and subsequent collision of these molecules with molecules in ground state. According to the authors [141] plasma chemical processes in active medium can make a significant contribution to the energy pumping into vibrational modes of molecules and allow achieving the efficiency pumping up to 18%. The use of argon as a buffer gas instead of helium should increase 1.5-fold the efficiency of pumping in vibrational levels and reduce by an order the threshold energy for active medium pumping. It should be noted that a record efficiency of electron beam-controlled laser operating on carbon monoxide—63% [142]—was achieved due to Ar:CO = 10:1 mixture.

Presently achieved low parameters of NPL operating on CO, the need to cool down an active medium reaching cryogenic temperatures, apparently, makes carbon monoxide medium insufficient for creation of nuclear-pumped lasers.

CO$_2$. The possibility of CO$_2$-laser pumping (λ = 10.6 μm) by nuclear radiation was consid- ered in the earliest stages of NPL study [2], and gas-discharge laser operating on carbon dioxide had the highest output parameters for that time. Numerous attempts to create NPL on CO$_2$ have yield negative results. Experiments on pumping ^{3}He-CO$_2$ and ^{3}He-CO$_2$-N$_2$ mixtures with products of ^{3}He(n,p)T reaction showed no gain in band of 10.6 μm at a wide variation of pressure (0.28–0.8 atm) and composition of mixture [143]. Moreover, these experiments showed probe laser radiation absorption, which indicates preferential population of the lower laser level of CO$_2$ when excited by ionizing radiation. Calculations of kinetic processes in CO$_2$- N$_2$-He mixture also support the conclusion on ineffectiveness of direct pumping of CO$_2$-laser with nuclear radiation [144].

Apparently, the most promising method of nuclear energy conversion into radiation on vibrational transitions of CO or CO$_2$ is the creation of nuclear power plant with electroioniza- tion laser based on thermionic converter reactor [145, 146].

4.4. Radiation of heteronuclear ionic molecules of inert gases

Molecular bands in radiation spectra of pair inert gas mixtures were first discovered more than half a century ago. The [147] recorded the band of 507–550 and 496–508 nm in Ar-Xe mixture, and the authors attributed the presence of these bands with emission of heteronuclear molecules or ions. When pumping Ar-Xe mixture by electron beam, the molecular band at 510 nm was detected and the band of 495–460 nm in Kr-Xe was registered for the first time [148]. Two emission systems were observed, in 600–670 and 670–685 nm regions, when Xe was added to Kr flowing afterglow at a pressure of 30 Pa [149]. Kugler [150] obtained similar results for Ar-Xe mixture and discovered new band in Ar-Kr in the region of 605–642 nm. He explained these bands as transitions of neutral heteronuclear molecule formed in the processes of metastable atoms of argon.

In 1975, Tanaka et al. [151] have published data on radiation spectra of 10 pair mixtures of inert gases in the region of 100–700 nm. Molecular bands observed in radiation spectra in the discharge were identified as transitions between states of heteronuclear ionic molecules:

$$M^+N \rightarrow MN^+ + h\nu \tag{27}$$

where molecular states of $M^+ N$ asymptotically correspond to states of M^++N, and MN^+ to the state of $M+N^+$; here M, N—atoms of inert gases, and N—a heavier atom. If the plasma of low pressure in an electric discharge in paired mixtures of inert gases has up to 5 similar bands [151], there are no transitions from levels corresponding to the states of atomic ions $^2P_{3/2}$ [152, 153] when excited by ionizing radiation of medium- and high-pressure mixtures.

Kinetics of Ar-Kr, Ar-Xe, and Kr-Xe mixtures' excitation by low activity ^{241}Am alpha particles was studied in [152, 154]. Constants of processes rates in these mixtures were identified; however, constants' values of a number of processes are underestimated: ~10^{-15} cm^3/s for two-particle and ~10^{-34} cm^6/s for three-particle processes. Emission of Ar-Xe and Kr-Xe mixtures when excited by ^{210}Po alpha particles with activity of ~0.5 Cu investigated in [153, 155–157] determined the rate constants of processes of formation and destruction of levels of heteronuclear ionic molecules. In [153] includes first noted high luminescence efficiency of (ArXe)$^+$, (KrXe)$^+$ at pumping by ionizing radiation. Luminescence of Ar-Xe mixture pumped by powerful electron beam was studied in [158], attempts to obtain lasing on transitions of (ArXe)$^+$ at pumping by an electron beam had yield negative results [155, 158].

In [42] was built kinetic model of Ne-Ar mixture relaxation pumped by a hard ionizer with regard to the possibility of lasing on transition Ne$^+$Ar→NeAr$^+$. When using typical rates of plasma chemical reactions, calculations show that lasing is only possible at high pressure (above 16 atm) and powerful pumping (1 MW/cm^3), and lasing efficiency should not exceed 0.05–0.25%. In this work were considered the triple (with Kr) instead of the binary mixtures of inert gases, as the authors suggested that the constant of deactivation rate of the lower level in exchange processes:

$$NeAr^+ + Ar \rightarrow Ar_2^+ + Ne \tag{28}$$

may occur negligible. It was assumed that the lower laser level will be deactivating in the processes with Kr atoms:

$$NeAr^+ + Kr \rightarrow Kr^+ + Ne + Ar \tag{29}$$

$$NeAr^+ + Kr + M \rightarrow Kr^+ + Ne + Ar + M \tag{30}$$

5. Conclusions

Nuclear-pumped lasers are of great interest as the way to extract high-quality energy from a nuclear reactor core. Presently achieved pulse power of NPL in quasi-continuous mode exceeds 1 MW. However, the most promising active media on transitions of inert gas atoms have a number of disadvantages: relatively low efficiency, radiation in IR region, low operating temperature. There is no clarity as to the basic mechanism of upper laser level population: direct excitation by secondary electrons, excitation transfer from buffer gas atoms, electron-ion recombination of molecular ions (dimers or heteronuclear ions). This area requires further research.

Acknowledgements

This work has been supported by the Ministry of Education and Science of the Republic of Kazakhstan (Grant No. 0681/GF4). Author is grateful to staff of the WWR-K reactor and DC-60 accelerator for their assistance in conducting experiments.

Author details

Mendykhan U. Khasenov[*]

Address all correspondence to: mendykhan.khassenov@nu.edu.kz

National Laboratory Astana, Nazarbayev University, Astana, Kazakhstan

References

[1] Herwig L.O. Concepts for direct conversion of stored nuclear energy to laser beam power. Trans Am Nucl Soc. 1964;7:131–132.

[2] Thom K., Schneider R.T. Nuclear pumped gas lasers. AIAA J. 1972;10:400–406.

[3]   Mel'nikov S.P., Sizov A.N., Sinyanskii A.A., Miley G.H. Lasers with nuclear pumping. New York: Springer; 2015. 455 p. DOI: 10.1007/978-3-319-08882-2.

[4]   Ebert P.J., Ferderber L.J., Koehler H.A., et al. Amplified spontaneous emission in xenon pumped by gamma rays. IEEE J Quantum Electron. 1974;QE-10:736.

[5]   Dondes S., Hartek P., Kunz C. A spectroscopic study of alpha-ray-induced luminescence in gases. Radiat Res. 1966;27:174–210.

[6]   Thiess P.E., Miley G.H. New near-infrared and ultra-violet gas proportional scintillation counters. IEEE Trans Nucl Sci. 1974;21:125–145.

[7]   Voinov A.M., Konak A.I., Melnikov S.P., Sinyanskiy A.A. Feasibility of developing a cw laser with a radionuclide pump source. Quantum Electron (Sov J). 1991;21:1179–1181.

[8]   Bigio I.J. Preionization of pulsed gas laser by radioactive source. IEEE J Quantum Electron. 1978; QE-14:75–76.

[9]   Batyrbekov G.A., Batyrbekov E.G., Tleuzhanov A.B., Khasenov M.U. Electrodischarge laser with radioisotope pre-ionization. Tech Phys (Sov J). 1987;57:783–785.

[10] Nemez O.F., Hofman J.V. Handbook on nuclear physics. Kiev: Naukova dumka; 1975. 415 p. (in Russian).

[11] Wilson J.W., DeYoung R.G. Power density in direct nuclear-pumped ^{3}He lasers. J Appl Phys. 1978;49:980–988.

[12] Pikulev A.A. Energy deposition in helium-3 based nuclear-pumped lasers. Tech Phys. 2006;51:1344–1350.

[13] Wilson J.W., DeYoung R.G. Power deposition in volumetric ^{235}UF$_6$–He fission-pumped nuclear lasers. J Appl Phys. 1978;49:989–993.

[14] Pikulev A.A., Vlokh G.V., Limar' Y.M., et al. Determination of energy deposition into the cells of nuclear-pumped lasers. Tech Phys. 2012;57:1127–1134.

[15] Gudzenko L.I., Slesarev I.S., Yakovlenko S.I. Proposed nuclear laser reactor. Tech Phys (Sov J). 1975;20:1218–1220.

[16] De Yong R.G., Shiu Y.J., Williams M.D. Fission-fragment nuclear lasing of Ar(He)–Xe. Appl Phys Lett. 1980;37:679–681.

[17] Voinov A.M., Vorontsov S.V., Krivonosov V.N., et al. Studies performed at VNIIEF, on investigating the possibility for creating a reactor laser. In: Proceedings of the 4th International Conference "The physics of nuclear-pumped lasers and pulsed reactors". Obninsk: FEI; 2009. vol. 1. pp. 17–33 (in Russian).

[18] Karelin A.V., Sinyanskii A.A., Yakovlenko S.I. Nuclear-pumped lasers and physical problems in constructing a reactor-laser. Quantum Electron. 1997;24:375–402.

[19] Budnik A.P., Kosarev V.A. Kinetic model of the helium nuclear inducted plasma containing nanoclusters. Phys Chem Kinet Gas Dyn. 2010;9:art. 130.

[20] Gudzenko L.I., Yakovlenko S.I. Plasma lasers. Moscow: Atomizdat; 1978. 256 p. (in Russian).

[21] Starodubzev S.V., Romanov A.M. Passage of the active particles through matter. Tashkent: Uzbek SSR Academy of Sciences Printing House; 1962. 228 p. (in Russian)

[22] Moratz T.J., Kushner M.J. Fission fragment pumping of a neon plasma. J Appl Phys. 1988;63:1796–1798.

[23] Moratz T.J., Saunders T.D., Kushner M.J. Heavy-ion versus electron-beam excitation of an excimer laser. J Appl Phys. 1988;64:3799–3810.

[24] Tyukavkin A.V. Electron energy distribution in helium excited by ions. Plasma Phys Rep. 1999;25:90–93.

[25] Dymshits Y.I., Neverov V.G., Khoroshev V.G. Calculation of outputs of primary products of irradiation of rare gases by fast electrons. High Energy Chem (Sov. J). 1982;16:201–208.

[26] Pikaev A.K. Modern radiation chemistry, vol. 2. Moscow: Nauka; 1987. 439 p. (in Russian).

[27] Dyachenko P.P. Experimental and theoretical works performed by the Institute of Physics and Power Engineering on the physics of nuclear-induced plasmas. Laser Part Beams. 1993;11:619–634.

[28] Budnik A.P., Dobrovol'skaya I.V. Characteristics of the kinetics of the active media of gas lasers excited by fission fragments. Quantum Electron. 1997;27:492–496.

[29] Karelin A.V., Simakova O.V. Kinetics of the active medium of a multiwave IR xenon laser in hard-ioniser-pumped mixtures with He and Ar. II. Nuclear pumping. Quantum Electron. 1999;29:687–693.

[30] Khasenov M.U. On the mechanism of populating 3p levels of neon under pumping by a hard ionizer. Quantum Electron. 2011;41:198–201.

[31] Poletaev E.D., Dorofeev Y.B., Dyachenko P.P., et al. Emission characteristics of pure neon and He–Ne mixture excited by a high-pressure nuclear particles. Tech Phys. 1992;37:114–121.

[32] Svelto O. Principles of lasers. 5th edition. Heidelberg: Springer; 2010. 620 p. DOI: 10.1007/978-1-4419-1302-9.

[33] Khasenov M.U. Mechanisms of population of the levels in gas lasers pumped by ionizing radiation. Laser Part Beams. 2014;32:501–508.

[34] Kolokolov N.B., Kudryavtsev A.A., Romanenko V.A. A spectroscopic investigation of recombination populating of the $5p^56p$ and $5p^55d$ states of the Xe Atom. Opt Spectrosc (Sov J). 1989;67:292–296.

[35] Ivanov V.A. Dissociative recombination of molecular ions in noble-gas plasmas. Sov Phys Uspekhi. 1992;35:17–36.

[36] Batyrbekov G.A., Dolgikh V.A., Khasenov M.U., et al. Luminescence of mixtures of mercury and inert gases containing molecular additives with excitation by ionizing radiation. J Appl Spectrosc (Sov J). 1988;49:1139–1143.

[37] Fedenev F.V., Tarasenko V.F. Simulation of NPL in experiments with e-beam pumping. Laser Part Beams. 1998;16:327–380.

[38] Ulrich A. Light emission from the particle beam induced plasma: an overview. Laser Part Beams. 2012;30:199–205.

[39] Karelin A.V., Yakovlenko S.I. Kinetic model of an He–Ne–Ar–H_2 laser pumped by hard ionising radiation. Quantum Electron. 1995;25:739–745.

[40] Voinov A.M., Melnikov S.P., Sinyanskiy A.A. A kinetic model of recombination lasers at xenon atom transitions. Tech Phys (Sov J). 1990;35:1172–1182.

[41] Miskevich A.I. A kinetic model of a nuclear-pumped laser operating on cadmium vapors. Tech Phys (Sov J). 1987;32:1056–1063.

[42] Boichenko A.M., Yakovlenko S.I. The possibility of lasing in Ne^+Ar ionic molecules pumped by a hard ionizer. Quantum Electron. 2000;30:681–686.

[43] Dem'yanov A.V., Dyatko N.A., Kochetkov I.V., Napartovich A.P. Simulation of excimer lasers with nuclear pumping. In: Proceedings of the Specialist Conference "Physics of Nuclear-Excited Plasma and Problems of Nuclear-Pumped Lasers", Obninsk: FEI; 1992. vol. 1. pp. 252–261.

[44] Magda E.P. Analyses of experimental and theoretical research of nuclear pumped lasers at the Institute of Technical Physics. Laser Part Beams. 1993;11:469–476.

[45] Miley G.H. Overview of nuclear pumped lasers. Laser Part Beams. 1993;11:575–581.

[46] Hebner G.A., Hays G.N. Reactor-pumped laser experimental results. Proc SPIE. 1994;2121:10–20.

[47] Prelas M. Nuclear-pumped lasers. New York: Springer; 2016. 417 p. DOI: 10.1007/978-3-319-19845-3.

[48] Yang C., Chen H., Zheng C., Zhao X., Han H. The progress of nuclear pumped laser in CFBR-II reactor. Chin Opt Lett. 2003;1:292–293.

[49] Xingxing J., Kaisu W., Iluaming Z., Hande C. Gain test of nuclear-pumped ^{3}He–Ne laser. Chin J Lasers. 1992;19:N 7. (Chinese, Engl. resume).

[50] Mis'kevich A.I. Visible and near-infrared direct nuclear pumped lasers. Laser Phys. 1991;1:445–481.

[51] Khasenov M.U. Optical emission of the nuclear-induced plasmas of gas mixtures. Int J Opt. 2014;ID748763. 16 p.

[52] Batyrbekov E. Converting nuclear energy into the energy of coherent optical radiation. Laser Part Beams. 2013;31:673–687.

[53] Khasenov M.U. Nuclear-induced plasma of gas mixtures and nuclear pumped lasers. Almaty, Kazakhstan; 2011. 187 p. (in Russian).

[54] Batyrbekov G.A., Khasenov M.U., Soroka A.M., et al. Feasibility of construction of a quasi-cw laser utilizing 7s–6p transitions in mercury pumped by ionizing radiation. Quantum Electron (Sov J). 1987;17:774–775.

[55] Zdorovets M., Ivanov I., Koloberdin M., et al. Accelerator complex based on DC-60 cyclotron. In: Proceedings of the 24th Russian Particle Accelerator Conference. Obninsk: FEI; 2014. pp. 287–289.

[56] Bennett W.R. Optical spectra excited in high pressure noble gases by alpha impact. Ann Phys. 1962;18:367–420.

[57] Dmitriev A.B., Il'yashenko V.S., Mis'kevich A.I., Salamakha B.S. Luminescence of neon and some of its mixtures at high pressures. Opt Spectrsc (Sov J). 1977;43:687–688.

[58] Khasenov M.U. Kinetics of CO first negative system excitation by ionized radiation. Proc SPIE. 2004;5483:14–23.

[59] De Young R.J., Weaver W.R. Spectra from nuclear-excited plasmas. J Opt Soc Am. 1980;70:500–506.

[60] Gorbunov V.V., Grigor'ev V.D., Dovbysh L.E., et al. The luminescence spectra in the 350–875 nm range of the dense gas excited by uranium fission fragments. Proc RFNC-VNIIEF, 2004;iss.6:148–185 (in Russian).

[61] Abramov A.A., Gorbunov V.V., Melnikov S.P., et al. Luminescence of nuclear-induced rare-gas plasmas in near infrared spectral range. Proc SPIE. 2006;6263:279–296.

[62] Sinyanskii A.A., Melnikov S.P. Research on development of continuous nuclear-laser setups in VNIIEF. Proc SPIE. 1998;3686:43–55.

[63] Helmick H.H., Fuller J.I., Schneider R.T. Direct nuclear pumping of helium–xenon laser. Appl Phys Lett. 1975;26:327–328.

[64] McArthyr D.A., Tollefsrud P.B. Observation of laser action in CO gas excited only by fission fragments. Appl Phys Lett. 1975;26:187–190.

[65] Zagidulin A.V., Bochkov A.V., Mironenko V.V., Sofienko G.S. A 500-J nuclear-pumped gas laser. Tech Phys Lett. 2012;38:1059–1062.

[66] Alford W.J., Hays J.N. Measured laser parameters for reactor-pumped He/Ar/Xe and Ar/Xe lasers. J Appl Phys. 1989;65:3760–3766.

[67] Melnikov S.P., Sinyanskii A.A. Ultimate efficiency of nuclear-pumped gas lasers. Laser Part Beams. 1993;11:645–654.

[68] Voinov A.M., Zobnin V.G., Konak A.I., et al. Quasi-cw low-threshold lasing and line competition in the nuclear-pumped lasers based on atomic xenon transitions. Tech Phys Lett (Sov J). 1990;16:297–300.

[69] Hebner G.A., Shon J.W., Kushner M.J. Temperature dependent gain of the atomic xenon laser. Appl Phys Lett. 1993;63:2872–2874.

[70] Barysheva N.M., Bochkov A.V., Bochkova N.V., et al. On the possible mechanism of overheating of the active medium of an NPL operating on ir transitions of the xenon atom. In: Proceedings of the Specialist Conference "The Physics of Nuclear-Excited Plasma and the Problems of Nuclear-Pumped Lasers". Obninsk: FEI: 1992. vol. 1. pp. 374–380.

[71] Konak A.I., Melnikov S.P., Porkhaev V.V., Sinyanskii A.A. Nuclear-pumped gas lasers at temperatures up to 800°C. Laser Part Beams. 1993;11:663–668.

[72] Kryzhanovskii V.A., Mavlyutov A.A., Miskevich A.I. Lasing characteristics of a nuclear-pumped Ar–Xe laser at high temperatures. Tech Phys Lett. 1995;21:535–536.

[73] Mel'nikov S.P. Mechanisms of generation of nuclear-pumped lasers on IR transitions of inert gas atoms. In: Proceedings of the 4th International Conference "The physics of nuclear-pumped lasers and pulsed reactors". Obninsk: FEI; 2009. vol. 1. pp. 167–176 (in Russian).

[74] Shon J.W., Kushner M.J. Excitation mechanism and gain modeling of the high pressure atomic Ar laser in He/Ar mixture. J Appl Phys. 1994;75:1883–1890.

[75] Apruzese J.P., Giuliani J.L., Wolford M.F., et al. Experimental evidence for the role of the Xe_2^+ in pumping of the Ar–Xe infrared laser. Appl Phys Lett. 2006;88:121120.

[76] Denezhkin I.A., D'yachenko P.P. Population and relaxation kinetics of $5d[3/2]_1$ level upon pulsed electron-beam excitation of pure xenon. Quantum Electron. 2009;39:135–138.

[77] Carter D.D., Rowe M.J., Schneider R.T. Nuclear pumped CW lasing of the ^{3}He–Ne system. Appl Phys Lett. 1980;36:115–117.

[78] Prelas M.A., Schlapper G.A. Comments on nuclear pumped CW lasing of the ^{3}He–Ne system. J Appl Phys 1981;52:496–497.

[79] Batyrbekov G.A., Khasenov M.U., Tleuzhanov A.B., et al. Investigation of the active media of lasers operating in the nuclear reactor. Final Scientific Report, no. GR 81032078. Institute of Nuclear Physics. Almaty. 1986 (in Russian).

[80] Batyrbekov G.A., Batyrbekov E.G., Danilychev V.A., Khasenov M.U. Efficiency of populating neon 3p-levels under ionized pumping. Opt Spectrsc (Sov J). 1990;68:1241–1245.

[81] Smirnov B.M. Excited Atoms. Moscow: Energoizdat; 1982. 232 p. (in Russian).

[82] Schmieder D., Brink D.J., Salamon I.J., Jones E.G. A high pressure 585.3 nm neon–hydrogen laser. Opt Commun. 1981;36:223–226.

[83] Bunkin F.V., Derzhiev V.I., Mesyaz G.A., et al. Plasma laser emitting at the wavelength of 585.3 nm with Penning clearing of the lower level in dense mixtures with neon excited by an electron beam. Quantum Electron (Sov J). 1985;15:159–160.

[84] Aleksandrov A.Yu., Anan'ev V.Yu, Basov N.G., et al. Efficient visible laser based on neon 3p–3s transitions. Sov Phys Dokl. 1985;30:875–879.

[85] Karelin A.V., Tarasenko V.F., Fedenev A.V., Yakovlenko S.I. Ultimate efficiency of a Penning neon plasma laser. Quantum Electron. 1996;26:291–294.

[86] Basov N.G., Baranov V.V., Danilychev V.A., et al. High-pressure power laser utilizing 3p–3s transitions in NeI generating radiation of wavelengths 703 and 725 nm. Quantum Electron (Sov J). 1985;15:1004–1006.

[87] Voinov A.M., Krivonosov V.N., Mel'nikov S.P., et al. Quasicontinuous lasing on the 3p–3s transitions of a neon atom in mixtures of inert gases excited by uranium fission fragments. Sov Phys Dokl. 1990;35:568–572.

[88] Kopai-Gora A.P., Mis'kevich A.I., Salamakha B.S. Emission of laser radiation at a wavelength of 585.2 nm in a dense ^{3}He–Ne–Ar plasma. Tech Phys Lett (Sov J). 1990;16:411–414.

[89] Hebner G.A., Hays G.N. Fission-fragment-excited lasing at 585.3 nm in He/Ne/Ar gas mixtures. Appl Phys Lett. 1990;57:2175–2177.

[90] Shaban Y., Miley G.H. A practical visible wavelength nuclear-pumped laser. In: Proceedings of Specialist Conference on Physics of Nuclear Induced Plasma and Problems of Nuclear Pumped Lasers. Obninsk: FEI; 1993. vol. 2. pp. 241–247.

[91] Konak A.I., Mel'nikov S.P., Porkhaev V.V., Sinyanskii A.A. Characteristics of nuclear-pumped lasers based on the 3p–3s transitions in the neon atom. Quantum Electron. 1995;25:209–214.

[92] Bochkov A.V., Kryzhanovskii V.A., Magda E.P., Mukhin S.L. Quasi-cw lasing at λ = 585.2 nm in an Ne–H$_2$ mixture. Tech Phys Lett. 1993;12:750–751.

[93] Hebner G.A. Fission-fragment excitation of the high-pressure atomic neon laser at 703.2 and 724.5 nm. J Appl Phys. 1993;74:2203–2207.

[94] Aleksandrov A.Y., Dolgikh V.A., Kerimov O.M., et al. Basic mechanisms of inversion of the 3p–3s transitions of neon. Quantum Electron (Sov J). 1987;17:1521–1526.

[95] Shon J.W., Rhodes R.L., Verdeyen J.T., Kushner M.J. Short pulse electron beam excitation on the high-pressure atomic Ne laser. J Appl Phys. 1993;73:8059–8065.

[96] Andriyakhin V.M., Vasil'tsov V.V., Krasilnikov S.S., et al. On emission of the gas mixture Hg–He^3 irradiated by neutron flux. JETP Lett (Sov J). 1970;12:58–60.

[97] Akerman M.A., Miley G.H., McArthur D.A. A helium–mercury direct nuclear pumped laser. Appl Phys Lett. 1977;30:409–412.

[98] Batyrbekov G.A., Khasenov M.U., Soroka A.M., et al. Kinetics of excited states of Hg pumping by ionizing radiation. Preprint Inst Nucl Phys. 1987;3/87, Alma-Ata, Kazakhstan, (in Russian).

[99] Yakovlenko S.I., Karelin A.V. Kinetics of the active media of high-pressure metal-vapor lasers. Quantum Electron. 1993;23:545–563.

[100] Dmitriev A.B., Il'yashenko V.S., Mis'kevich A.I., Salamakha B.S. Spectroscopy of ^{3}He–Hg and ^{3}He–Kr–Hg high-pressure plasmas excited by ^{3}He(n,p)T reaction products. Opt Spectrosc (Sov J). 1979;47:34–36.

[101] Djeu N., Burnham R. Optically pumped CW Hg laser at 546.1 nm. Appl Phys Lett. 1974;25:1350–1351.

[102] Dmitriev A.B., Il'yashenko V.S., Mis'kevich A.I., et al. Excitation of laser transitions in parametallic gas mixtures using reaction products from nuclear reactions. Tech Phys (Sov J). 1982;27:1373–1374.

[103] Bochkov A.V., Kryzhanowskii V.A., Magda E.P., et al. Quasi-cw lasing on the 7^3S_1–6^3P_2 atomic mercury transition. Tech Phys Lett. 1992;18:241–243.

[104] Barysheva N.M., Bochkova N.V., Kosorukova A.A., Magda E.P. Kinetic model of He–Xe–Hg–H$_2$ laser with nuclear pumping. In: Proceedings of the 3rd International Conference on "Problems of Lasers with Nuclear Pumping and Pulsed Reactors". Snezhinsk: VNIITF; 2002. pp. 218–224 (in Russian).

[105] Rhoades R.A., Verdeyen J.T. Electron beam pumping of the 546.1 nm mercury laser. Appl Phys Lett. 1992;60:2951–2953.

[106] Mis'kevich A.I., Dmitriev A.B., Il'yashenko V.S., et al. Lasing of Cd vapor excited by the products of the nuclear reaction ^{3}He(n,p)T. Tech Phys Lett (Sov J). 1980;6;352–355.

[107] Mis'kevich A.I., Il'yashenko V.S., Salamakha B.S., et al. Lasing at wavelength 441.6 nm in a high-pressure ^{3}He–^{116}Cd mixture. Tech Phys (Sov J). 1982;27;260–262.

[108] Baltayan P., Peboy-Peyroula J.C., Sadeghi N. Determination of the rate constants for population of the individual Cd^{+*} levels in the thermal Penning and charge-transfer reactions of He*(2^{3}S) and He$^+$ with cadmium. J Phys B. 1985;18:3615–3628.

[109] Novoselov Yu.N., Tarasenko V.F., Uvarin V.V., Fedenev A.V. Influence of impurities and of pump power on the operational characteristics of a high-pressure He–Cd laser. Quantum Electron. 1996;26:205–210.

[110] Kopai-Gora A.P., Mis'kevich A.I., Salamakha B.S. Quasi-cw generation on the ZnII Beitler transition in dense ^{3}He–Zn plasma. Tech Phys Lett (Sov J). 1990;16:348–351.

[111] Bugaev S.P., Goryunov F.G., Nagornyii D.Y., et al. UV generation by electron beam pumping of He–Cd mixture. Opt Spectrosc (Sov J). 1988;65:442–445.

[112] Karelin A.V., Shirokov R.V. Kinetics of the active medium of a nuclear-pumped laser based on transitions in the cadmium atom. Quantum Electron. 1998;28:893–897.

[113] Khasenov M.U. Emission of the ^{3}He–Xe–Cd mixture in the active zone of a nuclear reactor. Quantum Electron. 2004;34:1124–1126.

[114] Basov N.G., Aleksandrov A.Y., Danilychev V.A., et al. Efficient high-pressure quasi-cw laser using the first negative system of nitrogen. JETP Lett (Sov J). 1985;42:47–50.

[115] Aleksandrov A.Y., Dolgikh V.A., Rudoi I.G., et al. Energy characteristics of visible-range and UV lasers based on the first negative system of nitrogen. Pis'ma v Zhurnal Tekhnicheskoi Fiziki. 1987;13:1370–1373 (in Russian).

[116] Barysheva N.M., Bochkov A.V., Bochkova N.V., et al. First nuclear-pumped ultraviolet laser. Tech Phys Lett. 1996;22:636–637.

[117] Dyuzhov Y.A., Poletaev E.D., Smol'skii V.N. Investigation of lasing on transitions of the first negative system of nitrogen (λ = 391.4, 428.1 nm) in the He–N$_2$–H$_2$-mixtures pumped by fission fragments from the pulse reactor BARS-6. In: Proceedings of the 4th International Conference "The physics of nuclear-pumped lasers and pulsed reactors". Obninsk: FEI; 2009. vol. 1. pp. 151–155 (in Russian).

[118] Budnik A.P., Kuznetsova E.A. Mathematical modeling of the lasing characteristics of a helium–nitrogen–hydrogen laser medium from the pressure of the mixture and the specific power of energy input. In: Proceedings of the 4th International Conference "The physics of nuclear-pumped lasers and pulsed reactors". Obninsk: FEI; 2009. vol. 1. pp. 160–166 (in Russian).

[119] Bochkov A.V., Zagidulin A.V., Magda E.P., et al. Effect of deuterium and neon on the laser parameters to the first negative system of nitrogen. In: Proceedings of the 4th International Conference "The physics of nuclear-pumped lasers and pulsed reactors". Obninsk: FEI; 2009. vol. 1. pp. 319–321 (in Russian).

[120] Aleksandrov Yu., Dolgich V.A., Kerimov O.M., et al. Effective collision lasers in the visible and UV regions of spectrum. Izvestija Akademii Nauk SSSR, Ser Phys. 1989;53:1474–1483 (in Russian).

[121] Khasenov M.U. Kinetics of the nitrogen first negative system excitation by ionising radiation. Quantum Electron. 2005;35:1104–1106.

[122] Waller R.A., Collins C.B., Cunningham A.J. Stimulated emission from CO^+ pumped by charge transfer from He_2^+ in the afterglow of an e-beam discharge. Appl Phys Lett. 1975;27:323–325.

[123] Khasenov M.U., Dolgich V.A., Soroka A.M. Kinetics of CO first negative system excitation by ionized radiation. Abstracts of Specialist Conference on "Physics of nuclear induced plasmas and problems of nuclear pumped lasers. Obninsk: FEI; 1992. p. 224.

[124] Khasenov M.U. Kinetics of CO first negative system excitation by ionized radiation. Proc SPIE. 2004;5483:14–23.

[125] Barysheva N.M., Bochkova N.V., Kosorukova A.A., Magda E.P. Kinetics of nuclear-pumped lasers on ultraviolet electronic transitions in molecular ions. In: Proceedings of the 3rd International Conference "Problems of Lasers with Nuclear Pumping and Pulsed Reactors". Snezhinsk: VNIIEF; 2002. pp. 51–55 (in Russian).

[126] Pearse R.W.B., Gaydon A.G. The identification of molecular spectra. 2nd ed., New York: Wiley; 1950. 276 p.

[127] Rhodes C.K., editor. Excimer lasers. 2nd ed. Berlin: Springer-Verlag; 1984. 271 p. DOI: 10.1007/3-540-13013-6.

[128] Yakovlenko S.I., Karelin A.V., Morovov A.P., et al. Investigation of an XeF laser pumped by gamma radiation from a nuclear explosion. Quantum Electron. 1996;26:410–412.

[129] Ulrich A., Adonin A., Jacoby J., et al. Excimer laser pumped by an intense, high-energy heavy-ion beam. Phys Rev Lett. 2006;97:153901.

[130] Hays G.N., McArthur D.A., Neal D.R., Rice J.K. Gain measurements near 351 nm in $^3He/Xe/NF_3$ mixtures excited by fragments from the $^3He(n,p)^3H$ reaction. Appl Phys Lett. 1986;49:363–366.

[131] Bochkov A.V., Kryzhanowskii V.A., Magda E.P., et al. Investigation of the characteristics of excimer laser media. In: Proceedings of the Specialist Conference "The Physics of Nuclear-Excited Plasma and the Problems of Nuclear-Pumped Lasers". Arsamas-16:VNIIEF; 1995. vol. 1. pp. 154–161.

[132] Yakovlenko S.I., Karelin A.V., Boichenko A.M. Calculation of the threshold characteristics of a nuclear-pumped $Ne–Xe–NF_3$ laser. Quantum Electron. 1995;25:521–524.

[133] Mavlyutov A.A., Miskevich A.I. Nuclear-pumped excimer laser with a wavelength of 308 nm. Tech Phys Lett. 1996;22:326–327.

[134] Khasenov M.U., Nakiskozhaev M.T., Syrlybaev A.S., Smirnova I.I. Emission of inert gas halides at excitation by alpha-particles. Atmos Ocean Opt. 2009;22:1057–1059 (in Russian).

[135] Mis'kevich A.I., Guo Jinbo, Dyuzhov Yu.A. Spontaneous and induced emission of $XeCl^*$ excimer molecules under pumping of $Xe–CCl_4$ and $Ar–Xe–CCl_4$ gas mixtures

with a low CCl_4 content by fast electrons and uranium fission fragments. Quantum Electron. 2013;43: 1003–1008.

[136] Batyrbekov G.A., Khasenov M.U., Kuzmin Yu.E., et al. Radiation resistance of elements of the laser system in the core of nuclear reactor. Izvestija Akademii Nauk KazSSR, Phys Math Ser. 1986;6:23–26 (in Russian).

[137] Batyrbekov G.A., Khasenov M.U., Kostriza S.A., et al. Feasibility of excimer lasers with ionization by radiation from a nuclear reactor. Tech Phys Lett (Sov J). 1982;8:789–791.

[138] Treanor C.E., Rich J.W., Rehm R.J. Vibrational relaxation of anharmonic oscillators with exchange-dominated collisions. J Chem Phys. 1968;48:1798–1803.

[139] Jalufka N.W., Hohl F. A direct nuclear-pumped ^{3}He–CO–laser. Appl Phys Lett. 1981;39:139–142.

[140] Gudzenko L.I., Malyshevskii V.S., Yakovlenko S.I. CO laser pumping with high-energy particles. Tech Phys. 1978;23:1228–1231.

[141] Zherebtsov V.A. Nuclear pumping of a carbon monoxide laser. Tech Phys. 1998;43:818–823.

[142] Mann M.M., Rice D.K., Eguchi R.G. An Experimental Investigation of High Energy CO lasers. IEEE J Quantum Electron. 1974;QE-10:682–685.

[143] Jalufka N.W. Direct nuclear excitation of a ^{3}He–CO$_2$ mixture. Appl Phys Lett. 1981;39:190–192.

[144] Hassan H.J. Kinetics of a CO_2 nuclear pumped laser. AIAA J. 1980;18:1221–1222.

[145] Batyrbekov G.A., Danilychev V.A., Kovsh I.B., Mardenov M.P., Khasenov M. Preionization CO_2 laser operating in the active zone of a stationary nuclear reactor. Quantum Electron (Sov J). 1977;7:667–668.

[146] Batyrbekov G.A., Danilychev V.A., Kovsh I.B., Khasenov M.U. Operation of a cooled electroionization CO laser in the active zone of a nuclear reactor. Tech Phys Lett (Sov J). 1979;5:345–346.

[147] Jongerius H.M., Van Koeveringe J.L., Oskam H.L. Argon–xenon bands. Physica. 1959;25:406–408.

[148] Friedl W. Krypton–Xenon Banden. Z Naturforsch. 1959;14A:848–848a.

[149] Tsuji M., Tanaka M., Nishimura Y. New emission spectra of $KrXe^+$ produced from Kr afterglow reactions of Xe. Chem Phys Lett. 1996;262:349–354.

[150] Kugler E. Über die Lumineszenze der Edelgasgemische Ar/Xe, Kr/Xe, Ar/Kr und der Gemische Xe/N$_2$ und Kr/N$_2$ bei Angerung mit schnellen Elektronen. Ann Phys Leipz. 1964;B14:137–146.

[151] Tanaka Y., Yoshino K., Freeman D.E. Emission spectra of heteronuclear diatomic rare gas positive ions. J Chem Phys. 1975;62:4484–4496.

[152] Millet P., Barrie A.M., Birot A., et al. Kinetic study of $(ArKr)^+$ and $(ArXe)^+$ heteronuclear ion emissions. J Phys Ser B. 1981;14:459–472.

[153] Batyrbekov G.A., Batyrbekov E.G., Tleuzhanov A.B., Khasenov M.U. Molecular band in an emission spectrum of Ar–Xe. Opt Spectrosc (Sov J). 1987;62:212–214.

[154] Millet P., Birot A., Brunet H., et al. Kinetic study of the $KrXe^+$ heteronuclear ion emissions. J Phys Ser B. 1983;16:1383–1392.

[155] Khasenov M.U. On the possibility of the creation of nuclear-pumped lasers on transitions of the heteronuclear ionic molecules of inert gases. In: Abstracts of Specialist Conference on "Physics of nuclear induced plasmas and problems of nuclear pumped lasers". Obninsk: FEI; 1992. pp. 351–352.

[156] Khasenov M.U. Emission of the heteronuclear ionic molecules $(ArXe)^+$ at excitation by a hard ionizer. Proc SPIE. 2006;6263:141–148.

[157] Khasenov M.U. Emission of ionic molecules $(KrXe)^+$ at excitation by a hard ionizer. J Appl Spectrosc. 2005;72:316–320.

[158] Laigle C., Collier F. Kinetic study of $(ArXe)^+$ heteronuclear ion in electron beam excited Ar–Xe mixture. J Phys Ser B. 1983;16:687–697.

Free-electron Laser

Undulators for Short Pulse X-Ray Self-Amplified Spontaneous Emission-Free Electron Lasers

K. Zhukovsky

Additional information is available at the end of the chapter

http://dx.doi.org/10.5772/64439

Abstract

We review the synchrotron type radiation sources with focus on undulator and free-electron laser (FEL) schemes, aimed on working in X-ray range and ultra-short time interval. Main FEL schemes, useful for generation of high frequency radiation, extending to X-rays, are presented. High harmonic generation is explored. The advantages and disadvantages of single pass and of multipass designs are discussed. The viable ways to reduce the duration of the pulse, with the goal to generate femtosecond pulses, are indicated. Future developments of X-ray FELs (X-FELs) and the ways to improve the quality of the FEL radiation in this context are discussed.

Keywords: undulator radiation, harmonics generation and broadening, homogeneous and inhomogeneous losses, free-electron laser

1. Introduction

Synchrotron radiation (SR) and undulator radiation (UR) have been attracting researcher's attention for more than half a century. The reasons for that varied with time passing as the challenges for the scientists evolved and the technical progress stepped forward. UR was predicted [1] and then discovered [2] in the middle of the twentieth century. During the following 70 years, the study of the radiation, emitted by ultra-relativistic electrons, was performed, and the SR theory was refined; extensive theoretical studies of the electron motion in periodic magnetic fields of various configurations have been performed [3–7]. Now UR is

again in focus due to the request for coherent X-ray sources [8], while free-electron lasers (FELs) extend to X-range [9]. Both, SR and UR are due to the radiation of relativistic electrons, executing curved trajectories [10]. The difference between them is in the length on which the radiation is formed: short part of the circle for UR and the full length of the undulator for the UR. This determines the fundamental difference in the quality of the radiation obtained from these two sources: short pulses with very broad spectrum for the SR and relatively long-lasting radiation bursts with narrow spectrum for the UR. Nowadays, the research frontier is represented by studies of ultra short attosecond time intervals and Röntgen range [11, 12]. To achieve these characteristics, the devices require extremely high quality and intense magnetic fields, long undulators with many periods. To obtain high frequency radiation, sometimes undulator periodic structures with double or even triple period are used [13–18], facilitating control over high harmonics and regulating their emission [6, 19]. Maintaining best quality UR line is important. Nowadays, electron accelerators provide ultra fast and high coherent electron beams and intense emission even in Röntgen range. Modern undulators allow for the harmonic emission regulation [19], which can be achieved by superimposing different periodic magnetic fields in many period undulators. It should be noted that in long undulators the distortions of the magnetic field and of unavoidable inhomogeneities of periodic magnetic field have very strong effect on the operation of the devices. High gain in undulators is essential in FEL with self-amplified spontaneous emission (SASE), with high-gain harmonic generation (HGHG), and in other modern schemes. The quality of the electron beams, fed into the undulators, is extremely important for such new FELs [20], in particular at high frequencies. The requirements for the periodicity of the field are particularly rigid in long undulators [21]. Deviation from the ideal oscillatory trajectories of electrons results in degradation of the beam quality and of the FEL output in terms of power, brightness FEL gain. Therefore, the emitted UR can be exploited for diagnosing the quality of the undulator itself.

The radiation spectrum lines from undulators inevitably broaden due to a number of reasons, first of all due to the electron beam energy spread and the beam divergency, as well as due to inhomogeneity of the periodic magnetic field in undulators. They may have internal or external origin [22–26], but their presence is eminent also due to the fact that the ideal $\vec{H} = H_0 sin(2\pi z / \lambda)$ periodic magnetic field simply does not satisfy Maxwell equations. The electron energy spread is the most common detrimental factor; some researchers even concluded that the spectral properties of higher UR harmonics should be limited only by the electron beam properties and not by the undulators [24]. The role of the divergence was underlined, for example, in Refs. [22, 27]. At the same time the constant magnetic field shifts the resonance frequencies and causes loss of intensity [28–31].

The demand for radiation with specific properties and high requirement to the UR and undulator quality stimulated analytical study of their spectral properties [32–39]. There appear generalized Bessel and Airy functions naturally, as well as in other mathematical problems of radiation, emitted by charges, executing complicated oscillating trajectories. The mathematical apparatus of inverse differential operators and orthogonal polynomials was developed for treating broad spectrum of physical problems, which include radiation and propagation of electron beams [40–45]. The contributions of all sources of broadening in various undulator schemes were analyzed by means of precise analytical treatment of the UR, employing extended forms of special functions of Airy and Bessel types [34–37, 39, 42]. In these works, the role of the various broadening terms, accounting for the real size and the emittance of the electron beam, for the energy spread and for the constant field component, was explored. It was shown that the undulator length has strong detrimental effect on the spontaneous harmonic emission; partial compensation of the beam divergences by constant magnets was also demonstrated.

2. Synchrotron radiation, undulator radiation, and free-electron lasers

Charged particle radiates energy in the form of electromagnetic radiation when it accelerates. Following relativistic Lorentz transforms $\tan\theta = \frac{\sin\theta'}{\gamma(\beta + \cos\theta')}$, it is easy to conclude that the radiation emission of the relativistic electron is focused in the narrow angle $\theta \approx \frac{1}{\gamma}$ (see, for example, [46]). While the charge is moving on a circular orbit of radius R, it emits SR in a narrow cone of emission, which illuminates the receiver for a very short period of time, while passing from the point A to the point B. Lorentz transforms is applied to the period of time in the reference frames, related to the electron and to the observer and yield the time of the SR pulse $\Delta t = \frac{m}{2eB\gamma^2} = \frac{R}{2c\gamma^3}$, where $R \cong \frac{\gamma mc}{eB}$. For the UR, it is not so, since it is gathered all along the undulator and the characteristic length then equals that of the undulator, by far exceeding the arc, from which the SR, reaching the user, is gathered. Denoting the unit vector $\vec{n} = \vec{R}/R$ and $\vec{n} \cong (\psi\cos\phi, \ \psi\sin\phi, \ 1-\psi^2/2)$, it is easy to obtain the wavelength of the radiation, emitted off the axis in the angle θ

$$\lambda_n = \frac{\lambda_u}{2n\gamma^2}\left(1 + \gamma^2\langle\theta^2\rangle + \gamma^2\psi^2\right) \tag{1}$$

from the simple condition of positive interference of the wavelengths, emitted on each magnetic poles of the undulator. Since the SR from a relativistic charge is emitted in a narrow cone, which includes the undulator axis all the time, the electron drifts in the undulator at relativistic speed, if the electrons transversal oscillations are small. At the exit of the undulator the intense radiation appears (see **Figure 1**).

The spectral range of SR and UR extends up to Röntgen band. However, due to the fact that the SR is perceived as a very short pulse, its spectral range is very broad, starting from the synchrotron frequency $\omega_0 = v/R$, while the UR has few harmonics and in some cases, such as that of a spiral undulator, it can contain a single harmonic.

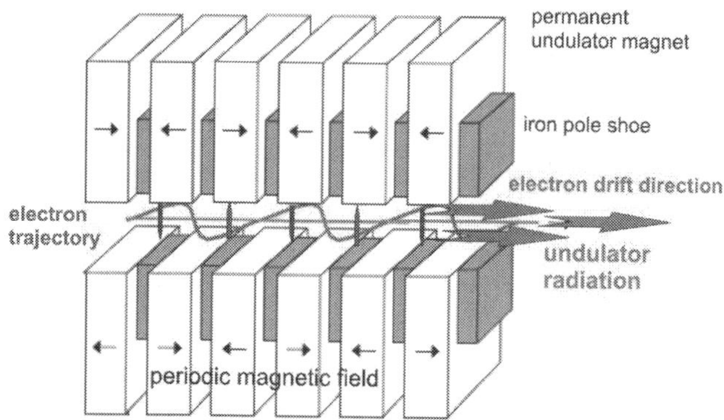

Figure 1. Schematic drawing of a planar undulator.

For spontaneous radiation, emitted by an electron, the undulator selects resonant UR wavelengths. The idea about it can be given by the simple consideration, demonstrated in **Figure 2**, where electric fields for two resonant wavelengths λ_n at the fundamental ($n = 1$; red) and at the third harmonics ($n = 3$; blue) are shown. A non-resonant electric field for the second harmonic is shown in green. The fundamental and the third harmonics are phase-matched with the electron after one undulator period as highlighted in **Figure 2**. The electron trajectory is drawn in gray. Such a phase matching of the radiation proceeds on each next undulator period. Thus, the radiation from one electron constructively interferes over many periods, and in this sense we obtain coherent radiation from one electron along the whole length of the undulator.

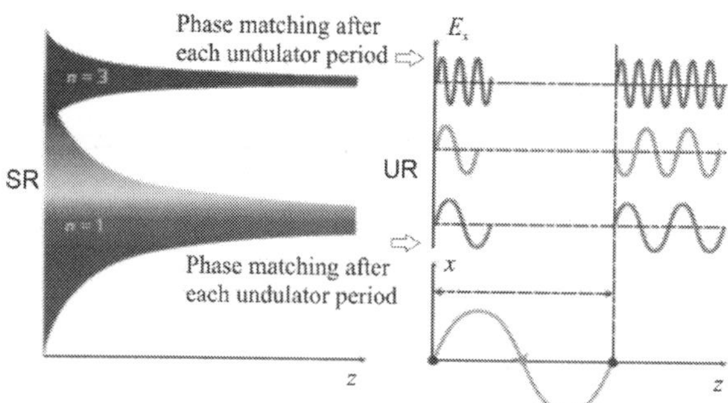

Figure 2. Undulator selects resonant frequencies for the emitted SR.

The following conditions are common in modern undulators: the electrons are ultrarelativistic, which is natural in contemporary accelerators, they have small transverse momentum, and the electric field is absent:

$$\gamma \gg 1, \beta_{\perp} \ll 1, H_{\parallel} \ll H_{\perp}, \vec{E} = 0 \tag{2}$$

The UR from a planar undulator with N periods of λ_0 with the undulator parameter $k = \dfrac{e}{mc^2}\dfrac{H_0}{k_\lambda}$, where $k_\lambda = 2\pi/\lambda_u$, λ_u is the undulator period, $H_y = H_0 sin(k_\lambda z)$ is the sinusoidal magnetic field, has the following peak frequencies:

$$\omega_n = n\omega_R = \frac{2n\omega_0\gamma^2}{1 + \dfrac{k^2}{2} + (\gamma\psi)^2}, \ \omega_{R0} = \frac{2\omega_0\gamma^2}{1 + k^2/2}, \ \omega_{n0} = n\omega_{R0}, \tag{3}$$

where $\omega_0 = k_\lambda \beta_z^0 c$, $\beta_z^0 = 1 - \dfrac{1}{2\gamma^2}\left(1 + \dfrac{k^2}{2}\right)$ is the average drift speed of the electrons along the undulator axis, $k \approx H_0 \lambda_0$ [Tcm] and ψ is off the undulator axis angle. The shape of the UR emission line is described by the $\dfrac{sin(v_n/2)}{v_n/2} = sinc\dfrac{v_n}{2}$ function, dependent on the detuning parameter

$$v_n = 2\pi N n\left(\frac{\omega}{\omega_n} - 1\right). \tag{4}$$

The homogeneous bandwidth, sometimes called half-width of UR spectrum line at its half-height or simply half-width, is $\dfrac{1}{2nN}$ and the half-width equals

$$\frac{\Delta\omega}{\omega_{n0}} = \frac{\omega - \omega_{n0}}{\omega_{n0}} = \frac{1}{nN}. \tag{5}$$

In real devices $\dfrac{1}{nN} \ll 1$. The emitted wavelength $\lambda_n = 2\pi/k_n$ can be expressed through the speed of the electrons in the undulator as follows:

$$\lambda_n = \frac{\lambda_1}{n} = \frac{\lambda_u}{n}\left(\frac{1 - u/c}{u/c}\right) \cong \frac{\lambda_u}{2n\gamma^2}\left(1 + \tilde{k}^2\right), \tag{6}$$

where λ_u is the undulator period, u is the electron speed

$$u = c\left(1 - \left(1 + \tilde{k}^2\right)/2\gamma^2\right), \tag{7}$$

The undulator parameter is k or $\tilde{k}$:

$$\tilde{k}^2 \equiv \frac{k^2}{2} = \gamma^2 \beta_\perp^2 = \gamma^2 \langle \theta^2 \rangle, \ k = eH_0 \lambda_u / 2\pi mc^2, \tag{8}$$

θ is the off-axis angle, which in essence indicates how much the electron deviates from the axis in its motion along the undulator due to transversal oscillations, caused by the periodic magnetic field. In real conditions for a weak undulator $k \sim 1$, while for wiggler or strong undulator $k \sim 10$. Thus, for a weak undulator the radiation is essentially directed all along the undulator axis, while for the wiggler, it is in much wider angle. Considering $n = 1$, i.e., the fundamental harmonic, we write $u = ck_1/(k_1 + k_u)$, $k_1 = 2\pi/\lambda_1$. Both SR and spontaneous UR are incoherent. In the full spectrum of the radiation, the components, whose wave length is longer than the bunch length, are coherent, while the radiation of the length, significantly exceeding the size of the bunch, is approximately coherent. The region of coherent radiation will enlarge as the bunch gets shorter. The difference between the radiations, emitted by various SR sources, is demonstrated in **Figure 3**. It shows how the radiation intensity and the degree of coherency of the emitted radiation varies from one device scheme to the other; N_f is the number of emitted photons, N_e is the number of the electrons in the beam, N is the number of undulator periods. For SR from a bending magnet, the intensity is roughly proportional to the number of electrons emitting photons in a bending magnet. Wiggler, or strong undulator, is essentially a number of bending magnets, where electrons deviate significantly from the axis; the character of the radiation remains that from the bending magnets, but the intensity is N times higher. The radiation is incoherent. In a weak undulator, where the electrons slightly oscillate in transversal to the axis plane, so that the cone of the emission of the radiation always includes the undulator axis, the radiation of one single electron is coherent along the undulator axis, but the radiation of the bunch of electrons is incoherent in between them. The intensity of the radiation is proportional to N^2. In free-electron laser, the electrons within a bunch emit largely coherent radiation, which transforms into a significant increase of the intensity $N_f \sim (N_e N)^2$.

New type of radiation source—X-ray free-electron laser (X-FEL)—provides a combination of sub-picosecond pulse duration of a conventional laser with the X-ray wavelength of a synchrotron radiation source. Practical research potential of short wave radiation in nature and in technology can hardly be overestimated. Indeed, wavelength about 200 nm allows study of viruses, the scale of an atomic corral is ~14 nm, and the wavelength of 1–2 nm gives the potential to resolve DNA helix width and carbon nanotubes, while shorter wavelengths could visualize the small molecules, such as water molecule, and atoms. On the other hand, studies of processes at mircoscale often require very short time resolution. Indeed, the processes of a water molecule dissociation takes as short time as 10 fs, Bohr period of valence electron is ~1 fs, shock wave propagates by 1 atom in 100 fs, and electron spin processes in the magnetic field of 1 T within 10 ps. Thus, it is of great practical importance to have a device, which produces radiation with a combination of very short wavelength and short duration. One of the most prominent devices of such a kind is a free-electron laser.

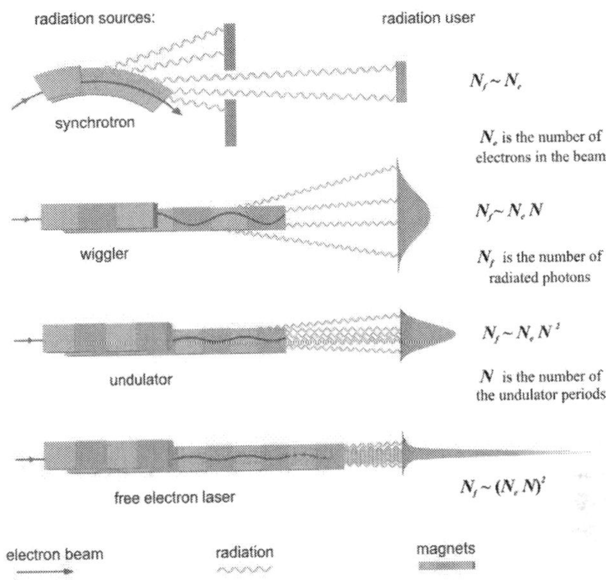

Figure 3. Different sources of SR: synchrotron, wiggler, undulator, and free-electron laser.

3. Keynotes on free-electron lasers

In 1971, Madey published a theory of the FEL [47], where he described a small gain process in a system: relativistic electron beam/undulator. In his study he hypothesized that it could generate coherent X-ray radiation. Few years later the first demonstration of FEL amplification and lasing was performed in a low-gain infrared oscillator FEL at Stanford. At approximately the same time, Colson and Hopf described classically the FEL interaction, which had god quantum description by Madey before. Since 1970s, extensive classical description of the high-gain regime of FEL operation has been developed. Since the first X-rays were discovered by Wilhelm Röntgen in Würzburg in 1895, the peak "brilliance" of X-ray sources has increased by ~16 orders of magnitude. The so-called first-generation synchrotron sources were in fact particle accelerators, designed for experiments in high-energy physics. The second-generation labs were custom-built facilities, while the third-generation sources at labs such as the European Synchrotron Radiation Facility (ESRF) feature undulators, emitting light in a very narrow cone. The fourth generation of synchrotron-radiation sources includes free-electron lasers.

To account for the processes, which occur in a free-electron laser, we need to understand the fundamental difference between the coherent and non-coherent radiation, emitted by many electrons, which pass through undulator at the same time. The power, emitted by electrons

$$P \propto \left| \sum_{j=1}^{N_e} E_j e^{i\varphi_j} \right|^2 = \sum_{j=1}^{N_e} E_j^2 + \left| \sum_{j=1, j \neq k}^{N_e} \sum_{k=1}^{N_e} E_j E_k e^{i(\varphi_j + \varphi_k)} \right|^2 , \tag{9}$$

where φ_j are the relative phases of the emitted radiation electric fields E_j in a system of big number of electrons $N \gg 1$, includes two terms. For a system with uncorrelated phases, the terms in the second double sum, which is of the order of $\sim N^2$, tend to destructively interfere. This happens in incoherent spontaneous UR sources. Total power emitted then approximately equals to the sum of the powers from the N independent scattering electrons, which originates from the first term in Eq. (9). For correlated phases of the electric fields, we have $\varphi_j \approx \varphi_k$ for all the electrons and then the coherent second term $\sim N^2$ contributes. For it to happen, the electron sources must be periodically bunched at the resonant radiation wavelength. **Figure 4** demonstrates how incoherent radiation from a bunch of electrons in an undulator becomes coherent toward the undulator's end. Indeed, at the beginning, the electrons in the bunch enter the undulator with initially random phases, which ensures that mostly incoherent radiation is emitted at the resonant radiation wavelength. Because of the electrons interact collectively with the radiation they emit, small coherent fluctuations in the radiation field grow along the undulator length and simultaneously begin to bunch the electrons at the resonant wavelength. This collective process continues until the electrons are strongly bunched toward the end of the undulator, where the process saturates and the electrons begin to de-bunch.

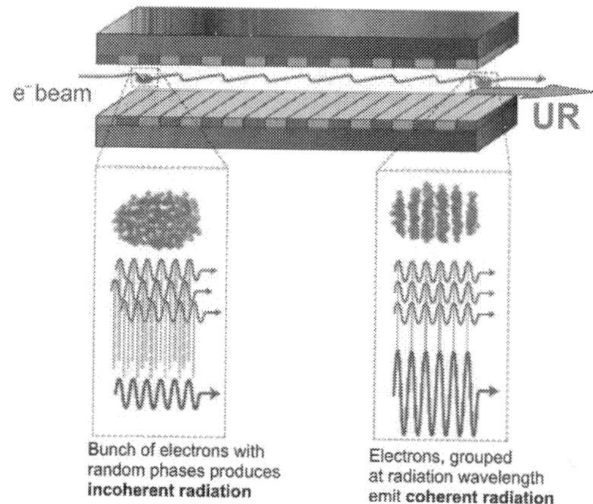

Figure 4. From incoherent to coherent emission along the undulator length.

Behind this phenomenon stands a simple physical mechanism, which is based on the fact that the speed of electrons, while being close to that of the radiation, is still smaller, and, therefore, the electrons appear behind its radiation, propagating in the undulator. It can be best illustrated in **Figure 5**.

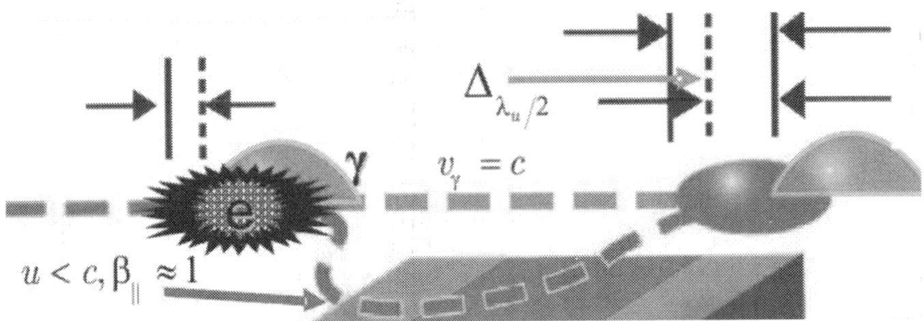

Figure 5. Slippage of the radiation and the electron bunch in FEL.

Indeed, the photons move at the speed of light $v_y = c$, while the electrons move at the speed $u < c$, $\beta_{\|} \approx 1$, i.e., slower than photons. Therefore a full slip between the electron bunch and the photon pulse accumulates along the length of the undulator and reads as follows:

$$\Delta = (1 - \beta_{\|}) N \lambda_u \cong N\lambda, \tag{10}$$

where λ is the emitted wavelength $\lambda = \frac{\lambda_u}{2\gamma^2}\left(1 + \frac{k^2}{2}\right)$ and the parallel component of the β is $\beta_{\|} \cong 1 - \frac{1}{2\gamma^2}\left(1 + \frac{k^2}{2}\right)$. Thus the slip on the whole undulator length is N times the emitted wavelength. Respectively, the slip on a distance, equal half an undulator period equals half of the emitted wavelength:

$$\Delta_{\lambda_u/2} = N\lambda/(2N) = \lambda/2. \tag{11}$$

In other words, as the radiation wave travels over a distance $\lambda/2$ in a time $\lambda/(2c)$, the electron travels over a smaller distance $u\lambda/(2c)$, and on one undulator period the electrons slip, respectively, to the photons by one emitted wavelength. The whole photon packet, having a higher velocity that the electrons, slips over the electron packet and the wave emitted by the electrons on a certain undulator period comes in phase with the wave, emitted on the next undulator period. To illustrate the formation of microbunches, which lead to coherent radiation of the electrons within a bunch, we present the following explanation in **Figure 6** (see [48]).

Suppose we have the magnetic field B_w of the already existing electromagnetic wave. Then its interaction with the electron transverse velocity v_t creates Lorentz force F_{bunch}, which we denote as f for brevity (see **Figure 6**). This force pushes the electron toward a wave node as seen in **Figure 6**. After the electron travels over one-half undulator period, its transverse velocity becomes reversed. In the meantime the electromagnetic wave travels ahead of the electron by one-half wavelength as follows from (10) and (11). Its field B_w is now also reversed, so that the

Lorentz force keeps its direction and the microbunching continues. For the charge, which is ahead with respect to the microbunch, which is in the wave node (see **Figure 6**), the Lorentz force F_{bunch} or f pushes the charge back to the node, thus grouping electrons in a bunch at the wave node. So it proceeds on other periods, because while the electrons move through a full oscillation period λ_w, the electromagnetic wave propagates by λ_u plus one wavelength λ. Consequently, the transverse movement of each electron has a constant phase with respect to the electromagnetic field.

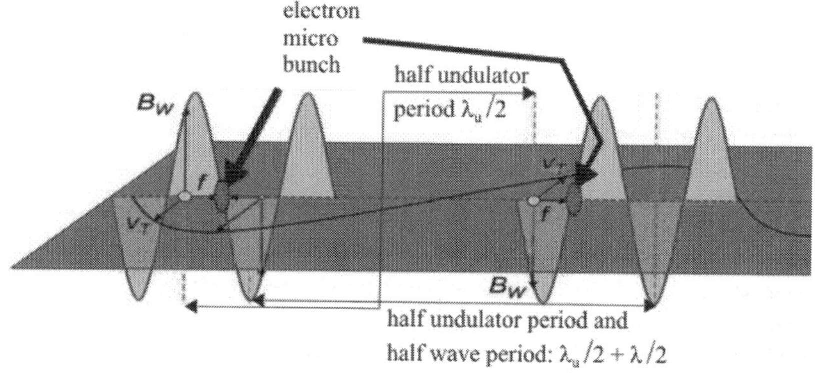

Figure 6. Interaction of the radiation with the electrons and formation of microbunches.

4. Correlated amplification in FEL

Amplification in FEL occurs because of the energy is transferred from the electrons to the previously emitted waves. This effect is due to the negative work of the force, produced by the transverse electric field of the wave, since the magnetic field of the wave does not work. The time rate of the energy transfer for a single electron is proportional to the product $E_W v_t$, where E_W is the electric field of the radiation wave and v_t is the electron transverse velocity.

Then $E_W \propto I^{1/2}$, where I is the wave intensity and $dI/dt \propto I^{1/2} v_t$. Note that uncorrelated combination of the effects of individual electrons would not lead to an exponential increase of the intensity with the distance (or time), but to a quadratic law: $I(z) \propto z^2$. Transverse velocity and the field B produce longitudinal Lorentz force $F_{bunch} = v_t B_W$, which pushes the electrons and forms microbunches. This force is proportional to the transverse electron velocity, and it is also

orthogonal to the wave field B of the strength B_W. Since $B_W \propto I^{1/2}$, the microbunching force is

also proportional to $I^{1/2}$: $B_{bunch} \propto I^{1/2}$. We now assume that this force enhances the correlated emission by a factor, proportional to the microbunching force. Multiplied by the energy transfer rate for each electron, this factor gives $\frac{dI}{dt} = AI$, A = constant. Proceeding on the supposition that $A = u/L_G$, where L_G is the gain length, we obtain the exponential growth:

$$I = I_0 \exp\left(\frac{ut}{L_G}\right) = I_0 \exp\left(\frac{z}{L_G}\right). \tag{12}$$

The exponential gain only occurs after bunching is established. This process continues until the saturation is reached. Provided at the beginning, the initial position of the electron, respectively, the existing wave, is favorable for the energy transfer from the electron to the wave and the direction of the electron transverse velocity, respectively, to the wave electric field E result in negative work, the electron energy is successfully transferred to the wave. This results in the decrease in the electron energy and of the longitudinal speed of the electron u, which becomes $u - \Delta u$. This decrease in longitudinal speed in its turn changes the above conditions, making them less favorable for the energy transfer from the electron to wave. At a certain point, as this process continues and the electron speed reduction Δu becomes more and more significant, the electrons can give no more energy to the wave and the wave starts giving its energy to the electrons of the beam. It increases the electron speed u and restores the favorable for the energy transfer from the electrons to the wave conditions. Thus, in this regime the exponential wave energy growth is over and instead the energy oscillates between the wave and the electrons.

To find the amplification value, we must evaluate v_t and the degree of bunching. The transverse speed of the electron reads as follows:

$$v_t = \left(\frac{euH_0}{\gamma m_0}\right)\left(\frac{\lambda_u}{2\pi u}\right)\cos\frac{2\pi ut}{\lambda_u}. \tag{13}$$

The energy transfer rate by a single electron to preexisting wave is proportional to $I^{1/2}H_0\lambda_u/\gamma$; the field B_W of the electromagnetic wave is proportional to the square root of the wave intensity $B_W \propto I_0^{1/2}\exp(ut/2L_G)$. Then the bunching force, which is the longitudinal force, can be written as follows:

$$F_{bunch} = \text{const} \times \left(\frac{H_0\lambda_u}{\gamma}\right)I_0^{1/2}\exp\left(\frac{ut}{2L_G}\right), \tag{14}$$

which, in turn, yields the following equation of motion for longitudinal mass $\gamma^3 m$:

$$\gamma^3 m \frac{d^2 \Delta x}{dt^2} = F_{\text{bunch}} = \text{const} \times \left(\frac{H_0 \lambda_u}{\gamma} \right) I_0^{1/2} \exp\left(\frac{ut}{2L_G} \right). \tag{15}$$

The bunching force F_{bunch} induces a small longitudinal electron displacement Δx:

$$\Delta x = \text{const} \times \frac{1}{\gamma^3} \frac{H_0 \lambda_u}{\gamma} L_G^2 I_0^{1/2} \exp\left(\frac{ut}{2L_G} \right) = \frac{H_0 \lambda_u L_G^2}{\gamma^4} I^{1/2}. \tag{16}$$

As discussed above, the electrons are concentrated in narrow slabs, separated from each other by a distance of the wavelength λ. The degree of microbunching, corresponding to the fraction of electrons that emit in a correlated way, can be assumed to be proportional to $(\Delta x/\lambda)$. The corresponding number of electrons is proportional to $N_e(\Delta x/\lambda)$. Their contribution to the wave intensity reads as follows:

$$j\frac{\Delta x}{\lambda} \propto j\frac{I^{1/2}H_0\lambda_u L_G^2 / \gamma^4}{\lambda_u / \gamma^2} = jI^{1/2}H_0 L_G^2 / \gamma^2, \quad j \equiv \frac{i}{\Sigma} = \frac{\text{current}}{\text{cross-section}}. \tag{17}$$

Microbunching effects correspond to a factor, proportional to the longitudinal microbunching force F_{bunch} and, therefore, proportional to $\sim I^{1/2}$, where I is the electromagnetic wave intensity. Multiplying this factor by the energy transfer rate for electrons, we obtain the total transfer rate as follows:

$$\frac{dI}{dt} = \text{const} \times j\frac{H_0 L_G^2}{\gamma^2} I^{1/2} \left(I^{1/2} \frac{H_0 \lambda_u}{\gamma} \right) = \text{const} \times j\frac{H_0^2 \lambda_u L_G^2}{\gamma^3} I. \tag{18}$$

The solution of this equation reads $I = I_0 \exp(z/L_G)$ and for ultrarelativistic electrons $u/L_G \cong c/L_G \propto jH_0^2 \lambda_u L_G^2 / \gamma^3$. Thus we obtain the FEL gain length

$$L_G = \text{const} \times j^{-1/3} H_0^{-2/3} \lambda_u^{-1/3} \gamma. \tag{19}$$

Gain length is related to the Pierce parameter as follows:

$$\rho = \frac{\lambda_u}{4\pi\sqrt{3}L_G} \propto j^{1/3} H_0^{2/3} \lambda_u^{4/3} \gamma^{-1} \tag{20}$$

which can be also written as $\rho = \frac{1}{2\gamma}\left(\frac{I_e}{I_A}\left(\frac{\lambda_u \check{k} f_B}{2\pi\sigma_r}\right)^2\right)^{1/3}$, where $I_A \cong 17$ kA is the Alfven current, I_e is the electron beam current, σ_r is the beam radius, $f_B(J_n)$ is the bunching coefficient, and J_n are the Bessel functions. The arguments and the varieties of the Bessel functions, in charge of the description of the harmonics of the UR, will be explored in what follows.

5. Cavity-based FELs

There are essentially two main types of design for FEL: cavity-based FELs and single-pass FELs. Cavity-based FELs exploit many passes of the UR in the undulator and use mirrors. Relativistic electron beam passes through periodic magnetic field of an undulator and the mirrors feed spontaneous emission back onto the beam. Consequently, the spontaneous emission is enhanced by the stimulated emission.

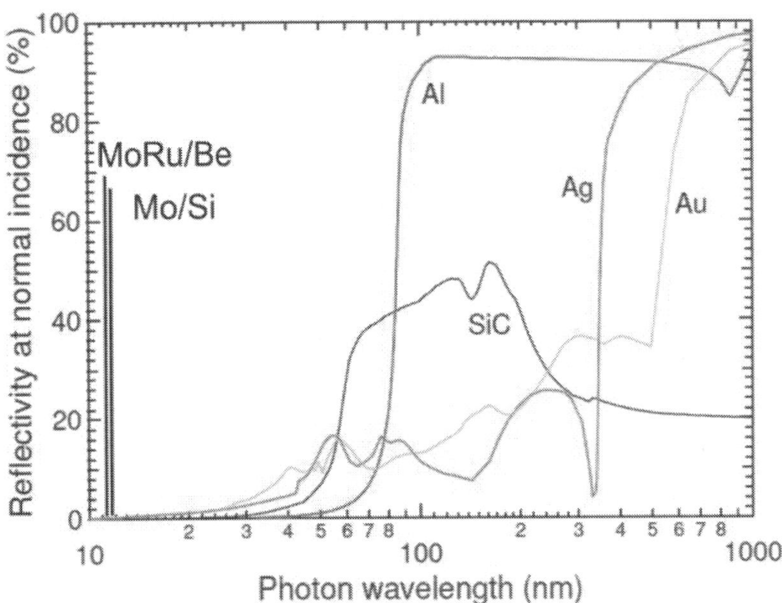

Figure 7. FEL mirror limitations.

Mirror design imposes limitations on the wavelength due to the mirror material. Common mirror materials limit the wavelength to IR-UV (see **Figure 7**). Recent improvements in materials and technology result in new multilayer X-ray reflective mirrors, which have the efficiency up to 70%, such as MoRu/Be mirrors have 69.3% reflectivity maximums at 11.43 nm, Mo/Si mirrors have 67.2% reflectivity peak at 13.51 nm.

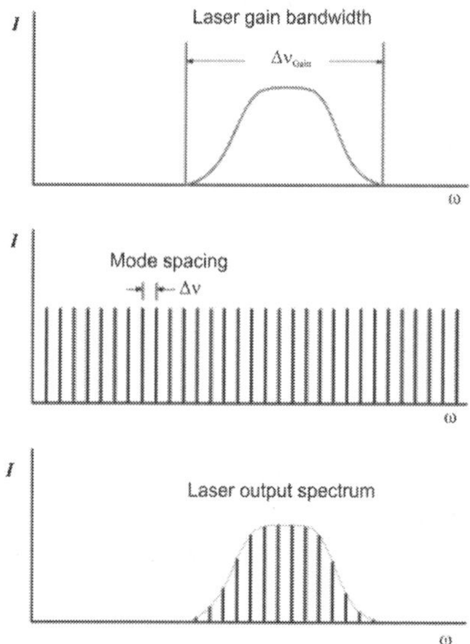

Figure 8. Longitudinal mode structure in mirror-based FEL design.

The mirror design has significant advantage: mirrors naturally select longitudinal modes, which are determined by the integer number of half-periods of the wavelength in between them. The longitudinal modes in optical cavity are equally spaced in frequency and time (see the middle plot in **Figure 8**):

$$\Delta v = \frac{c}{2L}, \Delta \tau = \frac{1}{\Delta v}. \tag{21}$$

The interval between them depends on the cavity length L (see the middle plot in **Figure 8**).

Taking into account the laser gain bandwidth

$$\Delta v_{Gain} \approx \frac{1}{\tau_{Pulse}} = \frac{c}{2\Delta} \tag{22}$$

where $\Delta = N\lambda$ (see (10), see the top plot in **Figure 8**) and, imposing it on the longitudinal picture of the middle plot in **Figure 8**, we obtain the laser spectrum as shown in the bottom plot of **Figure 8**. Inside the gain bandwidth, the longitudinal modes of the laser will periodically and constructively interfere with each other when in phase, producing an intense burst of light. These light bursts are separated by the period of time $\tau = 2L/c$, necessary for the light to make exactly one round trip of the laser cavity; the comb of equally spaced modes in the output spectrum with spacing

$$\Delta \nu_{\text{RoundTrip}} = \Delta \nu = \frac{1}{\tau_{\text{RoundTrip}}}, \text{ where } \tau_{\text{RoundTrip}} = \frac{2L}{c} \tag{23}$$

and L is the cavity size. In other words, only phase-matched wavelengths will constructively interfere and form the modes of the radiation field to create a comb of equally spaced modes in the output frequency spectrum. Such a laser is said to be mode-locked or phase-locked. Mode locking modifies the temporal envelope of the output field from a continuous wave to a series of short, periodically spaced pulses. The homogeneous gain bandwidth of a FEL is determined by the slippage length $\Delta = N\lambda$ (10). This last plays central role in FEL physics. Indeed, it is easy to see that only finite number of longitudinal modes with positive gain N_{Gain} exist: $\Delta \nu_{\text{Gain}} \approx \frac{1}{\tau_{\text{Pulse}}} = \frac{c}{2\Delta}$ (see (21)–(23)), where $\Delta = N\lambda$, λ is the radiated wavelength, and N is the number of the undulator periods. Therefore the number of gained modes reads as follows:

$$N_{\text{Gain}} = \frac{\Delta \nu_{\text{Gain}}}{\Delta \nu_{\text{RoundTrip}}} \approx \frac{L}{\Delta}, \tag{24}$$

and the duration of the pulse is evidently

$$\tau_{\text{Pulse}} \approx \frac{1}{\Delta \nu_{\text{Gain}}} = \frac{\tau_{\text{RoundTrip}}}{N_{\text{Gain}}}. \tag{25}$$

For a Gaussian-shaped pulse, we have $\tau_{\text{Gaussian Pulse}} \cong \frac{0.44}{\Delta \nu_{\text{Gain}}}$. Thus, a laser macro pulse of, to say, microsecond duration consist of a train of short micro pulses, which are picoseconds or less in length. Such a short duration of the pulses is useful for studies of ultrashort physical and chemical processes. It is used for ultrafast spectroscopy and in femto-chemistry. The macro pulses repeat at a repetition ratio, limited by the accelerator usually at 10–100 Hz. The micro pulse repetition rate can vary from several MHz to even THz.

We omit in this work the field gain equations, which are well known and can be found elsewhere, and without derivation we just state the formulation of the Madey's theorem [47], which claims that the gain in the so-called Compton or low-gain regime is proportional to the slope of the spontaneous UR spectrum $f(\nu)$: $G(\nu) \propto \frac{\partial f(\nu)}{\partial \nu}$, $f(\nu) = sinc^2(\nu)$, and, therefore,

$$G(\nu) = -\frac{j}{2}\partial_\nu \left[\sin\frac{\nu}{2} \Big/ \frac{\nu}{2} \right]^2, \tag{26}$$

where j is the current density, ν is the detuning parameter (4). **Figure 9** demonstrates the FEL gain; the homogeneous bandwidth is given by $\frac{\Delta\omega}{\omega_0} = \frac{1}{2N}$. In the presence of constant magnetic field, the line shape is described by the Airy-type function $S(\nu_n,\beta) = \int_0^1 d\tau e^{i(\nu_n\tau+\beta\tau^3)}$ and the derivative modifies into $-\partial S^2/\partial\nu_n$. The commonly known shape $-\partial(sinc^2\nu_n)/\partial\nu_n$ is seen in **Figure 9** in the rear vertical plane, where $\beta = 0$. The constant magnetic field $H_d = \kappa H_0$, where H_0 is the amplitude of the undulator periodic field, produces the bending angle $\theta_H = \frac{2}{\sqrt{3}}\frac{k}{\gamma}\pi N \kappa_1$, resulting in nonzero values of $\beta = \frac{(2\pi n N + \nu_n)(\gamma\theta_H)^2}{1+(k^2/2)+(\gamma\theta_H)^2}$ [35, 37, 39] and shifts the maximum of the curve to lower frequencies (see **Figure 9**).

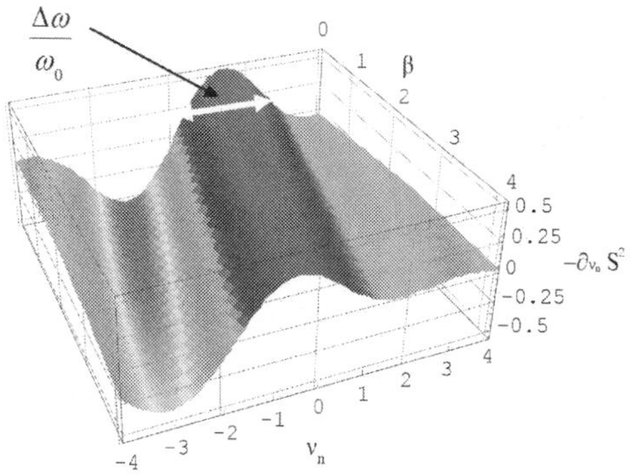

Figure 9. Function $-\dfrac{\partial(S^2(\nu_n))}{\partial\nu_n}$, which describes FEL gain in an undulator in external field.

6. High-gain single-pass FELs

There are several reasons why it is more difficult to build free-electron lasers for X-ray region that is for larger wavelengths. First of all, for small wavelengths we need high-energy electrons, but high electron energy also increases the gain length $L_G \propto \gamma$, and we must keep the gain length short, as required for an X-FEL. Then, the undulator parameters H_0 and λ_u must be maximized, keeping in mind, however, that λ_u also determines the radiation wavelength: $\lambda \propto \lambda_u/\gamma^2$. Then, the electron beam current i must be high and its transverse cross section σ small. However, the γ-factor cannot be freely decreased if we want to obtain X-ray wavelengths. These challenges can be better faced in single-pass FELs with typical high-gain regime.

In the high-gain regime the radiation power increases exponentially as the electron beam and radiation co-propagate along the FEL undulator, and this happens on a single pass of the radiation along the FEL. These kinds of FELs are sometime called amplifiers, since everything starts from an initially small source, which may originate as noise, and it can be amplified by many orders of magnitude before the process saturates. In such FEL, there are no mirrors to form an oscillator cavity and this is particularly good for X-ray region, where mirrors are the most compromised element of the cavity-based FEL. Such FEL essentially works as an amplifier, in which the radiation forms on the single pass through a very long undulator, reaching peak pulse power ~ 10^{10} W for few dozens of femto-seconds. Overwhelming majority of current X-ray FELs are based on this type of design, which has been made possible due to the twenty-first century advances in magnet technology, accelerator constructions, and electron beam production.

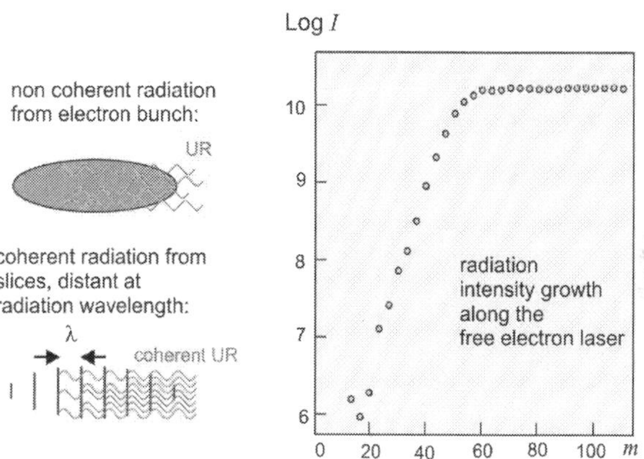

Figure 10. Radiation mechanism in a SASE FEL (amplifier).

The principle of the high-gain FEL interaction is based on the positive feedback process. The electrons emit undulator radiation, which corrects their position in space and their phase with respect to the electromagnetic wave; this groups the electrons on the radiation wavelength scale and thus the more and more coherent radiation is emitted along the undulator. First, the electrons in the bunch have random phases and produce incoherent emission. Already the first waves, emitted by these electrons, trigger formation of microbunches as discussed above (see **Figures 4, 6** and **10**). Contrary to non-micro-bunched electrons, which emit incoherent waves, the emission of electrons, collected in micro-bunches, which are separated from each other by one wavelength, is correlated (see **Figures 4** and **10**). This causes an exponential intensity increase with the distance that continues until saturation is reached (see **Figure 10**). The schematic drawing in **Figure 10** represents modeling of the performance of the Linac Coherent Light Source—LCLS, with the parameters of the undulator: L = 100 m, λ_u = 3 cm, L_G = 3.3 m, λ = 1.5Å = 0.15 nm. The whole installation has the length of L = 1 km, current I = 3 kA, E = 13.6 GeV, Linac and bunch compression: $\gamma\varepsilon_{x,y}$ = 0.4 μm (slice), σ_E/E = 0.01% (slice).

Simulation of the electron density in a bunch of the electron beam as it develops from the entrance toward the exit of the undulator is demonstrated in **Figure 11**. Left picture simulates the electron density in the bunch at the beginning of the undulator, middle picture represents the simulation in the middle of the undulator length, and the right picture demonstrates the electron density in the bunch at the end of the undulator.

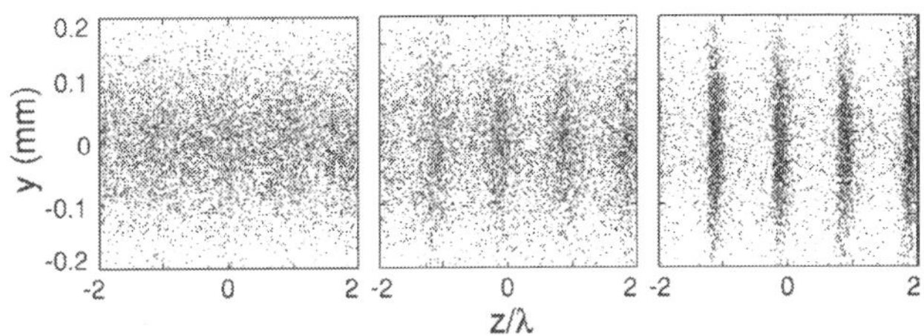

Figure 11. Simulation of the density modulation of the electron beam along the undulator: undulator beginning—left picture, undulator middle—middle picture, undulator end—right picture.

The transverse structure of the electron bunch is much larger than its longitudinal substructure. Note the length between the slices can be of the order of nm, and there can be up to hundreds of thousands of slices in a bunch.

7. Some advanced single-pass FEL schemes

Perhaps, the most common development of the basic SASE FEL scheme, where the amplification starts from noise with random phase, is represented by SASE FEL with seeding. In this case, already at the beginning of the undulator, unbunched electrons with random phase interact with coherent laser seed radiation, which bunches them accordingly. This scheme has the advantage of the stability of the phase, because the process is controlled by the seeding laser.

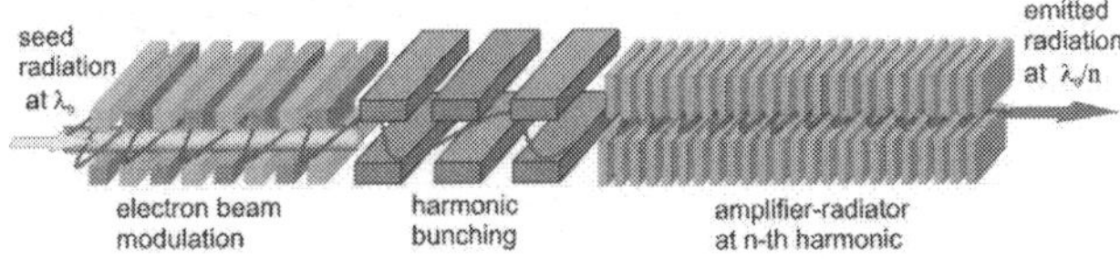

Figure 12. HGHG FEL with laser seed, harmonic generator, buncher, and amplifier.

To achieve extremely high frequencies, for example, those of the X-ray band, the following modification of SASE FEL with high-gain harmonic generation (HGHG) is sometimes employed (see schematic drawing in **Figure 12**).

It consists of the seed laser, harmonic generator, and the amplifier. The coherent seed laser radiation first passes the short undulator, called modulator, which is tuned to the seed frequency. The interaction with the electron beam gives the latter small longitudinal energy modulation. The following section of the installation converts this energy modulation into a beam density modulation in a magnetic dispersion unit. Then the modulated electron beam and the UR pass the second undulator, which is tuned to the nth harmonic of the modulator. When they pass the second undulator, the nth harmonic of the UR is fully amplified to saturation levels in this second undulator, frequently called radiator. At the exit the pulses with <20 fs duration can be obtained.

One of the most important advantages of the mirror FEL is the possibility to select optical modes with the help of the mirrors. SASE FEL in its classical scheme is lacking this ability: there are no mirrors, which, in turn, gives other advantages. Together the advantage of the cavity-based mirror design with those of SASE FEL, the magnetic chicanes (small blocks in **Figure 13**) can be introduced between a sequence of undulators (big blocks in **Figure 13**) to impose a sort of mode locking in mirrorless single-pass FEL. The proper scheme is given in **Figure 13**.

Figure 13. SASE FEL design with chicane mode-locking (big blocks are undulators, small blocks are chicanes).

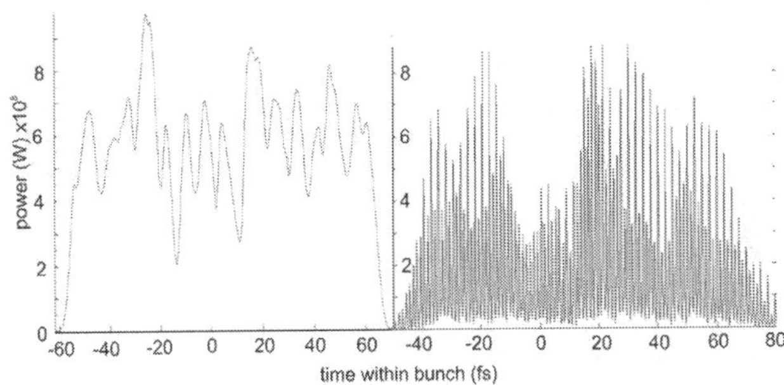

Figure 14. SASE output power after the undulator without (left) and with (right) chicanes.

The magnetic chicanes introduce an extra slippage of the radiation with respect to the electron bunch. Only those radiation wavelengths that have an integer number fit into the relative slippage of the radiation with respect to the electron bunch in one module will remain phase-matched. Only they will constructively interfere over many such modules. In this way form the modes of the radiation field, which create a comb of equally spaced modes in the output frequency spectrum. Such a mode locking can be achieved by the beam energy modulation or by a beam current modulation of the same period. The result is similar to that created by the optical mode locking. We present the example of the power output and the spectrum of a common SASE FEL (see **Figures 14** and **15** left, respectively) to compare it with the example of the power output and of the spectrum of a FEL with chicane mode locking (see **Figures 14** and **15** right, respectively). Note that instead of the relatively broad SASE FEL spectrum, we now see the series of sharp equidistant peaks. The same regards the output power. At the end of the undulator-chicane line we achieve higher spectral power from the FEL.

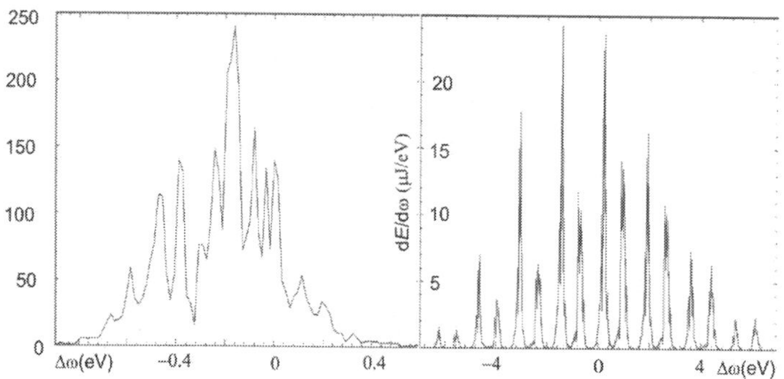

Figure 15. SASE spectrum after the undulator without (left) and with (right) chicanes.

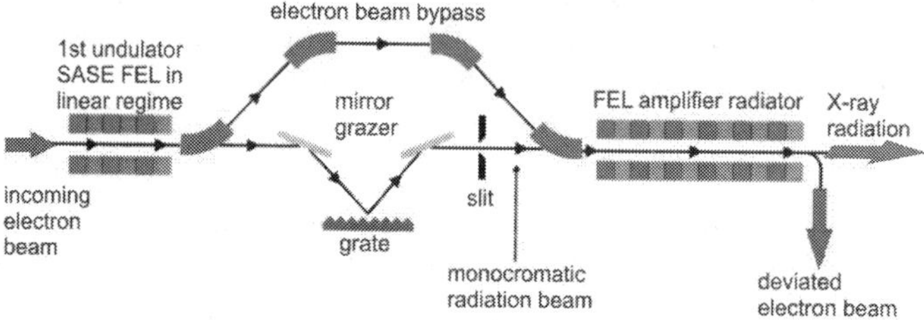

Figure 16. Schematic design of a self seed FEL.

Eventually, we touch on the functioning of the so-called self-seed FEL scheme, demonstrated in **Figure 16**. The advantage of this type of FELs consists in that they are independent of any external radiation source, which must be otherwise very stable and precisely matched to the electron beam in space and time. Both undulators in the self-seed design are tuned to the same frequency. First undulator is in essence a short SASE FEL, which operates in the linear gain regime far from the saturation level. It produces the radiation in the form of typical SASE FEL pulses at the power level approximately 1000 times below the saturation level. Then the electron beam is fed through a magnetic chicane, which eliminates the density modulation, introduced in the first undulator, and delays the electron bunch to match the UR pulse at the next undulator. The radiation from the first undulator, which works in linear regime, is spectrally filtered by a narrow-band grating monochromator. The latter stretches the radiation pulse, where the coherence length now exceeds that of the electron bunch. After all, the reshaped radiation pulse and the delayed electron bunch meet. The filtered UR becomes the seed for the second undulator—radiator—which amplifies the UR to saturation level.

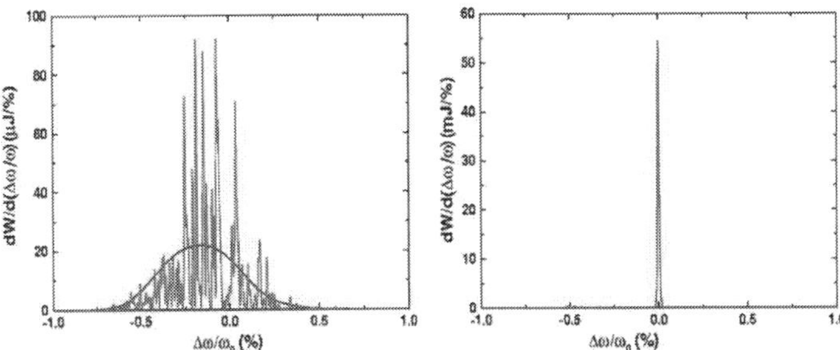

Figure 17. Spectrum after the first SASE FEL and after the second undulator-amplifier.

Not only this FEL is independent on any external radiation source, but due to its design it also produces remarkably good radiation pulses with given characteristics.

In **Figure 17** we demonstrate the example of the output characteristics of such a self-seed FEL. Note that the spectral power distribution after the first (left) and the second undulator (right) differ from each other. The common SASE FEL spectrum (left) is rather broad and consists of many spectral lines even far below the saturation level. The output radiation on the contrary represents a narrow spectral line with only a small background of spontaneous radiation (right). Spectral brightness has increased by almost two orders of magnitude.

8. Challenges and future developments of X-ray FELs

To conclude the review of some of the most prominent for X-FEL schemes, let us formulate main challenges for the X-ray FEL:

1. The radiation wavelength $\lambda = \frac{\lambda_u}{2\gamma^2}\left(1 + \frac{k^2}{2}\right)$ should be reduced to 1 Å. For small wavelengths we need high-energy electrons, but high electron energy also increases the gain length L_G. Longer undulator period λ_u also determines longer radiation wavelength λ and longer gain length L_G.

2. The electron beam energy has to be 10–20 GeV. There is a compromise to contain the gain length and obtain short wave radiation.

3. The gain length $L_G = \frac{1}{\sqrt{3}}\left(\frac{4mc}{\mu e}\frac{\gamma^3\lambda_u}{k^2}\frac{\sigma_r^2}{I}\right)^{1/3}$ has to be contained preferably about of 10 m or so.

    Note that $\frac{\sigma_r^2}{I} = \frac{\sigma_r^2 L_{bunch}}{qc}$, where L_{bunch} is the longitudinal size of the bunch and q is the electric charge.

4.   The electron beam current I must be high and its transverse cross section σ small. However, the γ-factor cannot be freely decreased if we want to obtain X-ray wavelengths.

5.   As follows from the expression for L_G high peak current is requested for it, such as $I \sim$ kA or more.

6.   Transverse electron beam size has to be small, possibly $\sigma_R \sim 10$ μm or so.

7.   The energy spread of the electron beam σ_e has to be as small as possible, preferably $\sigma_e \approx 10^{-4}$ or less.

8.   The electron and the photon beams have to be overlapped properly.

Note also that in the light of the above said the following requirements arise for the electron beam in transverse: low emittance of the beam and preservation of this low emittance; for the longitudinal dimension good compression and acceleration are required. The main negative factors, which affect the amplification, are the electron energy spread, the angular divergence, the transverse electron beam size, the diffraction of the wave and others. The electron energy spread has negative effect on both the amplification and FEL saturation level. Amplification mainly starts with the optimal electron energy, whose γ-factor determines the wavelength. As the energy is transferred from the electrons to the radiated electromagnetic wave, the energy of the electrons naturally decreases. The wave emissions from all the electrons differ from each other, because different electrons have different energies. After the wave-electron interaction, the electron beam energy spread increases and at a certain point it grows to a level, where no gain occurs. Moreover, well before the electrons loose a substantial portion of their energy, they slow down by emitting electro-magnetic energy and change their phase with respect to the wave. Thus they begin to take the energy from the wave rather than giving it.

In conclusion, let us state some areas, where the performance of the X-ray sources of coherent radiation can be further improved. First of all, the temporal coherence of SASE FELs can be improved. The improved temporal coherence would in turn improve the spectral brightness of the sources, which means the users will have more useful photons. The way to accomplish it could consist, for example, in seeding X-FEL from a radiation source with good temporal coherence.

An alternative approach to single-pass high-gain amplifier schemes is to use cavity feedback in a relatively low-gain system. The development of relatively high-reflectivity diamond crystal mirrors in the X-ray regime makes them feasible.

Reducing X-ray pulse durations to the attosecond regime will provide spatiotemporal resolution of atomic processes. Two techniques have so far been reported that can take pulse durations significantly below 100 as toward the atomic unit of time 24 as. The first technique employs a variation of an echo-enabled harmonic generation method and produces pulses of ~20 as duration at a wavelength of 1 nm with the power, which in peaks reaches ~200 MW. The second technique is based on mode-locking in conventional cavity lasers—oscillators. It could generate radiation at 1.5 Å wavelength in sequences of pulses with ~150 as intervals between them. The peak power in each pulse could reach ~5 GW and the duration of the pulse could be as short as ~20 as.

Eventually, γ-ray FEL would be extremely interesting for studying nuclear processes. This may be the future of X-FEL.

Author details

K. Zhukovsky

Address all correspondence to: zhukovsk@physics.msu.ru

Department of Theoretical Physics, Physical Faculty, M.V. Lomonosov Moscow State University, Leninskie Gory, Moscow

References

[1] Ginzburg VL. "On the radiation of microradiowaves and their absorbtion in the air". *Isvestia Akademii Nauk SSSR (Fizika)*, vol. 11, N 2, 947, 1651.

[2] Motz H, Thon W, Whitehurst RNJ. "Experiments on radiation by fast electron beams". *Appl. Phys.*, vol. 24, 826, 1953.

[3] Artcimovich AL, Pomeranchuk IJ. "Radiation from fast electrons in a magnetic field". *J. Phys. USSR*, vol. 9, 267, 1945.

[4] Ternov IM, Mikhailin VV, Khalilov VR. *Synchrotron Radiation and its Applications*, Harwood Academic Publishers, Chur, London, Paris, New York, 1985.

[5] Alferov DF, Bashmakov YuA, Bessonov EG. "Radiation of relativistic particles in an undulator". *Sov. Phys. Tech. Phys.*, vol. 17, N 9, 1540, 1973.

[6] Alferov DF, Bashmakov UA, Cherenkov PA. "Radiation from relativistic electrons in a magnetic undulator". *Uspehi Fisicheskih Nauk*, vol. 32, 200, 1989.

[7] Bordovitsyn VA (Ed.), *Synchrotron Radiation Theory and its Development: in the Memory of I.M.Ternov*. Series on Synchrotron Radiation Technique and Applications, vol. 5, World Scientific Publishing, Singapore, 1999.

[8] Bessonov EG, Gorbunkov MV, Ishkhanov BS, Kostryukov PV, Maslova Yu YA, Shvedunov VI, Tunkin VG, Vinogradov AV. Laser-electron generator for X-ray applications in science and technology, *Laser Part. Beams*, vol. 26, N 3, 489–495, 2008.

[9] Mcneil BWJ, Thompson NR. X-ray free-electron lasers, *Nat. Photon.*, vol. 4, N 12, 814–821, 2010. doi: 10.1038/nphoton.2010.239.

[10] Sokolov AA, Ternov IM. *Radiation from Relativistic Electrons* (edited by C.W. Kilmister), American Inst. of Physics, New York, 1986.

[11] Feldhaus J, Sonntag B. Strong field laser physics. *Springer Ser. Opt. Sci.*, vol. 134, 91, 2009.

[12] Zholents AA. "Attosecond X-ray pulses from free-electron lasers". *Laser Phys.*, vol. 15, N 6, 855, 2005.

[13] Bessonov EG. Light sources based on relativistic electron and ion beams. *Proc. SPIE*, vol. 6634, 66340X, 2007.

[14] Zhukovsky KV. "Harmonic radiation in a double frequency undulator with account for broadening". *Moscow Univ. Phys. Bull.*, vol. 70, N 4, 232, 2015.

[15] Zhukovsky K. "Undulator radiation in multiple magnetic fields". *Synchrotron: Design, Properties and Applications*, vol. 39, Nova Science Publishers Inc., USA, 2012.

[16] Zhukovsky K. "Harmonic Generation by ultrarelativistic electrons in a planar undulator and the emission-line broadening". *J. Electromagn. Wave*, vol. 29, N 1, 132–142, 2015.

[17] Tripathi S, Mishra G. "Three frequency undulator radiation and free electron laser gain". *Rom. J. Phys.*, vol. 56, N 3–4, 411, 2011.

[18] Mishra G, Gehlot M, Hussain J-K. "Spectral properties of bi-harmonic undulator radiation". *Nucl. Instrum. A*, vol. 603, 495, 2009.

[19] Dattoli G, Mikhailin VV, Ottaviani PL and Zhukovsky K. "Two-frequency undulator and harmonic generation by an ultrarelativistic electron". *J. Appl. Phys.*, vol. 100, 084507, 2006.

[20] D. Iracane and P. Bamas, "Two-frequency wiggler for better control of free-electron-laser dynamics", Phys. Rev. Lett. 67 (1991) 3086.

[21] V. N. Korchuganov, N. U. Sveshikov, N. V. Smolyakov, C. I. Tomin, "Special-purpose radiation sources based on the Siberia-2 storage ring", *J. Surf. Invest.: X-ray, Synchrotron and Neutron Tech.*, Vol.11, p. 22, 2010.

[22] Walker RP. "Interference effects in undulator and wiggler radiation sources". *Nucl. Instrum. Methods*, vol. A335, 328, 1993.

[23] Onuki H, Elleaume P. *Undulators, Wigglers and Their Applications*, Taylor & Francis, New York, 2003.

[24] Vagin PV, Englisch U, Müller T, et al.. "Commissioning experience with insertion devices at PETRA III". *J. Surf. Invest.: X-ray, Synchrotron Neutron Tech.*, vol. 6, N 5, 1055, 2011.

[25] Hussain J, Gupta V, Mishra G. "Effect of two-peak beam energy spread on harmonic undulator radiation and free-electron laser gain". *Nucl. Instrum. A*, vol. 608, 344, 2009.

[26] Reiss HR. "Effect of an intense electromagnetic field on a weekly bound system". *Phys. Rev.*, vol. A22, 1786, 1980.

[27] Smolyakov NV. "Focusing properties of a plane wiggler magnetic field". *Nucl. Instrum. A*, vol. 308, 83–85, 1991.

[28] Hussain J, Mishra G. "Harmonic undulator radiations with constant magnetic field". *Opt. Commun.*, vol. 335, 126, 2015.

[29] Dattoli G, Mikhailin VV, Zhukovsky KV. "Influence of a constant magnetic field on the radiation of a planar undulator". *Moscow Univ. Phys. Bull.*, vol 64, 507, 2009, c/c of Vestnik Moskovskogo Universiteta Ser. 3 Fizika Astronomiya, vol. 5, 33, 2009.

[30] Mikhailin VV, Zhukovsky KV, Kudiukova AI. "On the radiation of a planar undulator with constant magnetic field on its axis taken into account". *J. Surf. Invest.: X-ray Synchrotron Neutron Tech.*, vol. 8, N 3, 422, 2014.

[31] Dattoli G, Mikhailin VV, Zhukovsky K. "Undulator radiation in a periodic magnetic field with a constant component. "*J, Appl. Phys.*, vol. 104, 124507, 2008.

[32] K. Zhukovsky, "Emission and tuning of harmonics in a planar two-frequency undulator with account for broadening", *Laser Part. Beams*, 2016, doi: 10.1017/S0263034616000264

[33] Mirian NS, Dattoli G, DiPalma E, Petrillo V. "Production and properties of two-color radiation generated by using a Free-Electron Laser with two orthogonal undulators". *Nucl. Instrum. A*, vol. 767, 227, 2014.

[34] Zhukovsky K. "Inhomogeneous and homogeneous losses and magnetic field effect in planar undulator radiation". *Prog. Electromagn. Res. B*, vol. 59, 245, 2014.

[35] Zhukovsky K. "Analytical account for a planar undulator performance in a constant magnetic field". *J. Electromagn. Wave*, vol. 28, N 15, 1869, 2014.

[36] Zhukovsky KV. "A model for analytical description of magnetic field effects and losses in a planar undulator radiation". *J. Surf. Invest.: X-ray Synchrotron Neutron Tech.*, vol. 8, N 5, 1068, 2014.

[37] Zhukovsky K., High harmonic generation in undulators for FEL, *Nuclear Instrum. Methods Phys. Res. B*, vol. 369, 9–14, 2016, DOI: 10.1016/j.nimb.2015.10.041.

[38] Quattromini M, Artioli M, Di Palma E, Petralia A, Giannessi L. Focusing properties of linear undulators, *Phys. Rev. ST Accel. Beams*, vol. 15, 080704, 2012, DOI: 10.1103/PhysRevSTAB.15.080704.

[39] Zhukovsky K. "High harmonic generation in the undulators for free electron lasers". *Opt. Commum.*, 353, 35–41, 2015.

[40] Dattoli G, Srivastava HM, Zhukovsky K. "Orthogonality properties of the Hermite and related polynomials". *J. Comput. Appl. Math.*, vol. 182, 165–172, 2005.

[41] Dattoli G, Srivastava HM, Zhukovsky K. "A new family of integral transforms and their applications". *Integral Transform. Spec. Funct.*, vol. 17, N 1, 31–37, 2006.

[42] Dattoli G, Zhukovsky K. "Evolution of time dependant linear potentials and non-spreading airy wave packets". *Appl. Math. Comput.*, V 217, 7966–7974, 2011.

[43] Zhukovsky KV. "A method of inverse differential operators using ortogonal polynomials and special functions for solving some types of differential equations and physical problems". *Moscow Univ. Phys. Bull.*, vol. 70, N 2, 93–100, 2015, DOI: 10.3103/S00271349150201 37 .

[44] Zhukovsky K. "Solution of some types of differential equations: operational calculus and inverse differential operators". *Sci. World J.*, vol. 2014, 8 pages, 2014, article ID 454865.

[45] Zhukovsky KV. "Operational method of solution of linear non-integer ordinary and partial differential equations". *SpringerPlus*, vol. 5, 119, 2016, DOI: 10.1186/s40064-016-1734-3.

[46] Jackson JD, *Classical Electrodynamics*. 2nd ed., Wiley, New York, 1975.

[47] Madey JMJ. Stimulated emission of bremsstrahlung in a periodic magnetic field. *J. Appl. Phys.*, vol. 42, 1906–1913, 1971.

[48] Margaritondo G, Rebernik Ribic P. A simplified description of X-ray free-electron lasers. *J. Synchrotron Rad.*, vol. 18, 101–108, 2011.

Chapter 9

Phase Manipulation of Ultrashort Soft X-Ray Pulses by Reflective Gratings

Fabio Frassetto, Paolo Miotti and Luca Poletto

Additional information is available at the end of the chapter

http://dx.doi.org/10.5772/63416

Abstract

In this chapter, we discuss the use of reflective diffraction gratings to manipulate the phase of ultrashort pulses in the extreme ultraviolet (XUV) and soft X-ray spectral regions. Gratings may be used to condition the spectral phase of ultrashort pulses, e.g., to compensate for the pulse chirp and compress the pulse, similarly to what is routinely realized for visible and infrared pulses. The chirped pulse amplification technique has been already proposed for soft X-ray free-electron laser radiation; however, it requires the use of a compressor to compensate for the pulse chirp and get closer to the Fourier limit. There are fundamental differences when operating the gratings at wavelengths shorter than ≈40 nm on a broad band: (a) the gratings are operated at grazing incidence; therefore, the optical design has to be consequently tailored to this peculiar geometry; (b) the grating efficiency is definitely lower; therefore, the number of diffractions has to be limited to two. We discuss the different configurations that can be applied to the realization of a grating stretcher/compressor.

Keywords: diffraction gratings, ultrafast optics, extreme ultraviolet, soft X-ray optics, chirped pulse amplification

1. Introduction

XUV and X-ray radiations have been used for many fundamental discoveries and outstanding applications in natural sciences. It has played a crucial role in basic research, medical diagnostics, and industrial development. In particular, the impressive developments in laser technology over the last decades lead to the generation of XUV and X-ray coherent ultra-

short and ultra-intense pulses in the femtosecond and sub-femtosecond time scale (1 fs = 10^{-15} s) [1–3]. Ultrafast short-wavelength radiation offers the unique capability to access and measure the structural arrangement and electronic structure inside the nucleus [4, 5]. The main available tools to generate ultrashort coherent pulses are presently high-order laser harmonics generated in gas and free-electron lasers [6].

High-order harmonics (HHs) are generated through the interaction between an ultrashort laser pulse and a gas in a cell or in a jet. Because of the strong peak power of the femtosecond laser pulse, a nonlinear interaction with the gas takes place and produces odd laser harmonics that may easily extend well above the order of several tens. The HH spectrum is described as a sequence of peaks corresponding to the odd harmonics of the fundamental laser wavelength and having an intensity distribution characterized by a plateau whose extension is related to pulse intensity and frequency. The use of advanced phase-matching mechanisms and interaction geometries has made possible the generation of HHs in the water window region between 2.3 and 4.4 nm, while still using a table-top laser source [7–9]. The radiation generated with the scheme of the HHs using few-optical-cycle laser pulses is currently the main tool for the investigation of matter with attosecond resolution (1 as = 10^{-18} s) [10–13]. Both trains [14, 15] and isolated [16–19] bursts of attosecond pulses have been experimentally demonstrated. The physical background of ultrashort pulses originates from the model of HH generation, i.e., the phase-matched emission of radiation results from the recombination of a tunnel-ionized electron with its parent ion. Once the conditions for such recombination are realized in only one occurrence per laser pulse, an isolated pulse is generated. Both trains and isolated attosecond pulses are positively chirped, resulting from the different duration of the quantum paths that contribute to the emitted spectrum [20]. Due to the nonzero chirp, the pulse temporal duration is longer than the Fourier limit. Positively chirped pulses may be temporally compressed by introducing a system that gives a compensating negative chirp. The compression has been achieved using a thin metallic filter with negative group-delay dispersion (GDD) as discussed in Refs. [15, 17] or broadband multilayer-coated optics with aperiodic layers [21].

Free-electron laser (FEL) sources generate radiation in the XUV and X-ray spectral regions with high spatial coherence, ultrashort time duration, and an increase of 6–8 orders of magnitude on the peak brilliance with respect to third-generation synchrotrons. FEL operation relies on a relativistic electron beam as the lasing medium which moves freely through a periodic magnetic structure (i.e., the undulator) that induces radiation [22]. There are presently four FEL user facilities operated at short wavelengths and dedicated to user-defined experiments: FLASH in Germany, SACLA in Japan, LCLS in USA, and FERMI in Italy. Presently, FEL pulses as short as 3 fs have been characterized at LCLS [23]. In order to increase the temporal resolution in pump-probe experiments, several approaches have been proposed for the generation of ultrashort FEL pulses in the femtosecond and sub-femtosecond regime, as the time slicing [24–26] or the reduction of the electron bunch charge [27]. Most of these methods rely on the selection of a small portion of the electron beam which undergoes FEL amplification, with a reduction of the amount of charge that contributes to the light amplification. A different possibility is the optical compression of the radiation pulse generated by the whole electron beam that is required to have a nonzero energy chirp in order to generate a chirped pulse. As for optical lasers, where frequency chirping is introduced to stretch the pulse before its amplification and then compensated after amplification to recover the ultrashort duration and high peak power, chirped-pulse amplification (CPA) may also be applied to FELs [28, 29]. In case of seeded FELs, the seeding laser pulse has to be stretched in time before interacting with the electron beam. This solution allows the use of the whole electron beam charge obtaining a significantly higher number of photons.

Indeed, for both HH and FEL facilities, the availability of a compressor tunable in the spectral band of operation of the source, capable of changing the spectral phase of the pulses, is particularly attractive, either to compensate for the intrinsic chirp or to realize CPA.

Here we discuss the use of gratings at grazing incidence to realize a tunable device to manipulate the spectral phase of XUV and X-ray–chirped pulses. In designing instruments for photon handling in the XUV ultrafast domain, there are some basic differences with respect to the traditional optical schemes in the visible and infrared domain [30, 31]. The first is to exploit the very short duration of the pulse, the study of the optical length of the rays gathered by the pupil and their equalization is mandatory. Moreover, the very extended bandwidth of operation may be exploited only if the instrument has a rather flat spectral response. Finally, an overall high throughput of the instrument is often a crucial feature to maintain high peak intensity.

The use of gratings at grazing incidence to realize tunable monochromators for XUV ultra-fast pulses is well established. Several monochromatic beamlines are presently in operation both with HHs [32–35] and FELs [36–39].

When using a grating for ultrafast pulses, the main problem faced with is the pulse-front tilt that is introduced by diffraction [40]. Indeed, each ray that is diffracted by two adjacent grooves is delayed by $m\lambda/c$, where m is the diffraction order, λ is the wavelength, and c is the speed of light in vacuum. The pulse-front tilt is given by the total difference in the optical paths of the diffracted beam, that is $\Delta\tau_G = m\lambda N/c$, where N is the total number of the illuminated grooves. This effect, that is totally negligible for picosecond or longer pulses, is noticeable in the femtosecond time scale, since it can dramatically degrade the instrumental ultrafast re-sponse. To overcome this effect, a double-grating configuration has to be adopted to compen-sate for the pulse-front tilt [41, 42]. The first grating is demanded to spectrally disperse the beam on an intermediate focal plane, where a slit performs the spectral selection, while the second grating compensates for the pulse-front tilt of the diffracted beam by equalizing the length of the optical paths. Double-grating instruments have been demonstrated to be very effective for HHs, with time resolution well below 10 fs [43–47]. Double-grating configura-tions, which are able to preserve the ultrafast duration of the pulse, have been also proposed as beam splitters for ultrafast intense pulses [48] and as infrared (IR)-XUV beam separators for HHs [49].

Here, we focus on the use of gratings to realize devices able to manipulate the spectral phase of chirped pulses, in particular to be used as XUV compressors.

2. Grating geometries for ultrashort pulses

Grazing incidence reflective gratings may be used either in the classical diffraction geometry (CDG) or in the off-plane geometry (OPG) [50].

The CDG is shown in **Figure 1(a)**. The grating equation is $\sin\alpha + \sin\beta = m\lambda\sigma_{CD}$, where α and β are, respectively, the incident and diffracted angles and σ_{CD} the groove density.

The OPG is shown in **Figure 1(b)**. The grating equation is $\sin\gamma\,(\sin\mu + \sin\nu) = m\lambda\sigma_{OP}$, where γ is the altitude angle, μ and ν are the azimuth angles as defined in the figure, and σ_{OP} is the groove density. The OPG, although seldom used, gives higher throughput than the classical mount, since it has been theoretically demonstrated and experimentally measured that the peak diffraction efficiency is close to the reflectivity of the coating at the altitude angle [51, 52]. Therefore, the OPG is suitable for the design of XUV grating instruments with high efficiency [53, 54].

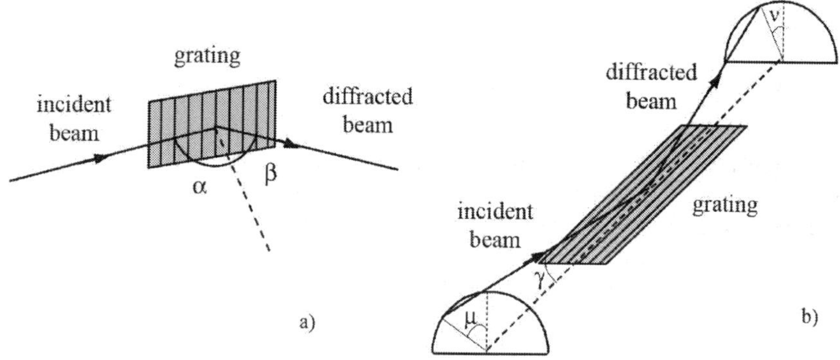

Figure 1. (a) Classical diffraction geometry; (b) off-plane geometry.

When realizing a grating compressor for ultrafast pulses, the main problem faced with is the pulse-front tilt given by the diffraction, as shown in **Figure 2** in the case of the CDG. Furthermore, different wavelengths are diffracted in different directions. The pulse-front tilt and the spectral angular dispersion have to be corrected by a second grating in a compensated configuration to fulfill the two following conditions: (1) the differences in the path lengths of rays with the same wavelength within the beam aperture that are caused by the diffraction from the first grating have to be compensated by the second grating; that is, the pulse-front tilt is corrected; (2) the angular spectral dispersion caused by the first grating has to be canceled by the second grating, that is, all the rays at different wavelengths exit the second grating with parallel directions. Both these conditions are satisfied by a scheme with two equal gratings mounted with opposite diffraction orders; that is, the incidence (incoming azimuth) angle on the second grating is equal to the diffraction angle (outcoming azimuth) from the first grating. The phase chirp introduced by the system is calculated as the difference in the optical paths of rays at different wavelengths. This principle is well known for the realization of stretchers and compressors in the visible and near infrared [55, 56].

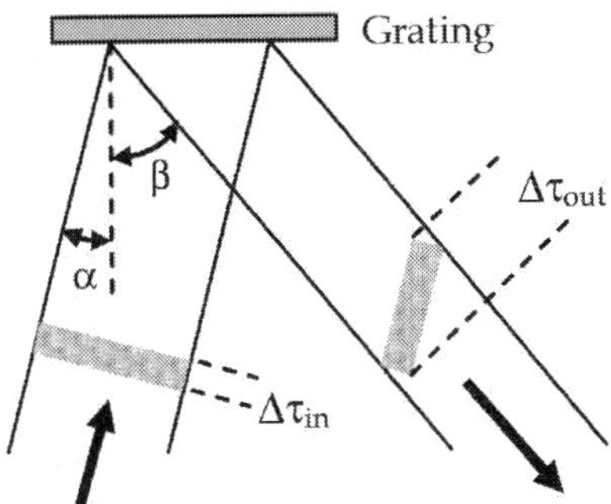

Figure 2. Pulse-front tilt of an ultrashort pulse diffracted by a grating in the CDG. $\Delta\tau_{in}$ and $\Delta\tau_{out}$ are the pulse duration at input and output, respectively. At the first diffracted order, the pulse-front tilt is $\Delta OP = N\lambda$, where N is the number of illuminated grooves.

3. Grazing incidence grating compressor

When applying the double-grating configuration to the phase manipulation of XUV pulses, all the optics have to be operated at grazing incidence. The simplest arrangement consists of two identical plane gratings mounted in the compensated configuration, as shown in **Figures 3 and 4**. Due to the symmetry of the configuration, the angular dispersion at the output is canceled, and the output rays are parallel to the input for all the wavelengths [57].

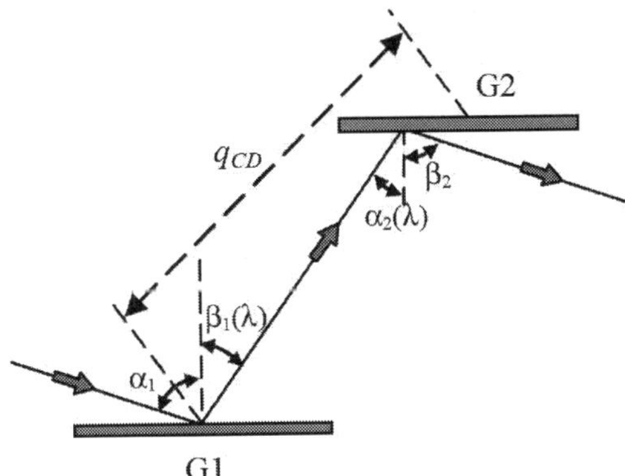

Figure 3. Double-grating compressor in the CDG. The diffraction angle from G2 is constant with the wavelength and equal to the incidence angle on G1, $\beta_2 = \alpha_1$.

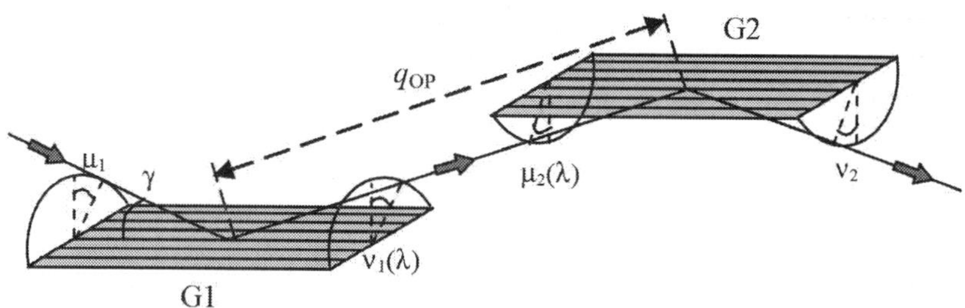

Figure 4. Double-grating compressor in the OPG. The outcoming azimuth from G2 is constant with the wavelength and equal to the incoming azimuth on G1, $v_2 = \mu_1$.

Since different wavelengths are diffracted by G1 at different angles, the rays do not make the same optical paths. In the case of the CDG, the optical path is analytically expressed (for less than a constant term) as

$$OP_{CD}(\lambda) = q_{CD}\frac{\cos\beta_c}{\cos\beta(\lambda)}[1 - \sin\alpha\sin\beta(\lambda)] \tag{1}$$

where α is the incidence angle on G1, $\beta(\lambda)$ and β_c are, respectively, the diffraction angles from G1 at the generic wavelength λ and at the central wavelength of the interval of operation λ_c, and q_{CD} is the G1-G2 distance. The bandwidth of the pulse $\Delta\lambda$ is limited between λ_{min} and λ_{max}, $\Delta\lambda = \lambda_{max} - \lambda_{min}$, and $\underline{\lambda}_c = (\lambda_{min} + \lambda_{max})/2$. In case of a narrow-band pulse with $\lambda/\Delta\lambda < 20\%$, Eq. (1) is linearized in λ as

$$OP_{CD}(\lambda) = q_{CD}\lambda_c\left(\frac{m\sigma_{CD}}{\cos\beta_c}\right)^2\lambda \tag{2}$$

Similarly, in the case of the OPM, the optical path is expressed as

$$OP_{OP}(\lambda) = q\frac{\sin^2\gamma\cos\mu}{\cos v}(1 + \sin\mu\sin v) \tag{3}$$

where μ is the incoming azimuth on G1 that has been chosen to have the central wavelength λ_c diffracted at $v_c = \mu$, i.e., $2\sin\gamma\sin\mu = \lambda_c\sigma_{OP}$, and q_{OP} is the G1-G2 distance. In case of a narrow-band pulse, Eq. (3) is linearized as

$$OP_{OP}(\lambda) = q_{OP}\lambda_c\left(\frac{m\sigma_{OP}}{\cos\mu}\right)^2\lambda \tag{4}$$

Note that in both cases the optical path increases with the wavelength, and this forces the group delay dispersion introduced by the double-grating configuration to be negative.

As usual, the group delay (GD) and the group delay dispersion (GDD) are expressed as a function of $\omega = 2\pi c/\lambda$: $GD(\omega) = \partial\varphi(\omega)/\partial\omega = OP(\omega)/c$ and $GDD(\omega) = \partial GD(\omega)/\partial\omega$. The central pulse frequency ω_c is defined as $\omega_c = 2\pi c/\lambda_c$.

For narrow-band pulses, the GD is also linear in frequency, and the GDD is constant and negative

$$GDD_{CD} = -\frac{q_{CD}c}{\omega_c^3}\left(\frac{2\pi\sigma_{CD}}{\cos\beta_c}\right)^2 \tag{5}$$

and

$$GDD_{OP} = -\frac{q_{OP}c}{\omega_c^3}\left(\frac{2\pi\sigma_{OP}}{\cos\mu}\right)^2 \tag{6}$$

Once the required GDD to manipulate the pulse has been defined, the above equations define the parameters of the grating compressor in both geometries.

Once the required GDD has been fixed, the two geometries give equivalent answer for $q_{OP}\,\sigma_{OP}^2/\cos^2\mu = q_{CD}\,\sigma_{CD}^2/\cos^2\beta_c$. In case of equal arms, i.e., $q_{OP} = q_{CD}$, since μ is typically below 20° and β_c above 80°, the groove density that would be required in the OPM is much higher than the CDM and may be not available from grating providers. Therefore, a compressor in the OPM is typically longer than the corresponding CDM, i.e., $q_{OP} > q_{CD}$.

The compressor introduces a spatial chirp of the pulse, i.e., rays with different wavelengths have the same output direction, but they are not exactly superimposed. In the conventional design of compressors for IR pulses, the spatial chirp is canceled by making the beam passing two additional times though the same gratings, so the output spatial dispersion is zero. This cannot be realized in grazing incidence, since it would require the insertion of two additional gratings that would make the configuration complex and inefficient. The spatial chirp $SC(\lambda)$ is expressed, in case of a narrow-band pulse, as

$$SC_{CD}(\lambda) = q_{CD}\sigma_{CD}\frac{\cos\alpha}{\cos^2\beta_c}\Delta\lambda \tag{7}$$

$$SC_{OP}(\lambda) = q_{OP}\sigma_{OP}\frac{1}{\cos^2\mu}\Delta\lambda \tag{8}$$

Since the rays are parallel, the spatial chirp does not influence the quality of the final spot size, since all the rays are focused on the same point.

In the following, we discuss the use of the double-grating configuration to the case of compression of FEL pulses and of attosecond pulses generated through HHs.

4. Grating compressor applied to FEL pulses

One of the main problems faced when manipulating intense FEL radiation is the use of robust optical components to be operated with the FEL pulses with minimum risk of damaging. From this point of view, grazing incidence elements are preferable. The double-grating compressor, whose optical elements are used at grazing incidence, is very suitable for FEL pulses. We assume FEL parameters already discussed in the literature [58] that may be a test case for the application to chirped-pulse amplification (CPA) of FEL pulses.

The scheme of the CPA applied to a seeded FEL is shown in **Figure 5**. The electron beam, being generated through a chirped laser seeding pulse, originates, after the radiator, a chirped optical pulse with positive GDD. The chirp is corrected by the grating compressor that introduces different optical paths for different wavelengths and shortens the pulse duration close to the Fourier limit.

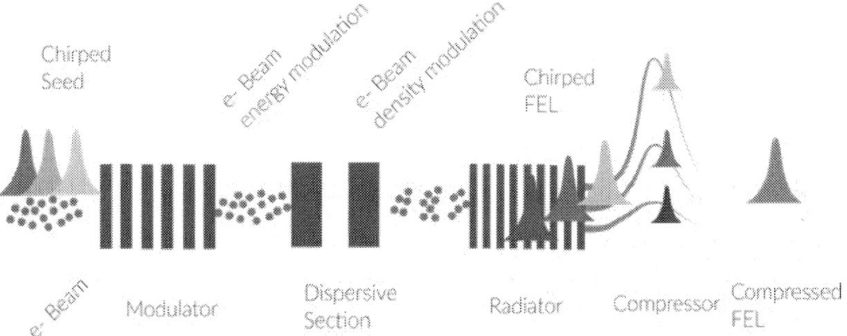

Figure 5. Schematic of the CPA applied to a seeded FEL.

Assuming a linear electron bunch energy spread at the entrance of the FEL radiator, the wavelength spread is evaluated as $\Delta\lambda/\lambda = -2\Delta E/E$. The photon chirp induced by the entrance energy spread is $2\alpha/E$, where α is the electron chirp. This has to be compensated to reduce the pulse time duration.

As a feasibility study, we want to define a configuration to compress an FEL pulse centered at 13.5 nm. **Table 1** resumes the FEL parameters used in the simulation. Using these parameters, the time to be compensated is calculated to be 310 fs for an FEL emission centered at 13.5

nm with a bandwidth $\Delta\lambda = 0.8$ nm. If the GDD introduced by the compressor is opposite to the intrinsic GDD of the chirped pulse, the pulse time duration is reduced.

Beam energy	0.96 GeV
Charge	0.7 nC
Beam current	1 kA
Bunch duration	270 fs
Energy chirp	–0.15 MeV/ • m

Table 1. FEL parameters used in the simulation.

The compressor parameters are summarized in **Table 2** for the two geometries. The GD of the configurations is shown in **Figure 6(a)**. The curve is almost the same for both geometries. As expected, for narrow-bandwidth pulses, the resulting GD is linear, and the GDD is constant: GDD ≈-37 fs^2.

Pulse central wavelength	13.5 nm
Bandwidth	0.8 nm
Stretching	310 fs
CDG	
Groove density	600 gr/mm
G1-G2 distance	285 mm
Subtended angle	164°
OPG	
Groove density	3600 gr/mm
G1-G2 distance	600 mm
Altitude angle	5°

Table 2. Parameters of the grating compressor in two geometries.

The spatial chirp calculated in the full-width-at-half-maximum (FWHM) bandwidth is 0.9 mm for both geometries. The compressor is typically inserted several tens of meters after the FEL source. As a typical angular divergence of the FEL source at 13.5 nm, 30 µrad (standard deviation) is assumed. The resulting beam diameter at the compressor input, that is assumed to be 50 m far from the source, is 3.6 mm FWHM, therefore much larger than the spatial chirp.

Note that the groove density required in the OPG is higher than the CDG and that the size of the instrument is longer for the OPG. On the basis of the efficiency measurements performed in the two geometries and already discussed in Ref. [54], the total efficiency in the OPG is expected to be higher than the CDG by a factor ≈2.5.

It can be shown that the FEL pulse duration is reduced by a factor of 10 at the output of the compressor, i.e., from 310 fs to about 30 fs. This gives a substantial increase in the temporal resolution of the FEL pulses when used for ultrafast experiments.

4.1. Tunability in wavelength and group delay

The compressor can be tuned in wavelength by rotating the gratings around an axis that is tangent to the surface, passes through the grating center, and is parallel to the grooves. The rotation changes the incidence angle α in the CDG at a constant subtended angle $K = \alpha + \beta$, or the azimuth angle μ in the OPM at constant altitude angle γ. The tuning in wavelength changes also the GD, since it depends on the incidence (azimuth) angles. The delays introduced in the bandwidth $\Delta\lambda = 0.8$ nm when the central wavelength is tuned in the 10–18 nm interval are shown in **Figure 6(b)**.

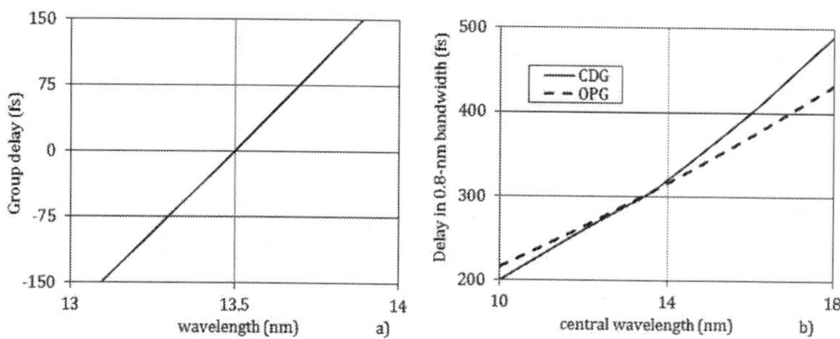

Figure 6. (a) GD of the compressor having the parameters of **Table 2**; (b) change of the delay in the bandwidth when the gratings are rotated to tune the wavelength in the 10–18 nm interval.

It is clear that the simple grating rotation is not sufficient to tune simultaneously the wavelength and the GD. An additional degree of freedom is required that may be the changing of the subtended angle K (the altitude angle γ) in the CDG (OPG), as shown in **Figure 6**. By acting simultaneously on grating rotation and subtended (altitude) angles, users can select simultaneously the wavelength and GD. This makes the design very flexible.

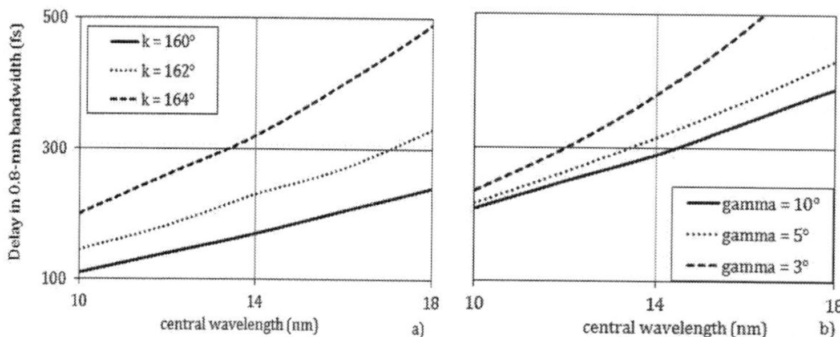

Figure 7. Delay in 0.8-nm bandwidth introduced by the compressor having the parameters of **Table 2** when the gratings are rotated to tune the wavelength in the 10–18 nm interval: (a) CDG, variable subtended angle; (b) OPG, variable altitude angle.

The optical setup of the compressor is shown in **Figure 7**. The instrument consists of two plane gratings and two plane mirrors. The two mirrors are used to deviate the FEL beam in the same direction as the input.

4.2. Operation with a diverging beam

The formulas calculated above assume to work with a collimated beam. In this case, the number of illuminated groove is the same for the two gratings when they are parallel, giving a corrected pulse-front tilt at the output. Indeed, in the real case of an FEL-divergent beam, this would require the use of an additional mirror at the input of the compressor to collimate the beam, namely a grazing-incidence parabola that makes the design complex. Indeed, the compressor, as presented in the previous paragraph, can be used in a divergent beam if the second grating is operated slightly out from the parallel condition, to have the same number of illuminated grooves.

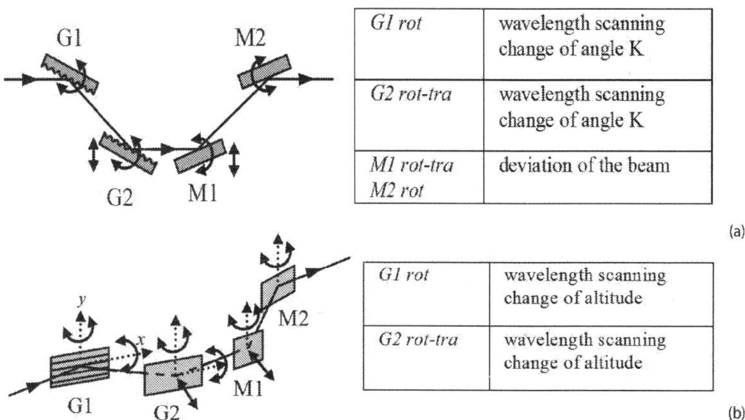

G1 rot	wavelength scanning change of angle K
G2 rot-tra	wavelength scanning change of angle K
M1 rot-tra M2 rot	deviation of the beam

(a)

G1 rot	wavelength scanning change of altitude
G2 rot-tra	wavelength scanning change of altitude

(b)

Figure 8. Optical setup of the compressor: (a) CDG; (b) OPG.

The geometry with a divergent beam is shown in **Figure 8** for CDG. The number of illuminated grooves is the same for both gratings if G2 is operated at a lower subtended angle, $k_2 < k_1$:

$$\cos\alpha_2 = \cos\beta_1 + \frac{q_{CD}\delta_1}{S_1}\frac{\cos^2\alpha_1}{\cos\beta_1} \qquad (9)$$

where δ_1 is the divergence of the incoming beam, S_1 is the beam cross section at G1, α_1 and β_1 the incidence and diffraction angles on G1, and α_2 the incidence angle on G2.

In the case of OPM, the compensation of the pulse-front tilt is expressed by

$$\cos \mu_2 = \cos \mu_1 + \frac{q_{OP}\delta_1}{S_1} \tag{10}$$

where μ_1 and μ_2 are the azimuth angles on G1 and G2.

The asymmetry between G1 and G2 depends on the actual divergence of the FEL beam and on the distance between the two gratings. Let us assume the parameters of **Table 2**, with 30-µrad divergence (standard deviation). The compressor stage is supposed to be installed 50 m far from the source. In the case of CDG, the residual pulse-front tilt at 13.5 nm is 5 fs FWHM if the gratings are operated parallel and the correction to be applied to α_2 is 0.1°. If the compressor is used at longer wavelengths, the asymmetry to be applied is more remarkable because of the higher FEL divergence.

The spatial chirp, that does not influence the quality of the final spot size in case of a parallel beam, has to be evaluated in case of a divergent beam, since different wavelengths are focused in different points in the direction of the spectral dispersion. This gives a slight asymmetry in the spot profile that is broadened in the direction of spectral dispersion. Let us define M as the total demagnification of the FEL beamline, $M \approx 50$–100. The spatial chirp SC gives a limit to the minimum focal spot that can be achieved in the direction of the spectral dispersion, as SC/M. Assuming the same parameters of **Table 2** and $M = 75$, the minimum spot size that can be achieved in the direction of the spectral dispersion is 12 µm FWHM. The broadening due to the spatial chirp is generally negligible for spot sizes in the 20–50 µm range. However, the use of the compressor may degrade the quality of the final focus if the beamline is tailored to give micro-focusing. In such cases, the insertion of the compressor has to be carefully evaluated.

4.3. Efficiency of the compressor

The efficiency of the compressor depends on the geometry adopted for the gratings, since it is well known that the OPG gives efficiency higher than the CDG [59].

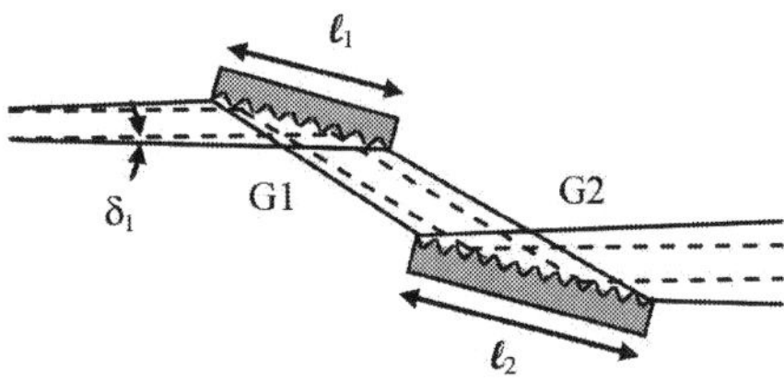

Figure 9. Grating compressor operated in divergent beam. The pulse-front tilt is corrected for $l_2 = l_1$.

To compare the two geometries, the first-order diffraction efficiency has been measured in the 25–35 nm (35–50 eV) region in the two different geometries. Both gratings are plane, gold-coated, and have 600 gr/mm groove density. The grating used in the CDG is blazed at 2°, and the grating used in the OPG is blazed at 7°. The results are shown in **Figure 9**. The efficiency in the OPG is a factor ≈2 higher than the CDG that gives a factor 4 in the total efficiency of the compressor. The latter is expected to be ≈5% in the CDG and ≈20% in the OPG [60].

5. Grating compressor for attosecond pulses

Attosecond pulses generated with the scheme of HHs by the use of laser pulses of few optical cycles are positively chirped as a result of the different duration of the quantum paths that contribute to the different portions of the emitted spectrum. They can be compressed by introducing a suitable device with a negative GDD.

The simplest device that exhibits negative GDD is a thin metallic filter. Anomalous dispersion near absorption resonances can be exploited to compensate the positive chirp of the generated attosecond pulses. Aluminum or zirconium is normally used in the XUV range, depending on the spectral range of operation [61]. Furthermore, the filter is also useful to block the IR laser beam since it is totally solar blind. The main drawback of the filter is the strong XUV absorption that may even exceed one order of magnitude.

Aperiodic multilayer mirrors have also been developed and successfully tested for the dispersion control in the XUV [62–65]. The multilayer coating is designed to compensate for the attosecond chirp and reduce the pulse duration. Pulse compression and focusing are demanded to the same optical element that makes the design simple and compact. For a center-photon energy range of 100–120 eV, the mirror reflectivity is approximately 10% and the bandwidth 10–13 eV. Recently, also mirrors for the water-window region have been tested, although with reflectivity lower than 1% [66]. A metallic filter has to be inserted anyway in the optical path to block the IR laser light.

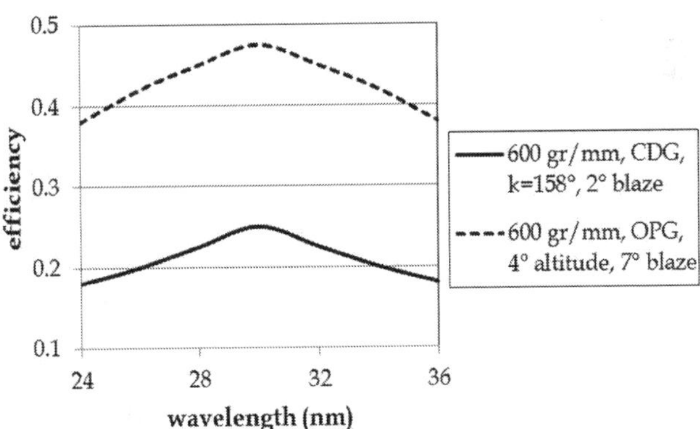

Figure 10. Comparison of efficiencies measured in CDG and OPG.

We discuss here the use of gratings to compress attosecond pulses by introducing a GDD that compensates for the intrinsic pulse chirp. Unfortunately, the configuration with plane gratings discussed above is not suitable, since the G1-to-G2 distance q that is required to give the necessary GDD is too small to be realized in practice. Therefore, the plane-grating configuration has to be modified as shown in **Figure 10** for the OPG [67, 68], by adding an intermediate focal point between the two gratings. The case of CDG is analogous.

The design consists of six optical elements: four identical grazing incidence parabolic mirrors (P1-P4) and two identical plane gratings (G1, G2). The XUV source is located in the front focal plane of P1, and the rays are collected at the focus of the last parabolic mirror P4. The parabolic mirrors are used to collimate and refocus the XUV radiation with negligible aberrations, and the gratings are illuminated in parallel light. A spectrally dispersed image of the source is obtained in the intermediate plane. The two focusing mirrors placed between the gratings act as a telescopic arrangement. Differently from the plane-grating compressor discussed above, this makes it possible to: (i) continuously tune the GDD from negative to positive values and (ii) achieve the exceedingly small grating separations necessary to compensate the attosecond chirp.

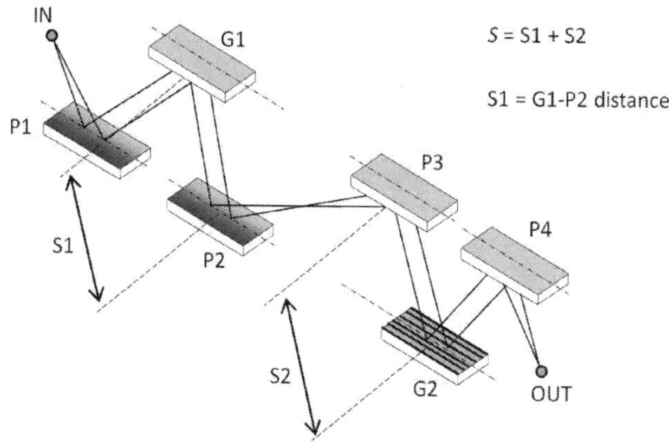

Figure 11. Grating compressor for attosecond pulses.

With reference to the symbols listed in **Figure 11**, the condition for zero GDD is to have G1 imaged on G2, that is realized when $S1 + S2 = 2f$. Since f is fixed, the GDD depends on $S1 + S2$: for $S1 + S2 < 2f$, G1 is imaged behind G2, and the resulting GDD is positive; for $S1 + S2 > 2f$, G1 is imaged before G2, and the resulting GDD is negative. Once the equivalent distance q that is required for compensation has been calculated, the effective displacement from the zero-dispersion case is $\Delta S = q/(\omega_0 \sin\gamma \cos\upsilon)$, where ω_0 is the center angular frequency. It can be noted that the design of the compressor is simplified if $S1$ is kept fixed, and only $S2$ is tuned to change the GDD. A suitable value for f is in the range of 200–300 mm, giving a total length of the compressor of ≈1.5 m.

As an application to ultrashort pulses, a compressor design for the 50–100 eV region is discussed here. The characteristics of the compressor are resumed in **Table 3**. The GD introduced with the distance $S = S1 + S2$ is shown in **Figure 12**. An example of compression of a pulse with a positive GDD is presented in **Figure 13**, adapted from [69]. Note that the chirp introduced by the compressor is able to compensate the pulse chirp down to a nearly single-cycle pulse (**Figure 14**).

Pulse spectral interval	50-100 eV, 12-24 nm
Mirrors	**Off-axis parabola**
Input/output arms	300 mm
Grazing angle	3°
Gratings	**Plane**
Groove density	200 gr/mm
Altitude angle	1.5°
Distances	S1+S2 > 600 mm for negative GDD

Table 3. Parameters of the compressor for the 50–100 eV region.

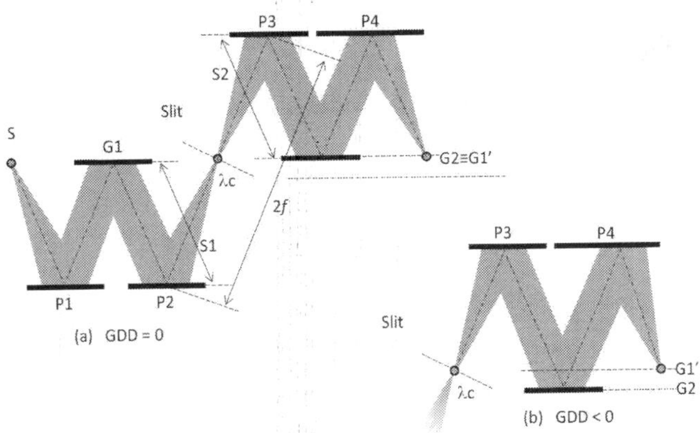

Figure 12. Operation of the attosecond compressor: (a) GDD = 0; (b) GDD < 0.

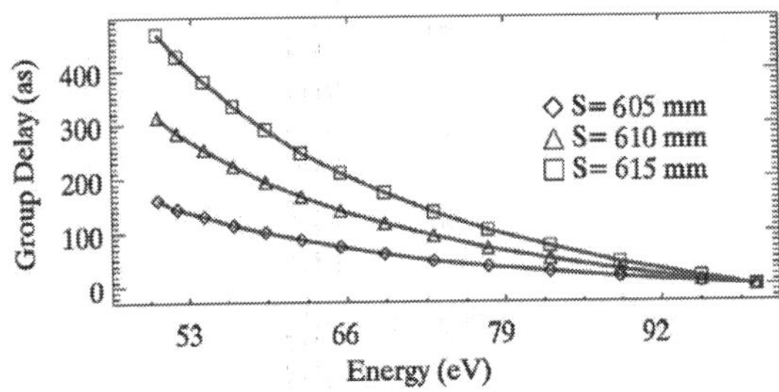

Figure 13. GD of the compressor with parameters listed in **Table 3**.

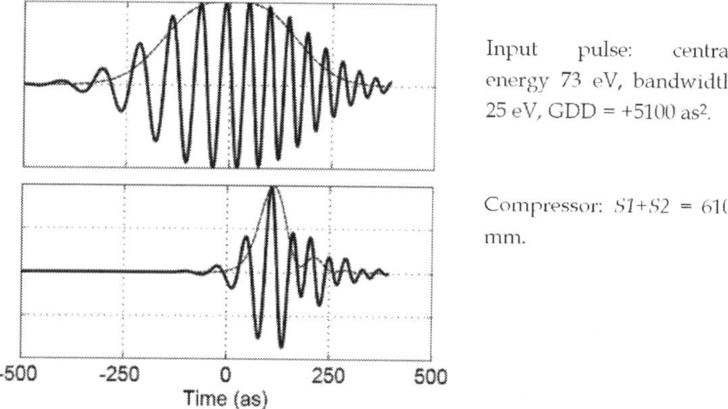

Figure 14. Simulation of the compression of a XUV pulse with the parameters of **Table 3**.

5.1. Example of application to attosecond pulses

A schematic view of an experiment using compressed attosecond pulses is shown in **Figure 15**. The XUV attosecond pulses are generated on a gas jet in a vacuum chamber and are intrinsically chirped. The XUV radiation is generated with the intrinsic attosecond chirp. Different wavelengths travel in different paths inside the compressor that compensates for the chirp and reduces the time duration of the pulse.

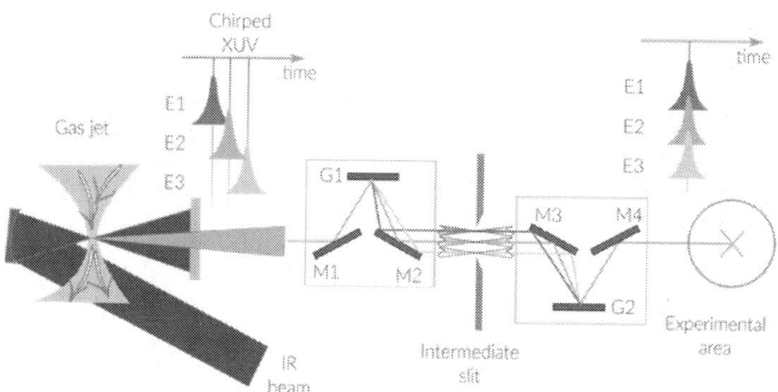

Figure 15. Schematic of the attosecond compressor.

Attosecond pulses are generated in different XUV spectral windows, depending on the interacting gas. Using argon and $\approx 2 \cdot 10^{14}$ W/cm² ultrafast laser intensity, radiation is generated in the 25–55 eV region, while with neon and $\approx 6 \cdot 10^{14}$ W/cm² intensity, radiation is generated in the 50–120 eV region. Attosecond pulses are generated from the short trajectory components, since this is the part of the generated radiation that survives propagation in the generating medium. Aluminum and zirconium filters can be used, respectively, in the low- (i.e., argon) and high-energy range (i.e., neon) to introduce negative chirp to compensate for the intrinsic chirp of the attosecond pulses and compress them close to the Fourier limit.

Here, we discuss the parameters to be used for a grating compressor in the two different XUV regions. For the lower intensity case, the parameters are σ = 100 grooves/mm, γ = 1.5°, μ = 3.7°, ΔS = 23 mm. For the higher intensity case, the parameters are σ = 200 grooves/mm, γ = 1°, μ = 4.1°, ΔS = 29 mm.

Simulations of the generated and compressed pulses have been performed starting from 25-fs driving pulses at 790 nm [70]. The generated pulses have a duration of 270 as in argon and 200 in neon, while the Fourier-limited duration would be, respectively, 115 and 35 as. At the output of the grating compressor, the pulse duration results, respectively, 160 and 60 as, much closer to the Fourier limit. It is also shown that the optimal Al filter would reduce the pulse duration to 130 as, therefore shorter than the grating compressor, while the optimal Zr filter would reduce the pulse duration to 90 as, therefore longer than the grating compressor.

In general, the grating compressor is more versatile than metal filters and can be continuous- ly tuned from negative to positive GDD with constant throughput.

6. Conclusions

The use of diffraction gratings to manipulate the spectral phase of XUV ultrashort pulses has been discussed. The system consists of two gratings to introduce the required phase chirp. Both the classical and the off-plane geometries have been discussed.

Both a simple design with two plane gratings working in collimated beam and a more complex design with two plane gratings and an intermediate focal point have been discussed. Both designs are tunable in wavelength range to cover the whole spectral extension of the source and operate at different wavelengths; furthermore, once the operating wavelength has been chosen, the GDD is tunable to compensate for the actual pulse chirp.

The use of elements at grazing incidence makes the system particularly suitable for the application to CPA of intense FEL pulses and to compression of attosecond pulses, playing an important role for the photon handling and conditioning of future ultrashort sources.

Acknowledgements

The results discussed here have been partially funded by Extreme Light Infrastructures (ELI) funds of the Italian Ministry for Education, University and Research.

Author details

Fabio Frassetto, Paolo Miotti and Luca Poletto*

*Address all correspondence to: luca.poletto@ifn.cnr.it

CNR-Institute of Photonics and Nanotechnologies, Padova, Italy

References

[1] Diels J-C, Rudolph W. Ultrashort Laser Pulse Phenomena. Oxford: Academic Press Inc.; 2006.

[2] Jaegle' P. Coherent Sources of XUV Radiation. Berlin: Springer; 2006.

[3] Marciak-Kozlowska J. From Femto-to Attoscience and Beyond. New York: Nova Science Publishers Inc; 2009.

[4] Vrakking M J J, Elsaesser T. X-Ray Photonics: X-rays Inspire Electron Movies. Nat. Photonics. 2012;6:645–647.

[5] Lépine F, Ivanov M Y, Vrakking, M J J. Attosecond Molecular Dynamics: Fact or Fiction?. Nat. Photonics. 2014;8:195–204.

[6] Canova F, Poletto L, editors. Optical Technologies for Extreme-Ultraviolet and Soft X-ray Coherent Sources. Berlin: Springer; 2015.

[7] Chang Z, Rundquist A, Wang H, Murnane M M, Kapteyn H C. Generation of Coherent Soft X Rays at 2.7 nm Using High Harmonics. Phys. Rev. Lett. 1997;79:2967–2970.

[8] Schnürer M, Spielmann Ch, Wobrauschek P, Streli C, Burnett N H, Kan C, Ferencz K, Koppitsch R, Cheng Z, Brabec T, Krausz F. Coherent 0.5-keV X-Ray Emission from Helium Driven by a Sub-10-fs Laser. Phys. Rev. Lett. 1998;80:3236–3239.

[9] Takahashi E J, Kanai T, Ishikawa K L, Nabekawa Y, Midorikawa K. Coherent Water Window X Ray by Phase-Matched High-Order Harmonic Generation in Neutral Media. Phys. Rev. Lett. 2008;101:253901.

[10] Corkum P, Krausz F. Attosecond Science. Nat. Physics. 2007;3:381–387.

[11] Krausz F, Ivanov M. Attosecond Physics. Rev. Mod. Phys. 2009;81:163–234.

[12] Sansone G, Poletto L, Nisoli M. High-Energy Attosecond Light Sources. Nat. Photonics. 2011;5:655–663.

[13] Krausz F, Stockman M I. Attosecond metrology: from electron capture to future signal processing. Nat. Photonics. 2014;8:205–213.

[14] Paul P M, Toma E S, Breger P, Mullot G, Auge F, Balcou P, Muller H G, Agostini P. Observation of a Train of Attosecond Pulses from High Harmonic Generation. Science. 2001;292:1689–1692.

[15] Lopez-Martens R, Varju K, Johnsson P, Mauritsson J, Mairesse Y, Salières P, Gaarde M B, Schafer K J, Persson A, Svanberg S, Wahlström C-G, L'Huillier A. Amplitude and Phase Control of Attosecond Light Pulses. Phys. Rev. Lett. 2005;94:033001.

[16] Kienberger R, Goulielmakis E, Uberacker M, Baltuska A, Yakovlev V, Bammer F, Scrinzi A, Westerwalbesloh T, Kleineberg U, Heinzmann U, Drescher M, Krausz F. Atomic Transient Recorder. Nature. 2004;427:817–822.

[17] Sansone G, Benedetti E, Calegari F, Vozzi C, Avaldi L, Flammini R, Poletto L, Villoresi P, Altucci C, Velotta R, Stagira S, De Silvestri S, Nisoli M. Isolated Single-Cycle Attosecond Pulses. Science. 2006;314:443–446.

[18] Zhao K, Zhang Q, Chini M, Wu Y, Wang X, Chan Z. Tailoring a 67 Attosecond Pulse Through Advantageous Phase-Mismatch. Opt. Lett. 2012;37:3891–3893.

[19] Chini M, Zhao K, Chang Z. The Generation, Characterization and Applications of Broadband Isolated Attosecond Pulses. Nat. Photonics. 2014;8:178–186.

[20] Mairesse Y, de Bohan A, Frasinski L J, Merdji H, Dinu L C, Monchicourt P, Breger P, Kovacev M, Auguste T, Carré B, Muller H G, Agostini P, Salières P. High-Harmonics Chirp and Optimization of Attosecond Pulse Trains. Laser Phys. 2005;15:863–870.

[21] Schultze M, Goulielmakis E, Uiberacker M, Hofstetter M, Kim J, Kim D, Krausz F, Kleineberg U. Powerful 170-Attosecond XUV Pulses Generated with Few-Cycle Laser Pulses and Broadband Multilayer Optics. New J. Phys. 2007;9:243–253.

[22] Saldin E L, Schneidmiller E A, Yurkov M V. The Physics of Free Electron Lasers. Berlin: Springer; 2000.

[23] Ding Y, Decker F-J, Emma P, Feng C, Field C, Frisch J, Huang Z, Krzywinski J, Loos H, Welch J, Wu J, Zhou F. Femtosecond X-Ray Pulse Characterization in Free-Electron lasers Using a Cross-Correlation Technique. Phys. Rev. Lett. 2012;109:254802.

[24] Emma P J, Bane K, Cornacchia M, Huang Z, Schlarb H, Stupakov G, Nalz D. Femtosecond and Sub-femtosecond X-Ray Pulses from a Self-Amplified Spontaneous-Emission–Based Free-Electron Laser. Phys. Rev. Lett. 2004;92:074801.

[25] Ding Y, Huang Z, Ratner D, Bucksbaum P, Merdji H. Generation of Attosecond X-Ray Pulses with a Multicycle Two-Color Enhanced Self-Amplified Spontaneous Emission Scheme. Phys. Rev. ST Accel. Beams. 2009;12:060703.

[26] Tanaka T. Proposal for a Pulse-Compression Scheme in X-Ray Free-Electron Lasers to Generate a Multiterawatt, Attosecond X-Ray Pulse. Phys. Rev. Lett. 2013;110:084801.

[27] Rosenzweig J B, Alesini D, Andonian G, Boscolo M, Dunning M, Faillace L, Ferrario M, Fukusawa A, Giannessi L, Hemsing E, Marcus G, Marinelli A, Musumeci P, O'Shea B,

Palumbo L, Pellegrini C, Petrillo V, Reiche S, Ronsivalle C, Spataro B, Vaccarezza C. Generation of Ultra-Short, High Brightness Electron Beams for Single-Spike SASE FEL Operation. Nucl. Instrum. Meth. A. 2008;593:39–44.

[28] Shu X, Peng T, Dou Y. Chirped Pulse Amplification in a Free-Electron Laser Amplifier. J. Elect. Spect. Rel. Phen. 2011;184:350–353.

[29] Feng C, Shen L, Zhang M, Wang D, Zhao Z, Xian D. Chirped Pulse Amplification in a Seeded Free-Electron Laser for Generating High-Power Ultra-Short Radiation. Nucl. Inst. Meth. Phys. Res. A. 2013;712:113–119.

[30] Poletto L, Tondello G, Villoresi P. High-order Laser Harmonics Detection in the EUV and Soft X-Ray Spectral Regions. Rev. Sci. Instr. 2001;72:2868–2874.

[31] Poletto L, Bonora S, Pascolini M, Villoresi P. Instrumentation for Analysis and Utilization of Extreme-Ultraviolet and Soft X-Ray High-order Harmonics. Rev. Sci. Instr. 2004;75:4413–4418.

[32] Frassetto F, Cacho C, Froud C, Turcu I C E, Villoresi P, Bryan W A, Springate E, Poletto L. Single-Grating Monochromator for Extreme-Ultraviolet Ultrashort Pulses. Opt. Express. 2011;19:19169–19181.

[33] Grazioli C, Callegari C, Ciavardini A, Coreno M, Frassetto F, Gauthier D, Golob D, Ivanov R, Kivimäki A, Mahieu B, Bučar B, Merhar M, Miotti P, Poletto L, Polo E, Ressel B, Spezzani C, De Ninno G. CITIUS: An Infrared-Extreme Ultraviolet Light Source for Fundamental and Applied Ultrafast Science. Rev. Sci. Instrum. 2014;85:023104.

[34] Poletto L, Miotti P, Frassetto F, Spezzani C, Grazioli C, Coreno M, Ressel B, Gauthier D, Ivanov R, Ciavardini A, de Simone M, Stagira S, De Ninno G. Double-Configuration Grating Monochromator for Extreme-Ultraviolet Ultrafast Pulses. Appl. Opt. 2014;53:5879–5888.

[35] Ojeda J, Arrell C A, Grilj J, Frassetto F, Mewes L, Zhang H, van Mourik F, Poletto L, Chergui M. Harmonium: A Pulse Preserving Source of Monochromatic EUV (30–110 eV) Radiation for Ultrafast Photoelectron Spectroscopy of Liquids. Structural Dynamics. 2016;3:023602.

[36] Martins M, Wellhöfer M, Hoeft J T, Wurth W, Feldhaus J, Follath R. Monochromator Beamline for FLASH. Rev. Sci. Inst. 2006;77:115108.

[37] Guerasimova N, Dziarzhytski S, Feldhaus J. The Monochromator Beamline at FLASH: Performance, Capabilities and Upgrade Plans. J. Mod. Opt. 2011;58:1480–1485.

[38] Heimann P, Krupin O, Schlotter W F, Turner J, Krzywinski J, Sorgenfrei F, Messerschmidt M, Bernstein D, Chalupský J, Hájková V, Hau-Riege S, Holmes M, Juha L, Kelez N, Lüning J, Nordlund D, Fernandez Perea M, Scherz A, Soufli R, Wurth W, Rowen M. Linac Coherent Light Source Soft X-Ray Materials Science Instrument Optical Design and Monochromator Commissioning. Rev. Sci. Inst. 2011;82:093104.

[39] Schlotter W F, Turner J J, Rowen M, Heimann P, Holmes M, Krupin O, Messersch-midt M, Moeller S, Krzywinski J, Soufli R, Fernández-Perea M, Kelez N, Lee S, Coffee R, Hays G, Beye M, Gerken N, Sorgenfrei F, Hau-Riege S, Juha L, Chalupsky J, Hajkova V, Mancuso A P, Singer A, Yefanov O, Vartanyants I A, Cadenazzi G, Abbey B, Nugent K A, Sinn H, Lüning J, Schaffert S, Eisebitt S, Le W-S, Scherz A, Nilsson A R, Wurth W. The Soft X-Ray Instrument for Materials Studies at the Linac Coherent Light Source X-Ray Free-Electron Laser. Rev. Sci. Inst. 2012;83:043107.

[40] Poletto L, Frassetto F. Time-Preserving Monochromators for Ultrafast Extreme-Ultraviolet Pulses. Appl. Opt. 2010;49:5465–5473.

[41] Villoresi P. Compensation of Optical Path Lengths in Extreme-Ultraviolet and Soft-X-Ray Monochromators for Ultrafast Pulses. Appl. Opt. 1999;38:6040–6049.

[42] Poletto L. Time-Compensated Grazing-Incidence Monochromator for Extreme-Ultraviolet and Soft X-Ray High-order Harmonics. Appl. Phys. B. 2004;78:1013–1016.

[43] Poletto L, Villoresi P. Time-Compensated Monochromator in the Off-Plane Mount for Extreme-Ultraviolet Ultrashort Pulses. Appl. Opt. 2006;45:8577–8585.

[44] Poletto L, Villoresi P, Benedetti E, Ferrari F, Stagira S, Sansone G, Nisoli M. Intense Femtosecond Extreme Ultraviolet Pulses by Using a Time-Delay Compensated Monochromator. Opt. Lett. 2007;32:2897–2899.

[45] Poletto L, Villoresi P, Frassetto F, Calegari F, Ferrari F, Lucchini M, Sansone G, Nisoli M. Time-Delay Compensated Monochromator for the Spectral Selection of Extreme-Ultraviolet High-order Laser Harmonics. Rev. Sci. Instrum. 2009;80:123109.

[46] Ito M, Kataoka Y, Okamoto T, Yamashita M, Sekikawa T. Spatiotemporal Characteri-zation of Single-order High Harmonic Pulses from Time-Compensated Toroidal-Grating Monochromator. Opt. Express. 2010;18:6071–6078.

[47] Igarashi H, Makida A, Ito M, Sekikawa T. Pulse compression of Phase-Matched High Harmonic Pulses from a Time-Delay Compensated Monochromator. Opt. Express. 2012;20:3725–3732.

[48] Poletto L, Azzolin P, Tondello G. Beam-Splitting and Recombining of Free-Electron-Laser Extreme-Ultraviolet Radiation. Appl. Phys. B. 2004;78:1009–1011.

[49] Frassetto F, Villoresi P, Poletto L. Beam Separator for High-Order Harmonic Radia-tion in the 3–10 nm Spectral Region. J. Opt. Soc. Am. A. 2008;25:1104–1114.

[50] Poletto L, Frassetto F, Villoresi P. Ultrafast Grating Instruments in the Extreme Ultraviolet. IEEE J. Sel. Top. Quantum Electron. 2012;18:467–478.

[51] Werner W. X-Ray Efficiencies of Blazed Gratings in Extreme Off-Plane Mountings. Appl. Opt. 1977;16:2078–2080.

[52] Petit R. Electromagnetic Theory of Gratings. Berlin: Springer; 1980.

[53] Werner W, Visser H. X-Ray Monochromator Designs Based on Extreme Off-Plane Grating Mountings. Appl. Opt. 2006;20:487–492.

[54] Pascolini M, Bonora S, Giglia A, Mahne N, Nannarone S, Poletto L. Gratings in the Conical Diffraction Mounting for an EUV Time-Delay Compensated Monochromator. Appl. Opt. 2006;45:3253–3562.

[55] Martinez O. 3000 Times Grating Compressor with Positive Group Velocity Dispersion: Application to Fiber Compensation in 1.3–1.6 μm Region. IEEE J. Quantum Electron. 1987;23:59–64.

[56] Martinez O. Design of High-Power Ultrashort Pulse Amplifiers by Expansion and Recompression. IEEE J. Quantum Electron. 1987;23:1385–1387.

[57] Frassetto F, Poletto L. Grating Configurations to Compress Extreme-Ultraviolet Ultrashort Pulses. Appl. Opt. 2015;54:7985–7992.

[58] Frassetto F, Giannessi L, Poletto L. Compression of XUV FEL Pulses in the Few-Femtosecond Regime. Nucl. Instr. Meth. A. 2008;593:14–16.

[59] Poletto L, Bonora S, Pascolini M, Borgatti F, Doyle B, Giglia A, Mahne N, Pedio M, Nannarone S. Efficiency of Gratings in the Conical Diffraction Mounting for an EUV Time-Compensated Monochromator. In: Proc. SPIE Vol. 5534 Fourth Generation X-Ray Sources and Optics II; August 2004; Denver (USA). SPIE Publisher; 2004. p. 144–153.

[60] Poletto L, Frassetto F, Miotti P, Gauthier D, Fajardo M, Mahieu B, Svetina C, Zangrando M, Zeitoun P, De Ninno G. Grating-Based Pulse Compressor for Applications to FEL Sources. In: Proc. SPIE Vol. 9512, Advances in X-ray Free-Electron Lasers Instrumentation; April 2015; Prague (Czech Rep.). SPIE Publisher; 2015. p. 951210.

[61] Sola J, Mevel E, Elouga L, Constant E, Strelkov V, Poletto L, Villoresi P, Benedetti E, Caumes J-P, Stagira S, Vozzi C, Sansone G, Nisoli M. Controlling Attosecond Electron Dynamics by Phase-Stabilized Polarization Gating. Nat. Phys. 2006;2:319–322.

[62] Morlens A-S, Lopez-Martens R, Boyko O, Zeitoun P, Balcou P, Varju K, Gustafsson E, Remetter T, L'Huillier A, Kazamias S, Gautier J, Delmotte F, Ravet M-F. Design and Characterization of Extreme Ultraviolet Broadband Mirrors for Attosecond Science. Opt. Lett. 2006;31:1558–1560.

[63] Suman M, Monaco G, Pelizzo M-G, Windt D L, Nicolosi P. Realization and Characterization of an XUV Multilayer Coating for Attosecond Pulses. Opt. Express. 2009;17:7922–7932.

[64] Hofstetter M, Schultze M, Fie M, Dennhardt B, Guggenmos A, Gagnon J, Yakovlev V S, Goulielmakis E, Kienberger R, Gullikson E M, Krausz F, Kleineberg U. Attosecond Dispersion Control by Extreme Ultraviolet Multilayer Mirrors. Opt. Express. 2011;19:1767–1776.

[65] Bourassin-Bouchet C, de Rossi S, Wang J, Meltchakov E, Giglia A, Mahne N, Nannarone S, Delmotte F. Shaping of Single-Cycle Sub-50-Attosecond Pulses with Multilayer Mirrors. New J. Phys. 2012;14:023040.

[66] Guggenmos A, Rauhut R, Hofstetter M, Hertrich S, Nickel B, Schmidt J, Gullikson E M, Seibald M, Schnick W, Kleineberg U. Aperiodic CrSc Multilayer Mirrors for Attosecond Water Window Pulses. Opt. Express. 2013;21:21728–21740.

[67] Frassetto F, Villoresi P, Poletto L. Optical Concept of a Compressor for XUV Pulses in the Attosecond Domain. Opt. Express. 2008;16:6652–6667.

[68] Poletto L, Frassetto F, Villoresi P. Design of an Extreme-Ultraviolet Attosecond Compressor. J. Opt. Soc. Am. B. 2008;25:B133–B136.

[69] Poletto L, Villoresi P, Frassetto F. Diffraction Gratings for the Selection of Ultrashort Pulses in the Extreme-Ultraviolet. In: M. Grishin, editor. Advances in Solid-State Lasers: Development and Applications. INTECH, Croatia; 2010. p. 413–438.

[70] Mero M, Frassetto F, Villoresi P, Poletto L, Varju K. Compression Methods for XUV Attosecond Pulses. Opt. Express. 2011;19:23420–23428.

Fiber-based Sources of Short Optical Pulse

Chapter 10

Fiber-Based High-Power Supercontinuum and Frequency Comb Generation

Qiang Hao, Tingting Liu and Heping Zeng

Additional information is available at the end of the chapter

http://dx.doi.org/10.5772/64209

Abstract

Ultrafast optics has been a rich research field, and picosecond/femtosecond pulsed laser sources seek many applications in both the areas of fundamental research and industrial life. Much attention has been attached to fiber lasers in recent decades as they offering various superiorities over their solid-state counterparts with compact size, low cost, and great stability due to the inherent stability and safety of the waveguide structures as well as high photoelectric conversion efficiency. Fiber-based sources of ultrashort and high-peak/high-average optical pulses have become extremely important for high-precision laser processing while sources whose carrier-envelop offset and repetition rate are stabilized can serve as laser combs with applications covering many research areas, such as precision spectroscopy, optical clock, and optical frequency metrology. For the application as laser combs, four parts as fiber laser, broadband supercontinuum, nonlinear power amplification, and repetition rate stabilization must be concerned. This chapter is intended to give a brief introduction about the achievement of the four technologies mentioned above with different experimental setups, recently developed such as divided-pulse amplification (DPA) in emphasize. Moreover, detailed descriptions of the experimental constructions as well as theoretical analyses about the phenomena they produced are also involved.

Keywords: fiber lasers, divided-pulse nonlinear amplification, four-wave mixing, frequency stabilized

1. Introduction

Ultrafast laser sources and their applications such as high-power supercontinuum and frequency comb have gained much attention in recent decades [1–7]. High-power fiber lasers

spur a rapid growth of industrial applications including laser cutting, laser marking, and so on [8]. Moreover, supercontinuum and frequency comb are considered as the breakthrough of laser field for their applications covering precision spectroscopy, astronomical observations, and optical frequency metrology [9, 10]. This chapter is intended to describe, from experimental point of view, the ultrashort pulse laser oscillators, high-power nonlinear fiber amplifiers, supercontinuum, and frequency combs. Section 2 shows the performance of two types of mode-locked lasers. The first one consisting of bulk and fiber optical components is mode-locked via nonlinear polarization rotation (NPR) mechanism at 1.03 µm. The other one, operating at 1.55 µm, is mode-locked by nonlinear amplified loop mirror (NALM) with polarization-maintaining (PM) fiber components in order to overcome environmental perturbation and thus maintain long-term operation. Section 3 introduces a practical method (spectral tailoring), which facilitates supercontinuum generation in single-mode fiber amplifier at 1.03 µm with a few picosecond laser pulses. The second part in this section introduced broadband supercontinuum generation (from 950 to 2200 nm) by injecting pulses with 72-fs temporal duration, 150-mW average power, and 60-MHz repetition rate at 1560 nm into 20-cm-long PM-HNLF. Section 4 gives a brief introduction of divided-pulse amplification (DPA). To generate transform-limited pulse at 1.55 µm, DPA with polarized pulse duplicating was employed to overcome the gain narrowing effect and control the nonlinear spectral broadening in anomalous dispersion Er-fiber amplifier. As high as 500-mW average power at 1560 nm is achieved by ×8 replicas. Moreover, the highest frequency-doubling conversion efficiency reached 56.3% by using a periodically poled lithium niobate (PPLN) crystal at room temperature. Section 5 discusses an all-optical control method via resonantly enhanced optical nonlinearity (or pump-induced refractive index change, RIC) for high-precision repetition rate stabilization. The standard deviation (SD) of repetition rate can be reduced to a record level of <100 µHz by using the RIC method in a PM figure-eight laser cavity.

2. Fiber laser

Fiber lasers offer several practical advantages, such as excellent spatial-mode quality, effective heat dissipation, and flexible optical path and, recently, are becoming attractive laser sources in both scientific researches and industry applications. Especially, mode-locked fiber lasers with ultrashort pulse duration and high-repetition rate have attracted a lot of attention for their applications in optical sensing, optical communication, optical metrology, and biomedical imaging and processing [11, 12]. Therefore, various femtosecond/picosecond mode-locked lasers have been constructed and developed. As mode-locked lasers are often affected by environmental perturbations (mechanical vibration and temperature fluctuation), robust and stable oscillator with compact design are urgently needed. In this section, we present a compact femtosecond fiber laser at 1.03 µm by using integrated fiber optical components. The shortest dechirped pulse duration reaches 81 fs for a net cavity dispersion value close to −0.001 ps^2. Another part of this section described a self-started Er-doped laser oscillator, which is mode-locked by NALM with PM-fiber configuration. By optimizing the net dispersion, the buildup time can be reduced from 8 min to ~ms order of magnitude.

2.1. Operation regime of mode-locked lasers

As well known, the main features of mode-locked fiber laser depend on the pulse evolution process, which is relevant to the group-velocity dispersion (GVD) and the nonlinearity in optical fibers. According to the net intra-cavity dispersion, the pulse-shaping process can be roughly distinguished into the four different regimes, such as soliton regime, stretch-pulse regime, parabolic-pulse regime, and giant-chirp pulse regime, corresponding to all-anomalous dispersion, normal-anomalous dispersion, all-normal dispersion, and large-normal dispersion, respectively. Due to the equilibrium between Kerr nonlinearity and GVD, pulses that propagate in all-anomalous dispersion laser cavity keep unchanged in the form of fundamental soliton [13, 14]. While in dispersion-managed laser cavities, the negative dispersion is compensated by positive dispersion and thus stretch pulse forms. When the net cavity dispersion is optimized to zero, significant variations on pulse duration could be observed [15, 16]. As pulse operates in the all-normal dispersion regime, where laser gain, self-phase modulation, and dispersion co-effect, spectral/temporal filtering effects force linear chirping in the pulse, so that similariton forms [17–19]. While in ultra-long laser cavity, giant-chirped oscillating can be realized with ultralow repetition rate but at high-pulse energy [20–23].

2.2. NPR mode locking at 1.03 μm

In this section, we designed a compact ultra-fast Yb-doped fiber laser with integrated optical components. By integrating wavelength division multiplexer and optical isolator with

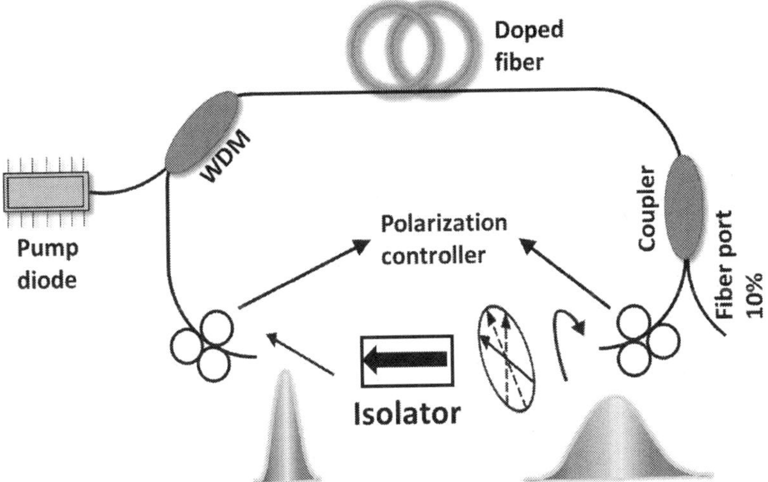

Figure 1. Schematic diagram of a passively mode-locked fiber laser via nonlinear polarization rotation mechanism.

collimators, the fiber loop was simplified. Self-started mode-locking could be realized by setting appropriate polarization angle of four intra-cavity wave plates. Due to the normal dispersion of fiber at 1.0 μm, transmission grating pair with 1250 l/mm was used to provide adjustable anomalous dispersion. As a result, 81-fs temporal duration with 65-MHz repetition rate and 0.5-nJ pulse energy was produced.

The mode-locking procedure can be explained in **Figure 1**. Two polarization controllers and a polarization-sensitive isolator (PSI) are used as the key elements for mode-locking. This combination acts as a virtual saturable absorber, which can absorb the low-intensity tail of pulse and transmit high-intensity part such that the pulse could be shortened. The pulse with linear polarization changes to elliptical polarization by twisting the polarization controller. As mentioned, self-phase modulation (SPM) or cross-phase modulation (XPM) can arouse energy coupling between two orthogonal polarizations. Moreover, serious nonlinear polarization rotation is produced by the high gain in the active fiber. Finally, another polarization controller is used to modify the polarization state to facilitate the central part of the pulse getting through the PSI [24].

In our experiment, a Yb-doped fiber laser shown in **Figure 2(a)** was firstly constructed without dispersion compensation elements. Three intra-cavity wave plates including two quarter-wave plates, QWP1 and QWP2, and one half-wave plate, HWP, were set with appropriate polarization angles to realize self-started mode-locking. The pigtails of fiber components are Hi1060 fiber with GVD of ~26 fs²/mm and TOD of ~41 fs³/mm, while the GVD for the active fiber is 39 fs²/mm. The repetition rate and pulse duration were measured to be 70 MHz and 13 ps, respectively. By external-cavity dechirping, the pulse can be compressed to 170-fs duration, but with obvious pedestal. The autocorrelation of pulse before and after dechirping is shown in **Figure 3(a)** and **(b)**, respectively.

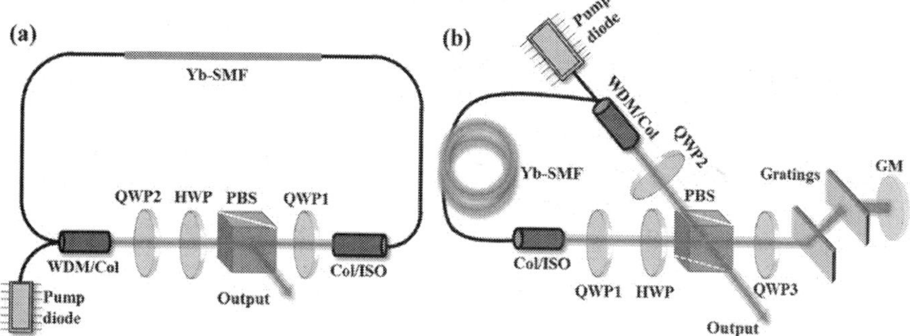

Figure 2. Structure of ultrafast Yb-doped fiber lasers without (a) and with (b) dispersion compensation. Pump diode: 400-mW laser diode at 976 nm; Yb-SMF: Yb-doped single-mode fiber; WDM/Col: the device combines WDM and collimator; Col/ISO: the device combines collimator and isolator; QWP1, QWP2, and QWP3: quarter-wave plate; HWP: half-wave plate; PBS: polarization beam splitter; GM: a gold-coated mirror.

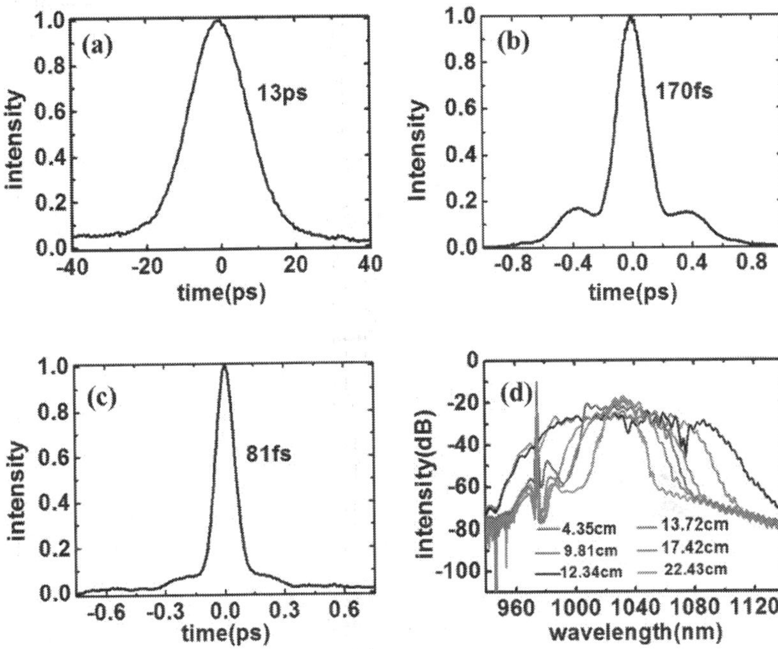

Figure 3. Autocorrelation trace of chirped pulses generated by picosecond fiber laser (a) and femtosecond fiber laser (c), (b) dechirped pulse of (a); (d) spectra generated by femtosecond fiber laser with different distances between gratings.

Secondly, a transmission grating pairs was used to manage the intra-cavity dispersion of the Yb-fiber laser, as shown in **Figure 2(b)**. The quarter-wave plate, QWP3, was used to impose 90° polarization rotation on laser pulses by double-passing the grating pairs. Soliton, stretch-pulse, and all-normal dispersion regime can be achieved by optimizing the distance between gratings. **Figure 3(d)** compares the various spectral shapes with different grating separations. As shown in **Figure 3(c)**, the shortest pulse duration was measured to be 81 fs. The black curve in **Figure 3(d)** represents the broadest spectra with a 10-dB bandwidth of 100 nm. The uncompensated phase was mainly caused by the accumulated high-order dispersion in fibers as well as intra- and extra-cavity grating pairs.

2.3. Polarization-maintaining figure-eight fiber laser at 1.55 μm

Compact size, low cost, and free maintenance fiber laser at 1.55 μm are desirable in many applications, such as eye surgery, Terahertz generation, and precision spectroscopy [25–28]. The standard repetition rate of commercial available fiber laser is typically 80 MHz with an optional design from 20 to 250 MHz. In order to combine both high pulse energy and high average power, a 10-MHz repetition rate is the best choice for applications. When the repetition rate is lower than 10 MHz, a pulse picker has to be used between the laser oscillator and the succeeding amplifier.

In this section, we introduce a PM figure-eight laser cavity which is the best option for oscillator operated at 10-MHz repetition rate. **Figure 4(a)** shows the experimental setup of the figure-eight laser cavity. The linear loop comprises of a 980/1550 nm wavelength division multiplexer, a segment of Er-doped fiber (PM-ESF-7/125, Nufern), an isolator, a 2-nm bandpass filter at 1550 nm, and an output coupler CP2 with a splitting ratio of 20:80. The active gain applied in the linear loop is to compensate the cavity loses and facilitate self-started mode-locking. The band-pass filter is used to block longer wavelength (Raman self-frequency shift) and reduce the temporal width of the pulse to be self-consistent. Pulses from the linear loop are coupled into NALM via CP1 with a splitting ratio of 45:55.

Over-pump with three LDs was applied to provide enough power for self-started mode-locking. Interestingly, the buildup time of mode-locking was found to be closely related to the net cavity dispersion. When the net dispersion was set to -0.115 ps^2, as long as 8-min buildup time was observed. After optimizing the net dispersion to about -0.062 ps^2, the time dramatically decreased to ~ms of magnitude.

Furthermore, we recorded the mode-locked pulse trains triggered by a square wave with 5-Hz modulation frequency which is simultaneously used to drive LD3. **Figure 4(c)** shows two

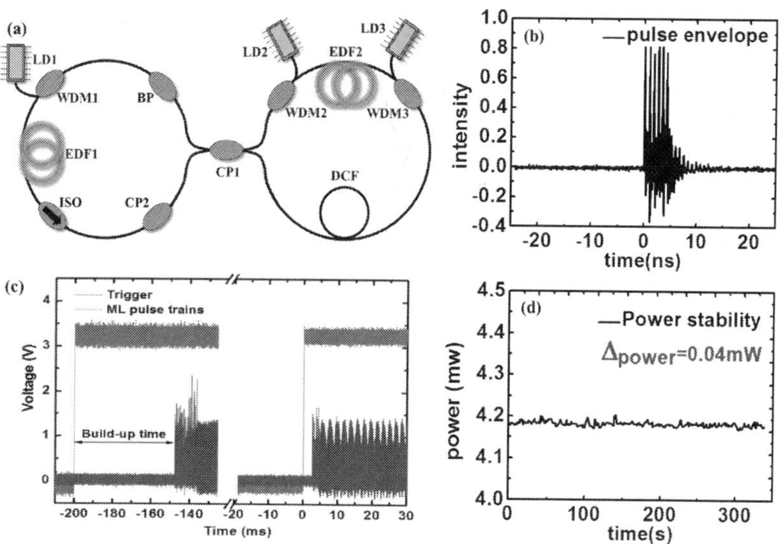

Figure 4. (a) Schematic of a polarization-maintaining figure-eight erbium-doped fiber laser. WDM1, WDM2, and WDM3: 980/1550 nm wavelength division multiplexers; ISO: isolator; EDF1 and EDF2: erbium-doped fiber; BP: 2-nm bandpass filter at 1550 nm; DCF: dispersion compensation fiber; CP1 and CP2: 1550 nm coupler with splitting ratio of 45:55 and 20:80. (b) The initial pulse that polarization-maintaining figure-eight erbium-doped fiber laser generated. (c) The buildup time measured by an oscilloscope. (d) The measurement of power stabilization once mode-locked.

adjacent periods with 50-ms pump duration, and the corresponding buildup time was measured to be 53 and 6 ms, respectively, exhibiting certain randomness. From experimental results point of view, the mode-locking buildup time is a random value in a certain range, which is related to the net cavity dispersion.

Interestingly, multiple-pulse operation was observed as the mode-locking is established, as shown in **Figure 4(b)**. Peak power clamping effect originating from sagnac mechanism resulted in the formation of pulse bunching [29]. Stable single pulse could be obtained by decreasing the pump power. In single-pulse regime, the 5-min power stability was measured to be 0.26%, as shown in the inset of **Figure 4(d)**.

3. Broadband supercontinuum

Recent years, supercontinuum generation (SC) has attracted much attention for its applications in optical coherence tomography, stimulated emission depletion microscopy, dense wavelength-division-multiplexing (DWDM) optical networks, and frequency comb generation [30–33]. In this section, several nonlinear optical effects such as SPM, XPM, four-wave mixing (FWM), and stimulated Roman scattering (SRS) that facilitate SC generation would be firstly discussed. Secondly, spectral filtering method is demonstrated to be an effective way for broadband supercontinuum generation in picosecond region [34]. By spectral filtering, a linear-chirped picosecond pulse with a 1-nm bandwidth filter installed between two Yb-doped single-mode preamplifiers, pulse shortening, and high peak power is achieved, so that an octave-spanning SC with bandwidth of 650 nm from 750 to 1400 nm and 10-dB peak-to-peak flatness was obtained at an output average power of 190 mW. Thirdly, SC covering from 950 to 2200 nm is generated in a 20-cm-long PM HNLF by injecting 72-fs pulse with 150-mW average power and 60-MHz repetition rate at 1.56 μm. Furthermore, an inline f-2f interferometer, including a PPLN for frequency doubling and a PM-fiber delay line, is used to generate carrier-envelop offset signal (f_{ceo}).

3.1. Nonlinear effects in optical fibers

Most of nonlinear effects in optical fibers attribute to nonlinear refraction, which refer to the intensity dependence of the refractive index. Especially, the lowest order nonlinear effects in optical fibers originate from the third-order susceptibility $\chi^{(3)}$, which governs the four-wave mixing, Raman effect, third-harmonic generation, and polarization properties [24].

This section does not thoroughly focus on the discussion of theoretical issues. In simple, the refractive index of the optical fiber can be described by the following equation

$$n = n_0 + n_1 \left| E(t) \right|^2 \tag{1}$$

where n_0 is the linear part and $n_1 |E(t)|^2$ is the nonlinear part.

An interesting phenomenon of the intensity dependence of the refractive index change in optical fiber occurs through SPM. When the input pulse is of low intensity, the corresponding refractive index remains a constant n_0. As the input pulse increases, the corresponding refractive index becomes nonlinear change with power intensity I. Hence, an additional phase shift is produced:

$$\delta\varphi(t) \propto |E(t)|^2 \tag{2}$$

This can be understood as an instantaneous optical frequency change from its central frequency:

$$\delta\omega(t) = -\frac{\partial}{\partial t}\delta\varphi(t) \tag{3}$$

Therefore, new spectral components are generated and time dependent frequency chirping is produced.

Another most widely studied nonlinear effect is XPM, which leads to asymmetric spectral and temporal changes for two co-propagating optical fields with different wavelength or orthogonally polarization. The contribution of the nonlinear phase shift induced by XPM is twice that of SPM. Therefore, the nonlinear part Δn_j induced by the third-order nonlinear effects is given by ($j = 1, 2$)

$$\Delta n_j \approx n_2\left(|E_j|^2 + 2|E_{3-j}|^2\right) \tag{4}$$

Eq. (4) shows the refractive index of the optical media seen by an optical field inside a single-mode fiber depends not only on the intensity of that field but also on the intensity of the other co-propagating fields [35]. As the optical field propagates inside the fiber, an intensity-dependent nonlinear phase shift shows up

$$\phi_j^{NL}(z) = n_2\left(\omega_j / c\right)\left(|E_j|^2 + 2|E_{3-j}|^2\right)z \tag{5}$$

The first term is related to SPM while the second term is related to XPM.

Stimulated Raman scattering (SRS) is an important nonlinear process that can produce red-shifted spectral components. Once the spectrum of the input pulse is broad enough, the Raman gain can amplify the long-wavelength components of the pulse with the short-wavelength components acting as pumps, and the energy appears red-shifted. The longer the propagating fiber, the more red-shifted spectral components can be generated. The red-shifted components are called Stokes wave. The initial growth of the Stokes wave can be described by

$$\frac{dI_s}{dz} = g_R I_p I_s \tag{6}$$

where I_s is the Stokes wave intensity, I_p is the pump-wave intensity, and g_R is the Raman-gain coefficient, which is related to the cross section of spontaneous Raman scattering.

The Raman-gain coefficient g_R (Ω) is the most important factor to describe SRS. Ω represents the frequency difference between the pump wave ω_p and Stokes wave ω_s. In the case of silica fibers, the Raman-gain spectrum is found to be very broad, extending up to approximately 40 THz. Assuming the pump wavelength is 1.5 μm and, peak gain is $g_R = 6 \times 10^{-14}$ m/W, the frequency downshift can be calculated to be 13.2 THz.

When supercontinuum is generating in an optical fiber, the SPM, XPM, and SRS are always accompanied by FWM. In optical fibers, FWM transfers energy from pump wave (ω_p) to two other waves in frequency domain, blue-shifted (anti-Stokes wave, ω_{as}) and red-shifted (Stokes wave, ω_s). Once the phase-matching condition $\Delta k = 2k(\omega_p) - k(\omega_s) - k(\omega_{as}) = 0$ is satisfied, the Stokes and anti-Stokes waves can be amplified from noise or an incident signal at ω_s or ω_{as}, respectively [36, 37]. Therefore, FWM process is used to produce spectral sidebands for supercontinuum generation.

3.2. Supercontinuum generation

SC is a powerful laser sources for many applications, such as nonlinear microscopy, optical coherence tomography, and frequency metrology [38–40]. Nowadays, more than one octave SC can be easily generated with a length of PCF, and the average power can reach tens of Watts [41, 42]. When ultrashort optical pulses propagate through a PCF fiber, the combination of SPM, XPM, SRS, and FWM is responsible for the spectral broadening. Generally speaking, the feature of SC depends on whether the incident laser wavelength λ located is below, closed to, or above the Zero-dispersion wavelength λ_D of the PCF. In the anomalous-dispersion regime ($\lambda > \lambda_D$) where $\beta_2 < 0$, soliton affects. If the λ nearly coincides with λ_D, $\beta_2 \approx 0$, β_3 dominant and the phase-matching condition of FWM are approximately satisfied. While in the normal-dispersion regime, $\beta_2 > 0$, GVD and SPM effects dominant SC generation. From the time domain of view, SPM and soliton effects dominant SC generation for femtosecond (typically <1 ps) pump pulses, while FWM and SRS contribute to spectral broadening for tens of picoseconds pulses.

3.2.1. Spectrally filtered seed for broadband supercontinuum generation in single-mode fiber amplifiers

There are several methods to extend the SC spectrum. Considerable spectral broadening could be observed with high-power incident laser. High-average/high-peak powers facilitate CW and pulse SC generation [43–45]. Besides, the SC bandwidth could also be increased by tapering PCFs. A flat (3 dB) spectrum from 395 to 850 nm was achieved in a tapered fiber with a continuously decreasing ZDW [46]. In this section, we demonstrate an effective method for broadband SC generation, which is valid in normal-dispersion fiber amplifiers. By spectral filtering of upchirped pulse at 1028 nm with 1-nm bandpass filter, as broad as 650-nm bandwidth from 750 to 1400 nm within 10-dB peak-to-peak flatness is obtained with an output power of 190 mW.

The experimental setup is shown in **Figure 5(a)**. The SC laser source is consisted of a picosecond mode-locked laser oscillator, a spectral filter, two-stage single-mode amplifiers, and 2-m-long PCF with ZDW at 1.02 µm. The laser oscillator operated in an all-normal-dispersion regime with repetition rate of 20 MHz. With 100 mW pumping power, 25 mW average output power laser is exported from the 30% port of the coupler. The pulse duration of highly up-chirped pulse was measured to be 10 ps.

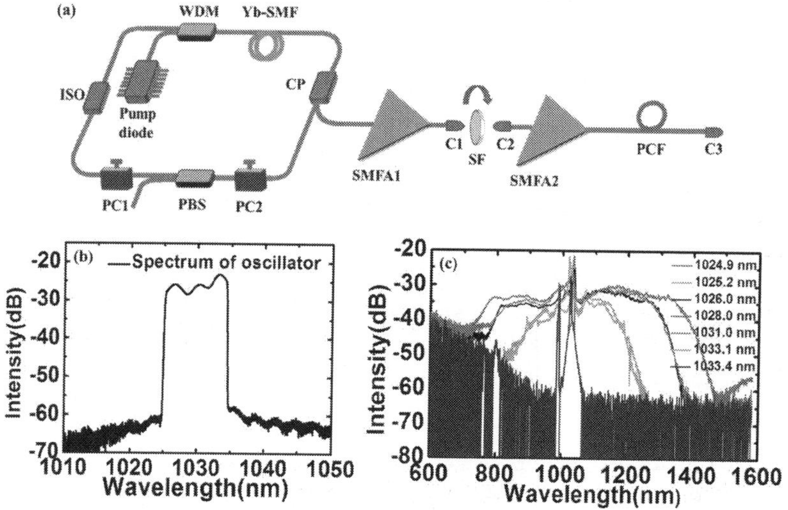

Figure 5. (a) Experimental setup for SC generation. Pump diode: 400 mW laser diode at 976 nm; WDM: 980/1040 nm wavelength division multiplexer; Yb-SMF: ytterbium-doped single-mode fiber; CP: 30:70 coupler; PC1 and PC2: polarization controller; PBS: polarization beam splitter; ISO: isolator; SMFA1 and SMFA2: single-mode fiber amplifiers; C1, C2, and C3: three collimators; SF: spectral filter; PCF: photonics crystal fiber. (b) The output spectrum of the laser oscillator. (c) SC with different filtering windows.

A bandpass spectral filter with 1-nm bandwidth at 1036 nm is installed between two single-mode fiber amplifiers. The transparent wavelength of the filter could be tuned from 1024 to 1036 nm by varying the incident deflection angle. For the large up-chirp with 10-nm spectral width (see **Figure 5(b)**) and 10-ps temporal duration, corresponding to a time-bandwidth product of 28.3, pulse can be greatly shortened by the filter. The shortest pulse duration of 2.9 ps was obtained with filtering window at 1028 nm. After the second-stage amplifier, the laser pulses could be amplified to an average power up to 190 mW with 400 mW pumping power.

A 2-m length of silica-based PCF with ZDW at 1024 nm is directly spliced to the fiber end of SMFA2. **Figure 5(c)** shows the output spectrum with different filtering window. The spectra keep unchanged when the filtering window is at the shoulder of the spectrum, shown as the red curve (1024.9 nm) and blue curve (1033.4 nm) in **Figure 5(c)**. When filtering window is located at the central wavelength of 1028 nm, the 10-dB bandwidth of SC is extended to 650 nm (from 750 to 1400 nm), shown as the pink curve in **Figure 5(c)**. Besides, filtering windows above or below the central wavelength produce a less broad SC.

3.2.2. One octave supercontinuum for frequency comb generation

Broadband supercontinuum of bandwidth up to 1250 nm can also be provided by HNLFs with spectral-tailored femtosecond pump pulses produced by erbium-doped power amplifiers. The schematic diagram of the experiment is shown in **Figure 6(a)**.

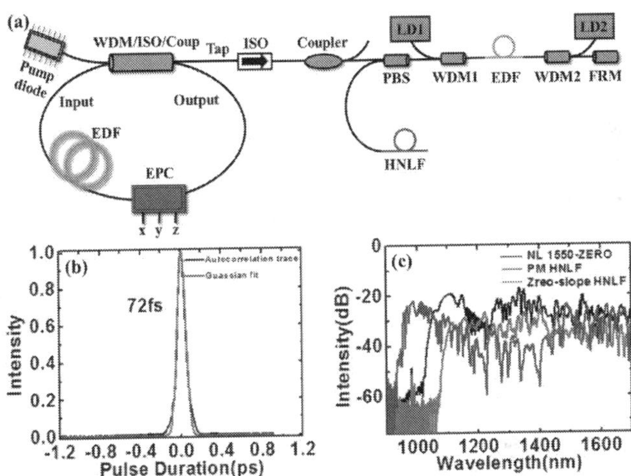

Figure 6. (a) Schematic diagram for SC generation. Pump diode: 400-mW pump at 976 nm; WDM/ISO/Coup: the device combines wavelength division multiplexer, isolator, and coupler; EDF: erbium-doped fiber; EPC: electric polarization controller; ISO: isolator; coupler: 30:70 polarization-maintaining coupler; PBS: polarization beam splitter; HNLF: high nonlinear fiber; LD1 and LD2: pumps at 976 nm; WDM1 and WDM2: 980/1550 nm wavelength division multiplexer; FRM: Faraday rotation mirror. (b) Autocorrelation trace of chirped pulses poured into HNLFs. (c) SC generated by different kinds of HNLFs on logarithmic coordinate.

The laser system consisted of an erbium-doped mode-locking fiber oscillator, a single-mode fiber amplifier (SMFA), and 20-cm-long PM-HNLF. To improve the mode-locking stability, an electric polarization controller (EPC) is utilized to replace the conventional mechanical polarization controller such that automatic and active control of mode-locking is accessible. By applying the voltage on three axes (x, y, and z) of EPC, accurate control of the temporal duration, spectral shape, f_{rep}, and f_{ceo} can be achieved [47, 48].

With the help of a PBS and a FRM, a dual-pass single-mode fiber amplifier with bidirectional pump configuration was used to boost the laser average power to more than 150 mW average power and reduce the environmental disturbance on SMFA. The pulse duration at the output port was measured to be 2.84 ps. Additional PM-1550 fiber was used to dechirp the pre-amplified pulse to 72 fs (shown in **Figure 6(b)**). Therefore, considering a repetition rate of 60 MHz, the pulse peak power achieved as high as 34.7 kW. Three types of HNLFs, such as NL 1550-ZERO, PM-HNLF, and Zero-slope HNLF, were applied to generate the supercontinuum by splicing the HNLFs to the dechirping fiber directly.

As shown in **Figure 6(c)**, 20-cm-long PM-HNLF with nonlinearity of 10.5 W^{-1} km^{-1} achieved the broadest spectrum, covering from 950 to 2200 nm, which is sufficient broad to produce f_{ceo} signal. The HNLF type should be taken into consideration as it influences the SC generation. We used three kinds of HNLFs: 25-cm-long NL 1550-ZERO with nonlinear coefficient of 10.4-W^{-1} km^{-1} and effective mode area of 13-μm^2, 20-cm-long PM-HNLF with nonlinear coefficient of 10.5-W^{-1} km^{-1} and effective mode area of 12.7-μm^2, 25-cm-long Zero-slope HNLF with nonlinear coefficient of 10.8-W^{-1} km^{-1} and effective mode area of 12.4-μm^2, and the corresponding SC was depicted in **Figure 6(c)**. Obviously, PM-HNLF produces broader spectrum than other two HNLFs.

As shown in **Figure 7**, a collinear setup was established for the detection of f_{ceo} signal. The SC generated by 20-cm-long PM-HNLF was coupled into free space via a lens (L1) with adjustable focal length. An inline f-2f interferometer, including a PPLN, several wave plates and lens, and a PM-fiber delay line, is used to produce the temporal overlapped components at 1.0 μm. The long-wavelength component of SC at 2092 nm was frequency doubled to match with the short-wavelength component at 1046 nm. After the PPLN, two lenses, L3 and L4, were used to couple the two components at 1046 nm back to PM-980 fiber. A half-wave plate, HWP2, is used to adjust the energy ratio on the fast and slow axes of PM-980 fiber. The pulse transmitted along the slow axis experiences a delay relative to the pulse on fast axis. With an optimized fiber length of 3.4 m, the differential delay between the fast and slow axes could be fully compensated [49]. Subsequently, a half-wave plate, HWP3, as well as a PBS were used to selected pulses to generate f_{ceo} signal on APD. Finally, with 28-dB signal-to-noise ratio was generated by using this setup.

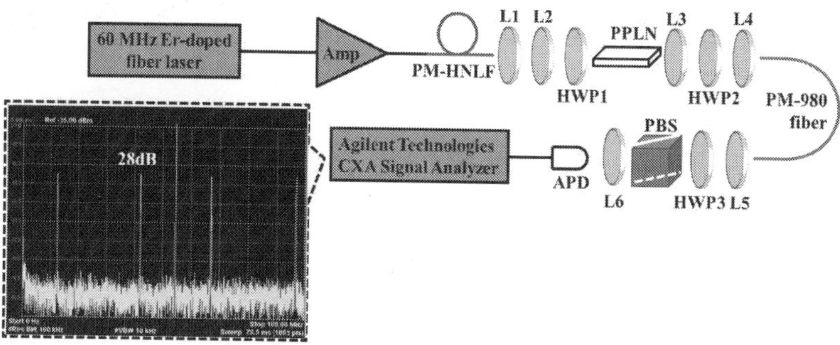

Figure 7. Setup for f_{ceo} detection. Amp: fiber amplifier; L1, L4, and L5: optical lens with adjustable focal length; L2, L3, and L6: optical lens with focal length of 50 mm; HWP1, HWP2, and HWP3: half-wave plates; PPLN: periodically poled lithium niobate; PBS: polarization beam splitter; APD: avalanche photodiode.

4. Nonlinear fiber amplifier

With the increasing applications in frequency metrology, THz generation, and in cataract surgery, the development for high-energy transform-limited pulse generation around 1.55 µm is still a fascinating area [50–53]. Owing to the limited available output power from laser oscillator, erbium-doped fiber amplifiers (EDFAs) are commonly used. Nevertheless, high-power amplification in EDFA is inevitably accompanied by several unwanted effects, such as SRS and amplified spontaneous emission (ASE), which would significantly deteriorate the temporal and spectral duration of pulse [54]. Chirped-pulse amplification (CPA) provides an effective way to decrease pulse peak power and avoid the nonlinearity in optical fibers [55–58]. In CPA, strong stretching and compression occurs to extract more energy and avoid nonlinear distortion as well as damage. However, CPA is inevitably accompanied by the gain-narrowing effect and therefore hardly produces pulse with temporal duration less than 400 fs [59].

Even though CPA has many advantages over the other techniques in amplifying pulses around 1.55 µm, ~100-fs pulse duration with above 10-nJ pulse energy is still a challenge because of the spectral gain-narrowing effects and nonlinear phase accumulation. Moreover, due to the involvement of bulk media, CPA is not suited for applications that require compact size and alignment-free laser source. A recent developed technique, divided-pulse amplification (DPA), opens up a new way for high-power laser pulse amplification [60–62]. In the configuration of DPA, the initial pulse is divided into a sequence of lower-intensity pulse with orthogonal polarization for successive replicas, and subsequently, the low-energy pulse is amplified and then recombined to create a high-energy pulse [61, 63].

In this section, we mainly focus on DPA at 1.56 µm where pulse amplification and compression can be simultaneous carried out so that a separate compressor is no longer necessary. The

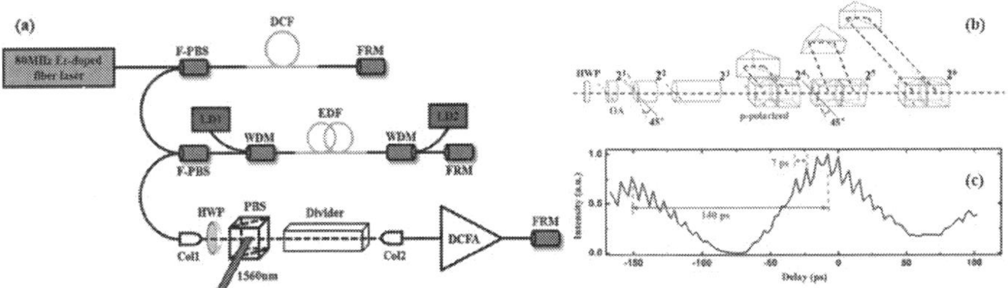

Figure 8. (a) Schematic diagram for laser system. F-PBS: fiber-coupled polarization beam splitter; DCF: dispersion compensation fiber; FRM: Faraday rotation mirror; LD1, LD2: pump diodes at 976 nm; WDM: wavelength division multiplexer; EDF: erbium-doped single-mode fiber; Col1, Col2: high-power collimators; HWP: half-wave plate; PBS: polarization beam splitter cube; DCFA: double-clad fiber amplifier. (b) Schematic of the pulse divider. The left-hand three cylinders (21, 22, and 23) represent YVO4-based dividers with given direction of crystal optical axes (OAs) shown as the red dash dot lines. The right-hand three parts (24, 25, and 26) represent PBS-based dividers with p-polarized direction shown as red dash dot lines. The red dot lines represent the horizontal plane, which is the same direction as the OA of 21 and 23 and the p-polarized direction of 24 and 26. (c) The measured autocorrelation trace of the divided replicas.

schematic diagram of DPA is shown in **Figure 8(a)**. The experimental setup is composed of a mode-locked fiber laser, a fiber stretcher, a single-mode fiber amplifier for preamplifying, and a pulse-divider as well as a double-clad fiber amplifier for main amplification. The Er-doped fiber laser with 80-MHz repetition rate shared the same configuration as **Figure 6(a)**, which takes the advantage of EPC to actively control the mode-locking. A photodiode and an electric loop were applied to monitor and feedback control the EPC for long-term stable operation. The fiber oscillator consisted of 1.74-m SMF28-e fiber with dispersion parameter of 19 ps/nm/km and 0.82-m Er-doped fiber with dispersion parameter of −51 ps/nm/km. There, the laser operated in the stretched-pulse regime and produced positively chirped pulses. As a result, with 200-mW pumping power at 976 nm, the laser oscillator produces 5-mW output average power with 1.5-ps pulse duration and 28-nm spectral bandwidth, corresponding to a time-bandwidth product of 5.2.

A fiber stretcher is spliced to the output of the fiber oscillator to stretch laser pulse and control the quantity of frequency up-chirp. However, over-long fiber could inevitably introduce too much high-order dispersion which is hardly compensated by the pulse-compressing stage. For the current configuration, a double-pass fiber stretcher consisting of a fiber-coupled PBS with PM fiber at input/output port and non-PM fiber at common port, a segment of non-PM dispersion compensation fiber with 6.0-μm-mode field diameter and −38 ps/nm/km dispersion at 1550 nm, and a FRM is used to reduce the environmental perturbation.

In our experiment, 6-m dispersion compensation fiber was applied to stretch pulses from the fiber oscillator. A dual-pass bidirectionally pumped single-mode fiber preamplifier was used to boost the average power to more than 100 mW to ensure efficient operation of the subsequent amplifiers. A FRM reflected the incident pulse to suppress ASE noise and rotated the polarization of the pulses by 90° to cancel all birefringence effects in the dual-pass amplifier. A fiber-based polarization beam splitter (F-PBS) was used to couple the seed laser to the preamplifier and reflected preamplified pulses to subsequent components. The output characters of the preamplifier were shown as the blue curves in **Figure 9(a)** and **(b)**. The FWHM temporal duration and spectrum bandwidth of the preamplified pulses is 4 ps and 15 nm, respectively, generating a time-bandwidth product of 7.4. Dramatic decrease in spectral bandwidth was

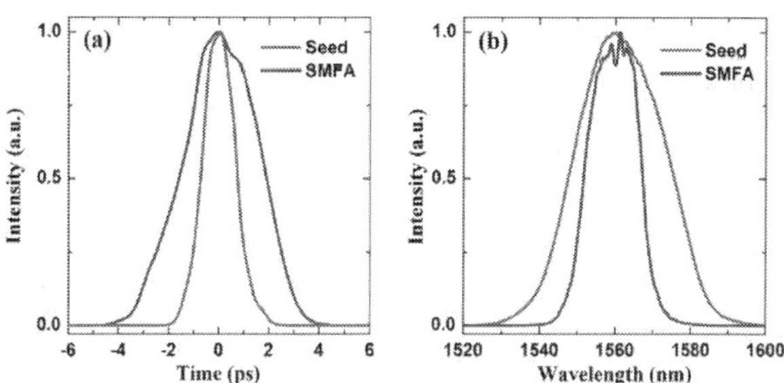

Figure 9. The temporal duration (a) and spectral bandwidth (b) of laser pulses from laser oscillator (red curves) and SMFA (blue curves).

observed due to the limited transmission bandwidth of WDM and FRM as well as spectral-narrowing effect in fiber amplifier.

Then, the concept of DPA was carried out to boost the laser to Watt-level average power. The preamplified laser is coupled into free space by collimator C1 and rotated to horizontal polarization to reach maximum transmission on PBS. The pulse division and combination were achieved by applying cascaded YVO_4-based and PBS-based dividers with the help of a FRM to reflect the replicas passing through the same divider but in the opposite direction. Each divider (YVO_4-based or PBS-based) can divide a single pulse into two cross-polarized replicas; hence, a single seed pulse could be temporally divided into 2^N (where N is the stage number of the divider) replicas. Ideally, each replica has identical pulse energy after division. As depicted in **Figure 8(b)**, three YVO_4 crystals with lengths of 10, 20, and 40 mm divided the initial pulse into $N = 8$ replicas. A half-wave plate (HWP) was used to produce the desired polarization of input pulses. The first (2^1) and third (2^3) YVO_4 crystals had their crystal optical axes (OA) oriented in the same direction as the horizontal plane, while the OA of the second (2^2) YVO_4 crystal oriented at a 45° angle to the horizontal plane. The polarization-mode delay between ordinary and extraordinary waves in YVO_4 is 0.7 ps/mm at 1560 nm. The shortest crystal length for our system was chosen to split the input pulse into replicas with 7 ps separation, about twice of the seed pulse duration.

To mitigate the nonlinearity in main amplifier, the string of pulse ($N = 8$) was further divided by three PBS-based dividers, resulting in a final pulse number of 64. For PBS-based divider, each incoming pulse was divided into an s-polarized beam and a p-polarized beam. All p-polarized components were directly transmitted the PBS, while the s-polarized components were reflected to the folded delay line. For the sake of simplicity, the second PBS-based divider (2^5) had its p-polarized direction oriented 45° to the direction of the horizontal plane, while the first (2^4) and third (2^6) PBS-based oriented in the same direction as the horizontal plane, such that separate half-wave plates were no longer necessary.

Owing to the delay length of 10, 20, 40, 26.8, 53.6, and 107.2 mm, the 2^1, 2^2, 2^3, 2^4, 2^5, and 2^6 stages approximately provided time delay of 7, 14, 28, 130, 260, and 520 ps, respectively. **Figure 8(c)** shows the measured autocorrelation trace of the pulse string which matches well with the designed time delays. The 7-ps interval between adjacent peaks in the same envelope was consistent with the expected time delay with 10-mm increment length of YVO_4, and the ~140-ps spacing between two adjacent envelopes was consistent with the expected time delay introduced by the PBS-based divider.

Intuitively, for simultaneous pulse amplification and compression in EDFAs, a positively pre-chirping seed pulse is desired. Numerical simulations show that there exists an equilibrium position that can not only restrict excessive nonlinear effects to ensure high-quality temporal integrity but also produce sufficient optical nonlinearity to broaden the spectrum around the wavelength of 1.55 μm. The generalized nonlinear Schrödinger equation (7) with the split-step Fourier method was used to carry out the simulation [24].

$$\frac{\partial A}{\partial z} = \frac{\alpha}{2} A + \sum_{n \geq 2} \frac{i^{n+1}}{n!} \beta_n \frac{\partial^n A}{\partial T^n} + i\gamma A \int_{-\infty}^{\infty} R(T') \left| A(z, T - T') \right|^2 dT' \tag{7}$$

where $A = A(z, t)$ is the complex amplitude of the pulse envelope of pulses, α is the laser gain coefficient, β_n is the dispersion parameter at ω_0 (1560 nm), and γ (3 W^{-1} km^{-1}) is the nonlinear coefficient. The right-hand side of Eq. (7) models laser gain, dispersion, and nonlinearity. Pulse of 2.5-ps temporal duration and 19.9-nm spectral width (corresponding to 0.16 ps^2 prechirping on 180-fs transform-limited pulse) and a pulse energy of 0.05 nJ were applied in the simulation.

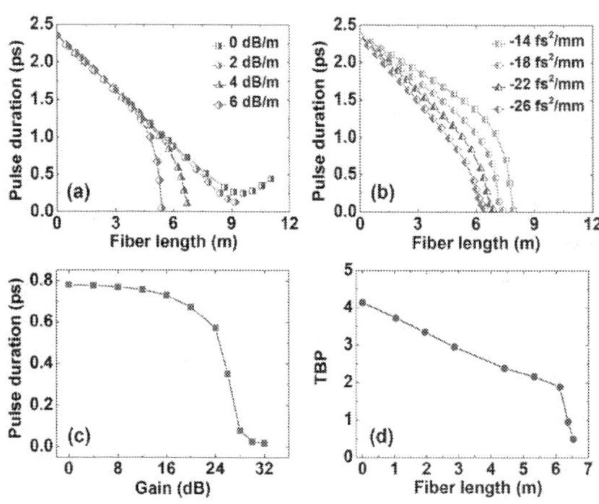

Figure 10. Amplified output pulse duration versus propagation length for the cases of different α (a) and β_2 (b). (c) Pulse duration versus the total gain provided by 6.5-m fiber. (d) Time-bandwidth product at different position of gain fiber.

The interplay of the SPM and group-velocity dispersion (GVD) as well as laser gain can lead to a qualitatively different behavior compared with that expected from them alone. SPM broadened the spectrum with increase in pulse energy, and simultaneously, the anomalous dispersion of the fiber compressed the new spectral components resulting in temporal shortening. **Figure 10(a)** compares the simulation results with different α but a fixed β_2 (−22 fs^2/mm). It is clear that pulse compression operates in linear regime when the laser gain is low, then it enters in nonlinear regime when the laser gain gradually increased. The shortest transform-limited pulse duration decreased from 180 fs at 7.0 m ($\alpha = 0$ dB/m) to 60 fs at 4.3 m ($\alpha = 3$ dB/m). **Figure 10(b)** compares the pulse compression with different β_2 but a fixed α (3 dB/m). The maximum pulse energies with respect to fiber length of 4.65, 5.07, and 5.53 m reached 1.24, 1.66, and 2.28 nJ, respectively. Therefore, higher α and smaller $|\beta_2|$ are benefit to overcome spectral bandwidth limitation for high-energy pulse amplification. For reference, the blue curves in **Figure 10(a)** and **(b)** present pulse evolution with the same parameters.

Next, we focus on pulse amplification and compression in a fixed fiber length by way of guiding the subsequent experiment. About 5.0-m-long fiber with $\beta_2 = -22$ fs²/mm was introduced to simulate the output pulse duration and the time-bandwidth product (TBP) at different position along the fiber. As shown in **Figure 10(c)**, when the total gain is smaller than 16 dB, the output pulse duration deceases linearly owing to the GVD and insufficient nonlinearity. As the total gain is greater than 24 dB, the output pulse duration dramatically decreases owing to strong nonlinear compression. Theoretically, pulse as short as 80-fs duration can be achieved with a total gain of nearly 28 dB. Although the pulse duration could be further decreased to 20 fs with 32-dB gain, considerable pedestal as well as wave breaking appears due to excessive nonlinearity. Meanwhile, the TBP of the pulse along the fiber gradually decreases from 4.1 at the input port to 0.5 at the output port.

Experimentally, the divider first operated with ×8 replicas. A dual-pass double-cladding fiber amplifier (DCFA) was used to boost the divided pulses. The DCFA consisted of 1.2-m 12/130 Er-doped double-clad fiber and 2.5-m SMF fiber. With a pump power of 4.3 W at 976 nm, the DCFA delivered 600 mW output power at 1560 nm, as measured at PBS output port. Along the first pass of DCFA, pulse evolution worked in the near-linear regime. Subsequently, as the pulse reflected by the FRM and passed through DCFA again, SPM and anomalous dispersion brought the pulse amplification into the moderate nonlinearity regime. **Figure 11(a)** and **(b)** shows pulse duration and spectral bandwidth of the combined pulse measured by an FROG (15-100-USB, Swamp Optics). The linear spectral phase indicates a nearly transform-limited pulse with the FWHM duration of 97 fs if a Gaussian pulse shape is assumed. The output power limitation was resulted from splicing losses of different type fibers and the insertion loss of FRM.

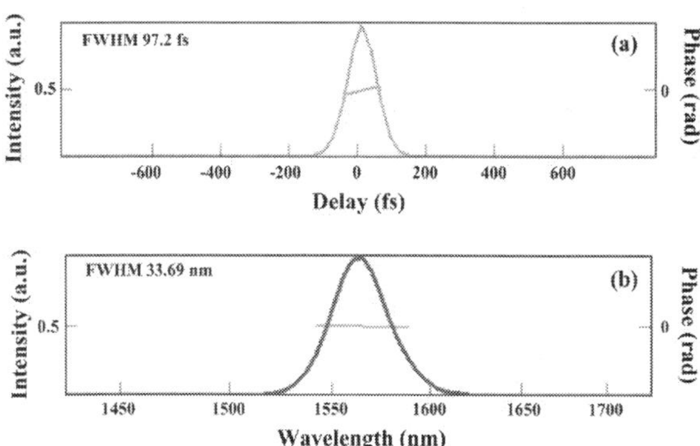

Figure 11. The measured pulse duration (a) and spectral width (b) of pulses from the nonlinear fiber amplifier.

Furthermore, a PPLN with 20.9-μm poling period and 0.3-mm length was used for frequency-doubling the amplified laser and checking the available peak power at 1560 nm. A pair of lens was used to focus and collimate the input and output beam on the PPLN, respectively. The output average power at 1560 nm and the corresponding SHG is shown in **Figure 12**. The highest SHG conversion efficiency was obtained as 56.3% with 302 mW incident power at 1560 nm. Further increasing the power at 1560 nm induced decay of conversion efficiency, shown as the black squares in **Figure 12**.

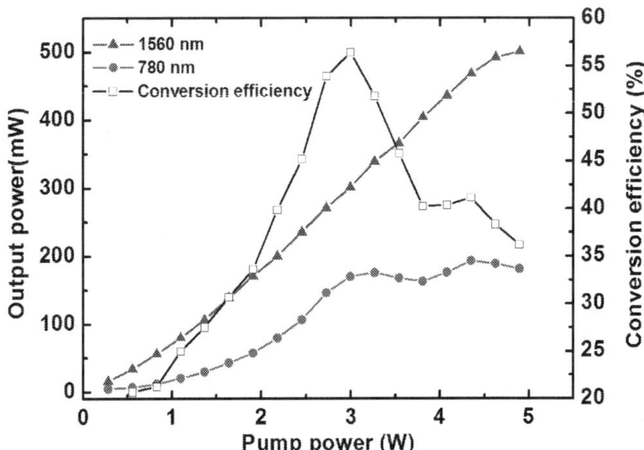

Figure 12. The output average power of 1560 nm (blue triangles) and 780 nm (red circles), and the SHG conversion efficiency.

To extract more energy from the double-pass amplifier, we will increase the number of the replicas from 8 to 16 and 32. The results for ×16 and ×32 replicas are still under investigation.

In conclusion, a divided fiber laser fiber amplifier delivering 500 mW average power at 1560 nm by the interplay between divided prechirped pulse amplification and nonlinear pulse compression. A small core double-clad erbium-doped fiber with anomalous dispersion carries out the pulse amplification and simultaneously compresses the laser pulses such that a separate compressor is no longer necessary. A numeric simulation reveals the existence of an optimum fiber length for producing a transform-limited pulse. Furthermore, frequency doubling to 780 nm with 170-mW average power is realized by using a PPLN at room temperature.

5. Repetition rate stabilization

Fiber-based frequency comb is recognized as the key breakthrough in the field of optics for it brings high accuracy in frequency domain as well as low jitter in time domain [49, 64–67]. Principally, as a frequency comb, two RF frequencies, f_{ceo} and f_{rep}, are required to be stabilized to external references. Therefore, the optical frequencies can be written as $v = m \times f_{rep} + f_{ceo}$, where m is a large integer of order 10^6 that indexes the comb line. Nevertheless, recent developments in adaptive dual-comb spectroscopy successfully employed free-running mode-locked lasers where the f_{ceo} instabilities could be compensated by data acquisition and electronic signal processing [68, 69]. Therefore, high accuracy f_{rep} stabilization of passively ML lasers is of great importance.

The relatively mature method for f_{rep} locking is to use a piezoelectric ceramic transducer (PZT) to control the geometrical length L of the laser cavity, and the best locking accuracy is in the range of ±0.5 mHz with the corresponding SD of 220 µHz [70]. However, the PZT-based stabilization encounters many limitations, such as significant positioning errors, hysteresis effect, bulky-design, and the need for time-consuming alignment.

In this section, we focus on the f_{rep} stabilization by using optical pumping scheme which can be achieved via resonantly enhanced optical nonlinearity or so-called pump-induced refractive index change (RIC) in doped fibers. In optical pumping scheme, the f_{rep} is stabilized by modulating the refractive index n, while keeping the geometrical cavity length L fixed. In the past, this method has been successfully applied in fiber switch where a low pump power and a short length doped fiber are sufficient for the switching [71]. Moreover, the validity of this concept has also been achieved in coherent combining and adaptive interferometry [72]. In 2013, Rieger et al. reported all-optical stabilized repetition rate by using the RIC-based method. With the help of thermos-electric element, over 12-h long-term stabilization was achieved in an NPE-mode-locked Er-doped fiber laser, while the SD of repetition rate drift was measured

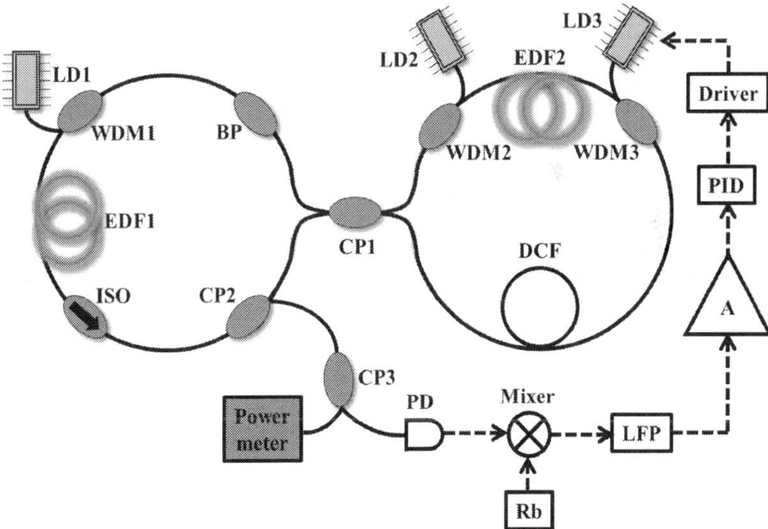

Figure 13. Experimental setup. LD1, LD2, and LD3: pump diodes at 976 nm; WDM1, WDM2, and WDM3: 980/1550 nm wavelength division multiplexers; EDF1, EDF2: erbium-doped fiber; BP: 2-nm bandpass filter centered at 1550 nm; CP1, CP2, and CP3: 1550 nm couplers with splitting ratio of 45:55, 30:70, and 50:50, respectively; DCF: dispersion compensation fiber; PD: photodiode detector; Rb: Rubidium clock; LFP: low-pass filter; A: electronic amplifier; PID: proportional–integral–derivative controller; Driver: precision current source.

to be 22 mHz. A recent experiment extends this concept to Yb-fiber laser and achieves 1.39-mHz SD of residual fluctuation in an hour measurement [73].

As reported in Ref. [74], a commercial available pump current supply can provide a minimum resolution of pump power as 1.5 µW and thus achieve a controlling accuracy of 0.05 Hz, which is more than two orders of magnitude than PZT-based method. Therefore, an interesting experiment worth to do is to use RIC-method to achieve high-precision f_{rep} stabilization. So far, the RIC method has been fully investigated in NPR mode-locked lasers, which applied non-PM fibers and components [73–75], and the locking accuracy limited to ~mHz. Considering the environmental perturbation on non-PM fiber, a straightforward idea is to implement RIC method in a PM fiber laser. Therefore, the following part will discuss high-precision repetition rate stabilization by using RIC method in a PM figure-eight laser cavity.

The laser setup shown in **Figure 13** is same as **Figure 4(a)**, except the net dispersion of laser cavity. In the current experiment setup for all-optical repetition stabilization, a 56-cm-long Er^{3+}-doped fiber (EDF2) is spliced asymmetrically in the NALM to act as a frequency controller, while the LD3, which is controlled by the error signal from frequency mixer, provides the feedback modulating pump power via WDM3 on EDF2. Besides, a segment of DCF38 is used to compensate the anomalous dispersion of PM1550 fiber. The dispersion of linear loop and the NALM was estimated numerically to be −0.208 and 0.025 ps^2, producing −0.183 ps^2 net dispersion for the whole cavity. Self-started mode-locking in multiple-pulse regime can be achieved by over-pumping method, and stable single-pulse operation can be obtained by decreasing the pump power of LD1 and LD2. At fundamental repetition rate of 11.9 MHz, the figure-eight laser cavity delivers 1.5-mW average power via CP2.

The repetition rate was detected by PD3 and successively compared with standard reference (Rb clock) in a frequency mixer to produce the error signal. Subsequently, the error signal was filtered and amplified by low-noise voltage preamplifier with frequency cutoff at 1 MHz and a maximum voltage gain of 5×10^4 and further processed by a proportional-integral-derivative (PID) controller.

The long-term stabilization was depicted in **Figure 14**. As low as 27-µHz accuracy is achieved within 16-h measurement. The inset of **Figure 14** magnifies the measured dates from 30,000 to 31,000 s and shows fluctuation range within ±0.1 mHz. Typically, thermal effect, Kerr nonlinear effect, pump-induced nonlinear effect, and random acoustic perturbations contribute to the precision of f_{rep} stabilization. For our experiment, a temperature-controlled incubator with a ripple of 0.2°C was used to take the laser cavity to isolate environmental perturbation. As for Kerr-nonlinearity, the RIC is proportional to the traveling power of resonant laser. Assuming 5-mW traveling power in NALM, the Kerr-induced RIC is estimated as 1.2×10^{-7}/mW, having the same order of magnitude of the pump-induced RIC (2.1×10^{-7}/mW). However, when the pump power of LD3 increased from 30 to 205 mW, only 1.6% of output power change was observed, which means little change on the dynamic process of pulse evolution in NALM. Thus, the Kerr-induced RIC is near ~1% of the RIC by pump-induced nonlinearity. Therefore, we postulate that the nonlinearity on the RIC of fibers owes to pump-induced nonlinear effect and thermal effect rather than Kerr effect.

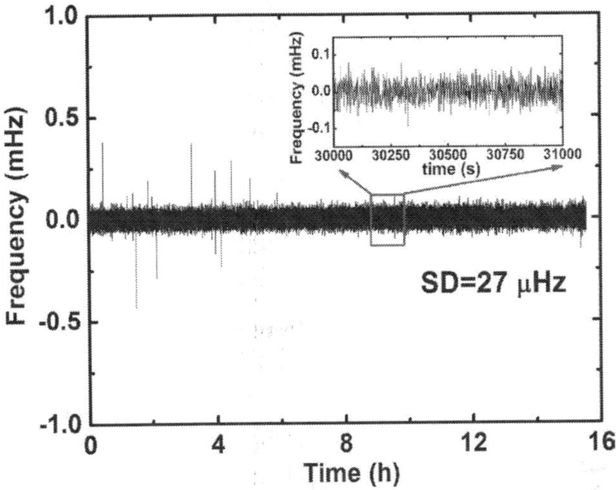

Figure 14. The long-time stabilization of repetition rate.

6. Conclusion

In this chapter, we first present several types of mode-locked fiber lasers, as well as their derivatives for SC generation. Second, an effective method named DPA was applied in Er-doped fiber laser system allowing simultaneous pulse amplification and compression so that additional pulse compressor is no longer needed. With ×8 replicas in DPA, as high as 500-mW average power was achieved and the highest SHG conversion efficiency was measured to be 56.3%. Third, an all-optical method, named as pump-induced nonlinearity, is applied to stabilize the repetition rate of a figure-eight Er-doped fiber laser, achieving as low as 27-µHz accuracy within 16-h measurement.

Author details

Qiang Hao, Tingting Liu and Heping Zeng*

*Address all correspondence to: hpzeng@phy.ecnu.edu.cn

Shanghai Key Laboratory of Modern Optical System, Engineering Research Center of Optical Instrument and System (Ministry of Education), School of Optical-Electrical and Computer Engineering, University of Shanghai for Science and Technology, Shanghai, China

References

[1] Hao, Q.; Li, W. & Zeng, H. (2007). Double-clad fiber amplifier for broadband tunable ytterbium-doped oxyorthosilicates lasers, *Opt. Express*, Vol. 15(25), 16754–16759.

[2] Xu, J.; Wu, S.; Liu, J.; Li, Y.; Ren, J.; Yang, Q. & Wang, P. (2014). All-polarization-maintaining femtosecond fiber lasers using graphene oxide saturable absorber, *IEEE Photon. Technol. Lett.*, Vol. 26(4), 346–348.

[3] Albert, A.; Couderc, V.; Lefort, L. & Barthelemy, A. (2004). High-energy femtosecond pulses from an ytterbium-doped fiber laser with a new cavity design, *IEEE Photon. Technol. Lett.*, Vol. 16(2), 416–418.

[4] Morin, F.; Druon, F.; Hanna, M. & Georges, P. (2009). Microjoule femtosecond fiber laser at 1.6 microm for corneal surgery applications, *Opt. Lett.*, Vol. 34(13), 1991–1993.

[5] Kieu, K.; Mehravar, S.; Gowda, R.; Norwood, R. A. & Peyghambarian, N. (2013). Label-free multi-photon imaging using a compact femtosecond fiber laser mode-locked by carbon nanotube saturable absorber, *Biomed. Opt. Express*, Vol. 4(10), 2187.

[6] Li, W.; Hao, Q.; Yan, M. & Zeng, H. (2009). Tunable flat-top nanosecond fiber laser oscillator and 280 W average power nanosecond Yb-doped fiber amplifier, *Opt. Express*, Vol. 17(12), 10113–10118.

[7] Xu, C. & Wise, F. W. (2013). Recent advances in fiber lasers for nonlinear microscopy, *Nat. Photonics*, Vol. 7(11), 875–882.

[8] Shi, W.; Fang, Q.; Zhu, X.; Norwood, R. A. & Peyghambarian, N. (2014). Fiber lasers and their applications [Invited], *Appl. Opt.*, Vol. 53(28), 6554–6568.

[9] Yang, K.; Li, W.; Yan, M.; Shen, X.; Zhao, J. & Zeng, H. (2012). High-power ultra-broadband frequency comb from ultraviolet to infrared by high-power fiber amplifiers, *Opt. Express*, Vol. 20(12), 12899–12905.

[10] Dudley, J. M. & Coen, S. (2002). Coherence properties of supercontinuum spectra generated in photonic crystal and tapered optical fibers, *Opt. Lett.*, Vol. 27(13), 1180–1182.

[11] Wan, P.; Yang, L. & Liu, J. (2013). All fiber-based Yb-doped high energy, high power femtosecond fiber lasers, *Opt. Express*, Vol. 21(24), 29854.

[12] Nicholson, J. W. & Andrejco, M. (2006). A polarization maintaining, dispersion managed, femtosecond figure-eight fiber laser, *Opt. Express*, Vol. 14(18), 8160–8167.

[13] Baumgartl, M.; Ortaç, B.; Limpert, J. & Tünnermann, A. (2012). Impact of dispersion on pulse dynamics in chirped-pulse fiber lasers, *Appl. Phys. B*, Vol. 107(2), 263–274.

[14] Iii, I. N. (1991). All-fiber ring soliton laser mode locked with a nonlinear mirror, *Opt. Lett.*, Vol. 16(8), 539–541.

[15] Tamura, K.; Ippen, E. P.; Haus, H. A. & Nelson, L. E. (1993). 77-fs pulse generation from a stretched-pulse mode-locked all-fiber ring laser, *Opt. Lett.*, Vol. 18(13), 1080.

[16] Tamura, K.; Doerr, C. R.; Nelson, L. E.; Haus, H. A. & Ippen, E. P. (1994). Technique for obtaining high-energy ultrashort pulses from an additive-pulse mode-locked erbium-doped fiber ring laser, *Opt. Lett.*, Vol. 19(1), 46–48.

[17] Herda, R. & Okhotnikov, O. G. (2004). Dispersion compensation-free fiber laser mode-locked and stabilized by high-contrast saturable absorber mirror, *IEEE J. Quantum Elect.*, Vol. 40(7), 893–899.

[18] Runge, A. F. J.; Aguergaray, C.; Provo, R.; Erkintalo, M. & Broderick, N. G. R. (2014). All-normal dispersion fiber lasers mode-locked with a nonlinear amplifying loop mirror, *Opt. Fiber Technol.*, Vol. 20(6), 657–665.

[19] Oktem, B.; Ulgudur, C. & Ilday, F. O. (2010). Soliton-similariton fibre laser, *Nat. Photon.*, Vol. 4(5), 307–311.

[20] Abdelalim, M. A.; Logvin, Y.; Khalil, D. A. & Anis, H. (2009). Properties and stability limits of an optimized mode-locked Yb-doped femtosecond fiber laser, *Opt. Express*, Vol. 17(4), 2264–2279.

[21] Zhao, L. M.; Tang, D. Y.; Cheng, T. H.; Tam, H. Y. & Lu, C. (2007). Bound states of dispersion-managed solitons in a fiber laser at near zero dispersion, *Appl. Opt.*, Vol. 46(21), 4768–4773.

[22] Chong, A.; Renninger, W. H. & Wise, F. W. (2007). All-normal-dispersion femtosecond fiber laser with pulse energy above 20 nJ, *Opt. Lett.*, Vol. 32(16), 2408–2410.

[23] Erkintalo, M.; Aguergaray, C.; Runge, A. & Broderick, N. G. (2012). Environmentally stable all-PM all-fiber giant chirp oscillator, *Opt. Express*, Vol. 20(20), 22669–22674.

[24] Agrawal, G. P. (2001). *Applications of nonlinear fiber optics* (1st edn): Academic Press; 525 B Street, Suite 1900, San Diego, California 92101–4495, USA.

[25] Senoo, Y.; Nishizawa, N.; Sakakibara, Y.; Sumimura, K.; Itoga, E.; Kataura, H. & Itoh, K. (2010). Ultralow-repetition-rate, high-energy, polarization-maintaining, Er-doped, ultrashort-pulse fiber laser using single-wall-carbon-nanotube saturable absorber, *Opt. Express*, Vol. 18(20), 20673–20680.

[26] Hofer, M.; Fermann, M. E.; Galvanauskas, A.; Harter, D. & Windeler, R. S. (1998). High-power 100-fs pulse generation by frequency doubling of an erbium ytterbium-fiber master oscillator power amplifier, *Opt. Lett.*, Vol. 23(23), 1840–1842.

[27] Takayanagi, J.; Kanamori, S.; Suizu, K.; Yamashita, M.; Ouchi, T.; Kasai, S.; Ohtake, H.; Uchida, H.; Nishizawa, N. & Kawase, K. (2008). Generation and detection of broadband coherent terahertz radiation using 17-fs ultrashort pulse fiber laser, *Opt. Express*, Vol. 16(17), 12859–12865.

[28] Gaponov, D. A.; Kotov, L. V.; Likhachev, M. E.; Bubnov, M. M.; Cabasse, A.; Oudar, J. L.; Fevrier, S.; Lipatov, D. S.; Vechkanov, N. N.; Guryanov, A. N. & Martel, G. (2012).

High power all-fibered femtosecond master oscillator power amplifier at 1.56 μm, *Opt. Lett.*, Vol. 37(15), 3186–3188.

[29] Tang, D. Y.; Zhao, L. M.; Zhao, B. & Liu, A. Q. (2005). Mechanism of multisoliton formation and soliton energy quantization in passively mode-locked fiber lasers, *Phys. Rev. A*, Vol. 72(4), 043816.

[30] Takayanagi, J.; Nishizawa, N.; Nagai, H.; Yoshida, M. & Goto, T. (2005). Generation of high-power femtosecond pulse and octave-spanning ultrabroad supercontinuum using all-fiber system, *IEEE Photonic. Technol. Lett.*, Vol. 17(1), 37–39.

[31] Chen, Y.; Räikkönen, E.; Kaasalainen, S.; Suomalainen, J.; Hakala, T.; Hyyppä, J. & Chen, R. (2010). Two-channel hyperspectral LiDAR with a supercontinuum laser source, *Sensors*, Vol. 10(7), 7057–7066.

[32] Kaminski, C. F.; Watt, R. S.; Elder, A. D.; Frank, J. H. & Hult, J. (2008). Supercontinuum radiation for applications in chemical sensing and microscopy, *Appl. Phys. B*, Vol. 92(3), 367–378.

[33] Rulkov, A.; Vyatkin, M.; Popov, S.; Taylor, J. & Gapontsev, V. (2005). High brightness picosecond all-fiber generation in 525–1800 nm range with picosecond Yb pumping, *Opt. Express*, Vol. 13(2), 377–381.

[34] Hao, Q.; Guo, Z.; Liu, Y.; Li, W.; Zhang, Q. & Zeng, H. (2014). Spectrally tailored supercontinuum generation from single-mode-fiber amplifiers, *Appl. Phys. Lett.*, Vol. 104(20), 201112.

[35] Tao, Z.; Yan, W.; Liu, L.; Li, L.; Oda, S.; Hoshida, T. & Rasmussen, J. C. (2011). Simple fiber model for determination of XPM effects, *J. Lightwave Technol.*, Vol. 29(7), 974–986.

[36] Kudlinski, A.; Pureur, V.; Bouwmans, G. & Mussot, A. (2008). Experimental investigation of combined four-wave mixing and Raman effect in the normal dispersion regime of a photonic crystal fiber, *Opt. Lett.*, Vol. 33(21), 2488–2490.

[37] Nodop, D.; Jauregui, C.; Schimpf, D.; Limpert, J. & Tünnermann, A. (2009). Efficient high-power generation of visible and mid-infrared light by degenerate four-wave-mixing in a large-mode-area photonic-crystal fiber, *Opt. Lett.*, Vol. 34(22), 3499–3501.

[38] Hao, Q. & Huang, Y. C. (2013). Two-octave polarized supercontinuum generated from a Q-switched laser pumped doubly resonant parametric oscillator, *Opt. Lett.*, Vol. 38(11), 1863–1865.

[39] Brown, W. J.; Kim, S. & Wax, A. (2014). Noise characterization of supercontinuum sources for low-coherence interferometry applications, *J. Opt. Soc. Am. A Opt. Image Sci. Vis.*, Vol. 31(12), 2703–2710.

[40] Klose, A.; Ycas, G.; Maser, D. L. & Diddams, S. A. (2014). Tunable, stable source of femtosecond pulses near 2 μm via supercontinuum of an erbium mode-locked laser, *Opt. Express*, Vol. 22(23), 28400.

[41] Wadsworth, W.; Joly, N.; Knight, J.; Birks, T.; Biancalana, F. & Russell, P. (2004). Supercontinuum and four-wave mixing with Q-switched pulses in endlessly single-mode photonic crystal fibres, *Opt. Express*, Vol. 12(2), 299–309.

[42] Holdynski, Z.; Napierala, M.; Mergo, P. & Nasilowski, T. (2015). Experimental investigation of supercontinuum generation in photonic crystal fibers pumped with sub-ns pulses, *J. Lightwave Technol.*, Vol. 33(10), 2106–2110.

[43] Cumberland, B. A.; Travers, J. C.; Popov, S. V. & Taylor, J. R. (2008). 29 W high power CW supercontinuum source, *Opt. Express*, Vol. 16(8), 5954–5962.

[44] Qiang, H. & Heping, Z. (2014). Cascaded four-wave mixing in nonlinear Yb-doped fiber amplifiers, *IEEE J. Sel. Top. Quant.*, Vol. 20(5), 900205.

[45] Chen, K. K.; Alam, S.; Price, J. H. V.; Hayes, J. R.; Lin, D.; Malinowski, A.; Codemard, C.; Ghosh, D.; Pal, M.; Bhadra, S. K. & Richardson, D. J. (2010). Picosecond fiber MOPA pumped supercontinuum source with 39 W output power, *Opt. Express*, Vol. 18(6), 5426–5432.

[46] Kudlinski, A.; George, A. K.; Knight, J. C.; Travers, J. C.; Rulkov, A. B.; Popov, S. V. & Taylor, J. R. (2006). Zero-dispersion wavelength decreasing photonic crystal fibers for ultraviolet-extended supercontinuum generation, *Opt. Express*, Vol. 14(12), 5715–5722.

[47] Shen, X.; He, B.; Zhao, J.; Liu, Y.; Bai, D.; Yang, K.; Wang, C.; Liu, G.; Luo, D.; Liu, F.; Hao, Q.; Li, W. & Zeng, H. (2015). Repetition rate stabilization of an erbium-doped all-fiber laser via opto-mechanical control of the intracavity group velocity, *Appl. Phys. Lett.*, Vol. 106(3), 31117.

[48] Xuling, S.; Wenxue, L.; Ming, Y. & Heping, Z. (2012). Electronic control of nonlinear-polarization-rotation mode locking in Yb-doped fiber lasers, *Opt. Lett.*, Vol. 37(16), 3426–3428.

[49] Sinclair, L. C.; Coddington, I.; Swann, W. C.; Rieker, G. B.; Hati, A.; Iwakuni, K. & Newbury, N. R. (2014). Operation of an optically coherent frequency comb outside the metrology lab, *Opt. Express*, Vol. 22(6), 6996.

[50] Lee, C.; Chu, S. T.; Little, B. E.; Bland-Hawthorn, J. & Leon-Saval, S. (2012). Portable frequency combs for optical frequency metrology, *Opt. Express*, Vol. 20(15), 16671–16676.

[51] Reichel, S. & Zengerle, R. (1999). Effects of nonlinear dispersion in EDFA's on optical communication systems, *J. Lightwave Technol.*, Vol. 17(7), 1152–1157.

[52] Mahran, O. & Aly, M. H. (2016). Performance characteristics of dual-pumped hybrid EDFA/Raman optical amplifier, *Appl. Opt.*, Vol. 55(1), 22–26.

[53] Ju, H. L.; You, M. C.; Young, G. H.; Haeyang, C.; Sang, H. K. & Sang, B. L. (2005). A detailed experimental study on single-pump Raman/EDFA hybrid amplifiers: static,

dynamic, and system performance comparison, *J. Lightwave Technol.*, Vol. 23(11), 3484–3493.

[54] Varallyay, Z. & Jasapara, J. C. (2009). Comparison of amplification in large area fibers using cladding-pump and fundamental-mode core-pump schemes, *Opt. Express*, Vol. 17(20), 17242–17252.

[55] Eidam, T.; Hanf, S.; Seise, E.; Andersen, T. V.; Gabler, T.; Wirth, C.; Schreiber, T.; Limpert, J. & Tünnermann, A. (2010). Femtosecond fiber CPA system emitting 830 W average output power, *Opt. Lett.*, Vol. 35(2), 94–96.

[56] Galvanauskas, A.; Cho, G. C.; Hariharan, A.; Fermann, M. E. & Harter, D. (2001). Generation of high-energy femtosecond pulses in multimode-core Yb-fiber chirped-pulse amplification systems, *Opt. Lett.*, Vol. 26(12), 935–937.

[57] Ancona, A.; Roser, F.; Rademaker, K.; Limpert, J.; Nolte, S. & Tünnermann, A. (2008). High speed laser drilling of metals using a high repetition rate, high average power ultrafast fiber CPA system, *Opt. Express*, Vol. 16(12), 8958–8968.

[58] Kuznetsova, L. & Wise, F. W. (2007). Scaling of femtosecond Yb-doped fiber amplifiers to tens of microjoule pulse energy via nonlinear chirped pulse amplification, *Opt. Lett.*, Vol. 32(18), 2671–2673.

[59] Sobon, G.; Kaczmarek, P.; Gluszek, A.; Sotor, J. & Abramski, K. M. (2015). µJ-level, kHz-repetition rate femtosecond fiber-CPA system at 1555 nm, *Opt. Commun.*, Vol. 347, 8–12.

[60] Guichard, F.; Hanna, M.; Zaouter, Y.; Papadopoulos, D. N.; Druon, F. & Georges, P. (2014). Analysis of limitations in divided-pulse nonlinear compression and amplification, *IEEE J. Sel. Top. Quant.*, Vol. 20(5), 619–623.

[61] Kong, L. J.; Zhao, L. M.; Lefrancois, S.; Ouzounov, D. G.; Yang, C. X. & Wise, F. W. (2012). Generation of megawatt peak power picosecond pulses from a divided-pulse fiber amplifier, *Opt. Lett.*, Vol. 37(2), 253–255.

[62] Hao, Q.; Zhang, Q.; Sun, T.; Chen, J.; Guo, Z.; Wang, Y.; Guo, Z.; Yang, K. & Zeng, H. (2015). Divided-pulse nonlinear amplification and simultaneous compression, *Appl. Phys. Lett.*, Vol. 106(10), 101103.

[63] Kienel, M.; Klenke, A.; Eidam, T.; Baumgartl, M.; Jauregui, C.; Limpert, J. & Tünnermann, A. (2013). Analysis of passively combined divided-pulse amplification as an energy-scaling concept, *Opt. Express*, Vol. 21(23), 29031.

[64] Cingoz, A.; Yost, D. C.; Allison, T. K.; Ruehl, A.; Fermann, M. E.; Hartl, I. & Ye, J. (2012). Direct frequency comb spectroscopy in the extreme ultraviolet, *Nature*, Vol. 482(7383), 68–71.

[65] Hsieh, Y. D.; Iyonaga, Y.; Sakaguchi, Y.; Yokoyama, S.; Inaba, H.; Minoshima, K.; Hindle, F.; Araki, T. & Yasui, T. (2014). Spectrally interleaved, comb-mode-resolved spectroscopy using swept dual terahertz combs, *Sci. Rep.*, Vol. 4, 3816.

[66] Heinecke, D. C.; Bartels, A. & Diddams, S. A. (2011). Offset frequency dynamics and phase noise properties of a self-referenced 10 GHz Ti:sapphire frequency comb, *Opt. Express*, Vol. 19(19), 18440–18451.

[67] Locke, C. R.; Ivanov, E. N.; Light, P. S.; Benabid, F. & Luiten, A. N. (2009). Frequency stabilisation of a fibre-laser comb using a novel microstructured fibre, *Opt. Express*, Vol. 17(7), 5897–5904.

[68] Ideguchi, T.; Poisson, A.; Guelachvili, G.; Picque, N. & Hansch, T. W. (2014). Adaptive real-time dual-comb spectroscopy, *Nat. Commun.*, Vol. 5, 3375.

[69] Schliesser, A.; Picqué, N. & Hänsch, T. W. (2012). Mid-infrared frequency combs, *Nat. Photon.*, Vol. 6(7), 440–449.

[70] Zhang, W.; Han, H.; Zhao, Y.; Du, Q. & Wei, Z. (2009). A 350 MHz Ti:sapphire laser comb based on monolithic scheme and absolute frequency measurement of 729 nm laser, *Opt. Express*, Vol. 17(8), 6059–6067.

[71] Pantell, R. H.; Sadowski, R. W.; Digonnet, M. J. & Shaw, H. J. (1992). Laser-diode-pumped nonlinear switch in erbium-doped fiber, *Opt. Lett.*, Vol. 17(14), 1026–1028.

[72] Fotiadi, A. A.; Zakharov, N.; Antipov, O. L. & Megret, P. (2009). All-fiber coherent combining of Er-doped amplifiers through refractive index control in Yb-doped fibers, *Opt. Lett.*, Vol. 34(22), 3574–3576.

[73] Yang, K.; Hao, Q. & Zeng, H. (2015). All-optical high-precision repetition rate locking of an Yb-doped fiber laser, *IEEE Photon. Technol. Lett.*, Vol. 27(8), 852–855.

[74] Bo, N.; Dong, H.; Peng, D. & Jianye, Z. (2013). Long-term repetition frequency stabilization of passively mode-locked fiber lasers using high-frequency harmonic synchronization, *IEEE J. Quantum Electron.*, Vol. 49(6), 503–510.

[75] Rieger, S.; Hellwig, T.; Walbaum, T. & Fallnich, C. (2013). Optical repetition rate stabilization of a mode-locked all-fiber laser, *Opt. Express*, Vol. 21(4), 4889–4895.

Chapter 11

High-Energy and Short-Pulse Generation from Passively Mode-Locked Ytterbium-Doped Double-Clad Fiber Lasers

Yuzhai Pan

Additional information is available at the end of the chapter

http://dx.doi.org/10.5772/63900

Abstract

Mode-locked ytterbium-doped fiber lasers capable of producing nanosecond-, picosecond- or femtosecond-level pulses with high energy or power have many advantages for various applications such as material processing and laser surgery. Firstly, in this chapter, the principles and methods used in passively mode-locked fiber lasers are briefly described. Secondly, mathematical modeling of all normal dispersion ytterbium-doped fiber lasers for analyzing the pulse generation and propagation has been established and simulated with the generalized nonlinear Schrödinger equation. Thirdly, short pulses generated from passively mode-locked fiber lasers have been demonstrated with carbon nanotube– deposited D-shaped fiber as the saturable absorber. Different pulse width can be realized with different parameters of the laser cavity. Finally, the main amplification methods for short laser pulses have been discussed, and a broad prospect for applications of various technologies using short-pulse fiber lasers is further introduced.

Keywords: Ultra-short pulse, passively mode-locked fiber laser, carbon nanotube saturable absorber, double-clad ytterbium-doped fiber

1. Introduction

Optical fiber lasers gained by rare-earth-ions-doped fibers with broad gain spectrum of tens of nanometers make them very attractive for ultrashort-pulse generation via the mode-locking mechanisms [1, 2]. They are under a growing interest because of their unique features of high efficiency and low consumption, good beam quality, high stability, naturally fiber coupled

and providing a powerful tool for high-speed optical communications, precise micromachining, biomedical imaging, and other applications [3].

Mode locking refers to phase locking of many different frequency modes in a laser cavity, which induces a laser to produce a continuous train of extremely short pulses. Unlike the Q-switched pulses, the mode-locked pulses are phase coherent with each other. Active and passive mode locking are two different methods of mode locking. Active mode-locking methods typically involve using external modulators, for example, electro-optical modulator, which induce a phase or amplitude modulation of the intra-cavity light including lots of longitudinal modes with a periodic duration according to the cavity length to generate the mode-locked pulses, whereas in passive mode locking, the generation of the mode-locked pulse is controlled by the saturable absorber (SA), that is a nonlinear optical element whose loss depends on the laser pulse intensity and causes self-modulation of the light.

Optical pulses generated from passive mode locking have the phase locking being carried out automatically in the cavity without external electrical components required, which results in extremely short and high stability. A number of potential operating and applying characteristics of passively mode-locked fiber lasers have been attractively demonstrated that produce laser pulses with durations from nanosecond to femtosecond, ultrawide bandwidth (~100 nm) at repetition rates ranging from several kHz to hundreds GHz. The research field in passive mode-locking fiber laser goes beyond the generation mechanisms and pulse behaviors that can be found in their operation. A lot of attempts and explores have been made to optimize the operation of the laser to suit for the particular application.

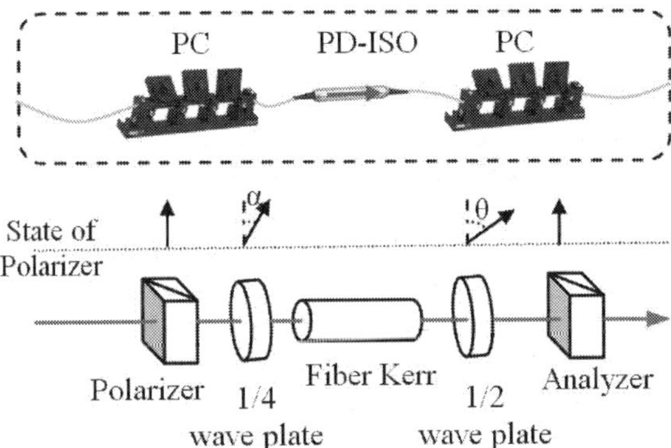

Figure 1. Basic configuration of fiber-based nonlinear polarization rotation. Top, basic configuration; bottom, principle model. PC, polarization controller; PD-ISO, polarization-dependent isolator.

The first passively mode-locked fiber laser was demonstrated in 1983 [4]. An all-fiber, unidirectional, mode-locked ring laser was first constructed using a type of artificial saturable absorber, which uses the effect of intensity-dependent polarization mode coupling in the fiber, that is, nonlinear polarization rotation effect [5]. Since self-phase modulation and other nonlinearity effects contribute to changing the refractivity index by field intensity, as well the change of the state of polarization as shown in **Figure 1**. Another type of common saturable absorber is based on the nonlinear interference between two polarization modes, that is, the so-called nonlinear optical loop mirror [6] and nonlinear amplifying loop mirror [7]. Nonlinear optical loop mirror relies on the nonlinear interference of the fields that counter-propagate, so the intensity of the pulse is determined by the product of loop length, peak power, and splitting ratio. The longer the loop length, the smaller the peak power required to reach the first transmission maximum. For passively mode-locked fiber lasers operating at large normal dispersion, a short loop length is preferred for ultrashort-pulse generation, while nonlinear amplifying loop mirror is designed with a gain medium placed asymmetrically in the Sagnac loop that is a ring-cavity interferometer. To date, this artificial saturable absorber continues to be an effective approach to generate ultrashort pulses from passively mode-locked fiber lasers [8].

Other methods of the use of new materials have been extensively investigated. Most commonly passive mode-locking devices used in research laboratories and commercial fiber or solid-state lasers are semiconductor saturable absorber mirror (SESAM) [9]. A semiconductor absorber mirror consists of semiconductor heterostructures embedded by a multiple-quantum-well structure like GaInNAs/GaAs. Such a semiconductor absorber mirror can achieve a recovery time of less than 1ns, but has some restrictions on the relatively narrow bandwidth operation, the lower antidamage threshold. Even so, semiconductor absorber mirrors have become widely commercially available and popularly utilization for environmentally robust and stable mode locking.

Various kinds of low-dimensional materials exhibiting the advantages of ultrafast recovery time and broadband saturable absorption have been presented for mode-locked fiber lasers, including carbon-based nanomaterials, such as carbon nanotubes [10], graphenes [11]. Recently, many two-dimensional (2D) layered materials have been investigated as broadband saturable absorbers for the mode locking [12–14]. Almost at the same time, another kind of nanomaterials, that is, the so-called topological insulators (TI), are characterized by a linear dispersion band structure with the Dirac point similar to graphene, which possess inherent features of broad response with a flat broadband wavelength absorption as well as high flexibility. This type of material includes bismuth telluride (Bi_2Se_3) and antimony telluride (Sb_2Te_3) [15, 16]. The manufacturing technique of these low-dimensional materials may have more simple process, easy integration, higher modulation depth, higher damage threshold, a broadband wavelength operation.

Most of passively mode-locked fiber lasers are based on a ring-cavity configuration, which conventionally contains a wavelength division multiplexer for the delivery of the pump energy to the cavity, a segment of gain fiber with the core doped by rare-earth ions, an isolator that provides the unidirectional travelling wave inside the cavity, a beam splitter for the output, a polarization controller, and a saturable absorber for the mode locking. In addition, a spectral filter or other optical elements may be inset into the cavity for the lasing stabilization. In an all-fiber scheme, the cavity contains the active medium and few fiber elements, which offer lower loss for laser pulses. For mode-locked ring-cavity fiber lasers, the fundamental repetition rate is determined by its cavity length L, the relation expression is as follows: *repetition rate =c/nL*, where c and n represents the speed of light, and refractive index respectively. On the state of the so-called harmonic mode locking (HML), the repetition rate can be two or more integer times of the fundamental repetition rate.

To improve the performance of the interaction of nanomaterials and light in fibers, various types of nanomaterial-based saturable absorbers have been demonstrated. For instance, a transmission-mode film-like saturable absorber is fabricated by nanomaterial-polymer composites sandwiched between two fiber ferrules in a standard fiber connector [10, 11]. **Figure 2(a)** shows the typical configuration of a transmission-type saturable absorber. This composite can be constituted by several easy fabrication and integration methods of sputtering, direct synthesis or deposition on the end surfaces of optical fibers, whereas there is a direct physical contact, that is, the laser light is directly transmitted through the nanomaterial film. It is noticed that there may be thermal and mechanical damages within the limited interaction length for high-energy pulsed fiber lasers. A promising alternative is lateral interaction with the evanescent waves of the fiber. The saturable absorbers with the evanescent wave interaction have been demonstrated by several fiber structures including tapered fiber [11], D-shaped fiber [13], etc. The structures and lateral interaction process of the tapered and D-shaped fibers are shown in **Figure 2(b)** and **(c)**. The evanescent wave is generated by total internal reflection of rays at the boundary of the fiber core with a lower index of clad medium. The nanomaterials like carbon nanotubes that contained in the lower-index region can raise the nonlinear reflection coefficient due to evanescent wave absorption along the fiber in a longer nonlinear interaction length in a centimeter scale. This configuration is compatible with the fiber format and easy to add the saturable absorber in the cavity by using simple fiber fusing splicing technique.

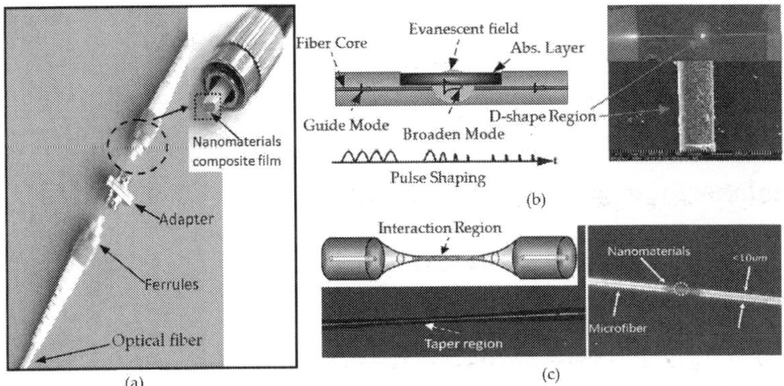

Figure 2. Typical configuration of transmission-type saturable absorber (a), evanescent field interaction of the D-shaped fiber (b), and tapered fiber (c).

Mode-locked silicate-based fibers lasers doped by rare-earth ions have been demonstrated directly operating around 1 µm, 1.5 µm, and 2 µm with very high optical efficiencies. There have also been reported that mode-locked fiber lasers based on Raman cascading, frequency conversion, and other nonlinear processes. The group velocity dispersion of silica fiber is normally positive at 1 µm. Many studies have demonstrated that all-normal dispersion passively mode-locked fiber lasers can generate various kinds of pulses [17], like dissipative solitons, similaritons, noise-like solitons, and soliton rains.

Dissipative soliton pulses refer to those confined wave packets of light in nonlinear optical systems with the balance of nonlinear gain, loss mechanisms. Dissipative solitons generally show large linear chirp with high pulse energies due to the large dispersion experienced in the fiber cavity and possible realization of larger compression ratio by means of simple chirp pulse compression techniques. Dissipative solitons offer highly desirable properties for some direct application and as the seed laser for pulse amplifier system, such as short light pulses of high pulse energies and the improved output stability with compactness, efficiency, and reliability.

An effective method of reducing the repetition rate of passively mode-locked fiber laser is to elongate the cavity by simply adding fiber lengths, and at this time, high-energy ultrashort pulses can be obtained by this effective approach. This method is well suited for fiber lasers where the resonator cavity may reach length in excess of one mile and generate higher energy pulse while maintaining a compact structure [18]. All-normal dispersion passively mode-locked ytterbium-doped fiber laser in a sense offers an ideal laser source of low repetition rate, long duration, and high-energy pulses suitable for a range of applications.

Passively mode-locked fiber laser yields a relatively lower pulse energy in the ultrashort duration because of mode confinement of conventional single-mode fiber. Also, the longer fiber enhances nonlinearities like stimulated Raman scattering and self-phase modulation, which lead to the distortions of the pulse and instability of the mode locking. Today, for the sake of the high-energy laser pulses with available pump power in many application fields like micromachining, many researchers pay their attentions to the use of double-clad fiber for high-power fiber amplifiers and lasers. The double-clad fiber can be pumped by high power laser diodes to get the higher gain [19–21], where the pump light is coupled into the larger inner cladding with a higher numerical aperture.

2. Simulation of passively mode-locked ytterbium-doped fiber laser

Ultrashort-pulse propagation in optical fiber can be accurately modeled by one or more coupled partial differential equations. Various of simulation methods with different theoretical models have been introduced to study laser pulse phenomenon and dynamic processes in the cavity with the parameters of dispersion, nonlinearity, gain, loss, etc. The well-known master mode-locking equation first proposed by H. A. Haus [22], which being a perturbation from the nonlinear Schrödinger equation, has the capacity of describing both the energy saturation and the pulse stabilization process. To prove the possibility of stable pulse generation in the presented cavity, numerical simulation of the generalized nonlinear Schrödinger equation is performed here, which provides an insights into the mode-locking dynamics of ytterbium-doped fiber lasers and directs to the performance optimization for the pulses.

The whole cavity is schematically shown as an analytical model in **Figure 3**. The main parameters of each fiber can be found in **Table 1**. Firstly, we have numerically simulated the pulse propagation and formation in the cavity governed by the following equation [23]:

$$\frac{\partial A}{\partial z} = \frac{g-\alpha}{2} A + \frac{g}{2\Omega_g^2}\frac{\partial^2 A}{\partial \tau^2} + i\gamma |A|^2 A - i\frac{\beta_2}{2}\frac{\partial^2 A}{\partial \tau^2} + i\frac{\beta_2}{2}\frac{\partial^3 A}{\partial \tau^3} \tag{1}$$

where A is the slowly varying envelope of the optical field, z is the axial distance, τ is the local time, α accounts for the loss, γ is the nonlinear coefficient of fiber, which accounts for the self-phase modulation effect, and β_2 is the second-order derivative of the propagation constant. During each cavity round-trip time, the pulse goes through different cavity components, and the output from one component is used as the input to the other, as described in **Figure 3**. Also, the pulse goes through the saturable absorber with a nonlinear loss and through the coupler with a fraction of the pulse energy outputted. The fiber parameters were chosen to match the measured or specified parameters of different elements used in the experiment.

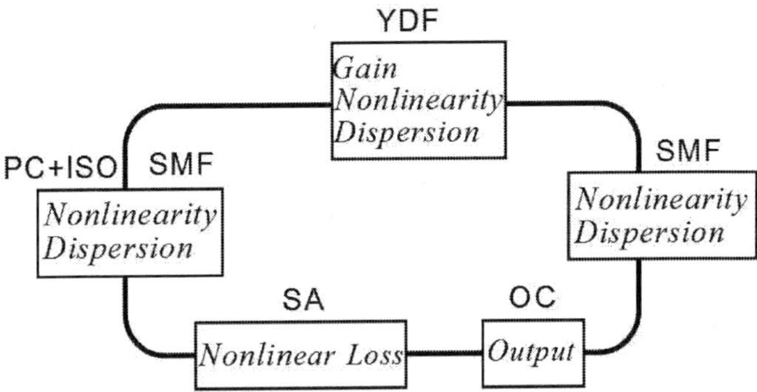

Figure 3. Proposed model of passively mode-locked fiber laser. YDF, ytterbium-doped fiber; SMF, single mode fiber; OC, output coupler; SA, saturable absorber.

Fiber type	β_2 (ps² m⁻¹)	β_3 (fs³ m⁻¹)	γ (W⁻¹ m⁻¹)	Length (m)
Ytterbium-doped fiber	0.021	0.0254	0.0048	10
Other fiber elements	0.022	0.0254	0.0044	5
Hi1060 fiber	0.022	0.0254	0.0044	20 or Var.

Table 1. Summary of the fiber parameters used in the simulation.

The parameter g denotes the gain coefficient of ytterbium-doped fiber, which can be described as the gain function and approximately expressed by:

$$g_i = \frac{g_0}{1 + \dfrac{E_{pulse}}{E_{sat}}}$$

(2)

where E_{sat} is the saturation energy due to the limited pump power, which is defined as $E_{sat}=(h\nu/\sigma)A_{eff}$ with the dependence of pump power. The pulse energy E_{pulse} is given by $E_{pulse} = \int_{T_R/2}^{-T_R/2} |A(z, \tau)|^2 d\tau$, where T_R is the cavity round-trip time. The same small signal gain g_0, depending on the doping concentration, can be assumed to be constant if only a small fraction of the pump light is absorbed provided an approximation of uniform pumping. Ytterbium-doped fiber is modeled with a total unsaturated gain of 30 dB, corresponding to these parameters: $g_0=6.9m^{-1}$ and $E_{sat}=1nJ$. Ω_g is the gain bandwidth of ytterbium-doped fiber, which is related to the bandwidth $\Delta\lambda$ through $\Omega_g = |2\pi c/\lambda^2| \Delta\lambda$, where $\Delta\lambda$ is chosen to be 55nm bandwidth.

The parameter β_2 represents dispersion of the group velocity contributing to time-domain broadening of laser pulse, that is, the so-called group velocity dispersion, which is commonly used by physicists in units of $ps^2\Delta m^{-1}$. For optical fibers, the group velocity dispersion usually refers to the chromatic dispersion parameter D that defined as a derivative $d\beta_1/d\lambda$, which is also used in practice with the relation of β_2 and n as: $\beta_2=-(\lambda^2/2\pi c)\Delta D$, where λ is the operating wavelength. The higher order dispersion and higher order nonlinear effects were ignored in simulations. Fiber nonlinear parameter γ relates the wavelength λ and effective area A_{eff} to the nonlinear index n_2 with an expression as: $\gamma=2\pi n_2/\lambda_0 A_{eff}$, when the radial field distribution is known.

Nonlinear transmission of the saturable absorber can be described by $T=\exp[-(\alpha_1+\alpha_{nl})]$, where α_1 is nonsaturable absorption loss, and α_{nl} denotes power-dependent nonlinear absorption loss, which is given by $\alpha_{nl}=\alpha_0/(1+P(\tau)/P_{sat})$, where α_0 is the saturable loss due to the absorption, that is, the modulation depth. $P(\tau)$ is the instantaneous pulse power and P_{sat} is the saturation power of the saturable absorber. The saturable loss, which acts as an equivalent-filtering effect, has a significant impact on the pulse duration and bandwidth of laser pulse. It is expected, for high power/energy pulse, that the saturable absorber with a larger modulation depth can be used in a fiber laser with large normal dispersion and strong nonlinearity. The further increase of the modulation depth should be carefully designed due to the limitation of the nonsaturable loss. One probable way for the increase of modulation depth is to reduce the evanescent field leaking of D-shaped fiber and enlarge the evanescent field interaction with the saturable absorber by lengthening the fiber D-shaped domain. Here, the parameters of the saturable absorber in the simulation model are as follows: $\alpha_1=45\%$, $\alpha_0=27\%$, and $P_{sat}=1000W$.

2.1. Numerical simulation and results

All optical fibers in the model above have normal dispersion within the laser spectral range. A 20 m length of Hi1060 fiber was inserted into the cavity aiming to increase the cavity length. The total dispersion is calculated to be about 0.75 ps^2. Eq. (1) has been solved with the standard split-step Fourier method. The simulation field is represented on a temporal grid (and via the Fourier transform on a frequency grid) consisting of 2^{11} points with a width of 0.2 ns. One initialized weak signal was introduced into the round-trip propagation in the cavity, and this pulse consecutively experiences each action of cavity components along the routes.

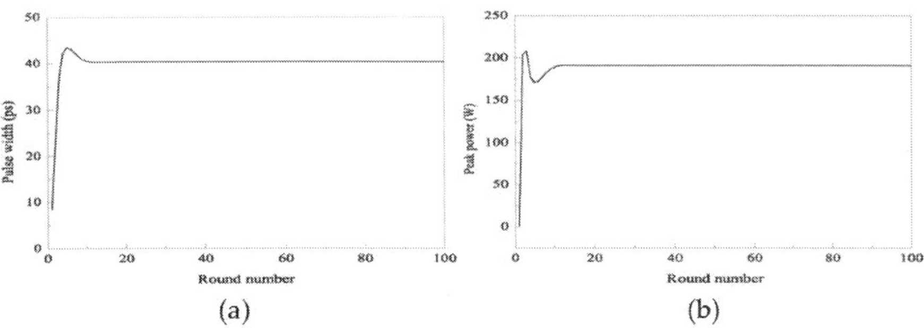

Figure 4. Transient evolution of the pulse width (a) and peak power (b).

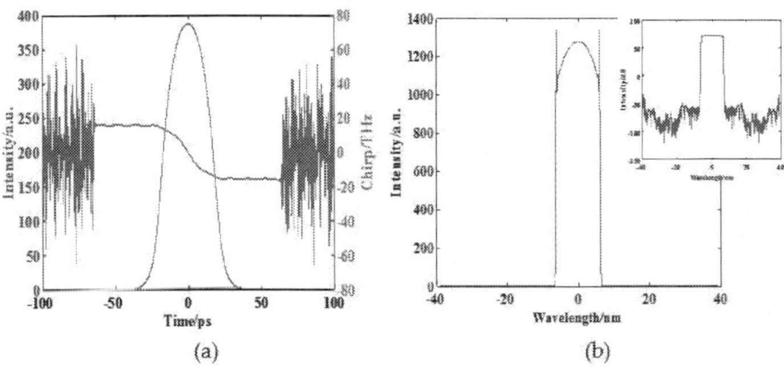

Figure 5. Pulse profile and chirp (a), and spectrum (b) of mode-locked laser. The inset is the spectrum in decibel scale.

After a finite number of round trips, the pulse started to converge into stable dissipative soliton solutions. The simulation results indicate that stable solutions do exist in such a laser, which can be confirmed from the peak power and pulse duration evolutions as shown in **Figure 4**. The pulse duration is ~40.6 ps and the spectral edge-to-edge bandwidth is 11.2 nm as shown in **Figure 5**, which both evidently indicate that the pulses are highly chirped. As shown in **Figure 5(b)**, the spectrum on a linear scale and a logarithmic scale is characterized by their steep edges, that is, the so-called M-shaped optical spectra. The higher peak power of the pulse after amplification by gain fiber induces a substantial nonlinear phase shift in the single-mode fiber, which results in sharp peaks on the spectrum edges. The gain and loss coexist in the dissipative system and play an essential role in the formation of dissipative solitons. Thus, dissipative soliton must be self-organized and its dynamics differ from that of the conventional soliton. It is noted that the saturable absorber has a nonlinear transmission depending on the light intensity, and the nonlinear phase shift is gradually varied in the process of gain saturation when an initial pulse is circulating in the cavity [24].

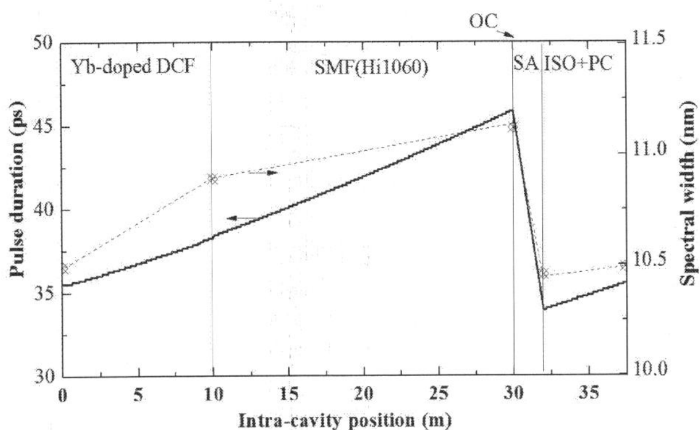

Figure 6. Evolution of the pulse width and spectral bandwidth in one cavity round trip.

The evolutions of the pulse duration and spectral edge-to-edge bandwidth in one cavity round trip are shown in **Figure 6**. It can be seen that the significant broadening of the duration and spectral width of the pulse after passing through the gain fiber. The large normal dispersion of a long segment of single-mode fiber induces large and positive chirp. The nonlinear phase shift broadens the spectrum of the pulse, while strong chirping induced by normal cavity dispersion enlarges the pulse pedestals, which are the lower intensity red-shifted and blue-shifted spectral weights located at both pulse edges in time-domain. When the pulse is travelling throughout the saturable absorber, pulse duration and spectrum width could be compressed. Here, the saturable absorber plays a key role of providing an effective filtering function to stabilize the mode locking.

When the cavity length was increased to 60m, the total cavity dispersion is up to ~1.3 ps^2. The laser works on a larger normal dispersion regime without dispersion compensation in this model as the same above. Simulation results showed that the stable pulse obtained with the pulse duration of 76.5 ps and the corresponding spectral width of 9.6 nm. With the larger positive dispersion, the stable pulses have been obtained again but highly chirp. Comparing the pulse characteristics of these two cavities, the broadened pulse induced mainly by self-phase modulation in the longer cavity has lower peak power, and both of the equivalent filter effect and lower self-phase modulation result in the narrower spectrum.

For the deliverable high energy of the pulse, the stabilization of mode locking should be ensured in a normal and large cavity dispersion accompanied with a high nonlinearity. A spectral filter, which is added in such a laser cavity, can be considered as an effective absorber in the spectral domain to cut off the temporal wings of the pulse and stable the mode locking. It is assumed that the spectral filter has a Gaussian profile, so the spectral filter is numerically implemented in the model by a function as: $T(\omega) = \exp[-(\omega/\Omega_f)^2]$, where ω is the angular frequency, and Ω_f is the bandwidth of the spectral filter. The influence of the spectra filtering on pulse shaping could be investigated for the stabilization of the mode locking, as well as the performance optimization of the pulses [25].

3. Mode-locked ytterbium-doped fiber laser with carbon nanotubes

Single-walled carbon nanotube is an enrolled two-dimensional graphene honeycomb sheet with a diameter of typically 0.6–2 nm and a length distribution ranging from tens of nanometers to several micrometers. Depending on their chirality [26], single-walled nanotubes exhibit two different electrical properties, metallic or semiconducting. Semiconducting nanotubes have an energy band gaps like those in ordinary semiconductors; thus, photons having corresponding wavelength are absorbed. Single-walled nanotubes are a kind of promising material as saturable absorber for passive mode locking because the bandgap energy can be controlled by the tube diameter, which would be applied for different spectral ranges.

The absorption loss and modulation depth of the nanotubes-based saturable absorber can be adjusted by changing the concentration of nanotubes, the interaction length of light with nanotubes, nonsaturable loss of fiber structure, and substrate materials. To study the effect of modulation depth on the mode locking in a larger normal dispersion cavity with high nonlinearity, the evanescent field used here is from the D-shaped fiber that was fabricated by ablating part of the cladding of single mode fiber by the femtosecond laser-induced water breakdown method [27]. Top- and side-view microscope images of the as-prepared D-shaped fiber are shown in **Figure 7(a)**. Averaged distance between the core and the flat face of the fiber is about 7 μm, and the entire length of D-shaped fiber is 900 μm. Measured insertion loss of the D-shaped fiber is about 0.6 dB which guarantees the lower loss in the cavity. Secondly, the commercial solution of single-walled carbon nanotubes with the purity of approximately 95% was used in the experiments. The diameter distribution of nanotubes is around 1.5 nm. The nanotube solution was further diluted to be a concentration of 0.05 wt% with a ten-hour ultrasonically agitated process to minimize bundling of nanotubes. Finally, the well-dispersed aqueous solution was sprayed on the D-shaped surface of the fiber. These devices were dried out in a vacuum oven at the temperature of 40 °C for 30 min carefully adjusted to reduce agglomeration and detaching of nanotubes. Simple encapsulation for these saturable absorbers was performed to avoid unexpected damage. A microscope image of the nanotube–deposited D-shaped fiber is shown in **Figure 7(b)**. The transmission values of the saturable absorber were recorded by using the ultrashort-pulse fiber laser with a MHz repetition rate at different average power. The results for both the D-shaped fiber with nanotubes and without nanotubes are presented in **Figure 7(c)**. The result shows that the transmissivity can be changed with the increase of the input power indicating the existence of saturable absorption. The power-dependent transmissivity induced by nanotubes can be decreased to ~28%. Nonsaturable loss induced by both D-shaped fiber and other impurities is about 21%. The modulation depth of about 51% can be provided, which is higher and sufficient.

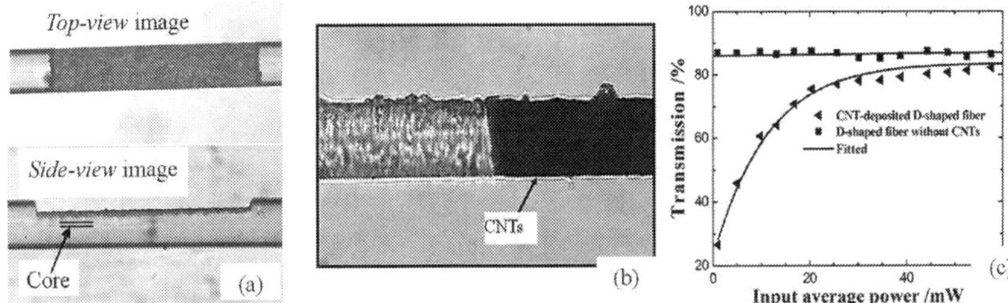

Figure 7. (a) Top-view and side-view microscope images of a D-shaped zone in fiber, (b) microscope image of carbon nanotube–deposited D-shaped fiber, and (c) transmissivity curves of saturable absorber and D-shaped fiber versus the input power.

3.1. Ultrashort-pulse generation

The presented experimental setup is schematically shown in **Figure 8(a)**. A side-pumping scheme of ytterbium-doped fiber was taken as the gain fiber pumped by 915-nm laser diode through a fiber combined with a coupling efficiency of 90%.

A lower absorption coefficient of gain fiber could suppress thermal effects to some extent, a 10-m-long ytterbium-doped double-clad fiber (Nufern SM-YDF-5/130) with a cladding absorption coefficient of 1.16 m^{-1} at 915 nm. The dispersion and nonlinear coefficients of ytterbium-doped fiber are 0.02ps$^2\cdot$m^{-1} and 0.0048W$^{-1}\cdot$m^{-1}, respectively. A 20-m-long single-mode fiber was employed to extend the cavity length, which correspondingly decreased the repetition rate. An optical coupler (OC) provided a 20% output ratio. The dispersion and nonlinear coefficients of single-mode fiber are 0.022 ps$^2\cdot$ m^{-1} and 0.0047 W$^{-1}\cdot$m^{-1}, respectively. The total length of laser cavity is ~36.5 m with the all-normal cavity dispersion of ~0.76 ps^2.

Self-starting mode locking has been achieved at the pump power of ~0.6 W by appropriately adjusting polarization controller. The self-consistent pulse evolution and stable mode locking indicate that the saturable absorber performs a filtering-equivalent function by the loss depending on light intensity to promote and stabilize the mode-locking operation in all-normal dispersion cavity as we expected. The pulse train generated has been observed by the oscilloscope trace as shown in **Figure 8(b)**. The pulse sequence was traced up to 2.5 μs by an oscilloscope connected with a high-speed photodetector (3 GHz). The figure exhibits the round-trip time of ~178.7 ns corresponding to the repetition rate of 5.59 MHz, which is consistent with the cavity length.

The temporal profile and spectral of the pulse recorded at the pump power of 2 W are shown in **Figure 8(c)** and **(d)**. The spectrum was measured by a spectrometer with the resolution of 1 nm (Ocean Optics Inc., HR4000). The operation wavelength is around 1085 nm, which indicated nearly four energy-level lasing behavior of ytterbium ions. The pulse duration is 46.6 ps, and the spectrum has an approximately M-shaped profile on a linear scale with a bandwidth of ~12.8 nm, which implies that the mode-locking operation in the dissipative soliton regime. The pulse duration could be further decreased to 941 fs from the outside cavity simple compression by using a segment of single-mode fiber with negative dispersion. The near Gaussian fitting shape of the pulse suggests that linear chirp dominates across the pulse in the cavity.

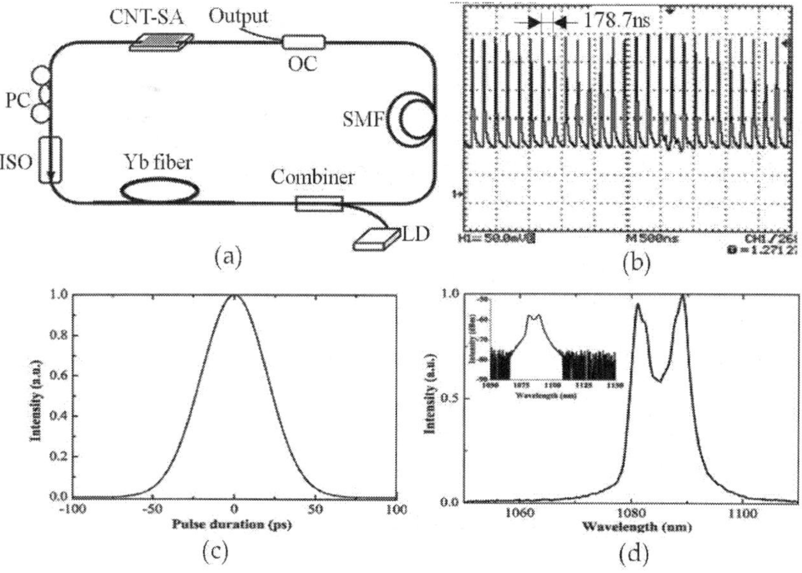

Figure 8. (a) Schematic diagram of mode-locked fiber laser, (b) oscilloscope traces of pulse train, (c) pulse profile trace of the pulse, and (d) output spectrum (the inset is the spectrum in a log scale).

When the pump power was increased to 3.5 W, the corresponding pulse duration was enlarged to 62.6 ps, and the spectral width was broadened to 16.3 nm. The output power was almost linearly increased to 162 mW, and the corresponding pulse energy was raised to ~29 nJ. The experimental results manifest that the evanescent-field interaction scheme and large modulation depth of the saturable absorber would be preferentially chosen for the achievement of high-energy pulses. Mode-locked fiber lasers could be robust against optical wave breaking due to the linear chirp across the pulse, showing a stretched pulse with the duration up to a few or even several hundred picoseconds.

Unlike nonlinear polarization rotation effect, various new nanomaterials hardly generate fs-level ultrashort pulses directly from the laser oscillators [28]. A robust self-starting picoseconds ytterbium-doped fiber laser is easy to be realized by using one of other 2D materials or topological insulators, at the aspect of characteristics of the pulses, those results are similar to these of carbon nanotubes yet. The pulse duration is limited by large normal dispersion, while the larger linear chirp dissipative solitons pulses are easy to be compressed. So the research interests are focused on various pulses dynamics of stable, self-starting mode locking of the fiber lasers in all-normal dispersion regime.

3.2. Nanosecond-level pulse generation

Carbon nanotubes have been demonstrated it suitable for stable long-duration pulse mode locked in all-normal dispersion regime [29]. Absence of nonlinear polarization evolution dynamics gives giant chirped pulses that can be suitable for compression. Here, an ultralong cavity ytterbium-doped fiber laser mode locked by nanotubes-based saturable absorber has been experimentally investigated. It is used a saturable absorber with the unsaturated loss of ~57.8% and the modulation depth of ~4.7%. The ring cavity was elongated by an one-kilometer-long single-mode fiber (YOFC, C1060), which results in an ultralong laser cavity with the length of 1021 m.

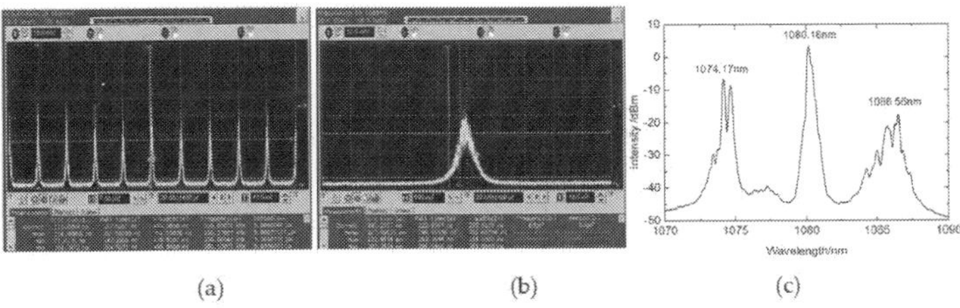

(a)   (b)   (c)

Figure 9. Characteristics of noise-like pulses. Train trace (a), single pulse (b) and optical spectrum (c).

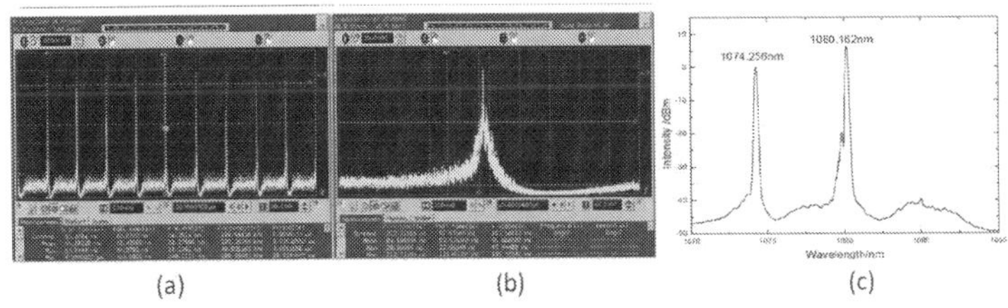

(a)   (b)   (c)

Figure 10. Characteristics of soliton rains. Train trace (a), single pulse (b), and optical spectrum (c).

Stable mode-locked pulses were obtained by slightly adjusting PC at the pump power of 0.81 W. The amplitude of the pulse was increased with the increase of pump power until the power of 1.81 W, which changing into a state of multipulses. The pulse sequence and single pulse, which is a typical noise-like pulse, have been measured and shown in **Figure 9(a)** and **(b)**, respectively. The repetition period of the pulse is approximately 5 μs. The pulse duration is approximately 292.6 ns. When the pump power was 1.81 W, another stable state, that is, soliton rains, could be obtained with carefully adjusting the polarization controller. **Figure 10** shows the pulse train, single pulse, and the spectrum of the fiber laser at the pumping power of 1.93 W. Within each pulse period, the pulse contains background noise, drifted pulse, and phase-condensed soliton. The intensity of the drifted pulse is about 10% of the phase-condensed soliton. The pulse width of phase condensation soliton is about 102.5 ns at 3 dB, as shown in **Figure 10(b)**. The steady soliton rains cannot be maintained once the pump power is above 1.93 W. The maximum output power is ~40.3 mW with single pulse energy of ~201.5 nJ. Output spectra of the noise-like pulses and soliton rains have been measured and shown in **Figures 9(c)** and **10(c)**. It can be seen that both spectra have several central wavelengths indicating that the presence of filtering effect in the cavity could be used as a multiwavelength short-pulse fiber laser.

Recently, it has been reported that the generated pulses of an ultralong cavity fiber laser can deliver microjoule-level energy in the nanosecond range [30]. In the all-normal dispersion fiber laser systems, the stable mode-locking pulses exhibits that the formation of pulse shaping is the product of complicated processes of energy conversion. Various nonlinear effects such as self-phase modulation, dispersion wave, peak clamping, which have strong influence on the stability of mode locking, and combining with high cavity dispersion can lead to complex pulsing phenomena, like wave-breaking of the soliton pulse as noise-like pulses in the results above. On the other hand, the Raman-induced noise-like pulses can be realized by the Raman effect in a fiber laser with high nonlinearity and dispersion [31].

4. Amplification of short-pulse fiber lasers

Because laser pulses that extracted from the master oscillator are generally of relatively low energies, an additional external amplifier is required for the enhancement of the pulse energy, which is of key importance of power scaling of fiber lasers for the wide applications. Ytterbium-doped fiber laser systems are excellently suited to generate and amplify ultrashort laser pulses due to their large amplification bandwidth supporting pulse durations of few hundred femtoseconds. This approach is benefited from the simple fiber connection between the oscillator and the amplifier. There are mainly two methods, that is, the chirped pulse amplification (CPA) and the master oscillator power amplification (MOPA).

For the CPA technique, ultrashort pulses are amplified by time stretching of the original pulses and later recompress them back into a short duration after the fiber amplifier [32]. The chirped pulse amplifier system, as schematically shown in **Figure 11**, is composed of a seed laser, a pulse stretcher, amplifier chains, and a pulse compressor. The duration of laser pulses is firstly increased temporally to a much longer duration of the order of 1ns, that is chirped by using a pulse stretcher, for example a grating pair, fiber chirped Bragg grating, etc., which reduces the peak power to a level so that the nonlinear effects in the gain medium can be avoided. The stretched pulse was amplified in next amplifier system, which typically consisted of large-mode area single-mode fiber or photonic crystal fiber (PCF) gained by multimode laser diodes, allowing more peak power generation below the limit of nonlinear optical intensity effects. Finally, a low-loss compressor is used to temporally compress the pulses to a duration similar to the input pulse duration. The pulse stretcher is necessary for the amplification of ultra-short pulses. Or otherwise, high nonlinearities in the fiber induced by high optical peak power in ultrashort pulses would affect the recompression to an ideal short pulse in the final compression part. Of course, the compressor also needs to tolerate high peak powers without introducing nonlinear distortions.

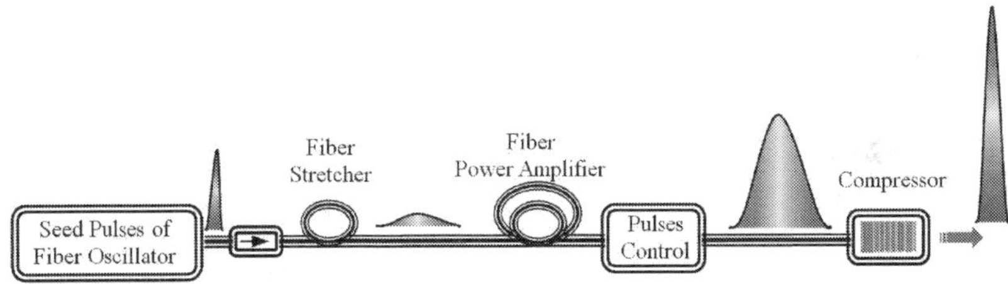

Figure 11. Schematic diagram of the chirped pulse amplifier for ultrashort-pulse fiber laser.

Femtosecond fiber amplifier systems have the potential for millijoule pulse energies at megahertz repetition rate. Fox example, a ytterbium-doped fiber amplifier system has delivered millijoule level pulse energy at repetition rates above 100 kHz corresponding to an average power of more than 100 W, the compressed pulse is as short as 800 fs [33], where a short-length PCF with 80-μm core diameter is employed, which allows the pulse energies up to 1.45 mJ with a stretched pulse duration of 2 ns. Scaling up of pulse energy in an ultrafast fiber laser has been demonstrated that the simultaneous generation of 60 W of compressed average power at 100 kHz, together with 320 fs and 600 μJ pulses [34].

Highly chirped pulse fiber oscillators may create powerful all-fiber generator-amplifier systems, that is, the so-called master oscillator power amplifier (MOPA), which is an attractive technology to achieve picoseconds-level laser pulses with higher average output power and peak power [35–38]. The master oscillator power amplifier, as shown in **Figure 12**, generally consists of a laser oscillator that produces the weak seed pulse and series of amplifiers that increase the laser power to the required level. Chirped pulses from the generator are directly fed to an amplifier without the use of a stretcher or a modulator and compressed after one or more amplifying stages. In addition, the chirped pulses coming out of the master oscillator in the normal dispersion regime, like dissipative solitons, whose energy exceeds the energy of classical solitons by tens or hundreds of times because of longer duration at the same peak power, can be further increased to the required level in one power amplifier, as well compressed by an external compressor.

Double-clad fibers have been extensively used to build fiber amplification systems, exhibiting desirable characteristics such as high gain, good efficiency, and excellent beam quality. A diode-pumped mode-locked ytterbium-doped fiber seed laser followed by two fiber amplifiers has been demonstrated [36], where both fiber amplifiers with the design of large-mode area were cladding pumped. The oscillator produced 30 pJ, 1.8 ps pulses. After two-scales amplifying, the output pulses compressed by 830 grooves/mm gratings produced high-quality 400 nJ pulses with a pulse duration of 110 fs at average power levels in excess of 25 W. A carbon-nanotube-based master oscillator power amplifier has also been reported that a compact picosecond-level pulse fiber laser with high average power [37]. The seed laser that is the nanotube-based mode-locked Nd:YVO$_4$ laser is further amplified with a single-stage fiber amplifier. An amplified pulse with a pulse width of 15.7 ps, pulse energy of 244 nJ, has been achieved with an average power of 20 W at a repetition rate of 82 MHz. A fiber amplifier contains a seed source and two-scale amplifiers in that the gain fibers are different size double-clad fibers to suppress the nonlinear effects, and single pulse energy of 4.56 µJ with a pulse width of 0.62 ns at 26.3 MHz has been realized [38]. These systems present simple and practical fiber-based solutions for high-average-power ultrashort-pulse laser applications.

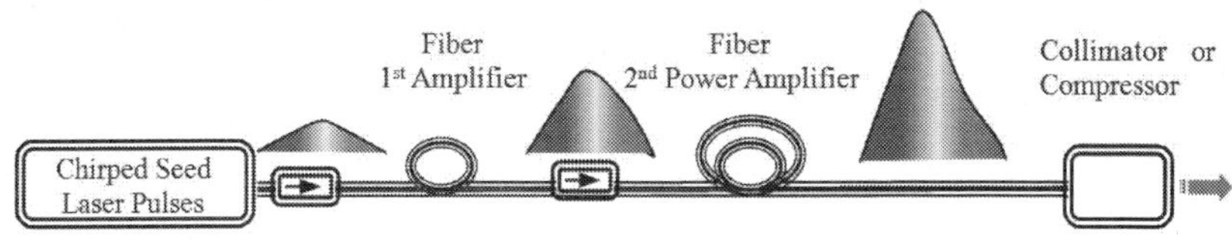

Figure 12. Schematic diagram of a master oscillator power amplifier system.

Fiber nonlinearity is proportional to the length of fiber and inversely proportional to the fiber core size. The developed large-mode area double-clad photonic crystal fiber can be considered as a possible approach to overcome many of the difficulties as mentioned above. These amplifiers enabled by advancements in photonic crystal fiber manufacturing technology can generate kilowatts to multigigawatts of peak power using direct amplification showing excellent conversion efficiency, diffraction-limited beam quality.

5. Typical applications of short-pulse mode-locked fiber lasers

Fiber laser sources with nanosecond pulses have great potential in a variety of applications requiring low temporal coherence, such as optical metrology, or sensor interrogation based on low-coherence spectral interferometry technology. Furthermore, the applications of fiber-based master oscillator power amplifier sources can also extend to industrial fields, such as laser marking, engraving, and other micromachining in various materials, particularly suitable for cutting high reflectivity materials like titanium, copper [39], silver [40]. Other successfully commercial examples include the high-resolution 3D imaging lidar system [41], nonlinear frequency conversion [42], etc. The pulse duration of ultralong cavity mode-locked fiber laser may be up to several or even hundreds of nanoseconds with higher energy. In contrast to the Q-switched lasers, obviously, there would be the limited range of direct applications. For this reason, the interests would focus on the high-chirp solitons or dissipative solitons resonance with the duration as long as nanoseconds that can further be compressed into a picosecond level.

All-fiber subnanosecond lasers have great potential application in the generation of wide-band supercontinuum (SC) source [38]. The supercontinuum laser source is the widely broadened spectrum generated by strong nonlinear effects using highly nonlinear fibers. The output power of this novel supercontinuum light source is high by using the small core fibers, so that the supercontinuum source is of interest to many kinds of applications. For example, an ultrahigh-resolution optical coherence tomography (OCT) has been investigated by using the ultrashort-pulse fiber supercontinuum source as fiber-based, high-power, wideband sources [43].

Laser beam at the output of the fiber can be easily focused in a spot with a radius of $\sim\mu m$, a few nanojoules of energy in a tens femtoseconds pulse resulting in intensities on the order of $\sim GW/cm^2$. So, ultrashort laser pulses would not induce heat diffusion during the fast interaction with objects such as various materials and living structures, that is, free from cracks and melting and other thermal effects [44]. For ultrashort laser pulses with the duration below several picoseconds, the pulses interacting period is generally shorter than the lattice heating time, which is necessary for energy diffusion processes for most materials. Ultrashort-pulse fiber lasers presently have provided a stable and reliable platform for many applications.

Material micro- and nano-machining has been for a long time identified as an important and the largest market for high-power/energy ultrashort-pulse fiber lasers [45]. High pulse intensities are widely used for permanent transparent material modification. In addition, these extremely high power densities of the pulses resulting in a highly localized disruption of the material matrix with very little energy deposited and few heat transferred to the surrounding material. This is a so-called laser cold process. A clear processing edge formed by laser ablation is of key importance for medical, photovoltaic, and semiconductor industry, especially for thermally sensitive material, like nitinol shape memory alloys, bio-absorbable polymers like polylactic acids, glass, etc. [46]. In comparison, nanosecond laser pulses micromachining in glass or other materials would leave an undesirable heat-affected zone, numerous stress fractures, and micro-cracks around the processing edge.

Although known as a cold ablation process, by precisely controlling localized heat accumulation to melt material accompanying with the inhibition of shrinkage stress by producing embedded molten pool by nonlinear absorption process, ultrashort laser pulses at high repetition rates (hundreds of kHz and above) have been applied into micro-welding of materials including glass and plastic, silicon and glass, and medical piece part [47].

Ultrashort-pulse laser make them ideal sources for time resolved measurement of the faster physical and chemical phenomenon. By controlling the optical carrier frequency and the carrier phase of ultrafast lasers, optical frequency combs, spectroscopy and precision metrology of optical frequency transitions and natural constants have been realized. Pump-probe measurements can use ultrashort laser pulses to measure and determine the evolution of a series of ultrafast processes in many kinds of materials, molecules or even in internal states of atoms, with the advantage of shortening the transient behavior resulting from the optical excitation [48]. Femtosecond laser pulses with modest energies generating the intensities above 10^{15} W/cm^2 are used to determine the elements of the sample in a technique called femtosecond laser-induced breakdown spectroscopy (fs-LIBS) [49].

Ultrashort fiber lasers are developing rapidly in the medical and biology applications. These laser pulses can be used as a laser scalpel directly to medical treatment on the one hand. The advantages of accuracy, absence of thermal interaction, and safety have been accepted by a wide customer. On the other hand, the indirect applications refer to high precision and high-quality medical devices, such as stents, implants, and catheters, requiring sophisticated manufacturing techniques, which are available from medical industrial manufacturing processes. An example has been reported in 2014, the multifunctional biochips for realizing high-performance biochemical analysis and cell engineering [50].

In conclusion, passively mode-locked ytterbium-doped fiber lasers operating in the normal dispersion regime have been firmly established in the field of various short-duration pulses, and attracted increasing attention due to their compactness, low cost, and widespread applications. Various technologies have been developed with the aim to realize short-pulse all-fiber laser sources with the desirable energies, durations, average powers and beam quality, as well the environmental stabilization and reliability. This chapter is intended to smoothly understand each topic on this field and has the interest to further read and explore high-energy short-pulse generation and their applications.

Acknowledgements

This work was partially supported by Shandong Graduate Teaching Innovation Project (SDYY15003) and HIT Graduate Teaching Reform Project(JGYJ201436 & WH2015008).

Author details

Yuzhai Pan*

Address all correspondence to: panyzh2002@163.com

Department of Optoelectronic Science, Harbin Institute of Technology at Weihai, Hi-Tech District, Weihai, China

References

[1] S M Swift. Q-switched and mode locked short pulses from a diode pumped, Yb-doped fiber Laser. BiblioScholar; 2012

[2] M J F Digonnet. Rare-earth-doped fiber lasers and amplifiers. 2nd edn. Boca Raton, FL: CRC Press; 2001

[3] H Endert, A Galvanauskas, G Sucha and R Patel. Novel ultrashort pulse fiber lasers and their applications. Proceedings of SPIE. 2002; 4426:483–488

[4] M I Dzhibladze, Z G Esiashvili, E S Teplitskii, S K Isaev and V R Sagaradze. Mode-locking in a fiber laser. Kvantovaya Elektron. 1983; 10:432–434

[5] V J Matsas and T P Newson. Self starting passively mode locked fibre ring soliton laser exploiting nonlinear polarization rotation. Electron Lett. 1992; 28(15):1391–1393

[6] Doran N J and Wood D, Nonlinear-optical loop mirror. Opt. Lett. 1988; 13(1):56–58

[7] M. E. Fermann, F. Haberl, M. Hofer, and H. Hochreiter, Nonlinear amplifying loop mirror. Opt. Lett. 1990; 15(13):752–754

[8] X Jin, X Wang, X Wang, and P Zhou, Tunable multiwavelength mode-locked Tm/Ho-doped fiber laser based on a nonlinear amplified loop mirror. Appl. Opt. 2015; 54(28): 8260–64

[9] U Keller, K J Weingarten, F X Kartner, et al. Semiconductor saturable absorber mirrors (SESAM's) for femtosecond to nanosecond pulse generation in solid-state lasers. IEEE J. Sel. Top. Quantum Electron. 1996; 2(3):435–453

[10] S Y Set, H Yaguchi, Y Tanaka, and M Jablonski, Ultrafast fiber pulsed lasers incorporating carbon nanotubes. IEEE J. Sel. Top. Quantum Electron. 2004; 10(1):137–146

[11] J Du, Q Wang, G Jiang, et al. Ytterbium-doped fiber laser passively mode locked by few-layer molybdenum disulfide (MoS_2) saturable absorber functioned with evanescent field interaction. Sci. Rep. 2014; 4(4):6346

[12] D Mao, Y Wang, C Ma, L Han, B Jiang, X Gan, S Hua, W Zhang, T Mei, and J Zhao, WS_2 mode-locked ultrafast fiber laser. Sci. Rep. 2015; 5:7965

[13] Y Chen, G Jiang, S Chen, et al. Mechanically exfoliated black phosphorus as a new saturable absorber for both Q-switching and Mode-locking laser operation. Opt. Express. 2015; 23(10):12823–12833

[14] Z Luo, Y Huang, J Weng, et al. 1.06 μm Q-switched ytterbium-doped fiber laser using few-layer topological insulator Bi_2Se_3 as a saturable absorber. Opt. Express. 2013; 21(24):29516–29522

[15] J Sotor, G Sobon, K Grodecki, et al. Mode-locked erbium-doped fiber laser based on evanescent field interaction with Sb_2Te_3 topological insulator. Appl. Phys. Lett. 2014; 104(25):251112

[16] K Kashiwagi and S Yamashita. Deposition of carbon nanotubes around microfiber via evanescent light. Opt. Express. 2009; 17(20):18364–18370

[17] Z Cheng, H Li, and P Wang. Simulation of generation of dissipative soliton, dissipative soliton resonance and noise-like pulse in Yb-doped mode-locked fiber lasers. Opt. Express. 2015; 23(5):5972–5981

[18] A Ivanenko, S Turitsyn, S Kobsev, and M Dubov. Mode-locking in 25-km fibre laser. in Optical Communication (ECOC), 2010 36th European Conf. and Exhibition on. 2010:1–3

[19] A Hideur, T Chartier, M Brunel, S Louis, C Özkul, and F Sanchez. Generation of high energy femtosecond pulses from a side-pumped Yb-doped double-clad fiber laser. Appl. Phys. Lett. 2001; 79:3389

[20] K Kieu, W H Renninger, A Chong, and F. W. Wise. Sub-100 fs pulses at watt-level powers from a dissipative-soliton fiber laser. Opt. Lett. 2009; 34(5):593–595

[21] Y Huang, Z Luo, F Xiong, Y Li, M Zhong, Z Cai, H Xu, and H Fu. Direct generation of 2 W average-power and 232 nJ picosecond pulses from an ultra-simple Yb-doped double-clad fiber laser. Opt. Lett. 2015; 40(6):1097–1100

[22] H A Haus. Mode-locking of lasers,. IEEE J. Sel. Top. Quant Electron. 2000; 6(6):1173–1185

[23] G P Agrawal. Nonlinear Fiber Optics. 4th ed. San diego, CA. Academic Press; 2009

[24] Z Zhang and G Dai. All-normal-dispersion dissipative soliton ytterbium fiber laser without dispersion compensation and additional filter. IEEE Photon. J. 2011; 3(6):1023–1029

[25] B Sonia, F Christophe, K Huseyin, and P Periklis. Pulse shaping in mode-locked fiber lasers by in-cavity spectral filter. Opt. Lett. 2014; 39(3):438–441

[26] S Yamashita. A tutorial on nonlinear photonic applications of carbon nanotube and graphene. J. Lightwave Technol. 2012; 30(4):427–447.

[27] Y Li and S L Qu. Femtosecond laser-induced breakdown in distilled water for fabricating the helical microchannels array. Opt. Lett. 2011; 36(21):4236–4238

[28] Z Sun, T Hasan, F Wang, A G. Rozhin, I H. White, and A C. Ferrari. Ultrafast stretched-pulse fiber laser mode-locked by carbon nanotubes. Nano Res. 2010; 3:404–411

[29] E J R Kelleher, J C Travers, Z Sun, A G Rozhin, A C Ferrari, S V Popov, and J R Taylor. Nanosecond-pulse fiber lasers mode-locked with nanotubes. Appl. Phys. Lett. 2009; 95(11):111108

[30] S Kobtsev, S Kukarin, and Y Fedotov. Ultra-low repetition rate mode-locked fiber laser with high-energy pulses. Opt. Express. 2008; 16(26):21936–21941

[31] T North and M Rochette. Raman-induced noiselike pulses in a highly nonlinear and dispersive all-fiber ring laser. Opt. Lett. 2013; 38(6):890–892

[32] D Strickland and G Mourou. Compression of amplified chirped optical pulses. Opt. Commun. 1985; 56(6):447–449

[33] M E Fermann and I Hartl. Ultrafast fibre lasers. Nature Photon. 2013; 7:868–874

[34] Y Zaouter, F Guichard, C Hoenninger, et al. High average power 600 uJ ultrafast fiber laser for micromachining application. Journal of Laser Applications. 2015; 27: S29301

[35] G A Ball, C E Holton, G Hull-Allen, and W W Morey. 60 mW 1.5 µm single-frequency low-noise fiber laser MOPA. IEEE J. Photon. Technol. Lett. 1994; 6(2):192–194.

[36] A Malinowski, A Piper, J H V Price, K Furusawa, Y Jeong, J Nilsson, and D J Richardson. Ultrashort-pulse Yb^{3+}-fiber-based laser and amplifier system producing>25-W average power. Opt. Lett. 2004; 29(17):2073–2075

[37] L Zhang, Y G Wang, H J Yu, et al. 20 W high-power picosecond single-walled carbon nanotube based MOPA laser system. J. Lightwave Technol. 2012; 30(16):2713–17

[38] J J Chi, P X Li, H Hu, et al. 120W subnanosecond ytterbium-doped double clad fiber amplifier and its application in supercontinuum generation. Laser Phys. 2014; 24(8): 085103

[39] http://www.ipgphotonics.com/apps_mat_lab_cutting.htm

[40] J Gabzdyl. Application versatility of nanosecond pulsed fiber lasers. Ind Laser Solut. 2014; 29 (1):p4

[41] J Yun, C Gao, S Zhu, C Sun, H He, L Feng, L Dong, et al. High-peak-power, single-mode, nanosecond pulsed, all-fiber laser for high resolution 3D imaging LIDAR system. Chin. Opt. Lett. 2012; 10(12):121402

[42] A Liu, M A Norsen and R D Mead. 60W green output by frequency doubling of polarized Yb-doped fiber laser. Opt. Lett. 2005; 30(1):67–69

[43] N Nishizawa, H Kawagoe and M Yamanaka. Highly functional ultrashort pulse fiber laser sources and applications for optical coherence tomography. CPMT Symposium Japan (ICSJ) 2015, IEEE. Kyoto. 2015: 85–87.

[44] M E Fermann, A Galvanauskas, and G Sucha. Ultrafast lasers: technology and applications. 1st edn. Marcel Dekker AG: CRC Press; 2002

[45] K Sugioka and Y Cheng. Ultrafast lasers—reliable tools for advanced materials processing. Light: Science & Applications. 2014; 3: e149

[46]   L Rihakova1 and H Chmelickova. Laser micromachining of glass, silicon, and ceramics. Adv Mater Sci Eng 2015; 2015:584952

[47]   I Miyamoto, K Cvecek, Y Okamoto, et al. Internal modification of glass by ultrashort laser pulse and its application to microwelding, Appl. Phys. A. 2014; 114:187–208

[48]   N A Inogamov, V V Zhakhovsky, S I Ashitkov, et al. Pump-probe method for measurement of thickness of molten layer produced by ultrashort laser pulse. AIP Conf. Proc. 1278. 2010; 590–599

[49]   J M Vadillo and J J Laserna. Laser-induced plasma spectrometry: truly a surface analytical tool. Spectrochim. Acta Part B Atomic Spectroscopy. 2004; 59(2):147–161

[50]   D Wu, S Z Wu, J Xu, et al. Hybrid femtosecond laser micro-fabrication to achieve true 3D glass/polymer composite biochips with multiscale features and high performance: the concept of ship-in-a-bottle biochip. Laser Photon. Rev. 2014; 8(3):458–467

Applications of Short Pulse Lasers

Effects of Different Laser Pulse Regimes (Nanosecond, Picosecond and Femtosecond) on the Ablation of Materials for Production of Nanoparticles in Liquid Solution

Abubaker Hassan Hamad

Additional information is available at the end of the chapter

http://dx.doi.org/10.5772/63892

Abstract

Ultra-short laser pulse interaction with materials has received much attention from researchers in micro- and nanomachining, especially for the generation of nanoparticles in liquid environments, because of the straightforward method and direct application for organic solvents. In addition, the colloidal nanoparticles produced by laser ablation have very high purity—they are free from surfactants and reaction products or by-products. In this chapter, nanosecond, picosecond and femtosecond laser pulse durations are compared in laser material processing. Due to the unique properties of the short and ultra-short laser pulse durations in material processing, they are more apparent in the production of precision material processing and generation of nanoparticles in liquid environments.

Keywords: lasers, laser ablation, nanoparticles, pulse duration, nanosecond lasers, picosecond lasers, femtosecond lasers

1. Introduction

Different laser pulse regimes (nanosecond, picosecond and femtosecond) were used not only to generate nanoparticles but also to manipulate them. Different laser wavelengths were selected to reduce the size of the nanoparticles and change their morphology. The study in this filed focuses on using different laser types and parameters to generate and manipulate of micro- and nanomaterials. Researchers have used different types of lasers not only to produce new materials

but also for precise micro- and nanomachining. Generating lasers with short and ultra-short pulse duration leads to high-precision laser processing. Lasers with pulse durations in the femtosecond to picosecond range demonstrate a significant development in quality for different materials in comparison with nanosecond or longer laser pulses [1]. In addition, using pulsed lasers to produce nanoparticles in liquid environments is a promising alternative to chemical methods for the production of totally ligand-free colloidal nanoparticles [2].

In general, for laser material processing, two different laser pulse duration regimes are used: long pulse duration, such as nanosecond pulse duration, which generates quite a significant heat-affected zone in the material because "the pulse duration is longer than the thermalisation time of most metals" [3]. This type of laser is suitable for removing materials or ablation. Short pulse duration (picosecond laser) and ultra-short pulse duration (femtosecond laser) yield better results, suited to the production of high-precision micro- and nanomachining. Typical laser pulse durations for precise laser material processing are 10 ps or less. It has been shown that suitable laser pulse durations for the micromachining of copper and stainless steel are in the range of 10–100 ps [3].

However, there is little evidence that researchers have approached the issue of lasers in nanotechnology in terms of precise and controllable ablation and their ability to generate nanoparticles from different materials. Consequently, the aim of this chapter is to provide an overview of how different laser pulses can be used in laser-material interaction and the production of nanoparticles.

Two important parameters used to describe lasers are their pulse duration (width) and pulse repetition rate (PRR). As shown in **Figure 1a**, laser pulse duration can also be described as full-width at half of the maximum (FWHM) amplitude of the laser pulse. Pulse repetition rate or pulse repetition frequency refers to the number of pulses emitted per second. For 1 kHz of PRR, the time period T would be 0.001 s (see **Figure 1b**).

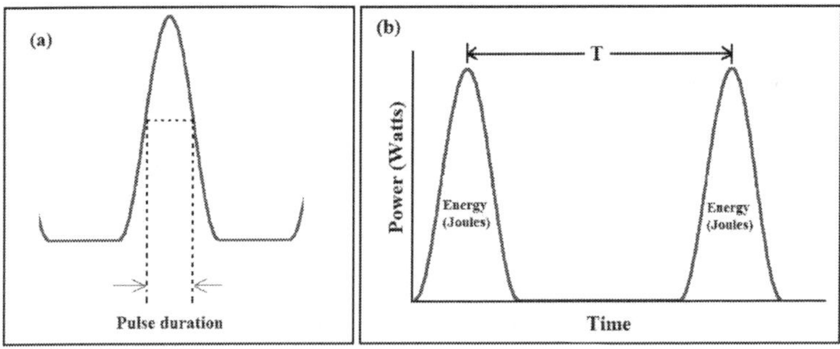

Figure 1. Pulse width or pulse duration (a) and pulse repetition rate (b) of a laser.

Two further parameters of pulsed lasers, which are especially relevant for ultrafast lasers, are the laser's peak power (P_{peak}), which is equal to the laser pulse energy divided by laser pulse duration ($P_{peak} = E_p/\tau$) and the average laser power, which is equivalent to the laser pulse energy multiplied by the pulse repetition rate ($P_{ave.} = E_p \times PRR$) [4].

In this chapter, the author shows the effects of different laser pulse interactions (nanosecond, picosecond and femtosecond lasers) with materials and the ablation of nanoparticles in a liquid environment in terms of their size, size distribution, morphology and production rate (productivity).

2. Pulsed laser ablation

Ultra-short laser pulse duration in the range of femtosecond laser pulses and a few picoseconds can be used to produce high-quality and precise material processing. Ultra-short pulse duration can only interact with electrons but longer pulse duration interacts with lattice. It is worth mentioning that during the interaction of an ultra-short laser pulse with materials, heat conduction is limited [5]. As a result of this, the material will be ablated within a spatial or well-defined area with minimised mechanical and thermal damage of the ablated area on the target. In contrast, longer (nanosecond) pulse duration irradiation on the materials leads to continuously heating the target material. The laser pulse energy will then be spread by heat conduction to an area outside the laser spot size, causing the irradiated target to boil and evaporate. Boiling and evaporation of the target material leads to the production of an uncontrollable melt layer [6]. In the case of nanosecond laser pulse duration, this problem may be caused by imprecise machining or marking.

There is existing research on laser-material interaction [6–8]; however, these show how differently pulsed lasers can be used for material processing, especially in nanotechnology. Furthermore, there are relevant findings concerning laser ablation of nanoparticles in liquid environments [9–12], and they are somewhat showing the effects of different laser pulse durations on the ablation of nanoparticles. Even most previous research on laser-material interaction and laser-generated nanoparticles tends to highlight laser beam parameters and experimental setups to produce small and well-distributed nanoparticles and to precision material processing, there is little emphasis on the optimal laser parameters to apply in laser material processing. Moreover, little attention has been given to the conceptualisation of the structure and phase of the nanoparticles produced by laser ablation in a liquid solution. Consequently, an understanding of the optimal pulse regime for these applications is critical in order to support and enhance the performance of laser material processing in nanotechnology.

3. Laser-material interaction at different laser pulse durations

Laser ablation of the materials starts with photon absorption, followed by the heating and photoionisation of the subjected area on the target by the laser beam. Subsequently, the ablated materials released from the target surface as solid fragments, vapours, liquid drops or as an expanding plasma plume. The amount of ablated material and phase depends on the absorbed energy by the target material [13]. After laser-material interaction with short pulses and low intensity, due to the inverse Bremsstrahlung, the laser beam energy will be absorbed by free electrons from the material followed by thermalisation within the electrons, and energy transfer to the lattice. Finally, energy will be lost due to electron heat transfer to the target material. The energy transfer from the laser beam to the target material can be described using 1D and 2D diffusion models when this is considered rapid thermalisation in the electron subsystem and if both lattice and electron subsystems are characterised by their temperatures (T_i lattice temperature and T_e electron temperature) [5]:

$$C_e \frac{\partial T_e}{\partial t} = -\frac{\partial Q(z)}{\partial z} - \gamma\left(T_e - T_i\right) + S \tag{1}$$

$$C_i \frac{\partial T_i}{\partial t} = \gamma\left(T_e - T_i\right) \tag{2}$$

$$Q(z) = -k_e\left(\frac{\partial T_e}{\partial z}\right) \tag{3}$$

$$S = I(t)\, A\alpha\, e^{-\alpha z} \tag{4}$$

where $Q(z)$ is the heat flux along the z-axis perpendicular to the target material surface, S is the laser heating source, $I(t)$ is the laser intensity as a function of time, A is the surface transmissivity ($A = 1 - R$), R is the reflectivity of the target material, α is the absorption coefficient of the target material, C_e is the heat capacity of the electrons per unit volume, C_i is the lattice heat capacity per unit volume, γ is the parameter characterising the electron-lattice coupling, and k_e is the thermal conductivity of the electrons. Two non-linear differential Eqs. (1) and (2) are used to model the cooling dynamics for T_e and T_i, which account for the electron-phonon coupling and thermal conductivity of the sample material [14]. In addition, these equations can be used to model the time evolution of the electron and lattice temperatures, T_e and T_i [15]. Eqs. (1)–(4) can be written as:

$$C_e \frac{\partial T_e}{\partial t} = k_e\left(\frac{\partial^2 T_e}{\partial z^2}\right) - C_i \frac{\partial T_i}{\partial t} + I(t)\, A\alpha e^{-\alpha z} \tag{5}$$

The lattice heat capacity (C_i) is considerably higher than the electronic heat capacity (C_e); in this case, the electrons have a very high temperature. When the Fermi energy is higher than the electron energy, the non-equilibrium thermal conductivity and heat capacity of the electron can be written as $k_e = k_o\,(T_i).T_e/T_i$ and $C_e = C'_e T_e$, respectively, where $k_o(T_i)$ is the conventional equilibrium thermal conductivity of a material and C'_e is a constant. In Eq. (5), a thermal conductivity in the lattice subsystem (phonon component) is neglected, and it has three characteristic timescales; τ_e is the electron cooling time ($\tau_e = C_e/\gamma$), τ_i is the lattice heating time ($\tau_e \ll \tau_i$) ($\tau_i = C_i/\gamma$), and τ_L is the laser pulse duration. In laser-material interaction, these parameters define three different interaction regimes: nanosecond, picosecond and femtosecond [5].

Zimmer [16] proposed a model for the analytical solution of the laser-induced temperature distribution across internal solid-liquid interfaces (see **Figure 2a**). It was shown that high solid surface temperature can be obtained with short laser pulse durations, sufficient interface absorption and high absorption liquids. As shown in **Figure 2b**, in the case of using a nanosecond laser (20 ns), the temperature of the liquid environment is quite higher than that of the transparent solid material.

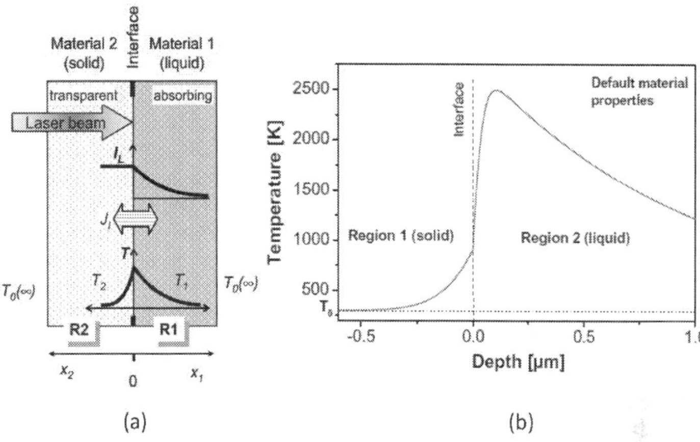

Figure 2. Diagram showing laser heating of a solid-liquid interface. I_L is the laser absorption, T is the temperature distribution in both materials, and J_i is the heat conduction across the interface [16].

3.1. Nanosecond laser

It was shown that the laser pulse duration has an effect on both the material ablation thresholds and penetration depths. Long pulse duration or increasing laser pulse duration increases the threshold fluence and decreases the effective energy penetration depth [1]. Low-intensity long laser pulse interaction with a target material firstly heats the surface of the target due to the absorbed energy, which leads to melting and vaporisation. It should be noted that vaporisation of the target requires much more energy than melting. "In case of low laser intensities the created vapour remains transparent for the laser radiation". The electron and lattice (ion) temperatures are equal ($T_e = T_i = T$) [6]. In other words, if the laser pulse duration is long in comparison with the electron-phonon energy-transfer time ($\tau_L \gg \tau_i$), the electrons and lattice temperatures will remain at the same thermal equilibrium point ($T_e = T_i = T$) [5, 17]; as such, Eq. (5) reduces to the heat equation:

$$C_i \frac{\partial T}{\partial t} = k_o \left(\frac{\partial^2 T}{\partial z^2} \right) + I_\alpha \alpha e^{-\alpha z} \tag{6}$$

In nanosecond laser beam interaction with material, the surface of the target material will be heated to melting point and then to vaporisation temperature. During the laser-material interaction, energy will be lost as heat conduction into the target material; the heat penetration depth (l) is given by $l \sim (D_t)^{1/2}$, where D is the heat diffusion coefficient and is given by ($D = k_o/C_i$). It can be noted that for this regime of lasers, the condition $D\tau_L \alpha^2 \gg 1$ is fulfilled, for example the thermal penetration depth is quite larger than the optical penetration depth [18]. The energy deposited in the target material per unit mass is given by $E_m \sim I_\alpha t/pl$; at a specific time ($t = t_{th}$), this energy becomes higher than the specific heat of evaporation Ω, at which point considerable evaporation will occur. When $E_m \sim \Omega$, the results are $t_{th} \sim D(\Omega\varrho/I)^2$. Consequently, for strong evaporation conditions, $E_m > \Omega$ or $\tau_L > f_{th}$ can be written for laser intensity as [5, 6]:

$$I > I_{th} \sim \frac{\rho\Omega D^{1/2}}{\tau_L^{1/2}} \tag{7}$$

and for laser fluence as:

$$F > F_{th} \sim \rho\Omega D^{1/2} \times \tau_L^{1/2} \tag{8}$$

The threshold laser fluence increases as $\tau_L^{1/2}$. In nanosecond laser ablation regimes, there is enough time for thermal waves to propagate into the target material and to create a relatively large layer of melted material target [5, 6]. Nanosecond laser pulses can ablate the target materials even at low laser intensities in both the vapour and liquid phases, so a recoil pressure that expels the liquid will be created due to the vaporisation process [6]. Evaporation occurrence makes challenge to precise laser processing with nanosecond laser pulses [18].

At long laser pulse duration, interaction with materials usually fulfils the condition $L_{th} \gg \alpha^{-1}$, L_{th} being the heat-penetration depth which is given by $L_{th} \approx (2D\tau_p)^{1/2}$, where $D = k/\rho c$, D is the heat-diffusion coefficient. So, long laser pulse duration creates sufficient time for thermal waves to propagate within the target material, and the absorbed energy will be stored in a layer with a thickness of about L_{th}. In this case, the target material needs much more energy to vaporise than to melt; in other words, evaporation will occur, while the energy absorbed per unit volume into the vaporised layer becomes higher than the latent heat of evaporation per unit volume, namely [19].

$$\Delta z_\upsilon \approx \frac{A(F_L - F_{th})}{\rho L_\upsilon} \tag{9}$$

where F_{th} is the laser fluence threshold which represents the minimum energy above which appreciable evaporation occurs from liquid metals. This figure is approximately given by the energy required to melt a surface layer of the target material of the order of L_{th} [19]:

$$F_{th} \approx \frac{\rho c \Delta T_m L_{th}}{A}$$ (10)

where $\Delta T_m = T_m - T_o$ and T_m and T_o are the melting and initial target temperature, respectively.

3.2. Picosecond laser

At low-intensity, short laser pulse interactions with a target material, due to the inverse Bremsstrahlung most of the laser energy will be absorbed by the free electrons of the target. This result can be described by the difference between the electron and lattice temperatures $(T_e > T_i)$ in a transient nano-equilibrium state. In spite of a small energy exchange between the lattice and the electron heat conduction, the electrons are cooled [6]. For picosecond laser ablation, time t is much greater than τ_e $(t \gg \tau_e)$, which is equivalent to $C_e T_e/t \ll \gamma T_e$. In addition, when the condition $\tau_e \ll \tau_L \ll \tau_i$ is fulfilled, Eq. (1) becomes quasi-stationary for the electron temperature [5]. In other words, when the laser pulse duration is shorter than the electron-phonon energy-transfer time, then the electron and lattice have different temperatures, meaning that they will be in a non-thermal equilibrium state. In this case, Eq. (5) becomes the following equations [5, 17]:

$$k_e\left(\frac{\partial^2 T_e}{\partial z^2}\right) - \gamma\left(T_e - T_i\right) + I_a \alpha e^{(-\alpha z)} = 0$$ (11)

$$T_i = T_o + \frac{1}{\tau_i}\int_0^t e^{\left(\frac{t-\theta}{\tau_i}\right)} T_e(\theta)\, d\theta$$ (12)

This method represents the lattice temperature in integral from. The above equations describe heating of metal targets by the laser pulses when the laser pulse duration $\tau_L \gg \tau_e$. By neglecting T_o and when $t \ll \tau_i$, because of the quasi-stationary character of the electron temperature, Eq. (12) can be reduced as follows [5, 18]:

$$T_i \simeq T_e\left(1 - e^{\left(\frac{-t}{\tau_i}\right)}\right) \simeq \left(\frac{t}{\tau_i}\right) T_e$$ (13)

It can be noted from the last equation that during picosecond laser ablation, the lattice temperature remains notably lower than the electron temperature, and thus, the lattice temperature in Eq. (11) can be neglected.

When $k_e T_e \alpha^2 \ll \gamma T_e$, from Eqs. (11) and (13), it can be concluded that the electron cooling is due to an exchange of energy with the lattice of the material target. Finally, both the lattice and electron temperature at the end of the laser pulse can be expressed by the following equation [5]:

$$T_e \simeq \frac{I_\alpha \alpha}{\gamma} e^{(-\alpha z)} \ \text{and} \ T_i \simeq \frac{F_\alpha \alpha}{C_i} e^{(-\alpha z)} \tag{14}$$

It can be noted that when the condition $\tau_e \ll \tau_L$ is fulfilled, at the end of the pulse, both lattice temperature and attainable lattice temperature are approximately equal.

3.3. Femtosecond laser

For femtosecond lasers, if the laser pulse duration τ_L is assumed to be shorter than the electron cooling time τ_e. ($\tau_L \ll \tau_e$.) and when $t \ll \tau_e$, it is equivalent to $C_e T_e / t \gg \gamma T_e$. In this case, the electron-lattice coupling can be neglected and Eq. (1) can be solved easily. When $D_e \tau_L < \alpha^2$ (where $D_e = k_e/C_e$ is the electron thermal diffusivity) is fulfilled, to simplify the solution of the equation, the electron heat conduction term can be neglected. Thus, Eq. (1) can be written as [5]:

$$C_e' \left(\frac{\partial T_e^2}{\partial t^2} \right) = 2I_\alpha \alpha e^{(-\alpha z)} \tag{15}$$

which gives

$$T_e(t) = \left(T_o^2 + \frac{2I_\alpha \alpha}{C_e'} t e^{(-\alpha z)} \right)^{1/2} \tag{16}$$

where $I(t) = I_o$ is assumed constant, and $I_\alpha = I_o A$, and $T_o = T_e(0)$ refer to the initial temperature.

It has been shown that heat conduction of the target material can be neglected at the very short timescales of picosecond and femtosecond laser pulse durations; thus the target temperature at the end of the pulse within the target material can be given by [19].

$$T(z,\tau_L) \approx \left(\frac{\alpha A F_\alpha}{\rho c} \right) e^{(-\alpha z)} \tag{17}$$

where $F_\alpha = I_o \tau_L$ is the laser pulse fluence, and τ_L is the laser pulse duration.

The evolution of the electron temperature (T_e) and lattice temperatures (T_i) after the laser pulse is described by Eq. (5) with S = 0. In addition, the electron temperature and lattice temperature initial conditions are given by Eq. (17) and $T_i = T_o$. Due to the energy transfer to the lattice and heat conduction of the bulk material, the electrons are rapidly cooled after the laser pulse. Since the electron cooling time is quite short, then Eq. (2) can be written as $T_i \simeq T_e(\tau_L)t/\tau_i$. It should be noted that the initial lattice temperature is neglected here. On the other hand, the attainable lattice temperature is determined by the average cooling time of the electrons $\tau_e^a = C_e^{'} T_e(\tau_L)/2\gamma$ and is given by the following equation [5]:

$$T_i \simeq T_e^2(\tau_L)\frac{C_e^{'}}{2C_i} \simeq \frac{F_\alpha \alpha}{C_i}e^{(-\alpha z)} \tag{18}$$

Fann et al. [20] and Wang et al. [14] have shown that the time scale for significant energy transfer and fast electron cooling is about 1 ps. In the case of $C_i T_i \gg \varrho\Omega$, where ϱ is the density and Ω is the specific heat of evaporation per unit mass, considerable evaporation will occur. From Eq. (18), the conditions for strong evaporation can be given in the form:

$$F_\alpha \geq F_{th}\, e^{(\alpha z)} \tag{19}$$

where $F_{th} = \varrho\Omega/\alpha$ is the threshold fluence laser evaporation by femtosecond laser pulses. Then, the ablation depth per laser pulse (or ablation rate) L can be written as [5, 21]:

$$L \simeq \alpha^{-1}\ln(\frac{F_\alpha}{F_{th}}) \tag{20}$$

The logarithmic dependence of the ablation depth on the laser pulse fluence is well known for the laser ablation of organic polymers and metal targets with femtosecond pulse duration [5].

It can be noted that Eqs. (14) and (18) give the same expressions for the lattice temperature in both picosecond and femtosecond laser regimes. Therefore, the condition for strong evaporation given by (19), the fluence threshold and the ablation depth per pulse given by (20) remain unchanged [5, 18]. Thus, in the picosecond laser range, it is possible that logarithmic dependence of the ablation depth on the fluence exists. Here, electron heat conduction inside the target material is neglected. In this case, laser ablation is accompanied by the electron heat conduction and production of a melted area in the target material. Even evaporation can be considered as a direct solid-vapour transition process, whereby the existence of a liquid phase in the target material reduces the precision of laser material processing. Femtosecond laser ablation effects a direct solid-vapour transition due to the short timescales in this laser regime; as a result of this, the lattice is heated on a picosecond timescale, leading to the production of vapour and plasma phases followed by a rapid expansion in vacuum. Here, in a first approximation, thermal conduction into the target material can be neglected during all of the aforementioned processes. Due to the advanced properties of picosecond laser ablation, highly precise and pure laser material processing can be achieved, as has been experimentally demonstrated by Chichkov et al. [5].

4. Comparison of different pulse durations for the ablation of materials and production of nanoparticles

A comparison of the characteristics of nano-, pico- and femtosecond lasers produced nanoparticles and materials processing have been studied [5, 22]. Even the comparison is not a like-for-like; the experimental work should be assumed as a fare comparison between these types of commercial lasers operating at their usual operating conditions [22]. Nanosecond laser ablation of materials occurs due to "melt expulsion driven by the vapour pressure and the recoil pressure of light". The melted material will solidify again because of the instability of this process, in which the fluid phase dynamics and the driving vapour conditions are very complex. As a result of this, the ablated area on the surface of the material target is not precise and uniform in comparison with that produced by a femtosecond laser (see **Figure 3**). Furthermore, nanosecond laser ablation creates a heat-affected zone (HAZ) [4]. It was shown that the HAZ of Al sample for the nanosecond laser ablation was about 40 mm wide, whereas for the femtosecond laser there was not observable because it was <2 mm wide [23].

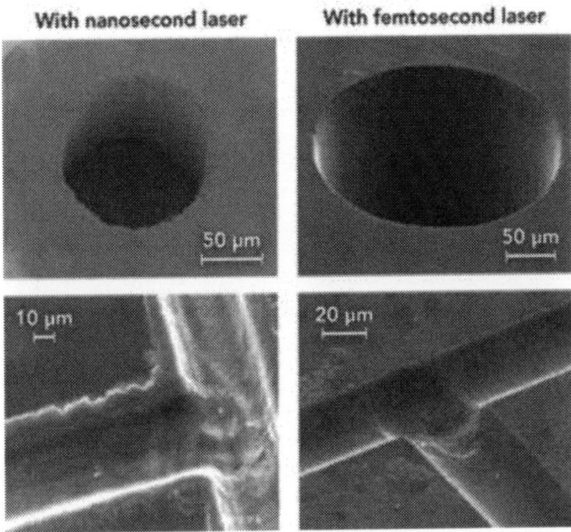

Figure 3. Laser material processing of a glass target by nanosecond laser (left) and femtosecond laser (right) ablation [4].

In the case of short (picosecond laser) and ultra-short laser pulses (femtosecond lasers), their pulse duration (τ) is considerably shorter than the timescale required for energy transfer between the lattice and the free electrons of the material target. As a result of this, very high temperatures and pressures are produced at a very shallow depth in the range of microns. Conversely, irradiated material is heated rapidly by pulsed laser and directly reaches the vapour phase with high kinetic energy without passing through the melting point temperature due to the absorbed energy. In other words, the material ablation will take place by vaporisation without producing a recast layer on the ablated area. As shown in **Figure 3**, the ablated area is very precise and smooth without forming any observable heat-affected zone (HAZ) [4]. The target materials with a high thermal conductivity are very important for the femtosecond laser ablation process because of the stable properties and chemical composition of the area ablated with femtosecond laser ablation [24].

For ultrafast lasers, laser beam energy deposition happens on a timescale that is quite short in comparison with atomic relaxation processes. After the laser energy is absorbed by the electrons, cold ions will be produced, leading to the occurrence of thermalisation at the end of the laser pulse. In addition, the femtosecond laser intensity is quite high and is sufficient to drive highly non-linear absorption processes in the target materials which the laser wavelength cannot absorb. At these high intensities, multi-photon ionisation becomes considerably strong [4]. Because of the very high flux of the femtosecond laser photons, several photons collide and become bound electrons; this is multi-photon ionisation. When the amount of total photon energies absorbed is higher than the ionisation potential, the bound electrons will absorb several photons. As a result, the bound electrons become free from the valence band. The multi-photon ionisation process is higher at very high laser intensities. When the laser intensity (photon flux) is above 10^{13} W/cm^2, multi-photon ionisation becomes very strong and seed electrons are not required to initialise ionisation in high-energy band gap materials or wide band gap materials (see **Figure 4a**) [4].

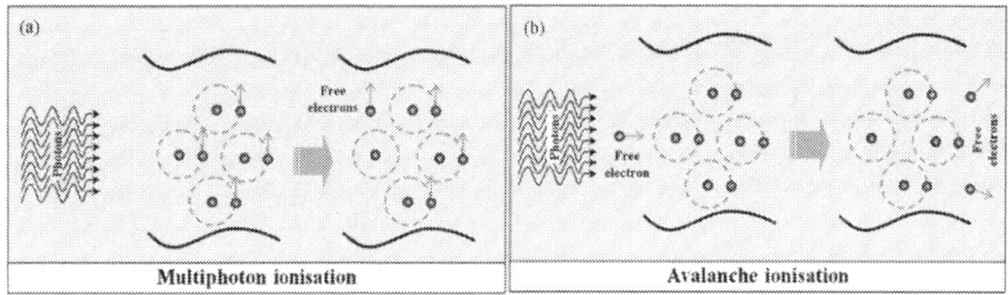

Figure 4. Non-linear ionisation processes of multi-photon absorption (a) and avalanche ionisation (b).

If the kinetic energy of the free electron is very high, some of the energy might be transferred to a bound electron in the target material by collision, thus overcoming the ionisation potential and generating two free electrons; this process is known as (collisional) impact ionisation. Thus, more free electrons will be produced from the bound electrons after the free electrons have absorbed photons. This is called avalanche ionisation, in which free electron density exponentially increases. It is worth mentioning that avalanche ionisation is highly dependent upon free electron density. Energy loss by phonon emission and energy gain by inverse Bremsstrahlung competition indicate the efficiency of this process. It is responsible for the high-energy gap transition materials at laser intensities below 10^{12} W/cm^2 (see **Figure 4b**) [4]. For femtosecond lasers, avalanche ionisation plays an essential role in the optical breakdown. Briefly, avalanche ionisation produces a macroscopic breakdown of the target materials and multi-photon ionisation produces seed electrons at a critical density of about 10^{20} cm^{-3} [25].

The amount of material removed during laser ablation depends on the amount of energy absorbed by the bulk material target. The dissipation of the absorbed laser energy will usually occur after the laser pulse duration. There are two major mechanisms to explain material removal by laser ablation: thermal vaporisation, where the local temperature increases to above the vaporisation point due to the electron-phonon collisions, and the occurrence of a Coulomb explosion, where the bulk materials release the excited electrons and produce a relatively strong electric field which revokes the ions inside the impact area [4, 26]. According to these mechanisms, femtosecond laser material removal can be classified in two ways: the first is "strong ablation dominated by thermal vaporisation at intensities significantly higher than the ablation threshold" and the second is "gentle ablation governed by the Coulomb explosion near the ablation threshold" [4]. As shown in **Figure 5**, the long-pulse lasers have more heat-affected zones and shock waves in comparison with the shorter picosecond and femtosecond lasers. The main differences between them are due to the mechanism or the basic principles of laser-induced target material removal processes [27].

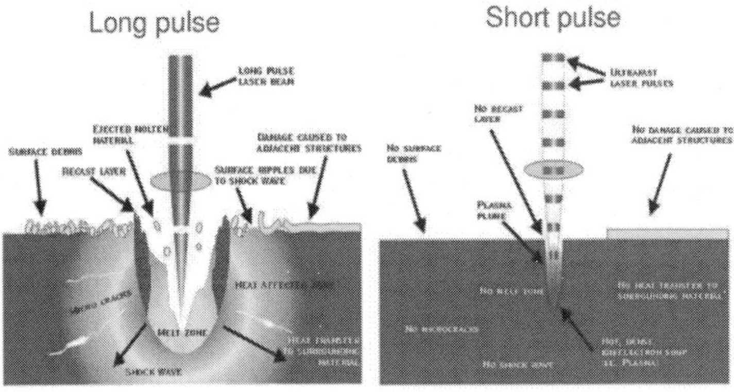

Figure 5. Long-pulse and ultrafast-pulse laser interaction with target material [27].

Laser pulse duration is an important parameter in the laser ablation process. There are quite large differences between the long pulse duration (nanosecond laser) and ultra-short laser pulse duration (femtosecond laser) for the ablation of materials [26]. As shown in **Figure 6**, during nanosecond laser ablation, plasma will be produced during the laser pulse, but during femtosecond laser ablation, it will be produced after the laser pulse has ended. During nanosecond laser ablation, plasma is a part of the pulse duration, so the pulse serves to reheat the plasma. This leads to higher persistence of the plasma for nanosecond laser ablation than for femtosecond laser ablation [26].

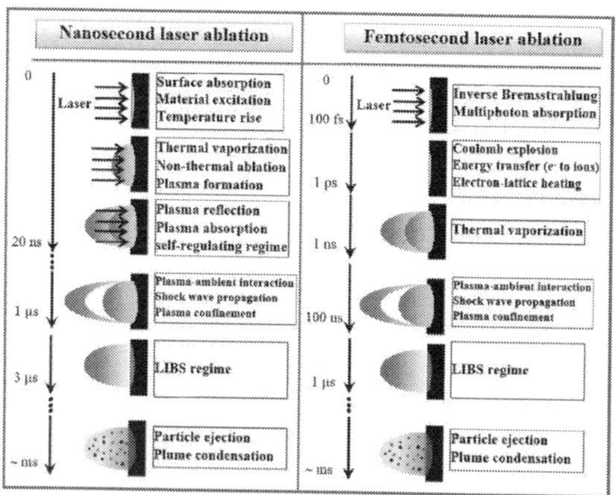

Figure 6. Approximate timescale comparison of pulsed laser energy absorption and ablation, along with the various processes, for nanosecond laser (10 ns) and femtosecond laser (50 fs) ablation in an ambient gas.

As shown in the schematic diagram in **Figure 7**, Lescoute et al. [28] showed that during subpicosecond laser ablation three phases occur behind the shock: liquid, vapour and plasma. The shock waves produced can compress the solid target material. The plasma plume effect is more predominant in picosecond laser ablation than in femtosecond laser ablation [22]. In general, laser-plasma interaction during laser-material ablation is strongly dependent on laser wavelength and the excitation wavelength is a very important parameter in nanosecond laser ablation, as in femtosecond laser ablation. The production of nanoparticles through plasma plume condensation occurs in the microsecond-millisecond timescales.

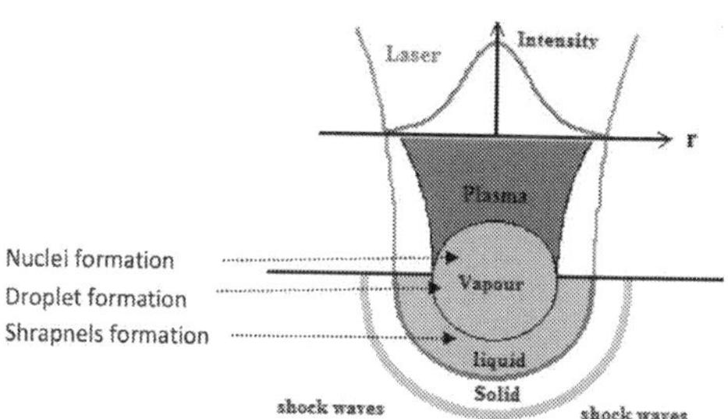

Figure 7. Schematic diagram of ablated matter depicted in an ultra-short, that is subpicosecond, laser-material interaction.

Another advantage of the femtosecond laser is its short timescale which prevents the material target from being heated; this leads to a reduction in the thermal effects in comparison with the nanosecond laser ablation. The duration of the energy transfer process from electron to ion (τ_{ei}) is approximately 1–10 s of a picosecond; this is in the same order as the heat conduction time (τ_{heat}). In an ideal femtosecond laser ablation case, this is considerably longer than the laser pulse duration (t_p). ($\tau_{ei} \sim \tau_{heat} \gg t_p$) [26]. This will depend on the type of material sample.

Femtosecond laser energy transfer in the material for ablation will occur through energy absorption by electrons, leading to ionisation and then redistribution of the laser energy to the lattice. Ionisation is either produced by multi-photon ionisation or collisional impact. These processes can be labelled multi-photon ionisation (MPI) and collisional impact ionisation (CII) [26]. Multi-photon ionisation (MPI) occurs when multiple photons provide sufficient energy to the electrons in the valance band; as a result, the electrons are free to reach metastable quantum states by excitation. Collisional impact ionisation (CII) will take place when bound electrons gain energy from free electrons by collision; as a result, the bound electrons will be released.

As shown in **Figure 8**, when a single high fluence fs laser pulse is incident onto a material, the timescale of the physical phenomena involved in laser-material interaction can take place as follows: after 1 fs, photo-ionisation, such as multi-photon absorption and/or tunnelling ionisation, will occur; after 50 fs, avalanche ionisation will occur; and after 100 fs, thermalisation of the electrons of the metal target will take place. Longer laser pulse durations lead to more physical phenomena; after 1 ps, energy transformation from electron to lattice will result, then after 10 ps, some thermodynamic processes such as thermal diffusion, fusion and explosion will occur. In much longer laser pulses, after 1 ns, photochemical processes such as chemical reactions and phase transformation will take place [29].

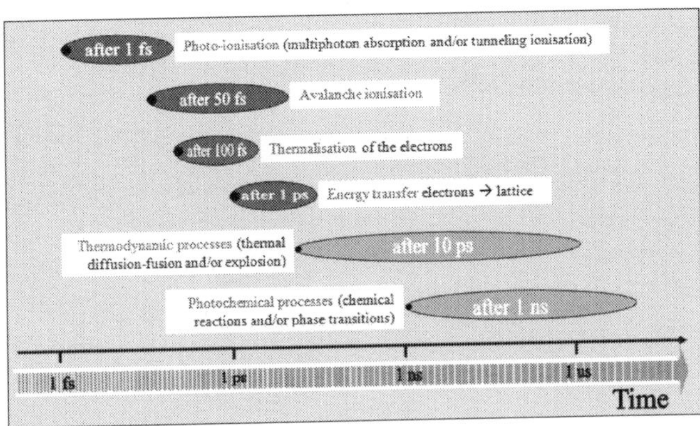

Figure 8. Occurrence of different physical phenomena during different timescales involved in laser-material interaction.

Different types of nanoparticles have been produced by nanosecond [30, 31], picosecond [32, 33] and femtosecond [28, 34, 35] lasers in different liquid environments, vacuum and gas media. A physical aspect of laser ablation in liquid environments is the liquid breakdown phenomena; it has been shown that the conversion efficiency of the radiation energy to a cavitation bubble during radiation focusing of different wavelengths in water is much lower for femtosecond lasers (at 100 fs, efficiency 1–2%) than for nanosecond lasers (at 76 ns, efficiency 22–25%). Femtosecond lasers (120 fs and 800 nm) can produce controllable size and size distribution of nanoparticles, while nanosecond laser produces relatively large, quite widely dispersed particles [36].

The generation of large nanoparticles in a liquid environment by nanosecond laser pulse durations and longer is due to the essential target material melting and laser pulse interaction with the cavitation bubbles that are then produced in the liquid environment [37]. The long pulse duration is sufficient to allow photon coupling with both the electronic and vibrational modes of the sample material. This case will be more predominant when the target material has low reflectivity at the laser wavelength, a large absorption coefficient, a low thermal diffusion coefficient and a low boiling point [38]. Long pulse duration heats the target material continually during the pulse duration. This causes the target material to start boiling and subsequently leads to evaporation, which produces a considerable melt layer. As a result, heat will be transferred to the target that prevents the production of very small structures and small nanoparticles. In nanosecond laser ablation, the laser power heats the target material to melting point and then vaporisation temperature; as such, this process can be considered an indirect solid-vapour transition but a solid-melt-vapour transition [19]. Nanosecond laser ablation ejects an ablation plume which creates a shielding on the surface of the material target, leading to a reduction in the laser power induced to the target. Chemistry material responses, cumulative changes in the target material's surface texture and morphology will occur due to surface melting and a combination of ablation and thermally activated processes [7].

However, the generation of small nanoparticles by short laser pulse duration is due to minimisation of both the thermal effects and the laser pulse interaction with the cavitation bubble [37]. Here, thermal conduction of the target material can be neglected because of the very short timescale and the heated area of the material being of the order of the laser absorption length α^{-1}, where α is the target material absorption coefficient [19]. At short and ultrashort laser pulse durations, because the laser interacts with materials due to very quick excitation of the electron distribution, electron-electron coupling leads to a rapid rise in the electron temperature, followed by lattice heating "at a rate dependent upon the electron-phonon coupling strength, and eventual vaporisation of the transiently heated target". Electronic contributions "can manifest themselves as unexpectedly large ion yields and/or supra-thermal propagation velocities within the expanding plasma plume" [38]. At short and ultra-short pulse durations, because the energy transfer from the electrons to the lattice happens on a longer timescale than for the short or ultra-short pulses, the pulses do not heat the target material continually. This limits the heat conduction within the target material and produces a superheated layer the size of the irradiated volume. This leads to ablation in a specific area, resulting in minimal thermal damage. Both picosecond and femtosecond laser ablation can be viewed as a direct solid-vapour transition [19].

At short and ultra-short laser pulse durations, the ablation threshold fluence of the target materials decreases and becomes very sharply defined. In addition, the excess energy in the target still can produce thermal effects on the ablation area after the pulse has ended. Furthermore, at ultra-short laser pulse duration, the optical absorption depth reduces due to an optical breakdown caused by the ultra-short laser pulse duration; this can also cause strong absorption in otherwise transparent wide-energy band gap materials. Another advantage of picosecond and femtosecond laser ablation is that they are "separated in time from material response and ejection" [7]. In contrast, for laser production of nanoparticles in a liquid solution, the picosecond laser performs better than the femtosecond laser [37]. It can be seen that only the femtosecond laser avoided the laser-bubble interaction [22].

The size and the morphology of the nanoparticles depend on the laser pulse duration, such as whether a nano-, pico- and femtosecond laser regime is being used and the nature of the target material [39]. Hamad et al. [22] showed that the picosecond laser (10 ps and 1064 nm wavelength) produced more spherical silver nanoparticles than those produced by both nanosecond (5 ns and 532 nm wavelength) and femtosecond (100 fs and 800 nm wavelength) lasers. Under the same conditions, smaller gold nanoparticles were produced in comparison with the silver and titania, with the latter being the largest. It was also concluded that the average size of the nanoparticles was smaller for the shorter wavelength (532 nm) nanosecond laser in relation to picosecond and femtosecond lasers, showing that the role of the laser wavelength is more significant than the laser pulse duration to control the size of the nanoparticles. It is worth mentioning that some silver nanoparticles produced by nanosecond laser have irregular (non-spherical) shapes. Dudoitis et al. [40] showed that the specific nanosecond and picosecond laser parameters can be used to produce controllable particle size distribution of nickel and aluminium in argon gas.

Lescoute et al. [28] observed two groups of particles of two different sizes after femtosecond laser ablation caused by the temperature achieved in the plasma plume: these could be classified into micrometer-size fragments and nanometre-size particles for temperatures lower than the critical temperature and for higher temperatures, respectively. Kabashin and Meunier [34] used femtosecond laser ablation to produce Au nanoparticles in deionised water. The researchers observed two different groups of nanoparticles: the small (about 3–10 nm) and monodispersed particles were produced due to a mechanism which, "associated with thermal-free femtosecond ablation, manifests itself at relatively low laser fluences F, 400 J/cm^2". The second group of particles was of a larger size and a broad size distribution, produced due to plasma-induced heating and ablation that occurs at high laser fluencies. Itina [41] showed that the ultra-short laser pulses produce small nanoparticles and a stable size distribution at optimal experimental parameters. This is because while using short and ultra-short laser pulse duration, laser radiation will not be absorbed by created plasma plume. This leads to transfer much more energy into the target material during very short time. Finally, the target ejected the cluster precursors directly to the liquid solution and the plasma plume is confined to a smaller volume. Chakravarty et al. [42] concluded that the smaller size of silver and copper nanoparticles can be produced by ultra-short laser pulse duration in comparison with the subnanosecond laser pulse duration. It was shown that broader surface plasmon resonance (SPR) and smaller reflection/transmission of silver and copper nanoparticles deposited by femtosecond laser in comparison with those deposited by subnanosecond laser indicating smaller size particle generation by ultra-short pulse durations. In contrast, Hubenthal et al. [43] concluded that the tailoring the size and shape of nanoparticles with nanosecond pulsed laser is more achieved in comparison with both continuous wave (CW) and femtosecond pulsed lasers.

5. Conclusions

Pulsed laser ablation for material processing and production of nanoparticles in liquid environments have been used due to the unique properties of the laser beam in marking and machining materials. There is a lack of knowledge of the optimal laser beam parameters in this field, as different laser pulse durations respond differently to different materials. A critical comparison of different laser pulse durations in micro- and nanomaterial processing was therefore sought in order to obtain higher quality and productivity. Femtosecond and picosecond lasers were found to be superior to the nanosecond laser for precision material processing. In addition, in spite of the short and ultra-short duration of their laser pulses, these lasers were observed to be more effective for the production of nanoparticles than those with longer pulse durations. Finally, laser wavelength was found to play significant role in the production of nanoparticles. The advantages of femtosecond laser ablation for micro- and nanotechnology are due to non-thermal interaction with solid target materials.

Author details

Abubaker Hassan Hamad*

Address all correspondence to: abubaker.hamad75@yahoo.co.uk

Laser Processing Research Centre, School of Mechanical, Aerospace and Civil Engineering, The University of Manchester, Manchester, UK

References

[1] Le Harzic, R., D. Breitling, M. Weikert, S. Sommer, C. Föhl, S. Valette, C. Donnet, E. Audouard and F. Dausinger, *Pulse width and energy influence on laser micromachining of metals in a range of 100 fs to 5 ps*. Applied Surface Science, 2005. 249(1): pp. 322–331. doi: 10.1016/j.apsusc.2004.12.027

[2]   Rehbock, C., J. Jakobi, L. Gamrad, S. van der Meer, D. Tiedemann, U. Taylor, W. Kues, D. Rath and S. Barcikowski, *Current state of laser synthesis of metal and alloy nanoparticles as ligand-free reference materials for nano-toxicological assays*. Beilstein Journal of Nanotechnology, 2014. 5(1): pp. 1523–1541. doi:10.3762/bjnano.5.165

[3]   Jaeggi, B., B. Neuenschwander, M. Schmid, M. Muralt, J. Zuercher and U. Hunziker, *Influence of the pulse duration in the ps-regime on the ablation efficiency of metals*. Physics Procedia, 2011. 12: pp. 164–171. doi:10.1016/j.phpro.2011.03.118

[4]   Lucas, L. and J. Zhang, *Femtosecond laser micromachining: a back-to-basics primer*. 2012, http://www.industrial-lasers.com/

[5]   Chichkov, B.N., C. Momma, S. Nolte, F. von Alvensleben and A. Tünnermann, *Femtosecond, picosecond and nanosecond laser ablation of solids*. Applied Physics A, 1996. 63(2): pp. 109–115. doi:10.1007/bf01567637

[6]   Momma, C., B.N. Chichkov, S. Nolte, F. von Alvensleben, A. Tünnermann, H. Welling and B. Wellegehausen, *Short-pulse laser ablation of solid targets*. Optics Communications, 1996. 129(1): pp. 134–142. doi:10.1016/0030-4018(96)00250-7

[7]   Brown, M.S. and C.B. Arnold, *Fundamentals of laser–material interaction and application to multiscale surface modification*, in *Laser Precision Microfabrication*. 2010, Springer. pp. 91–120. doi:10.1007/978-3-642-10523-4__4

[8]   Han, J. and Y. Li, *Interaction Between Pulsed Laser and Materials*. Lasers-Applications in Science and Industry, K. Jakubczak, Ed. InTech, 2011. doi:10.5772/25061

[9]   Sukhov, I.A., G.A. Shafeev, V.V. Voronov, M. Sygletou, E. Stratakis and C. Fotakis, *Generation of nanoparticles of bronze and brass by laser ablation in liquid*. Applied Surface Science, 2014. 302(0): pp. 79–82. doi:10.1016/j.apsusc.2013.12.018

[10]   Tyurnina, A.E., V.Y. Shur, R.V. Kozin, D.K. Kuznetsov and E.A. Mingaliev, *Synthesis of stable silver colloids by laser ablation in water*. Proceedings of SPIE 2013. 9065: p. 8. doi: 10.1117/12.2053557

[11]   Yan, Z. and D.B. Chrisey, *Pulsed laser ablation in liquid for micro-/nanostructure generation*. Journal of Photochemistry and Photobiology C: Photochemistry Reviews, 2012. 13(3): pp. 204–223. doi:10.1016/j.jphotochemrev.2012.04.004

[12]   Amans, D., C. Malaterre, M. Diouf, C. Mancini, F. Chaput, G. Ledoux, G. Breton, Y. Guillin, C. Dujardin and K. Masenelli-Varlot, *Synthesis of oxide nanoparticles by pulsed laser ablation in liquids containing a complexing molecule: impact on size distributions and prepared phases*. The Journal of Physical Chemistry C, 2011. 115(12): pp. 5131–5139. doi: 10.1021/jp109387e

[13]   Amendola, V. and M. Meneghetti, *Laser ablation synthesis in solution and size manipulation of noble metal nanoparticles*. Physical Chemistry Chemical Physics, 2009. 11(20): pp. 3805–3821. doi:10.1039/B900654K

[14] Wang, X., D.M. Riffe, Y.-S. Lee and M. Downer, *Time-resolved electron-temperature measurement in a highly excited gold target using femtosecond thermionic emission.* Physical Review B, 1994. 50(11): p. 8016. doi:10.1103/PhysRevB.50.8016

[15] Schoenlein, R., W. Lin, J. Fujimoto and G. Eesley, *Femtosecond studies of nonequilibrium electronic processes in metals.* Physical Review Letters, 1987. 58(16): p. 1680. doi:10.1103/PhysRevLett.58.1680

[16] Zimmer, K., *Analytical solution of the laser-induced temperature distribution across internal material interfaces.* International Journal of Heat and Mass Transfer, 2009. 52(1): pp. 497–503. doi:10.1016/j.ijheatmasstransfer.2008.03.034

[17] Fujimoto, J., J. Liu, E. Ippen and N. Bloembergen, *Femtosecond laser interaction with metallic tungsten and nonequilibrium electron and lattice temperatures.* Physical Review Letters, 1984. 53(19): p. 1837. doi:10.1103/PhysRevLett.53.1837

[18] Stafe, M., A. Marcu and N. Puscas, *Pulsed Laser Ablation of Solids: Basics, Theory and Applications.* Vol. 53. 2013: Springer Science & Business Media. doi:10.1007/978-3-642-40978-3

[19] Amoruso, S., R. Bruzzese, N. Spinelli and R. Velotta, *Characterization of laser-ablation plasmas.* Journal of Physics B: Atomic, Molecular and Optical Physics, 1999. 32(14): p. R131. doi:10.1088/0953-4075/32/14/201

[20] Fann, W., R. Storz, H. Tom and J. Bokor, *Electron thermalization in gold.* Physical Review B, 1992. 46(20): p. 13592. doi:10.1103/PhysRevB.46.13592

[21] Izawa, Y., Y. Setuhara, M. Hashida, M. Fujita and Y. Izawa, *Ablation and amorphization of crystalline Si by femtosecond and picosecond laser irradiation.* Japanese Journal of Applied Physics, 2006. 45(7R): p. 5791. doi:10.1143/JJAP.45.5791

[22] Hamad, A., L. Li and Z. Liu, *A comparison of the characteristics of nanosecond, picosecond and femtosecond lasers generated Ag, TiO$_2$ and Au nanoparticles in deionised water.* Applied Physics A, 2015. 120(4): pp. 1247–1260. doi:10.1007/s00339-015-9326-6

[23] Le Harzic, R., N. Huot, E. Audouard, C. Jonin, P. Laporte, S. Valette, A. Fraczkiewicz and R. Fortunier, *Comparison of heat-affected zones due to nanosecond and femtosecond laser pulses using transmission electronic microscopy.* Applied Physics Letters, 2002. 80(21): pp. 3886–3888. doi:10.1063/1.1481195

[24] Hirayama, Y. and M. Obara. *Molecular dynamics simulation of heat-affected zone of copper metal ablated with femtosecond laser.* in *Lasers and Applications in Science and Engineering.* 2005: International Society for Optics and Photonics. doi:10.1117/12.589513

[25] Musaev, O., A. Midgley, J. Wrobel and M. Kruger, *Laser ablation of alumina in water.* Chemical Physics Letters, 2010. 487(1): pp. 81–83. doi:10.1016/j.cplett.2010.01.011

[26] LaHaye, N.L., S.S. Harilal, P.K. Diwakar and A. Hassanein, *The effect of laser pulse duration on ICP-MS signal intensity, elemental fractionation, and detection limits in fs-LA-*

ICP-MS. Journal of Analytical Atomic Spectrometry, 2013. 28(11): pp. 1781–1787. doi: 10.1039/c3ja50200g

[27] *Machining with Long Pulse Lasers, Machining with Ultrafast Laser Pulses.* 2011 [cited 2016; Available from: http://www.cmxr.com/Education/Short.html, http://www.cmxr.com/Education/Long.html

[28] Lescoute, E., L. Hallo, D. Hébert, B. Chimier, B. Etchessahar, V. Tikhonchuk, J.-M. Chevalier and P. Combis, *Experimental observations and modeling of nanoparticle formation in laser-produced expanding plasma.* Physics of Plasmas (1994–present), 2008. 15(6): p. 063507. doi:10.1063/1.2936267

[29] Royon, A., Y. Petit, G. Papon, M. Richardson and L. Canioni, *Femtosecond laser induced photochemistry in materials tailored with photosensitive agents [Invited].* Optical Materials Express, 2011. 1(5): pp. 866–882. doi:10.1364/OME.1.000866

[30] Damian, V., C. Udrea, M. Bojan, C. Luculescu, A. Armaselu and I. Apostol. *Aluminum nanoparticles production by laser ablation in liquids.* in *International Conference on Applications of Optics and Photonics.* 2011: International Society for Optics and Photonics. doi: 10.1117/12.894525

[31] Gondal, M., T.F. Qahtan, M. Dastageer, T.A. Saleh and Y.W. Maganda. *Synthesis and characterization of copper oxides nanoparticles via pulsed laser ablation in liquid.* in *10th International Conference on High Capacity Optical Networks and Enabling Technologies (HONET-CNS).* 2013: IEEE. doi:10.1109/HONET.2013.6729774

[32] Hamad, A., L. Li, Z. Liu, X.L. Zhong and T. Wang, *Picosecond laser generation of Ag–TiO_2 nanoparticles with reduced energy gap by ablation in ice water and their antibacterial activities.* Applied Physics A, 2015. 119(4): pp. 1387–1396. doi:10.1007/s00339-015-9111-6

[33] Giorgetti, E., A. Giusti, S. Laza, P. Marsili and F. Giammanco, *Production of colloidal gold nanoparticles by picosecond laser ablation in liquids.* Physica Status Solidi (a), 2007. 204(6): pp. 1693–1698. doi:10.1002/pssa.200675320

[34] Kabashin, A.V. and M. Meunier, *Synthesis of colloidal nanoparticles during femtosecond laser ablation of gold in water.* Journal of Applied Physics, 2003. 94(12): pp. 7941–7943.

[35] Torres-Mendieta, R., R. Mondragón, E. Juliá, O. Mendoza-Yero, J. Lancis and G. Mínguez-Vega, *Fabrication of high stable gold nanofluid by pulsed laser ablation in liquids.* Advanced Materials Letters, 2015. 6(12): p. 6. doi:10.5185/amlett.2015.6038

[36] Perrière, J., E. Millon and E. Fogarassy, *Recent advances in laser processing of materials.* 2006, Elsevier.

[37] Intartaglia, R., K. Bagga and F. Brandi, *Study on the productivity of silicon nanoparticles by picosecond laser ablation in water: towards gram per hour yield.* Optics Express, 2014. 22(3): pp. 3117–3127. doi:10.1364/OE.22.003117

[38] Ashfold, M.N., F. Claeyssens, G.M. Fuge and S.J. Henley, *Pulsed laser ablation and deposition of thin films*. Chemical Society Reviews, 2004. 33(1): pp. 23–31. doi:10.1039/B207644F

[39] Kuzmin, P.G., G.A. Shafeev, G. Viau, B. Warot-Fonrose, M. Barberoglou, E. Stratakis and C. Fotakis, *Porous nanoparticles of Al and Ti generated by laser ablation in liquids*. Applied Surface Science, 2012. 258(23): pp. 9283–9287. doi:10.1016/j.apsusc.2011.08.108

[40] Dudoitis, V., V. Ulevičius, G. Račiukaitis, N. Špirkauskaitė and K. Plauškaitė, *Generation of metal nanoparticles by laser ablation*. Lithuanian Journal of Physics, 2011. 51(3): pp. 248–259.

[41] Itina, T.E., *On nanoparticle formation by laser ablation in liquids*. The Journal of Physical Chemistry C, 2010. 115(12): pp. 5044–5048. doi:10.1021/jp1090944

[42] Chakravarty, U., P. Naik, C. Mukherjee, S. Kumbhare and P. Gupta, *Formation of metal nanoparticles of various sizes in plasma plumes produced by Ti: sapphire laser pulses*. Journal of Applied Physics, 2010. 108(5): p. 053107. doi:10.1063/1.3475512

[43] Hubenthal, F., M. Alschinger, M. Bauer, D.B. Sánchez, N. Borg, M. Brezeanu, R. Frese, C. Hendrich, B. Krohn and M. Aeschliman. *Irradiation of supported gold and silver nanoparticles with continuous-wave, nanosecond, and femtosecond laser light: a comparative study*. in *Microtechnologies for the New Millennium 2005*. 2005: Proceedings of SPIE 5838, Nanotechnology II. doi:10.1117/12.608560

Chapter 13

Application of PLD-Fabricated Thick-Film Permanent Magnets

Masaki Nakano, Takeshi Yanai and
Hirotoshi Fukunaga

Additional information is available at the end of the chapter

http://dx.doi.org/10.5772/64235

Abstract

Isotropic Nd-Fe-B thick-film magnets have been prepared using a pulsed laser deposition (PLD) method with the control of laser energy density (LED) followed by post-annealing. The characteristics of the method are a high deposition rate up to several tens of microns per hour together with a reliability of magnetic properties due to the good transfer of composition from an Nd-Fe-B target to a film. Several micro-machines comprising the isotropic Nd-Fe-B films such as a miniaturized DC motor and a swimming machine in liquid were demonstrated. Furthermore, the deposition of isotropic Nd (or Pr)-Fe-B thick-film magnets on a Si or glass substrate was carried out to apply the films to various micro-electro-mechanical-systems (MEMS). We also introduced the preparation of isotropic Sm-Co, Fe-Pt, and nano-composite Nd-Fe-B+α-Fe film magnets synthesized using the PLD.

Keywords: pulsed laser deposition, thick-film magnet, isotropic, post-annealing

1. Introduction

Much research on anisotropic Nd-Fe-B film magnets thicker than several microns was carried out in order to advance miniaturized electronic devices including micro-electro-mechanical-systems (MEMS) [1–7]. It is generally known that the $(BH)_{max}$ value of an isotropic Nd-Fe-B film [8, 9] is smaller than that of an anisotropic Nd-Fe-B one; however, the flexibility of magnetization of the isotropic films is attractive in practical and various applications. Töpfer et al. demonstrated a multipolarly magnetized isotropic Nd-Fe-B film using a screen printing method

[10]. Yamashita et al. also reported that the torque of a milli-size motor comprising a multipo-larly magnetized isotropic thick film exceeded the torque of a motor using an anisotropic one [11].

In Sm-Co thick-film magnets, we have difficulty in overcoming the values of $(BH)_{max}$ for Nd-Fe-B films due to the low saturation magnetization. However, several researchers have demonstrated Sm-Co thick films because of their high Curie temperature and good corrosion resistance [12–19]. For example, Cadieu et al. reported in-plane anisotropic Sm-Co films with a thickness above 100 μm using sputtering together with pulsed laser deposition (PLD) methods [16, 17]. In addition, Budde and Gatzen demonstrated a magnetic micro-actuator comprising a sputtering-made 30-μm-thick Sm-Co film [19].

Fe-Pt magnet is a promising material to use in the medical field owing to its outstanding biocompatibility [20]. In order to advance medical MEMS, the miniaturization of Fe-Pt magnets is necessary [21]. Aoyama and Honkura [22] and Liu et al. [23] reported isotropic Fe-Pt film magnets thicker than several microns using a sputtering method from the medical application point of view.

Here, we show the properties of isotropic Nd-Fe-B [24–28] thick-film magnets prepared using the PLD method and several miniaturized machines [24, 25, 28] comprising the Nd-Fe-B thick film. Moreover, PLD-fabricated isotropic Sm-Co, Fe-Pt, and nano-composite Nd-Fe-B+α-Fe thick films were introduced.

2. Experimental

A target such as Nd-Fe-B, Pr-Fe-B, Sm-Co, and Fe-Pt was ablated with an Nd-YAG pulse laser with a wavelength of 355 nm together with a frequency of 30 Hz in a vacuum atmosphere. The chamber was evacuated to approximately 4×10^{-5} Pa with a vacuum equipment before the deposition. During the deposition, each target was rotated and the distance between the target and the substrate (T-S distance) varied from 5 to 20 mm. In the experiment, a laser power was measured with a power meter in front of the entrance lens of the chamber. The laser energy density (LED) varied by controlling the laser power (LP) together with a spot size of laser beam which could be changed by moving the distance between the focal lens and the target intentionally (see **Figure 1**). Here, the spot size was expressed as a defocusing rate (DF rate) = (TD − FD)/FD × 100(%), where TD is the distance between the condensing lens and the target and FD is the focal length. In the experiment, the control of the defocusing rate (DF rate) enabled us to change LED widely compared with the variation of LP.

In some experiments, the post-annealing of a conventional annealing (CA) and a pulse annealing (PA), respectively, was carried out as shown in **Figure 2**. After magnetizing each sample up to 7 T with a pulse magnetizer, M-H loops were measured by using a vibrating sample magnetometer (VSM) which could apply a magnetic field up to approximately 1800 kA/m reversibly. The average thickness was mainly measured with a micrometer. In some samples, the thickness was estimated from hysteresis loops of as-deposited films [29]. Surface observation and composition analysis were carried out by using a scanning electron microscope (SEM) and an SEM-energy-dispersive X-ray spectroscopy (EDX), respectively.

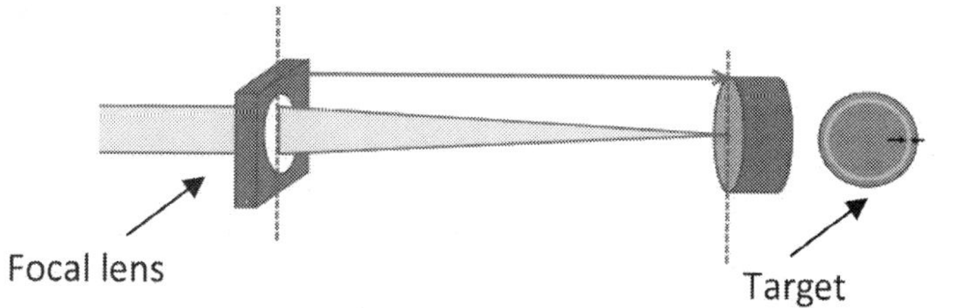

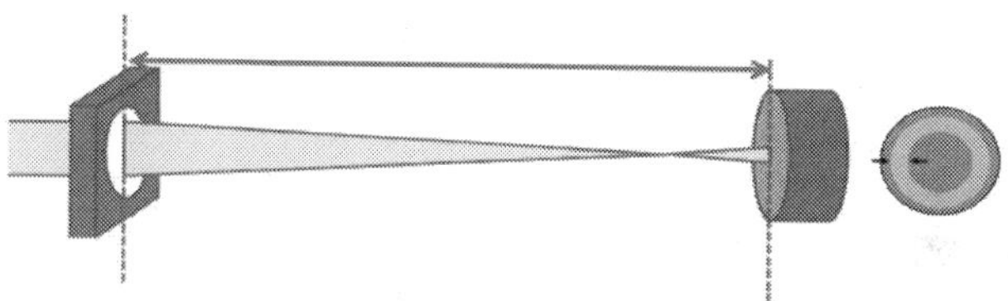

Figure 1. Schematic diagrams of deposition process with several values of the laser energy density (LED) which was controlled by changing the laser power and DF rate independently. (a) High laser energy density (DF rate of zero: just focus) and (b) low laser energy density.

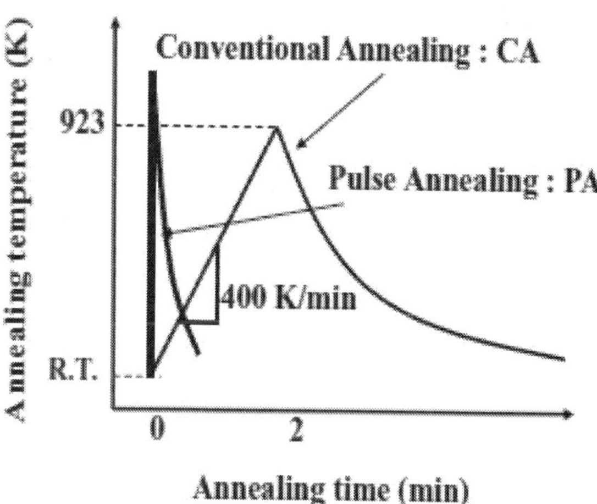

Figure 2. Schematic diagram of conventional annealing (CA) and pulse annealing (PA) methods.

3. Results

3.1. PLD-fabricated isotropic Nd-Fe-B thick-film magnets

Figure 3 shows the in-plane demagnetization curves of isotropic Nd-Fe-B thick-film magnets after the crystallization using two methods of CA and PA with each optimum condition. Both samples were prepared using an $Nd_{2.4}Fe_{14}B$ target under the deposition rate higher than 40 μm/h. The annealing duration of PA was approximately 1.8 s and the ramping rate of CA was 400 K/min (see **Figure 2**). The coercivity value of the annealed film using PA method was larger by approximately 300 kA/m than the annealed 673 K/min sample using CA method. On the other hand, the remanence and $(BH)_{max}$ values of the two samples were almost the same. Transmission electron microscopy (TEM) observation revealed that the use of PA method enabled us to reduce the size of $Nd_2Fe_{14}B$ grains compared with that of the sample annealed using CA method. We considered that the enhancement in coercivity is due to the increases in the domain pinning sites. From these results, a high-speed crystallization annealing process (PA method) is a promising method to refine $Nd_2Fe_{14}B$ grains of PLD-fabricated Nd-Fe-B thick-film magnets.

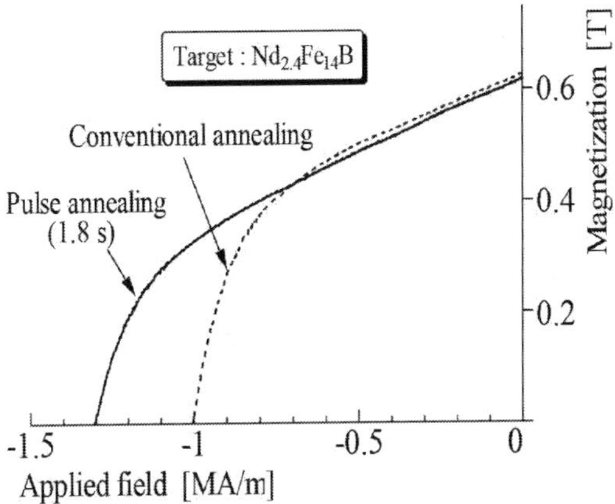

Figure 3. Demagnetization curves (in-plane) of PLD-made Nd-Fe-B thick films crystallized by two methods of CA and PA.

Figure 4 shows the relationship between the LED and magnetic properties of annealed Nd-Fe-B film magnets deposited on Ta substrates prepared using an $Nd_{2.6}Fe_{14}B$ target. All the samples were annealed using PA method. As the inset in each figure shows, the laser power LP and DF rate were controlled independently. In low energy density, high coercivity (H_c) and low remanence (M_r) were obtained, respectively. The samples with low H_c and high M_r were prepared in high energy density. In order to obtain the high deposition rate above 40 μm/h, LED less than 30 mJ/mm^2 was used. In this chapter, PA method was used in the results of

Sections 3.1–3.5. On the other hand, energy density above 200 mJ/mm^2 was used in the results of Sections 3.5 and 3.6.

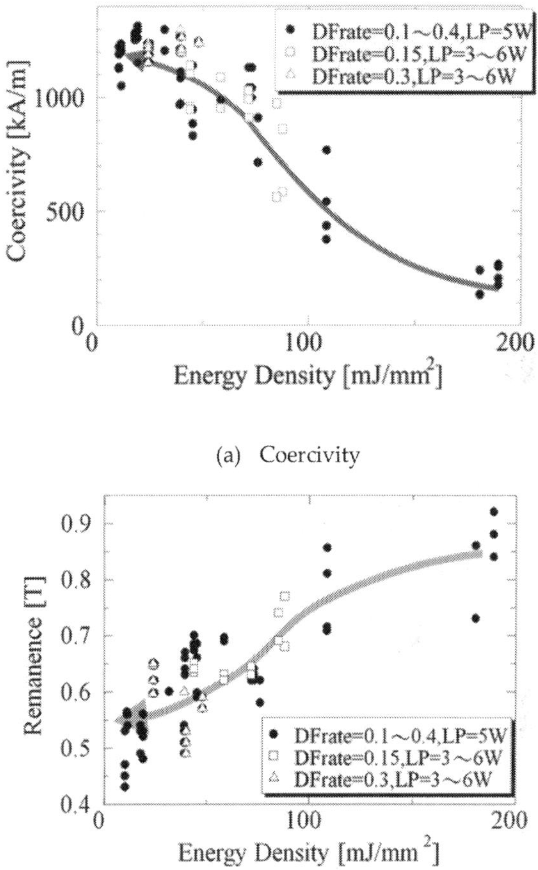

(a) Coercivity

(b) Remanence

Figure 4. Effects of energy density of laser beam on (a) coercivity and (b) remanence. The laser power LP and DF rate varied independently. (a) Coercivity and (b) Remanence.

3.2. Applications comprising isotropic Nd-Fe-B thick-film magnets

Several miniaturized devices comprising the above-mentioned isotropic Nd-Fe-B thick films were demonstrated [24, 25, 28]. **Figure 5(a)** shows a small DC brushless motor with a 200-μm-thick isotropic PLD-made Nd-Fe-B film magnet. The values of coercivity and remanence of the film deposited on an Fe substrate were approximately 970 kA/m and 0.6 T, respectively. It was confirmed that the motor with a thickness of 0.8 mm and diameter of 5 mm rotates at approximately 15,000 rpm under no-load test. The torque constant of 0.0236 mNm/A showed at the gap of 0.1 mm between a rotor and a stator.

(a)   DC brush-less motor

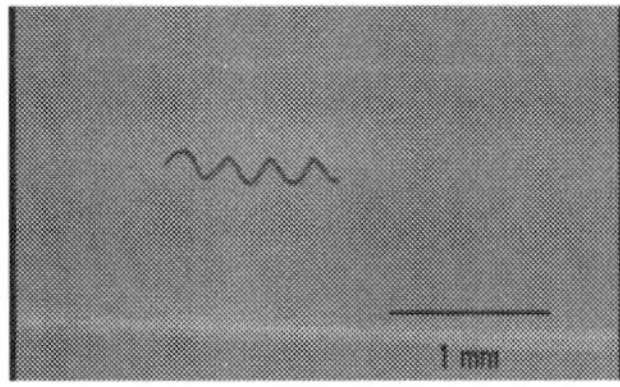

(b)   Swimming machine in liquid

(c)   Electromagnetic friction-drive micro-motor

Figure 5. Three micro-machines comprising PLD-made Nd-Fe-B thick films. (a) DC brushless motor, (b) swimming machine in liquid, and (c) electromagnetic friction-drive micro-motor.

A spiral-type micro-machine with 0.14 mm in outer diameter and 1.0 mm in length was fabricated as seen in **Figure 5(b)**. In the machine, an isotropic PLD-made Nd-Fe-B film magnet was deposited on a tungsten (W) wire. After magnetizing the film magnet in the circumferential direction, the machine rotated in sync with the rotating external magnetic field and the spiral structure generated the propellant force. In the experiment, three types of liquids with kinematic viscosity of 1, 10, and 100 mm^2/s, respectively, were used. In order to move the wireless micro-machine, an external magnetic field of 8 kA/m was applied under the frequency range between 2 and 10 Hz. It was confirmed that the machine swam at the speed of 0.2–1.6 mm/s under various conditions.

A micro-motor using a PLD-made Nd-Fe-B film magnet with a thickness of 384 μm deposited on a Ta substrate was demonstrated as shown in **Figure 5(c)**. The stator is a coil with a ferrite core in the center together with a ferrite disc at the bottom. As an alternating voltage was applied to the coil, pulling and reacting forces worked alternatively between the film and the ferrite core, and as a result the rotor of the film magnet vibrated. The rotor is magnetically mounted onto the stator without mechanical attachments. The motor rotated at approximately 300 rpm with a starting torque of approximately 2 μNm. We confirmed that the electromagnetic friction-drive motor had a large torque together with a low rotational speed.

3.3. Isotropic Nd-Fe-B thick-film magnets deposited on Si substrates

As mentioned earlier, increase in thickness of an Nd-Fe-B film magnet is indispensable to provide a sufficient magnetic field. Here, the deposition of isotropic Nd-Fe-B thick-film magnet on Si substrates was carried out in order to apply the film magnet to various MEMS. It is generally known that we had difficulty in suppressing the peeling phenomenon due to the different values of a linear expansion coefficient for a Si substrate and an Nd-Fe-B film. Even if a buffer layer such as a Ta film was used, the maximum thickness was less than 200 μm. Here, we reported that a control of microstructure of Nd-Fe-B thick films enabled us to increase the thickness above 100 μm without a buffer layer on Si substrates [30].

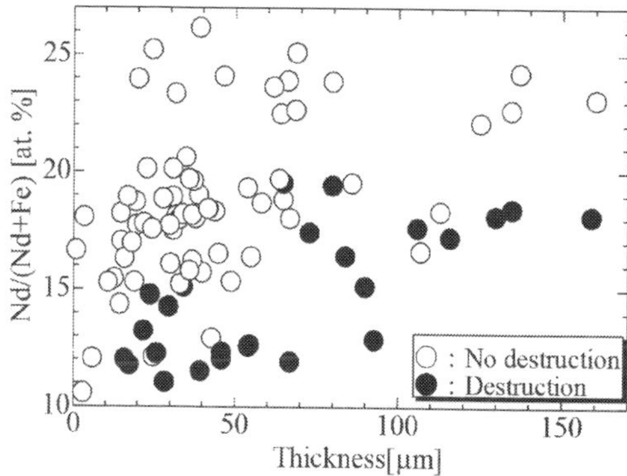

Figure 6. Relationship between thickness and Nd contents in isotropic Nd-Fe-B thick-film magnets deposited on Si substrates after an annealing process. Increase in Nd contents enabled us to increase the thickness up to 160 μm without mechanical destruction.

We investigated the relationship between thickness and Nd contents in annealed isotropic Nd-Fe-B thick films deposited on SiO$_2$/Si substrates as seen in **Figure 6**. After all the samples were annealed using PA method, many samples displayed by the symbol "○" could be prepared without mechanical destruction. All the other samples plotted as "●" were broken. As the Nd content exceeded by approximately 22 at.%, the thickness of the sample symbolized "○" could be enhanced up to approximately 160 μm. The thickness of Nd-Fe-B films deposited on Si substrates increased without the deterioration of mechanical properties. It was considered that the precipitation of Nd element at the boundary of Nd-Fe-B grains together with the triple junctions due to the composition adjustment of $Nd_2Fe_{14}B$ phase is effective to suppress the destruction of the samples through an annealing process.

3.4. Isotropic Pr-Fe-B thick-film magnets deposited on glass substrates

The laser beam was focused on the surface of a $Pr_XFe_{14}B$ (X = 1.8–2.4) target under a high deposition rate of approximately several tens of microns per hour (see **Figure 1(b)**). **Figure 7** shows the magnetic properties as a function of Pr contents in each film with a thickness above 10 μm. Coercivity increased and residual magnetic polarization decreased with an increase in the amount of Pr. As displayed in **Figure 6**, Nd(or Pr)-Fe-B thick films with rare-earth amount less than 15 at.% deposited on Si substrates were mechanically broken after a post-annealing process. On glass substrates, the Pr amounts could be reduced down to approximately 13 at.% without the deterioration of mechanical properties. It was also confirmed that an approximately 100-μm-thick Pr-Fe-B thick film with $(BH)_{max}$ of about 80 kJ/m³ could be deposited on a glass substrate.

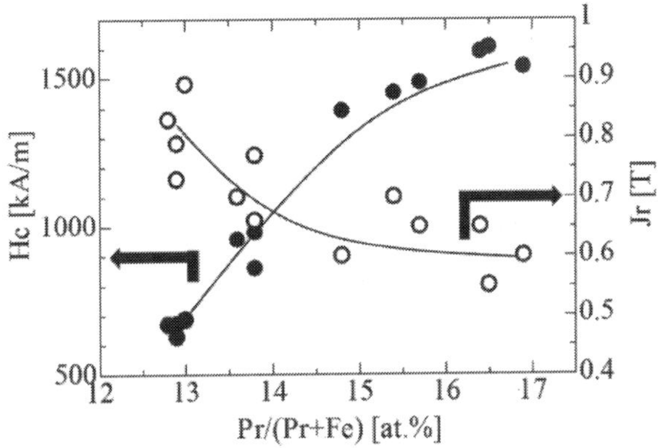

Figure 7. Remanence and coercivity as a function of Pr contents in Pr-Fe-B thick-film magnets deposited on glass substrates.

3.5. PLD-fabricated isotropic Sm-Co thick-film magnets

In this section, a high-speed PLD method with a deposition rate of approximately several tens of microns per hour was applied to fabricate Sm-Co thick-film magnets by using an $Sm_{1.2}Co_5$ target. In-plane and perpendicular M-H loops of a sample annealed by CA method are shown in **Figure 8**. The perpendicular M-H loop was corrected by a demagnetization factor of 1.0. The structure of as-deposited films prepared using an $SmCo_5$ target without a substrate heating system was amorphous as seen in **Figure 9**; therefore, the samples were post-annealed at a temperature of 973 K with a heating rate of 673 K/min (CA method). After the post annealing, not only $SmCo_5$ but also Sm_2Co_{17} phases were observed. In the present stage, we demonstrate the fabrication nano-composite Sm-Co/α-Fe multilayered film magnets using the PLD method [31].

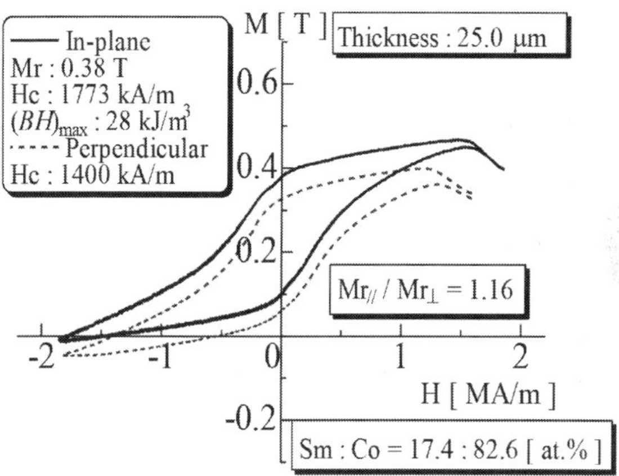

Figure 8. In-plane and perpendicular M-H loops of a PLD-fabricated Sm-Co thick film after a post-annealing process.

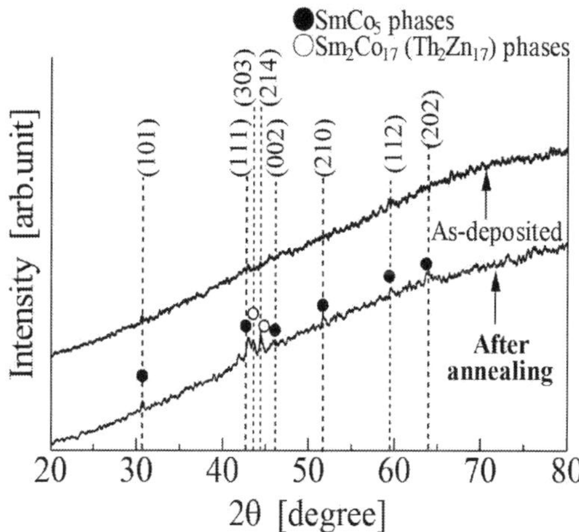

Figure 9. X-ray diffraction patterns of a PLD-fabricated Sm-Co thick film before and after a post-annealing process.

3.6. PLD-fabricated isotropic Fe-Pt thick-film magnets

In this section, a $Fe_{70}Pt_{30}$ target was ablated with an Nd-YAG pulse laser under LED above 200 mJ/mm^2 in a vacuum atmosphere (see **Figure 1(a)**). **Figure 10** shows coercivity values of the as-deposited films as a function of laser power. The values drastically increased at a power of 3 W and then gradually decreased with increase in power. We confirmed that the substrate temperature was proportional to the laser power due to the rise of a radiation heat from a target. As a result, it was found that $L1_0$ ordered phase together with relatively high coercivity could be obtained using a suitable laser power without using a substrate heating system and a post-annealing process. **Figure 11** shows in-plane M-H loop of a Fe-Pt film fabricated at a power of 3 W. The values of coercivity, remanence, and $(BH)_{max}$ were 378 kA/m, 0.94 T, and 104 kJ/m^3, respectively [32]. We, therefore, considered that remanence enhancement occurred in the sample because the saturation magnetization of 1.43 T for $Fe_{50}Pt_{50}$ ordered phase [33].

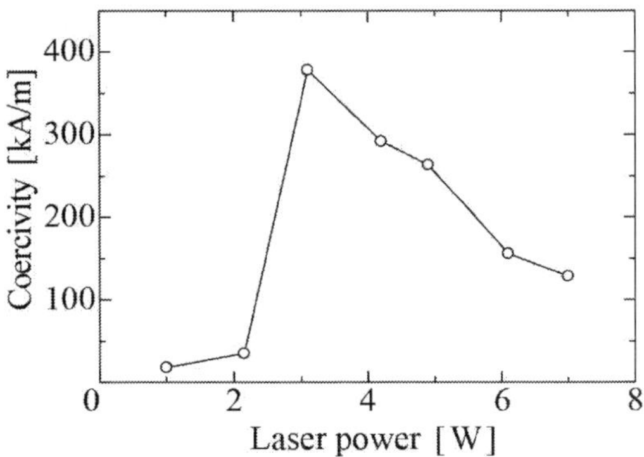

Figure 10. In-plane coercivity values for as-deposited Fe-Pt film magnets prepared from $Fe_{70}Pt_{30}$ target as a function of laser power.

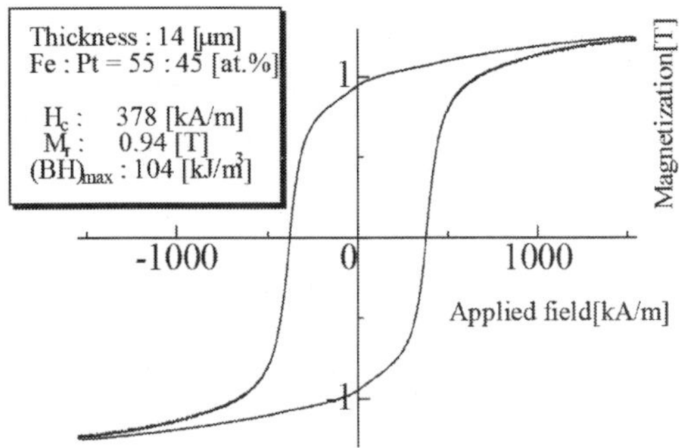

Figure 11. In-plane and perpendicular M-H loops of an as-deposited Fe-Pt film magnet prepared from a $Fe_{70}Pt_{30}$ target.

3.7. PLD-fabricated isotropic nano-composite Nd-Fe-B + α-Fe thick-film magnets

In this section, we focus on the use of high LED above 200 mJ/mm² in order to adopt a different deposition process by taking account of the explosively emitting process of atoms from a target. The as-deposited films had amorphous phase including α-Fe grains, and a nano-composite structure could be obtained after the pulse annealing [34].

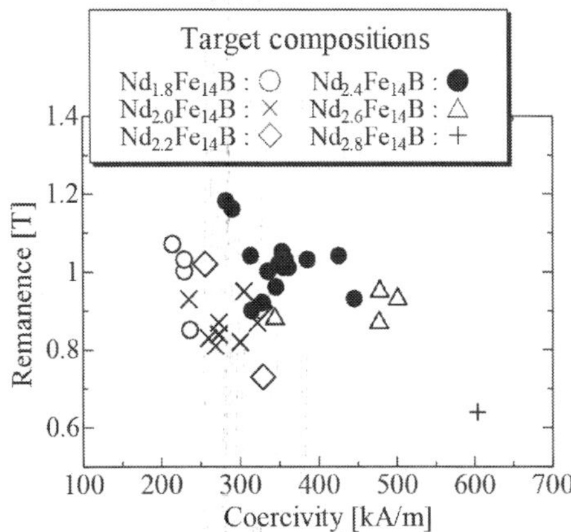

Figure 12. Relationship between remanence and coercivity values of films thicker than 10 μm by using six targets with various compositions. Use of an Nd$_{2.4}$Fe$_{14}$B target is effective to obtain good magnetic properties.

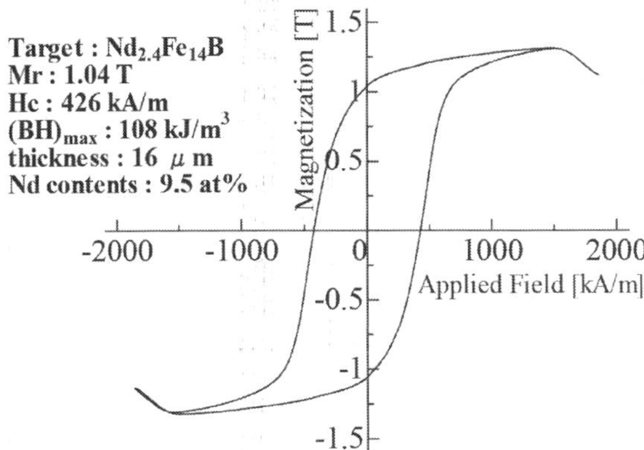

Figure 13. M-H loop of a 16-μm-thick Nd-Fe-B film magnet prepared using laser energy density above 200 mJ/mm² together with an Nd$_{2.4}$Fe$_{14}$B target.

Investigation on the relationship between coercivity and remanence of each sample thicker than 10 μm prepared by six Nd-Fe-B targets with various compositions under LED above

200 mJ/mm^2 (see **Figure 12**). In all the targets, the deposition rate was higher than 20 μm/h and the reduction in the Nd contents of each film by several atomic percentages was observed compared with that of the corresponding target. The samples prepared by using an Nd$_{2.4}$Fe$_{14}$B target had relatively large values of coercivity and remanence. **Figure 13** shows an in-plane M-H loop of a 16-μm-thick Nd-Fe-B + α-Fe thick-film magnet prepared using LED higher than 200 mJ/mm^2 together with an Nd$_{2.4}$Fe$_{14}$B target.

4. Conclusion

In this chapter, we introduced the preparation of rare-earth thick-film magnets using a PLD method and their applications. The PLD-made Nd-Fe-B thick-film magnets thicker than 10 μm deposited on metal substrates could be applicable to miniaturized devices by taking account of the relatively good magnetic and mechanical properties. In addition, Nd(or Pr)-Fe-B thick films with a thickness above 100 μm without a buffer layer on a Si or glass substrate. Preparation of Sm-Co, Fe-Pt, and nano-composite Nd-Fe-B + α-Fe thick-film magnets was also carried out.

Author details

Masaki Nakano*, Takeshi Yanai and Hirotoshi Fukunaga

*Address all correspondence to: mnakano@nagasaki-u.ac.jp

Graduate School of Engineering, Nagasaki University, Nagasaki, Japan

References

[1] F. Dumas-Bouchat, L. F. Zanini, M. Kustov, N. M. Dempsey, R. Grechishkin, K. Hasselbach, J. C. Orlianges, C. Champeaux, A. Catherinot, and D. Givord, *Appl. Phys. Lett.* 96, 102511 (2010).

[2] A. Walther, C. Marcoux, B. Desloges, R. Grechishkin, D. Givord, and N. M. Dempsey, *J. Magn. Magn. Mater.* 321, 590 (2009).

[3] N. M. Dempsey, A. Walther, F. May, D. Givord, K. Khlopkov, and O.Gutfleisch, *Appl. Phys. Lett.* 90, 092509 (2007).

[4] M. Uehara, *J. Magn. Magn. Mater.* 284, 281 (2004).

[5] L. K. B. Serrona, A. Sugimura, N. Adachi, T. Okuda, H. Ohsato, I. Sakamoto, A. Nakanishi, M. Motokawa, D. H. Ping, and K. Hono, *Appl. Phys. Lett.* 82, 1751 (2003).

[6] B. A. Kapitanov, N. V. Kornilov, Ya. L. Linetsky, and V. Yu. Tsvetkov, *J. Magn. Magn. Mater.* 127, 289 (1993).

[7] S. Yamashita, J. Yamasaki, M. Ikeda, and N. Iwabuchi, *J. Appl. Phys.* 70, 6627 (1991).

[8] T. Spelitos, D. Niarchos, V. Skumryev, Y. Zhang, and G. Hadjipanayis, *J. Magn. Magn. Mater.* 272–276, e877(2004).

[9] A. Melsheimer, and H. Kronmüller, *Phys. B.* 299, 251 (2001).

[10] J. Töpfer, B. Pawlowski, and B. Pawlowski, *Proceedings of the 18th Int. Workshop on High Performance Magnets and Their Applications*, 828 (2004).

[11] F. Yamashita, S. Nishimura, N. Menjo, O. Kobayashi, M. Nakano, H. Fukunaga, and K. Ishiyama, *IEEE Trans. Magn.* 46, 2012 (2010).

[12] S. Fahler, V. Neu, U. Hannemann, S. Oswald, J. Thomas, B. Holzapfel, and L. Schultz, Annual report 2001 in IFW Dresden, 21 (2001).

[13] K. Chen, H. Hegde, S. U. Jen, and F. J. Cadieu, *J. Appl. Phys.* 73, 5923 (1999).

[14] S. Takei, X. Liu, and A. Morisako, *Phys. Stat. Sol.(a)* 204, 12, 4166(2007).

[15] L. N. Zhang, J. S. Chen, J. Ding, and J. F. Hu, *J. Appl. Phys.* 103, 043911 (2008).

[16] F. J. Cadieu, H. Hegde, E. Schloemann, and H. J. Van Hook, *J. Appl. Phys.* 76, 6059 (1994).

[17] F. J. Cadieu, R. Rani, X. R. Qian, and Li Chen, *J. Appl. Phys.* 83, 6247 (1998).

[18] F. J. Cadieu, R. Rani, T. Theodoropoulos, and Li Chen, *J. Appl. Phys.* 85, 5895 (1999).

[19] T. Budde, and Hans H. Gatzen, *J. Appl. Phys.* 99, 08N304 (2006).

[20] T. Nakayama, M. Watanabe, M. Homma, T. Kanno, K. Kimura, and O. Okuno, *J. Magn. Soc. Jpn.* 21, 377 (1997). (in Japanese).

[21] A. Yamazaki, M. Sendoh, K. Ishiyama, K. I. Arai, R. Kato, M. Nakano, and H. Fukunaga, *J. Magn.Magn. Mater.* 272–276, e1741 (2004).

[22] H. Aoyama, and Y. Honkura, *J. Magn. Soc. Jpn.* 20, 237 (1996). (in Japanese).

[23] W. F. Liu, S. Suzuki, D. S. Li, and K. Machida, *J. Magn. Magn. Mater.* 302, 201 (2006).

[24] M. Nakano, S. Sato, H. Fukunaga and F. Yamashita, *J. Appl. Phys.* 99, 08N301 (2006).

[25] M. Nakano, S. Sato, F. Yamashita, T. Honda, J. Yamasaki, K. Ishiyama, M. Itakura, J. Fidler, T. Yanai, and H. Fukunaga, *IEEE Trans. Magn.* 43, 2672 (2007).

[26] M. Nakano, H. Takeda, F. Yamashita, T. Yanai, and H. Fukunaga, *IEEE Trans. Magn.* 44, 4229 (2008).

[27] H. Fukunaga, T. Kamikawatoko, M. Nakano, T. Yanai, and F. Yamashita, *J. Appl. Phys.* 109, 07A758 (2011).

[28] M. Nakano, T. Honda, J. Yamasaki, S. Sato, F. Yamashita, J. Fidler, and H. Fukunaga, *Sens. Lett.* 5, 48 (2007).

[29] M. Nakano, S. Tsutsumi, and H. Fukunaga, *IEEE Trans. Magn.* 38, 2913 (2002).

[30] M. Nakano, Y. Chikuba, M. Oryoshi, A. Yamashita, T. Yanai, R. Fujiwara, T. Shinshi, and H. Fukunaga, *IEEE Trans. Magn.* 51, #2102604 (2015).

[31] H. Fukunaga, A. Tou, M. Itakura, M. Nakano, and T. Yanai, *IEEE Trans. Magn.* 50, 2101504 (2014).

[32] M. Nakano, W. Oniki, T. Yanai, and H. Fukunaga, *J. Appl. Phys.* 109, 07A723 (2011).

[33] O. Gutfleisch, J. Lyubina, K. H. Müller, and L. Schultz, *Adv. Eng. Mater.* 7, 208 (2005).

[34] M. Nakano, K. Motomura, T. Yanai, and H. Fukunaga, *IEEE Trans. Magn.* 50, 2101404 (2014).

Obtaining a Thin and Flexible Dental Film of Hydroxyapatite

Lucia Marin Biolan, Andrei Bedros Agop and Doriana Forna

Additional information is available at the end of the chapter

http://dx.doi.org/10.5772/63955

Abstract

The deposition of hydroxyapatite thin films has become a topic of interest in medical applications. This dental film applied on the surface of the tooth may act as a highly resistant and flexible artificial enamel, protecting teeth and removing tooth sensitivity. Other possibilities include whitening and coating enamel-deficient structures. We obtained this flexible film of hydroxyapatite using laser ablation. The plasma plumes were generated by an Nd:YAG nanosecond laser in a vacuum chamber. We used the pulsed laser deposition (PLD) technique and conducted investigations using optical emission spectroscopy (OES), laser-induced breakdown spectroscopy (LIBS), and Raman spectroscopy. Initially, a thin film of HA was deposited on a soluble substrate and heated, followed by immersion into pure water to dissolve the substrates. The originality of our approach consists in the fact that the flexible HA film can be obtained in pure state, because it grows without a substrate, using just a base and lateral supports between, on which it will grow vertically. In order to verify the compatibility and the "stickiness" of HA on the teeth, we chose to grow the film between the roots of a tooth. In this case, besides the film, we also obtained HA microfibers. We tried to bind the film on an extracted tooth. A protocol must be established in order to allow the bonding of the film to the surface of the tooth, knowing that contaminants such as saliva or sulcular fluid increase bonding strength to enamel or dentin. We realized an efficient bonding as HA absorbs protein, the mineral also participates in this ionic exchange, and we strengthened the tooth structure. The main purpose of our research is to rebuild the dentine layer or enamel and close the dental channels. Our experiments led to the creation of an HA foil that has the role of protecting teeth against cariogenic bacteria and could even have cosmetic effects by teeth whitening. This dental plaster acts as an artificial HA enamel, very resistant and flexible, protecting the tooth and eliminating

dental sensitivity. Being very thin, it is invisible once applied on teeth and can be observed only by examination under a strong light.

Keywords: hydroxyapatite, biocompatibility, films, biomaterials, laser, tooth

1. Introduction

This chapter describes a unique way to obtain a flexible pure hydroxyapatite film. In the past years, the nanocomposites have been in the center of attention due to their unique physical and chemical properties.

Specifically, the biomaterials have been widely studied, motivated by their clinical applications. Among these, the hydroxyapatite has been studied due to its remarkable properties, such as biocompatibility, osteoconductivity, and bioactivity. This material is naturally found in the human body, being one of the major constituents of bones and teeth. As a consequence, HA has been widely used in many fields, including biomedical applications. Our experiments led to the creation of a thin HA film that has the role of protecting teeth against cariogenic bacteria and could even have cosmetic effects by teeth whitening. This dental plaster acts as an artificial HA enamel, very resistant and flexible, protecting the tooth and eliminating dental sensitivity. Being very thin, it is invisible once applied on teeth and can be observed only by examination under a strong light. The plaster, produced by pulsed laser deposition (PLD), can be manipulated with tweezers and applied on the tooth. The originality of our approach consists in the fact that the flexible HA film can be obtained in pure state, because it grows without a substrate, using just a base and lateral supports between which it will grow on vertical direction.

2. Experimental methods

The deposition of hydroxyapatite thin films has become a topic of interest in medical applications. This dental film applied on the surface of the tooth may act as a(n) artificial, highly resistant, and flexible enamel, protecting teeth and removing tooth sensitivity. There is also a possibility for whitening and covering deficient enamel structure. We obtained this flexible film of hydroxyapatite using laser ablation. We also tried to apply the film on an extracted tooth. The plasma plumes were generated by an Nd:YAG nanosecond laser in a vacuum chamber with 10^{-6} Torr. We used the pulsed laser deposition technique and for the investigations we opted for optical emission spectroscopy (OES) and Raman spectroscopy.

In this work, we studied the evolution of plasma plumes resulting from enamel and hydroxyapatite ablation. The plasma plumes were generated by an Nd:YAG nanosecond laser and the depositions were performed in a cylindrical stainless steel vacuum chamber (10 L volume, 30 cm height, 20 cm diameter) (**Figure 1**). The chamber is evacuated to a base pressure of 10^{-6}

Torr by a 550 L/s turbomolecular pump (Agilent Technologies TV-551) placed in a vertical position at the bottom of chamber. The target (HA) is placed on a micrometric precision XYZ stage and can be rotated with a motorized vacuum feedthrough (Caburn MDC). The target is placed on a metallic target holder, which is electrically isolated from the rest of the experiment by an alumina block. The substrate (salt) is placed parallel to the target. The target-substrate distance was 1.4 cm. The ablation laser beam was usually at 45° on the target surface (**Figure 2**).

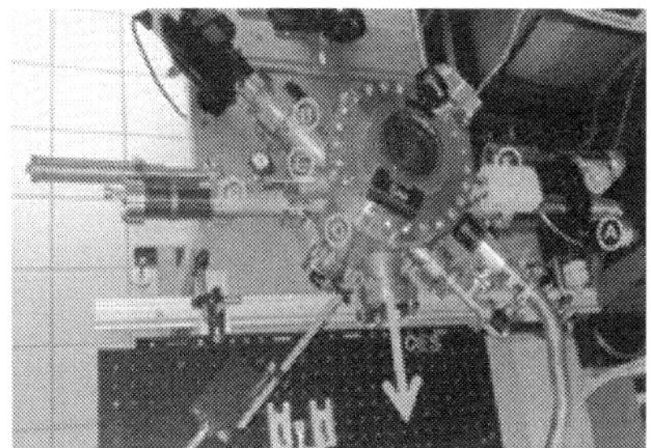

Figure 1. Vacuum chamber.

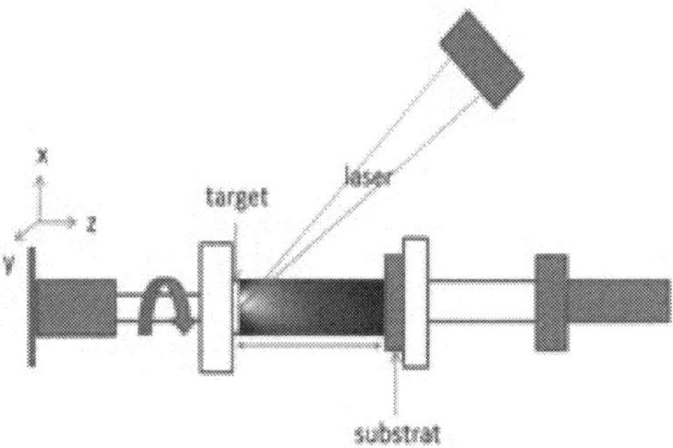

Figure 2. Experimental setup of an Nd:YAG nanosecond laser (Continuum Surelite III-10) (gw = 1 μs, gd = 25 ns).

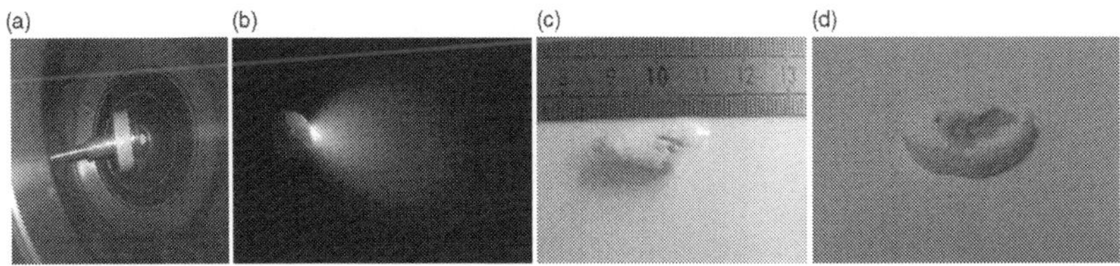

Figure 3. (a) Vacuum chamber with target; (b) 1 plasma tooth; (c) 1 tooth after ablation; (d) 1 HA after ablation.

2.1. Experimental data

We used λ = 266 nm, 10 ns, 10 Hz, E = 40 mJ/pulse, fluence 2.1 J/cm^2. The materials used in these studies were hydroxyapatite and a human tooth (**Figure 3a–d**).

2.2. Results and discussions

The dynamics of the plasma plume has been studied by means of a high-resolution monochromator (Acton SP 2500i) and intensified charge-coupled device (ICCD) camera (Roper Scientific PI MAX2-1003-UNIGEN2, 1024 × 1024 pixels). The monochromator has an alternative exit port fitted with a photomultiplier (PM, Hamamatsu) in order to record fast temporal profiles of a given spectral line. The PM output is sent to a fast digital oscilloscope (GHz, LeCroy). In order to obtain preliminary insight into the dynamics of the laser ablation plasma plume, ICCD sequential pictures of the spectrally unresolved plasma optical emission were recorded at various delays with respect to the laser pulse. We used the PLD technique to produce flexible films of hydroxyapatite deposited on glass or NaCl (salt) substrates, followed by a space-and-time-resolved optical emission spectroscopy (OES) investigation on selected spectral lines, the properties of the deposited films were investigated by Raman spectroscopy.

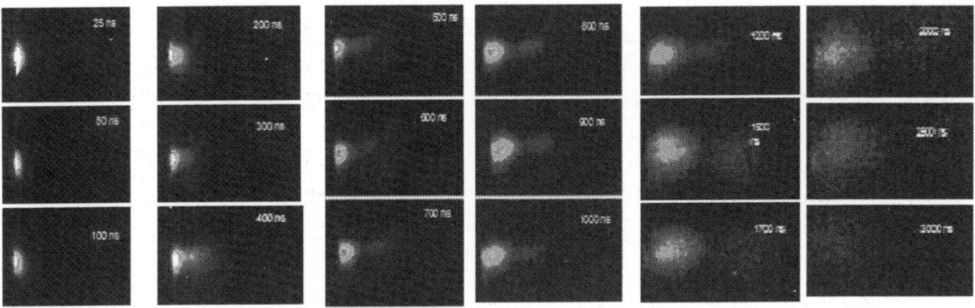

Figure 4. Temporal evolution of the plasma produced by irradiation of HA sample.

We used the ICCD camera to record the global template evolution of the plasma (**Figure 4**) and we observed the presence of two main structures: a fast one, represented by the spectral lines of ions, and a slower one, mainly due to the contribution of neutrals (the first one, plume-like-shaped, expands at a velocity of about 2 × 10^4 m/s; the second one, which looks like a small plasma cloud close to the target surface, shows an expansion velocity of about 2 × 10^3 m/s). OES allowed us to obtain information on the contribution of each species present in the plasma, the processes of formation and expansion of the plume. Identification of spectral lines observed in the OES was performed using the NIST database and CFA (**Figure 5**).

Emission is dominated by Ca I, Ca II ions but O II, H I, and contaminants such as Na I, Mg III, Hg I ions were also found. Raman analysis of HA deposition on glass and Raman analysis of HA deposition on salt are presented in **Figures 6** and **7**, respectively.

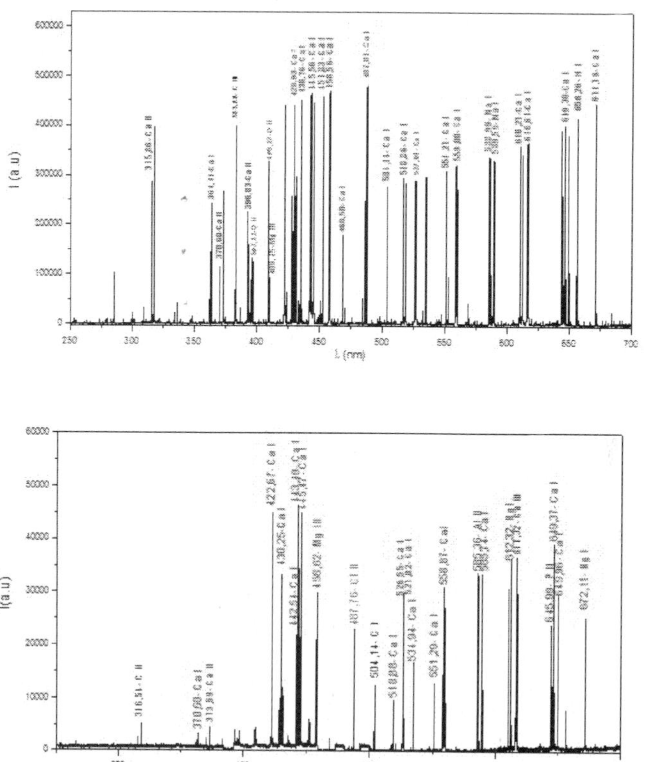

Figure 5. OES hydroxyapatite/OES enamel tooth.

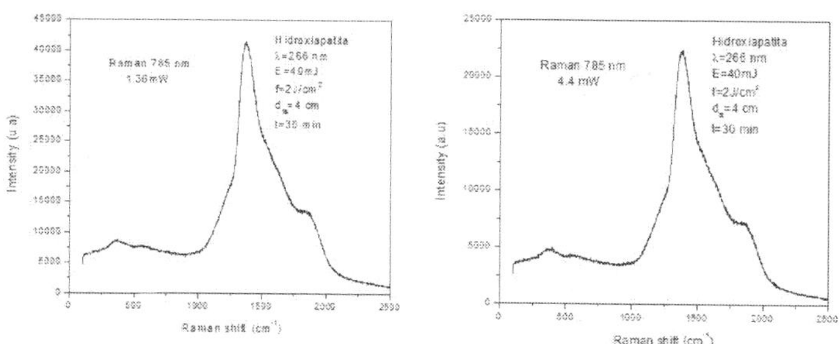

Figure 6. Raman analysis of HA deposition on glass. Experimental data: 785 nm, object 50×, 10 acc, 10 s, 1.36 mW/785 nm, 50×, 10 acc, 10 s, 4.4 mW.

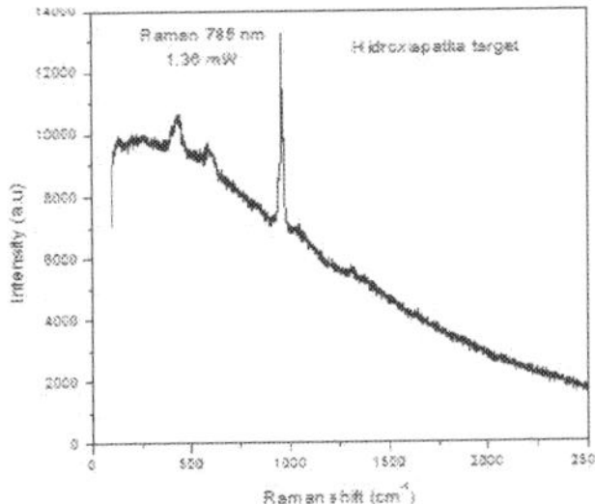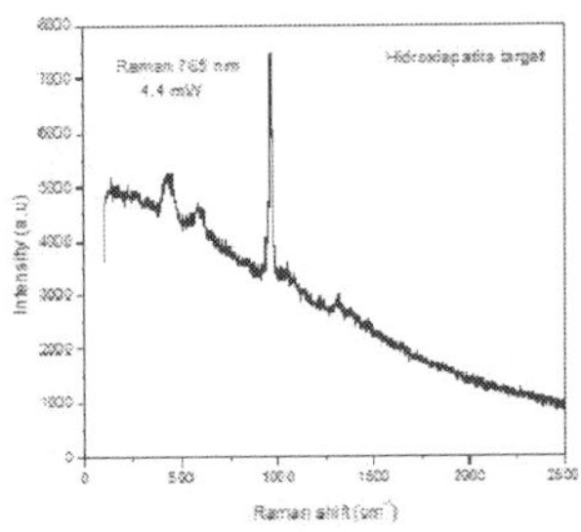

Figure 7. Raman analysis of HA deposition on salt. Experimental data: 785 nm, object 50×,10 acc, 10 s, 1.36 mW/785 nm, 50×, 10 acc, 10 s, 4.4 mW.

A starting point in the investigation of the thin HA film is to obtain information on the surface morphology of the deposited samples. For the study of the thin film discussed here, we used a confocal optical microscope (Olympus) coupled to the Raman spectroscopy setup (Renishaw). Our images were obtained using the 50× objective (**Figure 8**).

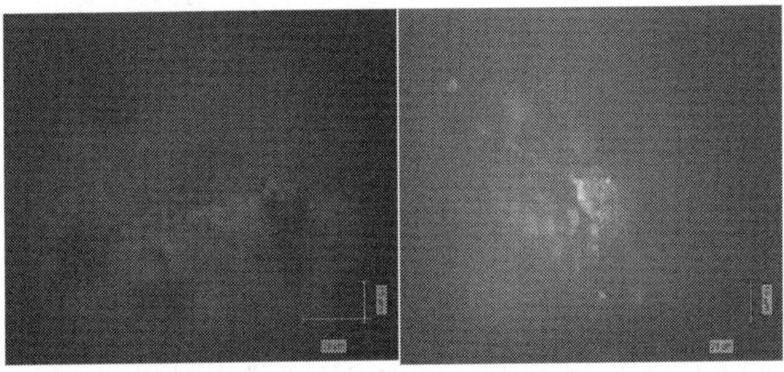

Figure 8. Optical microscopy for deposition: on glass/on salt.

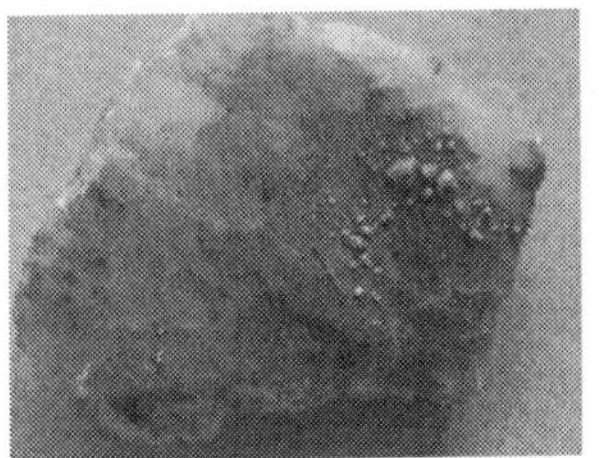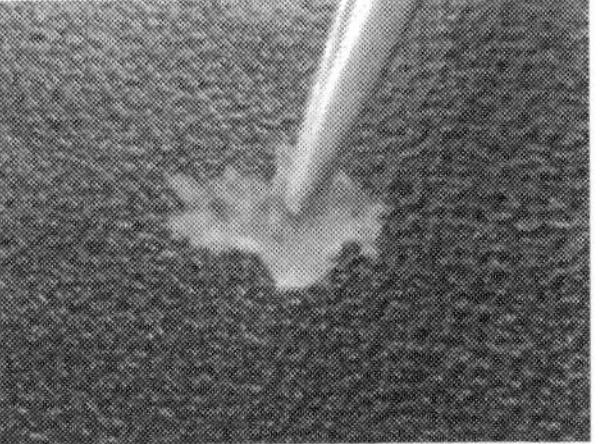

Figure 9. Thin film of hydroxyapatite on the salt after heated/thin and flexible film of HA.

A thin film of Hap was deposited on a soluble substrate by a pulsed laser deposition (PLD) technique. The substrates were then dissolved using a solvent and the thin Hap films were collected as freestanding sheets. The HA film was deposited on salt single crystals (**Figure 9**) and heated at 400°C, $t = 30$ min. These HA sheets were crystallized. Thereafter, the thin films were collected as freestanding sheets by immersing the NaCl substrates into pure water to dissolve the substrates. This procedure gave rise to flexible HA films.

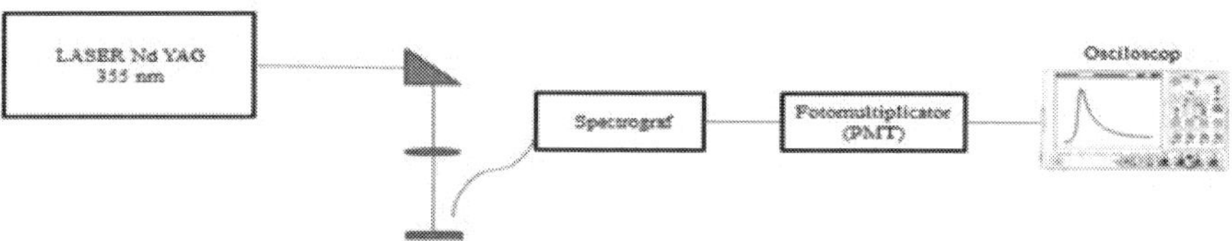

Figure 10. Schema of the LIBS setup.

Laser-induced breakdown spectroscopy (LIBS) uses a short laser pulse ($\approx$10 ns) focused on the surface of a solid sample to vaporize a very small quantity of material. The ejected material forms a plasma plume and the optical radiation emitted by the plasma species is collected through an optical fiber exactly where the plasma was produced, meaning that this method can be used for *in situ* analysis. Because the quantity of ejected material is in the order of 20–200 ng, this method is considered microdestructive because the crater formed on the surface of the sample is practically invisible to the naked eye. LIBS can be used for elemental analysis of materials and it allows the measurement of fluorescence lifetime of the species identified in the plasma. For the LIBS analysis, we used the third harmonic of an Nd:YAG (BMI LT-1233) laser (355 nm, 10 ns) focused by a 5 cm focal distance lens. The optical radiation emitted by the plasma plume was collected using an optical fiber and analyzed using a monochromator and detected with a photomultiplier (H9305-02 Hamamatsi) connected to a 500 MHz oscilloscope (Agilent Technologies) (**Figure 10**). This way, we were able to measure the intensity of the radiation emitted at various wavelengths and to record the time evolution of each signal.

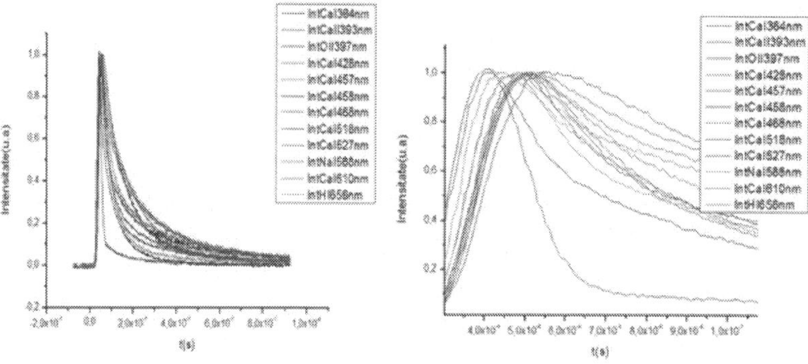

Figure 11. Fluorescence lifetime of HA species/normalized graph.

The measurements have been performed at IESL-FORTH (Institute of Electronic Structure and Laser-Foundation for Research and Technology-Hellas), Greece. The HA pellet was placed 1 cm before the focal point in order to avoid laser focusing in the air and air breakdown. The laser fluence could be adjusted by changing the area of the laser spot at the surface of the sample by modifying the lens-sample distance. Experiments were performed at 1.47 J/cm^2 (fluence) ($E = 7.4$ mJ, 10 Hz, $D = 0.8$ mm, $S = 0.50$ mm^2).

The main species identified are Ca I, then HI and OII, but also the main contaminant, Na I. The fluorescence lifetime of Ca I and oxygen is considerably larger (**Figure 11**).

2.3. Vertical growth methods for flexible HA films

In order to obtain flexible HA films, we used PLD starting from a solid pure HA pellet as a target. The powdered HA was processed in the form of solid pellets of 14–15 mm diameter and 4–6 mm thickness using a hydraulic press (15–25 tones) at "Gheorghe Asachi" Technical University of Iasi (**Figure 12**).

Figure 12. Manual press.

The hydroxyapatite thin films were obtained by nanosecond PLD. This method proved to be competitive for growing high-quality thin films because it has the capacity to preserve the stoichiometry of deposed compounds [4]. Obtaining thin films using PLD offers many advantages compared with other techniques, such as the laser source is external to the deposition chamber, most materials can be laser ablated and deposed in thin films, and the growth rate can be precisely controlled (10^{-2} to 10^{-1} nm/pulse)nm/pulse); the ablated material is localized in the volume of the generated plasma; the stoichiometry of the film is identical with that of the target and the high energy of ablated species allows one to obtain very adherent films [5–8]. The film depositions were performed using an Nd:YAG laser, BMI Industries, at IESL-FORTH (Institute of Electronic Structure and Laser-Foundation for Research and Technology-Hellas, Heraklion) Greece, at $\lambda = 266$ nm, repetition rate 10 Hz (**Figure 13**). Various fluencies were used and the target substrate distance was also adjustable (**Table 1**). As previously explained, the fluence can be varied by changing the distance between the focusing lens and the target surface.

Figure 13. PLD experimental setup.

Diameter (mm)	Area (mm²)	Area (cm²)	Measured power (mW)	Repetition rate (Hz)	Energy (mJ)	Fluence (mJ/cm²)	Fluence (J/cm²)
0.8	0.502655	0.005027	74	10	7.4	1472.183	1.472183
1	0.785398	0.007854	66	10	6.6	840.3381	0.840338
1.5	1.767146	0.017671	200	10	20	1131.768	1.131768

Table 1. Laser fluence.

Figure 14. HA ablation plasma/HA vertical film obtained on salt substrate.

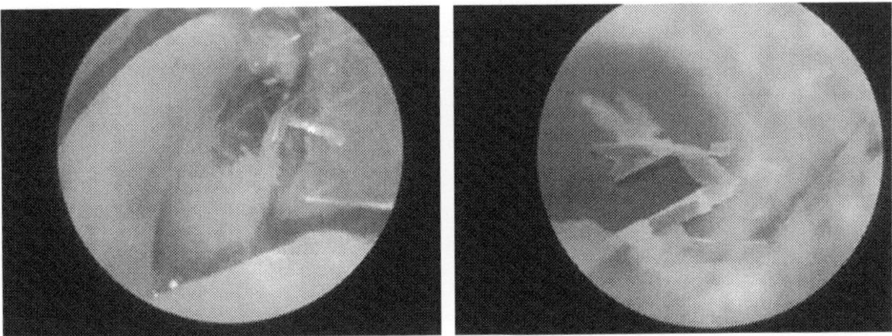

Figure 15. HA film formed between two lateral supports/vertical HA film attached only at the base.

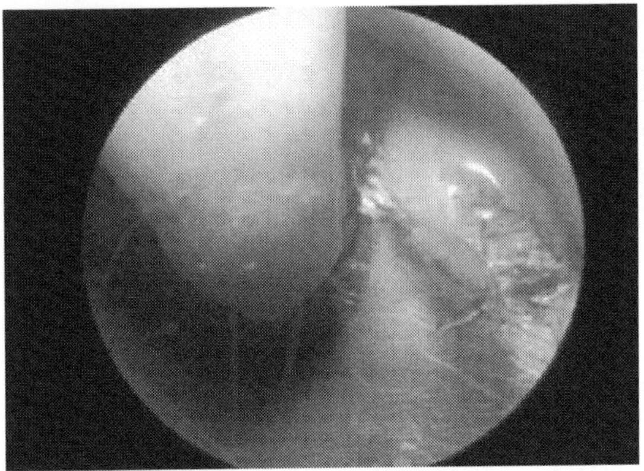

Figure 16. HA microfibers attached to the tooth surface.

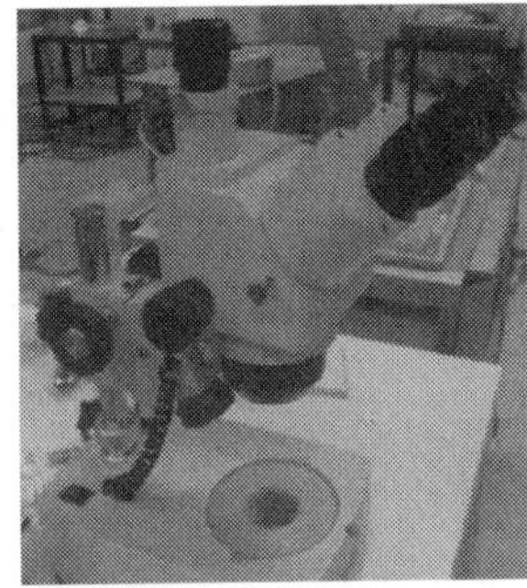

Figure 17. Optical microscope 15×, Brunel microscopes.

Figure 18. The oven used for thermal treatment, IESL, Greece.

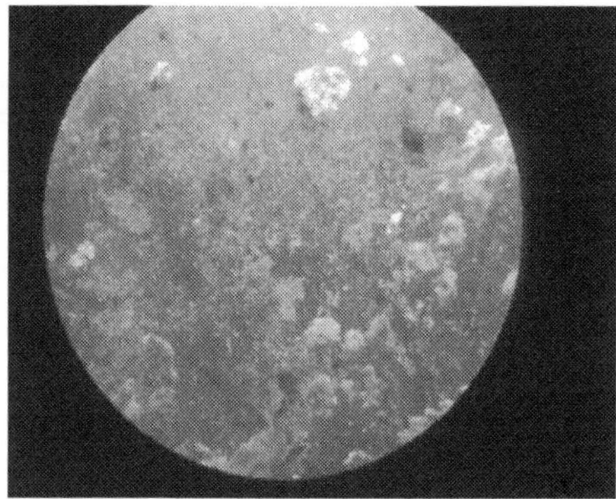

Figure 19. Free HA films, obtained by deposing on salt substrate, after thermal treatment.

Initially, we made a deposition on a salt (NaCl) substrate and it led to the formation of some HA vertical structures (**Figure 14**). The next step was to obtain films that were not completely attached to the substrate by growing them between two lateral supports. In order to verify the compatibility and the "sticking" of HA on the teeth as well, which corresponds to the intended application, we chose to grow the film between the roots of a tooth (**Figure 15**). In this case, besides the film, we also obtained HA microfibers attached only at one end on the tooth enamel (**Figure 16**). We used for these experiment Optical microscope (**Figure 17**).

In order to improve the crystallinity and biocompatibility of the films, they were thermally treated for 30 min at 400°C (**Figures 18** and **19**).

2.4. Methods of HA adhesion on enamel

This flexible HA film can be handled with tweezers and applied to the tooth. We tried to bind the film on an extracted tooth.

Adhesion to enamel is the hard process (**Figure 20**). A protocol must be established in order to allow the bonding of the film on the surface of the tooth knowing that contaminants such as saliva or sulcular fluid increase bonding strength to enamel or dentin. We realized an efficient bonding as HA absorbs protein, the mineral also participates in this ionic exchange and we strengthened the tooth structure.

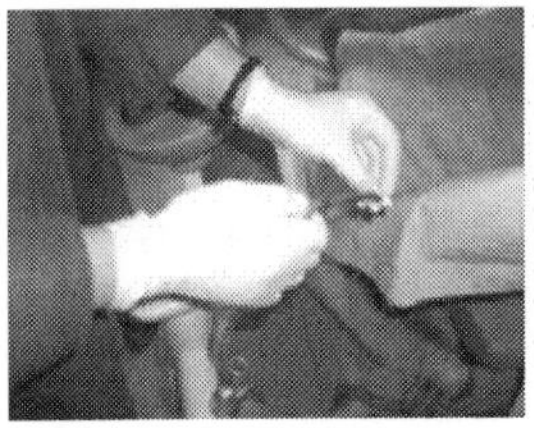

Figure 20. Methods of HA adhesion on enamel.

Using the PLD technique, we can obtain a flexible film of hydroxyapatite, a biocompatible dental material that can be immediately and directly attached to the tooth surface for restoration and conservation of teeth.

For the experimental part, we have attached a theoretical model [8–16].

We proposed a new approach for the analysis of dynamics in nanostructures. The dynamics of nanostructure quasiparticles takes place on continuous but nondifferentiable curves. Consequently, the standard properties of nanostructures, such as quasiparticles generation through self-structuring, interferential capacities through self-similar solutions of Kirchhoff type equations, etc., are controlled through nondifferentiability of motion curves.

The standard theoretical models of nanostructures dynamics are sophisticated and ambiguous. However, such situation can be simplified if we consider that complexities in interaction processes impose various time resolution scales and the evolution pattern leads to different freedom degrees. To develop new theoretical models, we have to admit that the nanostructures with chaotic behaviors can achieve self-similarity (space-time structures can appear) associated with strong fluctuations at all possible space-time scales. Then, for time scales that prove to be larger if compared with the inverse of the highest Lyapunov exponent, the deterministic trajectories are replaced by a collection of potential routes. In its turn, the concept of "definite positions" is replaced by that of probability density.

Thus, the nondifferentiability appears as a universal property of nanostructures and, moreover, it is necessary to create nondifferentiable physics of nanostructures. Under such circumstances, if we consider that the complexity of interactions in the dynamics of nanostructures is replaced by nondifferentiability, it is no longer necessary to use the whole classical "arsenal" of quantities from standard physics (differentiable physics).

This topic was developed using either the scale relativity theory (SRT) or scale relativity theory with arbitrary constant fractal dimension. According to these models, the dynamics of nanostructure quasiparticles takes place on continuous but nondifferentiable curves (fractal curves) so that all physical phenomena involved depend not only on space-time coordinates but also on space-time scale resolution. That is why physical quantities describing the dynamics of nanostructures can be considered as fractal functions. Moreover, according to geodesics in a nondifferentiable (fractal) space, the nanostructure quasiparticles may be reduced to, and identified with, their own trajectories (i.e., their geodesics) so that the nanostructure should behave as a special "fluid" lacking interactions—fractal fluid.

Various theoretical aspects of nanostructure dynamics (self-structuring, phenomena through quasiparticles generation, interferential capacities through self-similar solutions of Kirchhoff type equations, etc.) were analyzed using the SRT with arbitrary constant fractal dimension.

Any particle can take part in a permanent interaction with the "subfractal level" through the fractal potential, Q. The "subfractal level" is identified with a nonrelativistic fractal fluid described by momentum and state density conservation laws. The nondifferentiable hydrodynamics implies a quantum hydrodynamics model (QHM). Indeed, for motions described by fractal curves with fractal dimension $D_F = 2$, at Compton scale, the Non-differentiable hydrodynamics (NDH) reduces to quantum hydrodynamics model. Moreover, the "subfractal level" can be identified with "subquantum level." The fractal potential comes from the nondifferentiability and must be considered as a kinetic term and not as a potential one. Moreover, the fractal potential can generate a viscosity stress tensor type.

The main conclusions of the theoretical model are the following: (i) the nanostructure dynamics was theoretically analyzed assuming that the quasiparticle moves on continuous and nondifferentiable curves; (ii) a nondifferentiable hydrodynamic model containing the density and momentum conservation equations was built. The fractality is introduced via fractal potential; (iii) supposing that the fractal potential implies an isentropic-type behavior of the fractal fluid, the self-structuring phenomena are analyzed through numerical simulations; (iv) in the absence of convection, interferential properties are induced in nanostructures; (v) this chapter deals with the standard properties of nanostructures, such as quasiparticles generation through self-structuring or interferential capacities through self-similar solutions of Kirchhoff-type equations. In the literature, there are also other descriptions of the forms of organization of the matter. The quantum theory is used in each of these descriptions of the forms of organization of matter. However, the interpretation of modern quantum theory is still an open question as shown in the Reference [1–3].

3. Conclusions

We created a flexible HA film, a dental plaster grown vertically, thus improving its quality: there are no contaminations and impurities from the substrate. The crystalline structure can be improved by thermal treatment. The main purpose of our research is to rebuild the dentine layer or enamel and close the dental channels.

Author details

Lucia Marin Biolan[1*], Andrei Bedros Agop[2] and Doriana Forna[3]

*Address all correspondence to: luciamarin2015@yahoo.com

1 Faculty of Physics, "Alexandru Ioan Cuza" University of Iaşi, Iaşi, România

2 Faculty of Material Science and Engineering, "Gheorghe Asachi" Technical University of Iaşi, Iaşi, România

3 Faculty of Dental Medicine, "Grigore T. Popa" University of Medicine and Pharmacy of Iaşi, Iaşi, România

References

[1] Agop M, Forna N, Casian-Botez I, Bejenariu C. New theoretical approach of the physical processes in nanostructures. *Journal of Computational and Theoretical Nanoscience* 2008;5(4):483–489.

[2] Croca JR, de Lemos e Silva Cordovil JL. Nonlinear quantum physics. *Reviews in Theoretical Science* 2014;2(3):181–200.

[3] Casian-Botez I, Vrajitoriu L, Rusu C, Agop, M. *Interferential behaviors in nanostructures via non-differentiability.* Journal of Computational and Theoretical Nanoscience 2015;12:1–7.

[4] Mazilu N, Agop M, Axinte CI, Radu E, Jarcău M, Gârţu M, Răuţ, M, Pricop M, Boicu M, Mihăileanu D, Vrăjitoriu L. A Newtonian message for quantization. *Physics Essays* 2014;27:204–214.

[5] Landi E, Tampleri A, Celotti G, Spio S. Densigication behavior and mechanisms of synthetic hydroxyapatites. *Journal of European Ceramican Society* 2000;20(14–15):2377.

[6] Gurlui S, Agop M, Nica P, Ziskind M, Focsa C. Experimental and theoretical investigations of aluminium expanding laser-plasma. *Physical Review E* 2008;78:026405.

[7] Nica P, Vizureanu P, Agop M, Gurlui S, Focsa C, Forna N, Ioannou P, Borsos Z. *Japanese Journal of Applied Physics* 2009;48:066001.

[8] Baurle D. Laser Processing and Chemistry. Springer, Berlin; 2000.

[9] Serra P, Morenza JL. Imaging and spectral analysis of hydroxiapatite laser ablation plumes. *Applied Surface Science* 1998;127–129:662–667.

[10] Kokkinaki O, Mihesan C, Velegrakis M, Anglos D. Comparative study of laser-induced breakdown spectroscopy and mass spectrometry for the analysis of cultural heritage materials. *Journal of Molecular Structure* 2013;1044:160–166.

[11] Buonocore MG, Matsui A, Gwinnett AJ. Penetration of resin dental materials into enamel surfaces with reference to bonding. *Archives of Oral Biology* 1968;13:61–70.

[12] Raskin A, Michotte-Theall B, Vreven J, Wilson NH. Clinical evaluation of a posterior composite 10-year report. *Journal of Dentistry* 1999;27(1):13–19.

[13] Tesloianu D, Vrajitoriu L, Costin A, Vasincu D, Timofte D. Dispersive behaviours in biological fluids. Applications (II), *The Bulletin of the Polytechnic Institute of Jassy* 2014; Pg.12. Tomul LX(Fasc. 3), Sectia matematica, mecanica teoretica, fizica.

[14] Marin L, Mereuţă VD. Obtaining a thin and flexible dental film of hydroxiapatite using a PLD technique. Bulletin of the Polytechnic Institute of Iaşi; 2015.

[15] Marin L, Mihesan C, Velegrakis M. Obtaining hydroxiapatite flexible thin films without using a substrate. Bulletin of the Polytechnic Institute of Iaşi; 2015.

[16] Marin L, Mereuţă VD, Agop M. Nonlinear effects in complex systems. Publisher Ars Longa, Iasi, October 2015.

Chapter 15

High-Energy Nanosecond Laser Pulses for Synthesis of Better Bone Implants

Amirkianoosh Kiani and Mitra Radmanesh

Additional information is available at the end of the chapter

http://dx.doi.org/10.5772/63770

Abstract

The main objective of this chapter is to introduce high-energy nanosecond laser pulse treatment for enhancing the surface bioactivity of titanium for bone and tissue implant fabrication. Improvement to the implant performance could immensely benefit the human patient. Bioactivity enhancement of materials is currently an essential challenge in implant engineering. Laser micro/nano surface texturing of materials offers a simple, accurate, and precise method to increase the biocompatibility of materials in one single step. In this chapter, the effects of laser power, scanning parameters, and frequency on surface structure and topographic properties are studied. Through bioactivity assessment of treated titanium substrates, it was found that an increase in power and frequency increases the bioactivity of titanium, while a decrease in scanning speed of laser could lead to an increase in the cell adhesion ability of titanium.

Keywords: titanium, bone implant, laser surface treatment, micro- and nanotexturing, bioactivity, biocompatibility

1. Introduction: biomaterials and implant engineering

Biomaterials and implant engineering have become vital fields in the medical and surgical industries. These disciplines can enhance the quality and length of human life, and have an immense effect on the health of numerous individuals. The technologies associated with biomaterials and implant engineering, particularly in the fields of surgical implants such as dental, bone, and tissue implant applications, have led to the creation of numerous research opportunities [1–3]. In today's society, with the continuous growth in population and education, there is a preference for an improved lifestyle, better body functionality, and more appealing

aesthetics. This leads to ongoing expansion and discovery in the technology and science associated with biomaterials and implant engineering [3–6].

This chapter describes a research opportunity for the investigation and improvement of the effects of laser surface texturing on enhancing the biocompatibility and bioactivity of titanium. Titanium substrates are used to examine the effects of key laser process parameters, including power, scanning parameters, and frequency on their surface topography properties and biocompatibility.

1.1. Main challenge in implant engineering

Cell adhesion and biocompatibility are important parameters for implant fabrication and the production of biomedical devices. Improvements to an implant's performance in the implantation site could benefit the patient's quality of life. Low biocompatibility is often caused by poor integration of the implant with surrounding tissues. Cell behaviour on biomaterial surfaces depends upon implant-cell interactions, which are correlated with surface properties such as hydrophilicity, roughness, texture, chemical composition, charge, and topography properties [6–8]. Although there is a great range in the design and function of various implants, the one common factor for all of them is their biocompatibility, which affects overall implant performance. Biocompatibility enhancement of materials used in implants is currently an essential challenge in implant engineering [8]. This property is essential in order to avoid any infections and immune system rejection. It can also affect the healing process by reducing the healing time. A reduced healing time is desirable in implant applications, since the sooner the body accepts the implant organ, the sooner the user can function normally [7, 9, 10].

The implant's surface is the main area in contact with the body at an implantation site. Therefore, to increase the biocompatibility of a material, various methods of surface treatment are currently being used in industry [9, 11, 12].

1.2. Fabrication methods of biocompatible materials

One method to affect the surface properties of a material is to alter the surface topography properties. There are multiple conventional methods used commercially for processing and surface texturing of materials for bone and tissue implant applications [9, 11]. The most common mechanical methods for altering surface topography properties include sandblasting and machining, while acid etching and oxidation are common chemical methods. Although these methods are effectively used in industry, there are major disadvantages associated with them. Slow production time, complex control processes, and chemical contaminations are only a small number of the challenges presented by these commonly used processing techniques [13]. The newly developing method of laser surface texturing of materials addresses these disadvantages in a simple and effective manner. Lasers are able to deliver very low to high energy with extreme precision in dimension, spatial distribution, and temporal distribution. Lasers offer better control and precision, offer more feasibility, and are environmentally friendly [14–16]. Decreased processing time makes lasers particularly suitable for mass production, rapid prototyping, and custom-scale manufacturing for a wide variety of applications, such as microwelding, drilling, cutting, and heat treatment of metals and alloys. Laser treatment is known for its fast and precise manner in the processing of materials, and for the variety of scales offered by lasers, including micro-, submicro-, and nanofabrication [8, 15].

1.3. Physics of laser surface texturing

Laser processing can be applied in two categories based on the energy requirements: first, applications requiring relatively low energy with limited structural and physical changes; second, applications calling for high-energy transformations for significant structural changes over a large volume, such as welding. Energy transformation in applications involving lasers requires the coupling of the laser radiation with the electrons of the interacting surface, such as metals or semiconductors [17, 18]. In return, the speed of energy transformation from the laser beam to the surface becomes fully dependent on the nature of the interacting material and its chemical bonding. There have been significant improvements in energy transformation of lasers through the development of ultrashort lasers [19–23].

Ultrashort lasers decrease the interaction time (pulse duration) between the laser and the material, which reduces the effects on the bulk material. Despite these improvements, ultrashort laser systems are relatively expensive and cannot be utilized in industries. Hence, depending on the nature of the application, less expensive yet advanced laser systems, with pulse durations in the range of nanoseconds, are more popular among manufacturers because they are more widely available in manufacturing sectors. An understanding of the laser irradiation mechanism is required to utilize the more common lasers to their fullest capability, therefore, choosing the laser parameters is essential to the final quality of a particular application [24–27].

2. Laser-enhanced topography properties

Micro/submicro-treated surfaces are popularly used in scaffold systems for bone and tissue implant applications. The surface topography properties of a material, particularly roughness, are influential on the cell adhesion rate of bone-like apatite to surfaces. Increasing the cell adhesion rate to the surface of a material increases the biocompatibility of the material. Thus, surface texturing of materials to enhance their biocompatibility is an effective method in the fabrication of implant devices [26–28]. The effects of laser irradiation on the surface topography properties of materials are the main advantage of laser surface texturing. Laser parameters, including frequency, power, and laser scanning parameters, can influence these material properties greatly [27].

The irradiated surface area of a material has increased surface irregularities. In return, the exposed area is more readily available for cell attachment, which enhances the apatite-inducing ability and cell adhesion rate of the material [27–31]. Furthermore, laser treatment of the materials increases the surface temperature up to oxidation temperature, and results in the creation of thin layers of oxide upon the surface. An increase in the oxidation of the surfaces increases the wettability of surface of the material. Consequently, this leads to an increase in the apatite-inducing ability of the material and greatly improves the biocompatibility of the implant surfaces [30–35].

2.1. Laser system

The nanosecond laser used for obtaining all the results discussed in this chapter was a Nd:YAG pulsed laser system (SOL-20 by Bright Solutions Inc). The maximum output power is 20 W with a wavelength of 1064 nm and a repetition rate ranging between 10 and 100 kHz. This laser emits pulses of 6–35 ns pulse duration. The diameter of circular output beam from the laser is around 9 mm. The diameter of beam is reduced to 8mm by using an iris diaphragm before entering to galvo-scanner. A two-axis galvo-scanner (JD2204 by Sino Galvo) with the input aperture of 10 mm and beam displacement of 13.4 mm was used for beam scanning since it has a high scanning speed (to 3000 mm/s). In order to focus the normal beam to the surface of Ti, scan lens of a focal length of 63.5 mm was used. The theoretical focused laser spot diameter (d_0) is calculated to be 20 μm. During the experiment, the spot size may be bigger due to scatter and misalignment. The average laser fluence was 900 μJ at the frequency of 10 kHz.

The scanning parameters including scanning speed, and scanning configurations can be adjusted through the software operating the laser. When the combination of parameters is adjusted with this software, along with power and frequency set for laser irradiation, the desired pattern is irradiated across the surface of the selected material.

3. Effects of laser power

Power is one of the most influential laser parameters when surface treating a material. Adjusting the power increases the surface irregularities of the material and therefore enhances the topography properties of the substrate [25, 33]. In this chapter, the effects of four different powers on surface topographic and oxidation properties of titanium substrates are investigated, and the bioactivity of the treated substrates is examined through the use of simulated body fluid (SBF). Simulated body fluid (SBF), or formally known as hydroxyapatite, is a supersaturated insoluble calcium phosphate mineral, with the chemical composition of $Ca_{10}(PO_4)_6(OH)_2$. This fluid has a close similarity with human blood plasma and is the essential component of the biological hard tissues such as bones. Due to the high absorbance and catalytic properties of SBF, this fluid is commonly used to estimate the biocompatibility level of materials used in implant productions.

To treat the titanium substrates, a range of low to high average laser powers from 6 to 12W was used, while frequency and scanning speeds were kept constant in all cases. The effective number of pulses for all results was kept constant to be 25 pulses for frequency of 100 kHz [36, 37]. **Table 1** introduces the parameters used in more detail.

Average laser power (W)	Total number of laser pulses	Pulse energy (mJ) at 100 kHz	Pulse width (ns)	Total delivered energy (mJ)
6	25	0.06	35	1.5
8	25	0.08	35	2
10	25	0.10	35	2.5
12	25	0.12	35	3

Table 1. Laser parameters used in surface treating of Ti substrates.

3.1. Surface topography analysis

Once treated, there is often more available exposed surface area of the material because of the increase in surface roughness, which, in return, enhances the apatite-inducing ability and cell adhesion rate of that material. **Figure 1** displays the variation in the treated substrates using the stated powers.

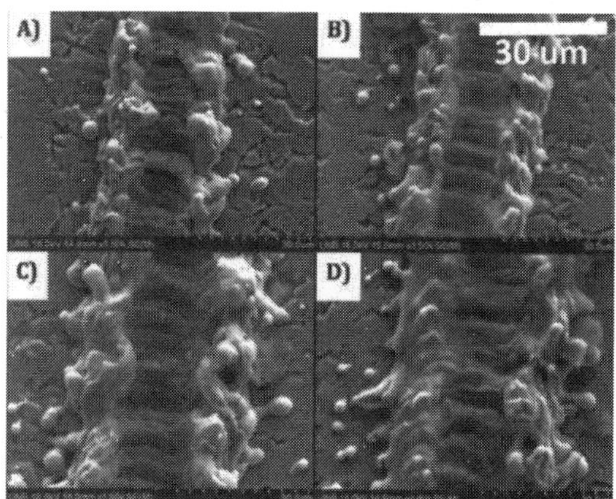

Figure 1. Surface of treated titanium (A) 6 W, (B) 8 W, (C) 10 W, (D) 12 W.

Although the number of pulses delivered to each of these substrates is the same, as indicated in **Table 1**, the images shown in **Figure 1** clearly indicate different surface topography after laser treatment. With an increase in power, the energy delivered to the surface of the material increases as well, hence, more ablation and topography changes occur. Substrates treated with a power of 12 W have a larger surface affected by laser irradiation as opposed to 6 W.

During laser irradiation, a plasma plume with radial surface tension is formed surrounding the irradiated area due to the high temperature gradient caused by laser energy transfer. Immediately following irradiation, a high temperature difference exists between the surface of the material and the generated plasma plume. Thus, a high cooling rate results, which causes rapid re-solidification of the ablated material. Consequently, the tension within the laser plume causes shooting of the molten titanium to outside of the ablated zones [37]. The micro re-solidified particles observed outside the crater in **Figure 1** have been formed due to the same reason.

Using 3D microscopy, the surface profile of the treated substrates can be compared; this is introduced in **Figure 2**. This study used the Zeta-20 optical profiler to scan the surface of the samples in order to obtain the surface topography results across each sample.

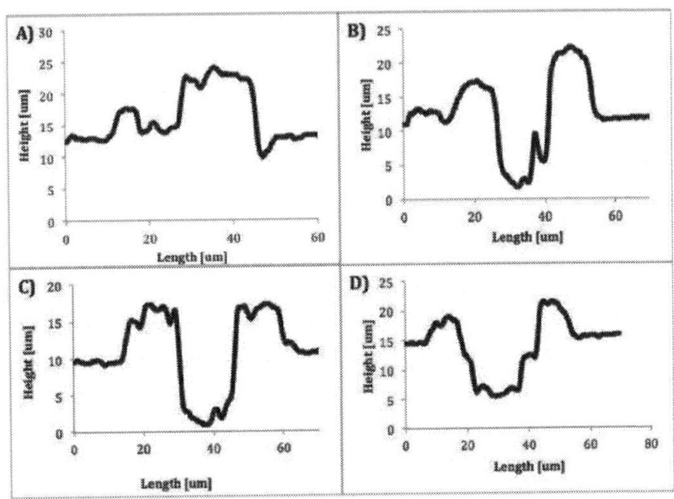

Figure 2. Surface profiles of Ti (A) 6 W, (B) 8 W, (C) 10 W, (D) 12 W.

As shown, an increase in laser power clearly affects the surface topography profile of titanium substrates. Different laser powers create different surface irregularities across the surface of the treated titanium substrates.

3.2. Surface temperature analysis

Different power levels deliver various amounts of energy to the surface of the substrate, and, therefore, increase the surface temperature of the titanium. This increase in the surface temperature could affect the surface structure, topographic properties, and oxidation of the titanium substrates. Oxidation is an important factor when considering bioactivity of a material. With higher oxidation, a higher surface energy is present, which results in more interaction between the implant and the body. Therefore, generating larger amounts of titanium oxide across the surface substrates consequently increases the biocompatibility of the treated work pieces. However, after a certain power threshold, the energy delivered to the surface of the titanium increases the surface temperature beyond the oxidation limit. Therefore, the material develops less titanium oxide as the surface undergoes larger ablation [32, 34, 38].

In order to calculate the surface temperature in this chapter, a theoretical calculation has been conducted according to the previous published results [25, 36, 39]; in this method, we assumed that the laser energy is absorbed in a much thinner layer compared to the penetration depth of the heat wave. Finally, the average surface temperature of the Ti substrate can be obtained as fully discussed in [25, 39]:

$$T_n = 2\alpha \frac{\left[1 - \frac{2}{3}\alpha\right]}{(1+\alpha^2)} \frac{T_m}{(1-\alpha)} \left[1 + \frac{\alpha^n - \alpha}{n(1-\alpha)}\right] \tag{1}$$

where T_m is the maximum temperature calculated at the end of first pulse, n is the pulse numbers, and α is the constant ratio for the previous maximum and the following minimum temperatures and equal to $\alpha = (t_p/t_{pp})^{1/2}$, where t_{pp} is the pulse interval and equal to $t_{pp} = 1/f$ (f is pulse repetition rate) and t_p is pulse duration. The analytical results obtained in this study are associated with possible errors due to multiple assumptions made throughout the study.

Upon completion of the analytical analysis, the temperature profile for both cases with 8 and 10 W powers, and a pulse number range of 2 pulses to 50 pulses is obtained.

The trend of average surface temperature of the treated titanium substrates is shown in **Figure 3**.

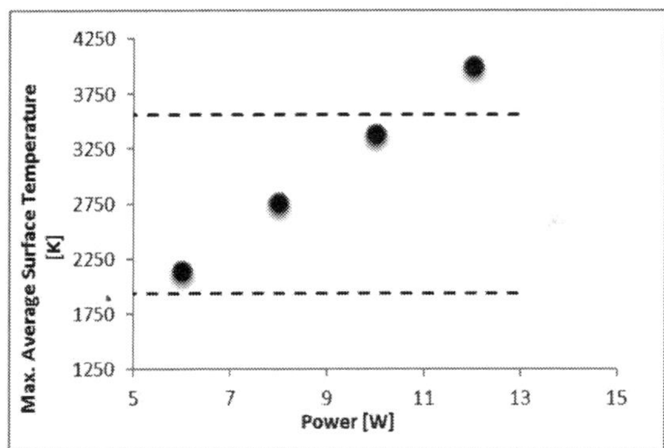

Figure 3. Maximum average surface temperature in each power.

As evident, with 12 W of power, the maximum average surface temperature of the titanium exceeds its evaporation point, and therefore less concentration of titanium oxide is generated along the surface of the substrates. Also, more ablation occurs, which agrees with the observations. Having a larger concentration of titanium oxide across the surface results in better cell interactions. To further investigate the oxidation of treated substrates, energy dispersive X-ray (EDX) analysis was conducted to detect the amount of oxygen available across the surface of substrates treated using the stated powers.

Figure 4 indicates the results.

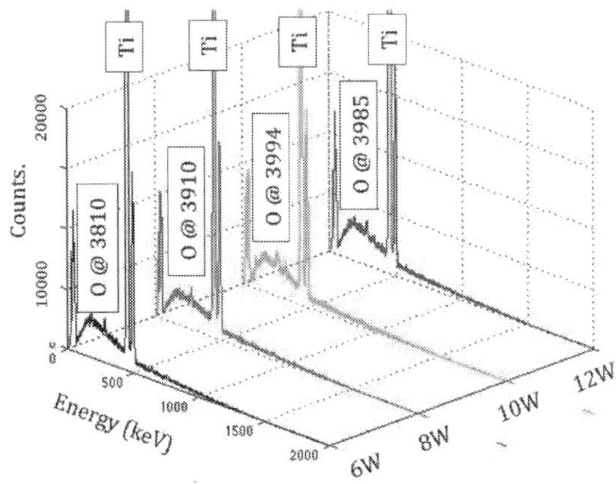

Figure 4. EDX analysis of the irradiated Ti at 6, 8, 10, and 12 W.

Comparing the trace of oxygen across the surface of all four substrates, it is observed that with an increase in power, the oxygen concentration increases slightly as well, until a power of 12 W is reached, which shows slightly a lower trace of oxygen. This is in agreement with the average temperature results observed in **Figure 3**. Considering the results shown in **Figures 4** and **5**, it is expected that powers of 8 and 10 W would result in a better biocompatibility compared to other two powers. This conclusion is assessed through the use of simulated bodily fluids (SBF).

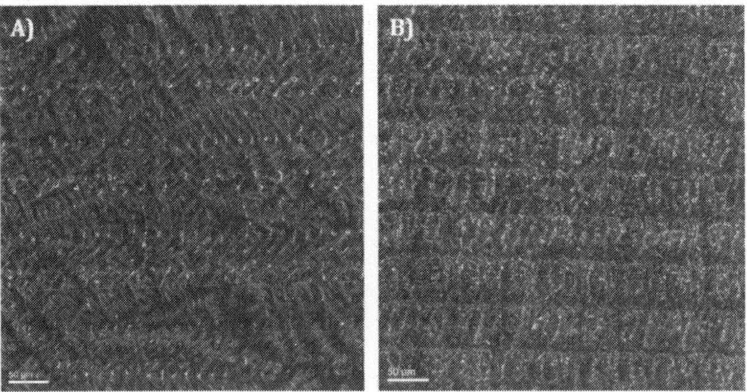

Figure 5. SBF-soaked substrates. (A) 8 W, (B) 10 W.

3.3. Biocompatibility assessment

To determine how the differences in surface topography affect the biocompatibility of titanium, the treated substrates using powers of 8 and 10 W were immersed in SBF for 7 days. **Figure 5** shows the scanning electron microscopy (SEM) photographs of the substrates after the completion of the assessment.

In **Figure 5**, the white layers across the induced line patterns indicate the bone-like apatite deposition on the surface of the titanium. These microscopy images show that generally with an increase in power, apatite-inducing ability across the surface of treated titanium increases as well. Looking closely into **Figure 5**, the EDX results for 10 W curve indicate a larger amount of apatite deposition compared to the 8 W curve. This is further observed in the EDX analysis conducted on these titanium substrates shown in **Figure 6**. Detecting larger amount of oxygen, calcium, and phosphorous elements indicates a higher apatite-inducing ability for the substrates.

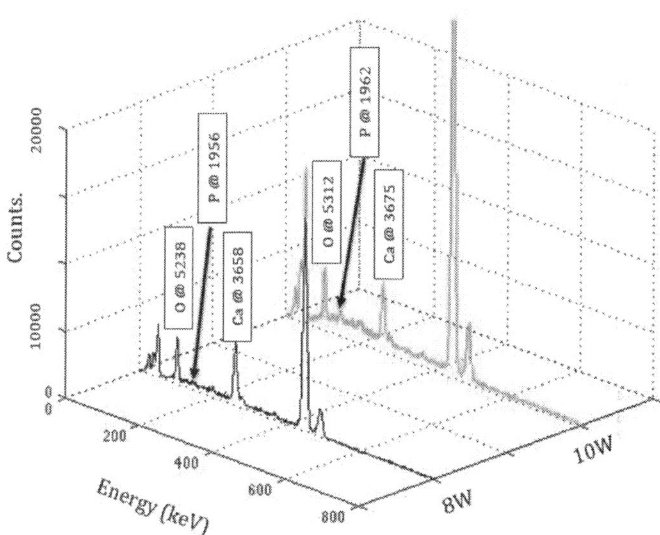

Figure 6. EDX analysis at 8 and 10 W.

As shown, the apatite-inducing ability of substrates treated with a power of 10 W is higher than the substrates treated with 8 W. This is in agreement with results observed with oxidation and SEM photography analysis.

Overall, the biocompatibility was highest for substrates treated with 10 W. This power provides adequate surface topography and the energy delivered is within the oxidation temperature range, hence creating a desirable environment for cell attachment to take place.

4. Effects of number of laser pulses

The total number of laser pulses delivered to the substrate is related to the scanning speed and laser frequency used while irradiation is taking place [40–42]. Effects of scanning speed on the number of laser pulses delivered to surface of the titanium substrates during laser irradiation are examined. Similar to power, scanning parameters have a direct effect on the surface topography and oxidation of treated titanium substrates. These effects are monitored using various scanning parameters, while keeping power and frequency constant. **Table 2** displays the parameters used.

Scanning speed (μm/ms)	Frequency (kHz)	Average power (W)	Total number of laser pulses	Total delivered energy (mJ)
50	100	10	167	16.7
200	100	10	41	4.1
400	100	10	21	2.1
500	100	10	17	1.7

Table 2. Scanning speed and pulse numbers used in surface treating of Ti substrates (pulse width, 35 ns; pulse energy, 0.1 mJ).

Scanning speed and the number of laser pulses are indirectly related. With an increase in scanning speed, the pulse number decreases. This is because with a lower scanning speed, more time is given to the laser to induce a pattern across the surface; hence, more pulses are delivered to the substrates.

4.1. Surface topography analysis

Similar to power, the effects of different scanning speeds on surface topographic properties of the treated substrates are examined using SEM photography. **Figure 7** displays these results.

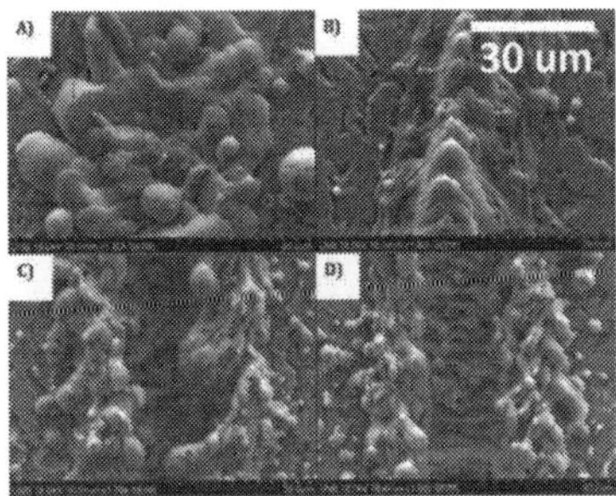

Figure 7. Surface of treated titanium (A) 50 μm/ms, (B) 200 μm/ms, (C) 400 μm/ms, (D) 500 μm/ms.

As shown in this figure, with an increase in scanning speed, the surface irregularities decrease in size; therefore, fewer topographic changes take place on those samples. This is in agreement with expectations, since **Figure 7(A)** with a scanning speed of 50 μm/ms has a larger number of pulses compared to **Figure 7(B)** with a scanning speed of 500 μm/ms.

The surface profile of these substrates is shown in more detail in **Figure 8**.

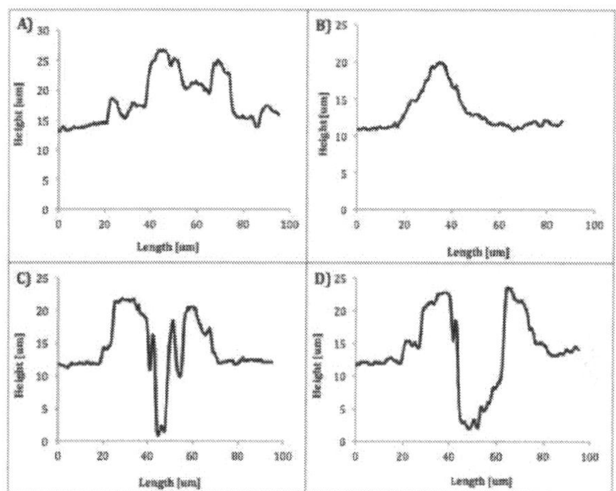

Figure 8. Surface profile of Ti (A) 50 μm/ms, (B) 200 μm/ms, (C) 400 μm/ms, (D) 500 μm/ms.

Figure 8(A) shows how irregularities cover the entire surface of the substrates, as opposed to only along the irradiation zone as shown in **Figure 8(D)**. This agrees with the expectations.

4.2. Biocompatibility assessment

To see how influential the number of pulses is on bioactivity enhancement of titanium, a wider range of scanning speeds is used to assess the apatite-inducing ability of treated titanium substrates. Scanning speeds of 300 and 700 μm/ms are used to treat these substrates in order to more clearly show the differences between the two substrates. All other parameters remain the same (**Table 3**).

Scanning speed (μm/ms)	Frequency (kHz)	Average power (W)	Total number of laser pulses	Total delivered energy (mJ)
300	100	10	28	2.8
700	100	10	12	1.2

Table 3. Scanning speed and pulse numbers used in surface treating of Ti substrates (pulse width, 35 ns; pulse energy, 0.1 mJ.

Figure 9 displays the substrates immersed in SBF for 7 days.

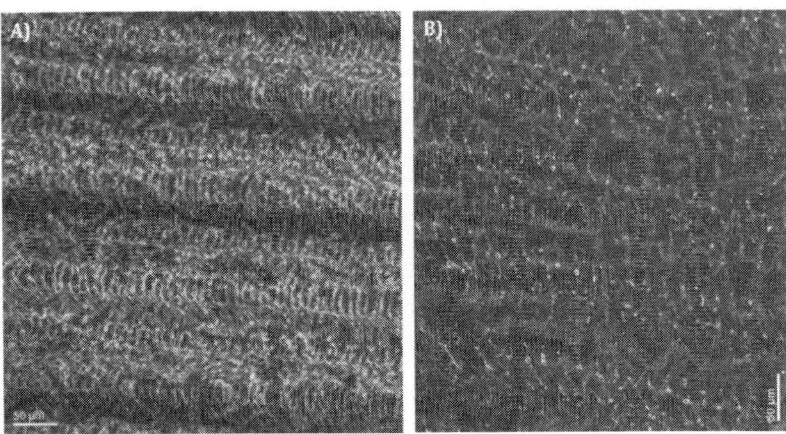

Figure 9. SBF-soaked substrates (A) 300 μm/ms, (B) 700 μm/ms.

As stated, a lower scanning speed delivers a larger number of pulses and therefore affects the surface topographic properties of the titanium more significantly. Additionally, with a larger number of pulses, the average surface temperature of the substrate increases, and therefore, a more ideal condition for the creation of titanium oxide is generated [25, 43]. **Figure 9** displays these effects clearly. Considering **Figure 9(A)**, more apatite deposition is observed across the surface as opposed to **Figure 9(B)**, which agrees with the expectations.

The apatite-inducing ability of these titanium substrates is further analyzed using EDX. **Figure 10** displays the results.

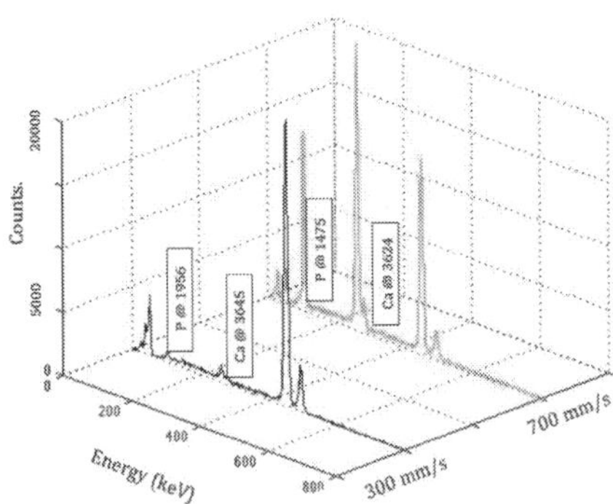

Figure 10. EDX analysis (A) 300 μm/ms, (B) 700 μm/ms.

It is evident in **Figure 10** that calcium and phosphorous concentrations detected across the surface are larger for the curve of 300 mm/s. This agrees with observations from **Figures 7** and **8**.

Overall, it was observed that with a larger number of laser pulses, more energy is delivered to the material, and therefore more surface topographic and oxidation changes take place across the surface of the treated substrates.

5. Effects of frequency

Lastly, the effects of frequency on surface of titanium substrates are examined.

A range of frequencies (25, 50, and 100 kHz) is used in surface treatment, while the average laser power and pulse width are held constant to 11 W and in the range of 15–35 nm, respectively, to ensure consistency of final results (**Table 4** introduces the parameters used in more detail).

Average laser power (W)	Laser frequency (kHz)	Pulse energy (mJ)
11	25	0.30
11	50	0.20
11	100	0.10

Table 4. Laser parameters used in surface treating of Ti substrates.

Figure 11 shows the untreated titanium substrates using laser frequencies of 25, 50, and 100 kHz.

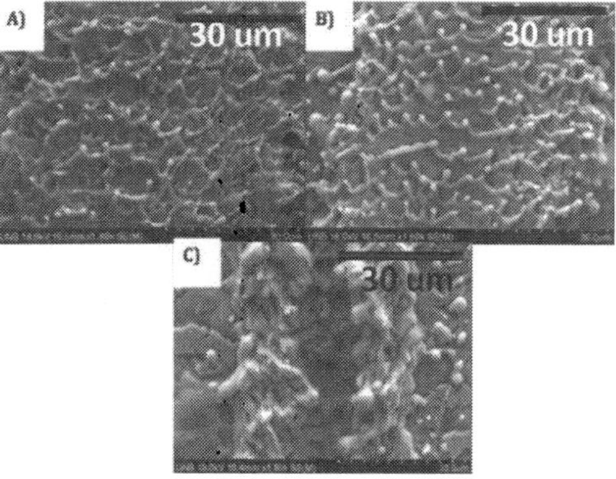

Figure 11. Non-SBF Ti substrates at laser frequency of (A) 25 kHz, (B) 50 kHz, (C) 100 kHz.

As evident in this figure, with a higher frequency, surface irregularities increase on the surface of the substrates. This enhances the surface topographic properties of the treated substrates and therefore increases the biocompatibility of the titanium [25, 41–43].

Furthermore, with a higher frequency, a larger amount of energy is delivered to the surface of the substrates, and the temperature increase causes more titanium oxide to form across the surface. This is similar to the results observed from the other parameters. Therefore, the substrates treated with a frequency of 100 kHz are expected to show the highest apatite-inducing ability across their surface.

5.1. Biocompatibility assessment

Figure 12 displays the EDX analysis of substrates immersed in SBF for 3 days.

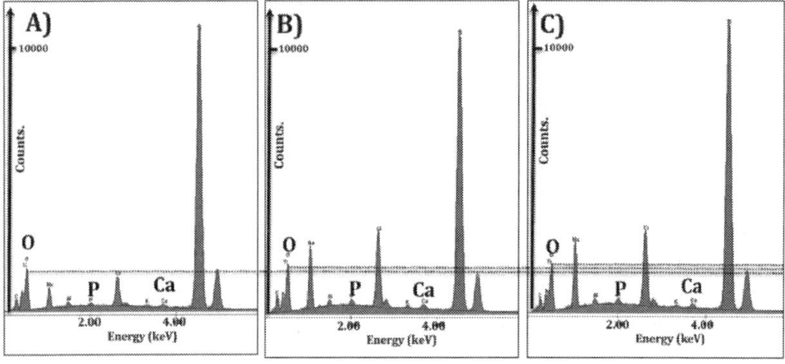

Figure 12. Ti substrates soaked in SBF for 3 days (A) 25 kHz, (B) 50 kHz, (C) 100 kHz.

As evident in **Figure 12**, oxygen concentration slightly increases with an increase in frequency, showing an increase in the concentration of titanium oxide. The higher oxidation level of titanium results in increased negative charge across the surface. Thus, positively charged CO_2, Ca^{2+}, NO_x, and H_2O atoms/ions are attracted to and absorbed by the negatively charged surface of the titanium (OH^-), and create an environment with excellent affinity to biomaterials, such as proteins. The attraction forces between the apatite and the negatively charged titanium oxides are increased through the laser treatment, which, in return, increases the biocompatibility of titanium [25, 42, 43].

This can be further observed by looking at the SEM photography of substrates immersed in SBF.

Figure 13 displays the result.

In this figure, the apatite deposition layer is evident across the irradiated area. As observed from the EDX analysis, the apatite-inducing ability of this substrate is high, which denotes the biocompatibility enhancement of the treated titanium.

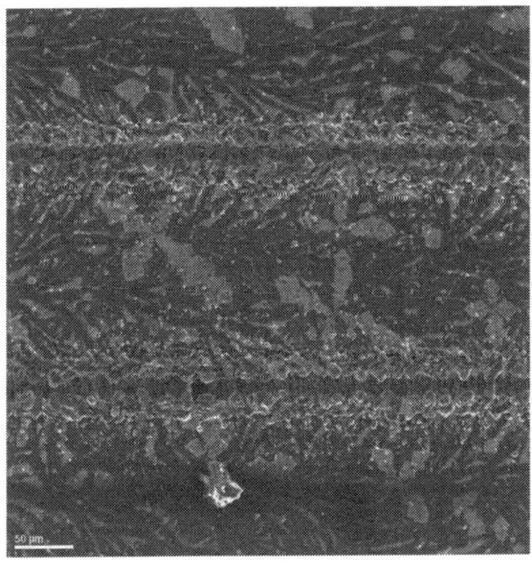

Figure 13. Substrate soaked in SBF for 3 days at 100 kHz.

6. Conclusion

In this chapter, a new method for micro/nano surface texturing of titanium was demonstrated using nanosecond laser irradiation. The application of this method in bone and tissue implant fabrication was explored. Systematic experimental and theoretical studies on effects of the laser parameters on biocompatibility and bioactivity of titanium were conducted. The effects of each parameter were explored individually while other parameters were held constant. Predetermined patterns were induced across thin sheets of titanium substrates to investigate the effects of power, scanning speed, and frequency on surface of the substrates.

Microscopy analysis determined that an increase in power increases the surface topography and ablation across the surface of the material. However, increasing the power within the oxidation limit of titanium can lead to generation of titanium oxide along the irradiation area. Using SBF and cell adhesion, it was found that bioactivity of titanium increases in areas with higher surface topography and higher concentration of titanium oxide.

Additionally, scanning parameters, including pulse number and scanning speed, were found to be influential on the biocompatibility enhancement of titanium substrates. Through a series of experimental and theoretical analysis, it was concluded that a higher pulse number increases the surface energy of titanium substrates, and hence increases the biocompatibility along the irradiation area.

Finally, a range of frequencies was used to examine the bioactivity along the irradiation areas. Microscopy photographs displayed the generation of structures with higher surface topography along the laser-induced patterns that have high oxidation and topography properties. Upon biocompatibility assessment, concentrations of calcium, oxygen, and phosphorous are observed on the surface, which shows the bioactivity enhancement of titanium.

This chapter introduced the use of commercially used nanosecond laser systems for biocompatibility enhancement of titanium substrates for fabrication of bone and tissue implants. The key features of this method are described as following:

- Easy controllability

- Simple procedure with rapid processing (single step)

- Feasibility for micro/nano scales

- High efficiency and accuracy

These key features make laser surface texturing desirable for rapid prototyping and fabrication of biomedical devices, and can lead to improvements in cost and durability.

Author details

Amirkianoosh Kiani* and Mitra Radmanesh

*Address all correspondence to: a.kiani@unb.ca

Department of Mechanical Engineering, Silicon Hall: Laser Micro/Nanofabrication Facility, University of New Brunswick, NB, Canada

References

[1] Geetha, M., Singh, A. K., Asokamani, R. and Gogia, A. K. (2009). Ti based biomaterials, the ultimate choice for orthopaedic implants–a review. *Progress in Materials Science*, 54(3), 397–425.

[2] Burg, K. J. L.; Porter, S.; and Kellam, J. F. (2000) Biomaterials developments for bone tissue engineering. *Journal of Biomaterials*, 21, 2347–2359.

[3] Geetha, M.; Asokamani, S.; and Gogia, A. K. (2009) Ti based biomaterials, the ultimate choice for orthopaedic implants – a review. *Journal of Progress in Materials Science*, 54, 397–425.

[4] Ponsonnet, L.; Reybier, K.; Jaffrezic, N.; Comte, V.; Lagneau, C.; Lissac, M.; and Martelet, C. (n.d.)., Relationship between surface properties (roughness, wettability) of titanium and titanium alloys and cell behaviour. *Materials Science and Engineering*.

[5]   Stevens, M. M. (2008) Biomaterials for bone tissue engineering. *Journal of Materials Today*, 11, 18–25.

[6]   Taddei, E. B.; Henriques, V. A. R.; Silva, C. R. M.; and Cairo, C. A. A. (2004) Production of new titanium alloy for orthopedic implants. *Journal of Materials Science & Engineering*, 24, 683–687.

[7]   Taimoor H.; Qazi, D.; Mooney, J.; Pumberger, M.; Geißler, S.; and Duda G. N (2015) Biomaterials based strategies for skeletal muscle tissue engineering: existing technologies and future trends. *Journal of Biomaterials*, 53, 502–521.

[8]   Gristina, A (1987) Biomaterial-centered infection, microbial adhesion versus tissue integration. *Journal of* Science, 237, 1588–1595.

[9]   Wang, H.; Liang, C.; Yang, Y.; and Li, C. (2010) Bioactivities of a Ti surface ablated with a femtosecond laser through SBF. *Journal of Biomedical Materials*, 5, 1–5.

[10]   Majumdar, J.; and Manna, I. (2011) Laser material processing. *Journal of International Materials Reviews*, 56, 341–388.

[11]   Pereira, A.; Cros, A.; Delaporte, P.; Georgiou, S.; Manousaki, A.; Marine, W.; and Sentis, M. (2004) Surface nano-structuring of metals by laser irradiation: effects of pulse duration, wavelength and gas atmosphere. *Journal of Applied Physics. A.*, 79, 1433–1437.

[12]   Ahmmed, T.; Yang, T.; Ling, E. J.; Servio, P.; and Kietzig, A. (2014) Introducing a new optimization tool for femtosecond laser-induced surface texturing on titanium, stainless steel, aluminum and copper. *Journal of Optics and Lasers in Engineering*, 66, 258–268.

[13]   Harooni, M.; Carlson, B.; Strohmeier, B. R.; and Kovacevic, R. (2014) Pore formation mechanism and its mitigation in laser welding of AZ31B magnesium alloy in lap joint configuration. *Journal of Materials & Design*, 58, 265–276.

[14]   Mukherjee, S.; Dhara, S.; and Saha, P. (2013) Enhancing the biocompatibility of Ti6A14V implants by laser surface microtexturing: an in vitro study. *Journal of International Journal of Manufacturing Technolog*, 76(1–4), 5–15.

[15]   Hutmacher, D. W. (2000) Scaffolds in tissue engineering bone and cartilage. *Journal of Biomaterials*. 21, 2529–2543.

[16]   Venkatakrishnan, K.; Stanley, P.; Sivakumar, N.R.; Tan, B.; and Lim, L.E.N. (2003) Effect of scanning resolution and fluence fluctuation on femtosecond laser ablation of thin films. *Journal of Applied Physics*, 77, 655–658.

[17]   Tong, W. Y., Y. M. Liang, Vivian Tam, H. K. Yip, Y. T. Kao, K. M. C. Cheung, K. W. K. Yeung, and Y. W. Lam. "Biochemical characterization of the cell-biomaterial interface by quantitative proteomics.(2010) Molecular & Cellular Proteomics 9 (10), 2089–2098.

[18] Tran, D. V.; Lam, Y. C.; Zhen, H. Y.; Murukeshan, V. M.; Chai, J. C.; and Hard, D. E. (2005) Femtosecond Laser Processing of Crystalline Silicon."http://dspace.mit.edu/bitstream/handle/1721.1/7449/IMST010.pdf

[19] Yilbas, B. S.; and Al-Aqeeli, N. (2009) Analytical Investigation Into Laser Pulse Heating and Thermal Stresses. *Journal of Optics & Laser Technology*, 41, 132–139.

[20] Kiani, A.; Venkatakrishnan, K.; and Tan, B. (2013) Optical absorption enhancement in 3D silicon oxide nano-sandwich type solar cell. Journal of Optics Express, 22, A120–A131.

[21] Ashby, M. F.; and Easterling, E. K. (1984) The transformation hardening of steel surfaces by laser beams; I. Hypo-eutectoid steels. *Journal of Acta Metallurgica*, 32, 1935–1937.

[22] Ion, J. C.; Scherclift, H. R; and Ashby, M. F. (1992) Diagrams for laser materials processing. *Journal of Acta Metallurgica et Materialia*, 40, 1539–1551.

[23] Kiani, A.; Venkatakrishnan, K.; and Tan, B. (2011) Enhancement of the optical absorption of thin-film of amorphorized silicon for photovoltaic energy conversion. *Journal of Solar Energy*, 85, 1817–1823.

[24] Woodard, p.; and Druden, J. (1998) Thermal analysis of a laser pulse for discrete spot surface transformation hardening. *Journal of Applied Science*, 85, 2488–2496.

[25] Radmanesh M.; Kiani A. (2015) Bioactivity enhancement of titanium induced by Nd: Yag laser pulses. *Journal of applied biomaterials & functional materials*, 0–0.

[26] Kokubo, T.; and Takadama, H. (2006) How useful is SBF in predicting in vivo bone bioactivity?. *Journal of Biomaterials*, 27, 2907–2915.

[27] Aza, De; Fernandez-Pradas, P. N.; and Serra, P. (2004). In vitro bioactivity of laser ablation pseudowollastonite coating. *Journal of Biomaterials*, 25, 1983–1990.

[28] Kurella, A., & Dahotre, N. B. (2005). Review paper: surface modification for bioimplants: the role of laser surface engineering. *Journal of biomaterials applications*, 20(1), 5–50.

[29] Egerton, R. F. (2005) Physical principles of electron microscopy: an introduction to TEM, SEM, and AEM. Springer, 202, New York, USA.

[30] Goldstein, J. (2012) Scanning Electron Microscopy and X-Ray Microanalysis, Springer, ISBN 978-0-306-47292-3, New York, USA.

[31] Albrektsson, T.; and Wennerberg, A. (2004). Oral implant surfaces: part 1 – review focusing on topographic and chemical properties of different surfaces and in vivo responses to them. *International Journal of Prosthodontics*, 17(5).

[32] Brady, J.; Newton, R.; and Boardman, S. (1995) New uses for powder x-ray diffraction experiments in the undergraduate curriculum. *Journal of Geological Education*, 43, 466–470.

[33] Trtica, M; Gakovic, B; Batani, D; Desai, T; Panjan, P; and Radak, B. (2006) Surface modifications of a titanium implant by a picosecond Nd: YAG laser operating at 1064 and 532nm. *Journal of Applied Surface Science,* 253, 2551–2556.

[34] Kiani, A.; Venkatakrishnan, K.; Tan, B.; and Venkataramanan, V. (2011) Maskless lithography using silicon oxide etch-stop layer induced by megahertz repetition femtosecond laser pulses. *Journal of Optics Express,* 19, 10834–10842.

[35] Colpitts, C.; and Kiani, A. (2016). Synthesis of bioactive three-dimensional silicon-oxide nanofibrous structures on the silicon substrate for bionic devices' fabrication. *Journal of Nanomater Nanotechnol,* 6:8. doi: 10.5772/62312

[36] Kiani, A.; Venkatakrishnan, K.; and Tan, B. (2010). Direct laser writing of amorphous silicon on Si-substrate induced by high repetition femtosecond pulses. *Journal of Applied Physics,* 108(7), 074907.

[37] Kiani, A.; Patel, N. B.; Tan, B.; and Venkatakrishnan, K. (2015). Leaf-like nanotips synthesized on femtosecond laser-irradiated dielectric material. *Journal of Applied Physics,* 117(7), 074306.

[38] Radmanesh, M.; and Kiani, A. (2015) ND:YAG laser pulses ablation threshold of stainless steel 304. *Journal of Materials Sciences and Applications,* 6, 634–645, doi: 10.4236/msa.2015.67065.

[39] Gamaly, E. G.; Rode, A. V.; and Luther-Davies, B. (1999). Ultrafast ablation with high-pulse-rate lasers. Part I: theoretical considerations. *Journal of Applied Physics,* 85(8), 4213–4221.

[40] Radmanesh, M.; and Kiani, A. (2015) Effects of laser pulse numbers on surface biocompatibility of titanium for implant fabrication. *Journal of Biomaterials and Nanobiotechnology,* 6, 168–175. doi:10.4236/jbnb.2015.63017.

[41] Zhao, G.; Schwartz, Z.; Wieland, M.; Rupp, F.; Geis-Gerstorfer, J.; Cochran, D. L.; and Boyan, B. D. (2005). High surface energy enhances cell response to titanium substrate microstructure. *Journal of Biomedical Materials Research Part A,* 74(1), 49–58.

[42] Tenner, F.; Brock, C.; Klampfi, F.; and Schmidt, M. (2014) Analysis of the correlation between plasma plume and keyhole behavior in laser metal welding for the modeling of the keyhole geometry. *Journal of Optics and Lasers in Engineering,* 64, 32–41.

[43] Radmanesh, M; and Kiani, A. (2015) "Enhancing biocompatibility of grade 4-titanium using laser surface texturing for bone and tissue transplant applications." 25th Canadian Congress of Applied Mechanics (CANCAM 2015), London, Ontario, Canada.

Chapter 16

Excimer Laser and Femtosecond Laser in Ophthalmology

Liang Hu, Yiqing Huang and Meng Lin

Additional information is available at the end of the chapter

http://dx.doi.org/10.5772/64238

Abstract

Laser technology is used in many basic and clinical disciplines and specialties, and it has played an important role in promoting the development of ophthalmology, especially corneal refractive surgery. We provide an overview of the evolution of laser technology for use in refractive and other ophthalmologic surgeries, mainly focusing on two types of lasers and their applications. First, we discuss the characteristics of the excimer laser and its application in corneal refractive surgery treating ametropia (e.g., photorefractive keratectomy (PRK), laser epithelial keratomileusis (LASEK), epipolis laser in situ keratomileusis (Epi-LASIK), and transepithelial photorefractive keratectomy (Trans-PRK) and presbyopia surgery). Second, we discuss the characteristics of the femtosecond laser and its application in corneal refractive surgery (e.g., femtosecond laser in situ keratomileusis (FS-LASIK), insertion of intracorneal ring segments, small-incision lenticule extraction (SMILE), and femtosecond lenticule extraction (FLEx)) and other ophthalmologic surgeries (e.g., penetrating keratoplasty (PKP), deep anterior lamellar keratoplasty, Descemet's stripping endothelial keratoplasty (DSEK), and cataract surgery). The patients studied received many benefits from the excimer laser and femtosecond laser technologies and were satisfied with their clinical outcomes.

Keywords: excimer laser, femtosecond laser, corneal refractive surgery, ophthalmology

1. Introduction

In the field of ophthalmology, laser technology is used in many basic and clinical disciplines and specialties. It has played an important role in promoting the development of ophthalmology. Advancements in technology have allowed measurable improvements in the surgical safety, efficacy, speed, and versatility of the laser, creating more operation methods to treat eye diseases, especially corneal refractive surgery. Because myopia is one of the most prevalent

ocular disorders, and high myopia may result in comorbidities associated with significantly increased risks of severe and irreversible loss of vision, it is always an important topic of research worldwide. Laser technology, including the excimer laser and the femtosecond laser, has brought an era of laser corneal refractive surgery. According to the location of ablation, corneal refractive surgery can be divided into two types: laser corneal surface refractive surgery and laser corneal lamellar refractive surgery. Because of the increasing numbers of applications in ophthalmology and their successful implementations, ophthalmic use of laser technology is expected to continue flourishing.

The aim of this chapter is to review the evolution of laser technology in refractive and other ophthalmologic surgeries, mainly focusing on the characteristics of two types of lasers and their applications: the excimer laser applied in laser corneal surface refractive surgery and presbyopia surgery and the femtosecond laser applied in laser corneal lamellar refractive surgery and other ophthalmologic surgeries.

2. Excimer laser in ophthalmology

2.1. Characteristics of the excimer laser

The *excimer* (comprising the terms *excited* and *dimer*) was named by the Russian, Nikolay Basov, in 1970, based on his work with a xenon dimer gas [1]. An *excimer* is a short-lived dimeric or heterodimeric molecule formed from two species (a noble gas and a halide), at least one of which has completely filled the valence shell by electrons. Excimers are only formed when one of the dimer components is in the excited state. When the excimer returns to the ground state, its components dissociate. The wavelength of an excimer's emission depends on the noble gas, such as ArF (193 nm), KrF (248 nm), XeCl (308 nm), or XeF (351 nm). The ultraviolet laser (193 nm) is commonly used to ablate tissue through ablative photodecomposition. The process of ablative photodecomposition involves three main components: absorption, bond breaking, and ablation. A large number of experiments have shown that the ArF excimer laser (193 nm) is the most optimal for corneal absorption and ablation because of its sufficient photon energy (6.4 eV) and precision (only penetrating the superficial layer; 0.3 μm). The tissue-ablation depth is positively correlated with the logarithm of laser density; 1-J/cm^2 energy can ablate approximately 1-μm corneal tissue. In addition, as a cold laser, the excimer laser can ablate the tissue accurately without thermal damage.

2.2. Photorefractive keratectomy (PRK)

Professor Jose I. Barraquer described his coined technique of "keratomileusis" to correct myopia in 1949, and this could be the original form of photorefractive keratectomy (PRK). A few years later, many researchers designed similar surgical procedures. In 1983, Stephen Trokel first started using the ArF (193 nm) excimer laser as a precise and safe tool of corneal shaping in calf eyes. He found that the excimer laser not only accurately ablated central corneal tissues but also did not do excessive damage to the peripheral corneal tissues. After a large number of animal experiments, in 1989, Marguerite McDonald, who first applied the technology to human eyes, presented the first patency of using the excimer laser as a corneal refractive tool, and it was accepted. Since then, PRK has become the classic refractive surgery, and its safety and efficacy have been proven by abundant studies. China introduced this surgery in 1993, but the presence of postoperative complications has influenced its development [1].

Compared with the current, more popular lamellar corneal refractive surgery, PRK has shown similar postoperative visual quality but more intraoperative and postoperative complications, such as haze. Nevertheless, Wagoner et al. [2] proposed that PRK had better visual results compared with lamellar corneal refractive surgery. O'Brart [3] also thought PRK showed better cornea curvature and excellent visual quality compared with those of lamellar corneal refractive surgery.

2.3. Laser epithelial keratomileusis (LASEK)

Laser epithelial keratomileusis (LASEK), also called laser sub-epithelial keratectomy, was first proposed and named by Italian doctor Massimo Camellin in 1998. The first case of LASEK was performed by Azar at Massachusetts Eye and Ear Infirmary in 1996. It is a modified operation based on the PRK, which enabled the epithelium to be preserved as a flap (about 50–70 μm) using 20% ethanol to infiltrate and release the connection between the corneal epithelium and Bowman layer and then overturn the flap. The flap is then reset after ablating the stroma using the excimer laser.

Therefore, LASEK can be considered a kind of PRK that is "wearing flap." The key to success throughout the surgical procedure is the activity and good adhesion of the flap. Owing to the viable flap, it combines the advantages of laser in situ keratomileusis (LASIK) and PRK. Because LASEK is essentially a kind of surface ablation evolved from PRK; consequently, it reserves the features of safety, validity, simplicity, and stability in low to moderate ametropia and presbyopia correction. Moreover, its flap design, which is similar to that of LASIK, has several merits: postoperative discomfort can reduce within 2–8 h after the procedure; the epithelium of the optical region in the slit lamp after surgery is as complete and clear as it is before surgery within 12–24 h; there is no edema or postoperative haze; and early complications, such as corneal epithelium necrosis, are less common compared with PRK. These outstanding characteristics are likely to have provided the inspiration for the advent of LASIK.

Furthermore, LASEK preserves the corneal biomechanical integrity and results in good clinical outcomes. Scholars have indicated that LASEK shows better postoperative outcomes including postoperative corneal topography and contrast sensitivity, and faster postoperative recovery rate in low to moderate myopia correction compared with PRK. In addition, LASEK has a unique advantage for patients with retinal diseases, high myopia, or blepharophimosis. Lu Xiong conducted a study on the clinical outcomes in low to moderate myopia after LASEK treatment, and the results showed that LASEK has a good effect and better postoperative experience compared with PRK.

According to the principles of LASEK, it not only optimizes PRK but also avoids some disadvantages of laser lamellar corneal refractive surgery such as LASIK. First, it avoids the risk associated with the corneal flap (e.g., free flap, broken flap, and button flap) made by microkeratome in LASIK. Second, it has less effect on the corneal nerve and less serious dry eye syndrome than LASIK. Third, LASEK creates less surgically induced wavefront aberration because of its thinner flap. In addition, it saves the cost of the microkeratome or femtosecond laser used in LASIK.

Nevertheless, LASEK has some common risks of surface ablation in high myopia correction, such as postoperative haze and side effects of corticosteroid eye drops required after surgery. What is more, LASEK is a complex surgical procedure that requires high surgical skill with a long learning curve for beginners.

2.4. Epipolis laser in situ keratomileusis (Epi-LASIK)

Epipolis laser in situ keratomileusis (Epi-LASIK) was first reported by the Greek doctor Pallikaris in 2003. Different from LASEK, Epi-LASIK uses a microkeratome instead of ethyl alcohol to bluntly separate the corneal epithelial from the Bowman layer. Therefore, Epi-LASIK avoids direct stimulation of alcohol and reserves the intact epithelial basement membrane, which results in good subjective feelings in patients as well as quick recovery and haze reduction.

Epi-LASIK surgery takes the activity of the flap as the core, which is crucial to the therapy effect. Compared with PRK, Epi-LASIK has an extra flap covering the corneal stromal bed, which can effectively protect the stroma below and also promotes corneal tissue healing. Epi-LASIK has different methods of flap creation than those of LASEK. Its flap basement membrane is intact, continuous integrity segments of stratum lucidum and a longer compact layer. The postoperative healing time of Epi-LASIK is shorter without alcohol stimulation or chemical damage to the corneal epithelium. Even though the flap of Epi-LASIK is closer to corneal natural state, the postoperative biomechanical change of the corneal flap and the effect on corneal healing are yet to be determined [4].

2.5. Transepithelial photorefractive keratectomy (Trans-PRK)

Transepithelial photorefractive keratectomy (Trans-PRK), where both the epithelium and stroma are removed in a single step, is a relatively new procedure of laser refractive error correction. It has become the focus in the refractive surgery field recently and the first choice to treat ametropia with or without an irregular cornea. The excimer laser has been developed to the sixth-generation lasers, targeting the goal of minimally invasive laser refractive surgery during the past 30 years. The latest laser system delivers more laser spots per second so as to reduce the treatment time [5]. Tran-PRK has brought a new revolution to excimer laser techniques for its bladeless surgery process [6] and superiority in efficiency and safety. Hence, it has been considered as the representative operation of the laser corneal surface refractive surgery, and it is a step closer to the perfect refractive surgery. Trans-PRK has been indicated in low to moderate myopia patients and a small number of high myopia patients (−8D below), but not in patients with a very low-degree myopia. Trans-PRK is currently regarded as an optimal safety choice for patients with a thin cornea. In addition, it is also the best choice for combat athletes, high-risk workers, and patients with ocular surgical history. For the second operation and patients with irregular corneas, who need synergistic or customized surgery, Trans-PRK may be the only choice.

2.5.1. Advantages and disadvantages

Compared with other corneal refractive surgeries, Trans-PRK has the advantages of no chemical toxicity or mechanical damage, no corneal incision, no negative pressure suction, and less risk of infection, and it avoids the potential damage caused by negative pressure suction.

Trans-PRK was considered as a kind of minimally invasive surgery for its optimal safety. Without creating a corneal flap, Trans-PRK significantly reduces the postoperative corneal biomechanical change, and it has little effect on the structure of the cornea, without corneal flap mark and flap-related complications. In addition, Tran-PRK offers faster epithelial healing, lower postoperative pain, and significantly less haze formation. Kaluzny et al. [7] used optical coherence tomography to supervise the epithelial recovery time and found that Trans-PRK (3 days) has a significantly shorter recovery time than PRK does (4 days). Aslanides et al. randomly selected 60 eyes of 30 myopic patients who had undergone conventional alcohol-assisted PRK in one eye and Trans-PRK in the other eye. The postoperative follow-up showed that Trans-PRK offers faster epithelial healing and 64% lower average pain scores. The haze level was consistently lower after Trans-PRK from 1 to 6 months [8].

Trans-PRK is considered a minimal complication, maximum security single-laser surgery at present. There is no significant different between Trans-PRK and other laser surgery in final visual acuity [9, 10]. Patient satisfaction is not as high with Trans-PRK compared with the femtosecond laser-assisted laser in situ keratomileusis (FS-LASIK) and small-incision lenticule extraction (SMILE) soon after surgery because of the slightly higher incidence of haze, pain, and slower recovery of visual acuity. However, Trans-PRK has good long-term postoperative satisfaction.

2.5.2. Visual outcomes and visual quality

Many studies have shown that [8, 9, 11] Trans-PRK has high-precision visual outcomes and good stability in ametropia correction. A large retrospective comparison of transepithelial PRK with LASEK, Epi-LASIK, and LASIK detected better visual outcomes with Trans-PRK for high myopia [12]. In hyperopia correction, the effectiveness showed a substantial increase over the previous study, but there are still difficulties, such as the relatively high rate of secondary operation, residual refractive error, surgically induced negative spherical aberration, and astigmatism. In astigmatism correction, static cyclorotation component and dynamic cyclorotation component techniques greatly improve the effectiveness.

With the development of refractive surgery technology, its safety is high. Increasingly, researchers have focused on improving postoperative visual quality. The assessment of the visual quality is divided into two main aspects. The subjective part includes vision acuity (near, intermediate, and distance) and contrast sensitivity, whereas the objective part includes the objective wavefront aberration, point spread function, modulation transfer function, visual quality scale, and so on. Research shows that topography-guided Trans-PRK can effectively correct the irregular astigmatism and improve the postoperative contrast sensitivity in patients. On the contrary, Trans-PRK simplifies the process, which reduces the potential higher order wavefront aberration caused by irregular ablation and decreases postoperative glare and night vision loss. Therefore, increasingly ophthalmologists choose Trans-PRK to improve postoperative visual quality.

Trans-PRK is the easiest laser refractive surgery for refractive surgeons to learn, and it is stable in techniques and cost-effective. As a result, it is considered an ideal corneal refractive operation for patients.

2.6. Excimer laser in presbyopia correction

Presbyopia is an age-related condition in which accommodation gradually decreases and the eyes are unable to focus and obtain clear near vision. There is one approach to correct presbyopia through laser refractive surgery commonly called PresbyLASIK. Synonymous with LASIK, PresbyLASIK contains the process of making a special ablation profile to either perform a multifocal cornea procedure or increase the depth of field.

PresbyLASIK has been described in three different approaches: central PresbyLASIK, peripheral PresbyLASIK, and transitional multifocality. In central PresbyLASIK, the central area is shaped for near vision and the mid-peripheral cornea is shaped for distant vision. On the contrary, in peripheral PresbyLASIK, the central area is shaped for distant vision and the mid-peripheral corneal area for near vision. Both central and peripheral techniques reportedly obtained adequate spectacle independence in both myopia and hyperopia. In addition, a neuroadaptation process is needed in peripheral techniques. The third approach combines depth of field increase and micro-monovision, which induces a certain degree of spherical aberration to each eye to increase the depth of field while making the nondominant eye slightly myopic [13].

Presbyopia remains the biggest challenge to be corrected: the mechanism of accommodation and the cause of presbyopia are complex to understand fully. Therefore, the efficacy of presbyopia correction even with the latest platform is still in dispute. However, there is enough scientific evidence to consider PresbyLASIK as a useful tool in presbyopia correction [14]. Epstein and Gurgos [15] reported that 89% hyperopia (25/28) patients and 91.3% myopia (94/103) patients who underwent peripheral PresbyLASIK were completely spectacle independent and with distance unaided visual acuity of 20/20 in 67.9% (19/28) in hyperopia patients and 70.7% (53/75) in myopia patients.

3. Femtosecond laser in ophthalmology

3.1. Characteristic of femtosecond laser

Femtosecond laser technology was first introduced by Dr. Kurtz in the early 1990s, and it has developed rapidly over the past two decades. "Laser power" is defined as energy delivered per unit time. The laser-related damage will reduce with the decrease of pulse duration. The femtosecond laser for ophthalmology works at 1053-nm wavelength with a very short pulse duration of 10^{-15} s, minimizing the collateral damage [16]. The accuracy is 5 μm, allowing high precision in ophthalmic operations [17]. The use of the femtosecond laser has revolutionized the modern ophthalmic surgery.

3.2. Femtosecond laser in situ keratomileusis (FS-LASIK)

The concept of the lamellar refractive procedure was first introduced by Barraquer in the early 1960s. In the 1990s, an excimer laser ablation-assisted lamellar procedure was developed, as the foundation of modern laser in situ keratomileusis (LASIK). Compared with photorefractive keratectomy (PRK), visual recovery is faster and visual outcome is rapidly stable after LASIK. Flap-associated complications and increased incidence of dry eye after surgery, however, affect the quality of life. The safety, precision, and predictability of the femtosecond laser have changed LASIK over recent years.

Flap formation is critical to a successful LASIK surgery. Improper flap geometry, decentration, irregular cut, and epithelial damage lead to a large number of complications. Over the past decades, mechanical microkeratome has been performed in LASIK because of its reliability and safety. However, complications such as incomplete flap, free flap, and buttonhole continue to plague surgeons. Furthermore, because of the instability of mechanical microkeratome, corneas may be too steep, too flat, or too thin even after a successful operation [18].

The femtosecond laser became available for LASIK flap formation approximately 10 years ago. It reduced the risk of the above-mentioned complications. With mechanical microkeratome, the flap is thinner centrally and thicker peripherally (meniscus-shaped flap), which increases the incidence of buttonhole perforation. The femtosecond laser allows thin and uniform flaps, which improves the stability, safety, and precision of the flaps. It can also create thinner flaps with minimum effects on stromal architecture. Flap centration, diameter, and thickness are also more precise in femtosecond-created flaps. Another advantage of the femtosecond laser is that it allows the surgeon to select the cutting angle, position, and diameter of the hinge, as well as the flap diameter and flap thickness, which may provide better flap stability and reduce clinical epithelial ingrowth.

Though femtosecond lasers reduce the incidence of complications such as buttonhole perforation, incomplete flap, free cap, and irregular cuts, there are still some specific limitations in FS-LASIK. FS-LASIK requires two laser platforms—one for flap creation (femtosecond laser) and another for stromal bed ablation (excimer laser)—which increases the time and cost of the laser procedure. Because of the response of corneal keratocytes to the energy and inflammatory responses of adjacent tissues to gas bubbles, patients may encounter photophobia, called transient light-sensitivity syndrome (TLSS), early after FS-LASIK. With the development of the femtosecond laser, it needs less energy for flap formation, and thus reduces the incidence of TLSS [19]. The presence of cavitation gas bubbles during FS-LASIK, which originate from stray laser pulses into the aqueous humor, can impede the eye tracker of the excimer laser. An increased rate of diffuse lamellar keratitis, a sterile inflammatory reaction, was also observed in FS-LASIK because of the higher flap interface inflammatory response to laser energy and gas bubbles. Revolution in femtosecond laser energy is expected to reduce the specific complications.

In our previous study, patients treated with a femtosecond laser showed better corneal regeneration than those with a microkeratome did. Because of its precision and predictability, the femtosecond laser makes a smoother flap and causes less damage to the corneal nerve [20]. However, different from others' opinions, our data showed that there were no significant differences in the tear meniscus parameters between the microkeratome and femtosecond laser [21].

3.3. Femtosecond lenticule extraction (FLEx)

A new approach called femtosecond lenticule extraction (FLEx) was introduced to correct myopia and astigmatism. FLEx does not require a microkeratome or an excimer laser. It uses only the femtosecond laser, which is more convenient than other procedures that require both excimer and femtosecond lasers. Two cuts (posterior and anterior) in the cornea are involved in the procedure, which thus create a lenticule that is ultimately removed. There is no significant difference between FLEx and conventional LASIK, both in efficacy and safety, which promotes FLEx to be a promising new corneal refractive procedure to correct refractive errors.

However, it has been reported that the visual outcome after FLEx was stable early after surgery, but visual recovery was slow [22]. As a corneal flap is necessary before lenticule extraction, associated complications such as dry eye and compromise of corneal biomechanical strength are inevitable. Therefore, the technique evolved into SMILE.

3.4. Small-incision lenticule extraction (SMILE)

SMILE, passing the Conformite Europeenne (CE) certification in 2009, is a novel technique to correct refractive errors. The procedure involves passing a dissector through a small incision to separate the lenticular interfaces and allow the lenticule to be removed, thus eliminating the need to create a flap [23]. Early or late complications associated with flaps, such as dislocations and buttonholes are avoided; therefore, patients' experience and visual outcomes improve.

The absence of flap creation minimizes the disruption of the stromal architecture because the corneal lenticule is extracted from the mid-stroma [24]. Fewer corneal nerve branches are disrupted compared with FS-LASIK, which preserves corneal biomechanical strength and maintains sensitivity. Thus, the risk of dry eye and patients' discomfort is reduced after surgery [25]. The minimal disruption of the anterior corneal surface epithelium, Bowman's layer, and anterior stroma is also associated with less risk of dry eye [26]. Laser fluence and difference in stromal hydration, which may affect stromal ablation, are avoidable in SMILE. Prospective and retrospective studies of SMILE have shown that in the terms of efficacy, predictability, and safety SMILE is similar to FS-LASIK.

However, there are still some difficulties in SMILE. Similar to a flap in FS-LASIK, a cap whose uniform regularity is essential to optimal visual outcome is created using a femtosecond laser.

In addition, the surface quality of the corneal lenticule can be irregular, causing tissue bridges, cavitation bubbles, scratches, or incomplete extraction of stromal lenticules [27]. At present, SMILE is mainly applied in mild myopia. It still needs more attempts and experience to be used in hyperopic eyes.

Complications such as epithelial erosion, suction loss, cap perforation, and lenticule extraction difficulty can all occur. Corneal haze, dry eye syndrome, keratitis, and interface inflammation have also been reported.

SMILE is a promising new technique for refractive error correction. With further development of the femtosecond laser, SMILE may gain greater acceptance in the future. However, we still have to pay more attention to its complications to verify its safety.

3.5. Femtosecond laser in presbyopia correction

3.5.1. Corneal inlay implantation

Corneal inlay implantation is performed for presbyopic correction. It changes the anterior corneal surface curvature, cornea refractive index, and depth of focus by placing a small inlay of suitable biocompatible material within the stroma. The benefits of inlays are the reversibility of the procedure by removal of the implant, implantation simplicity, and implant repositioning. The ability to perform ad hoc refractive procedures allows the simultaneous correction of ametropia. The most common complications after the surgery are glare and dry eye. The femtosecond laser advances flaps and tunnel creation to implant the inlays accurately on the line of sight, and thus result in remarkable improvements in uncorrected visual acuity (near and intermediate) and minimal change in uncorrected distance visual acuity. This procedure also provides good nonspectacle-corrected near vision for average daily activities [18, 28].

3.5.2. IntraCor surgery

IntraCor surgery is a new technique applicable for ametropic or low-degree hyperopic eyes (+0.5 to +1.5 D). It changes the topographic and refractive characteristics of the central portion of the cornea selectively with femtosecond laser. The procedure involves concentric intrastromal ring creation in the central portion of the cornea at different corneal depths (between the Bowman's and Descemet's boundaries). Other than the treatment failure in presbyopic correction, no major complications have been reported. IntraCor surgery is promising in the field of presbyopia correction [18].

3.6. Astigmatic keratotomy (AK)

To correct low to high astigmatism, astigmatic keratotomy (AK) is performed. The accuracy of length, optical zone, and incision depth are crucial in visual outcomes. However, the limitation of AK is the unpredictability of manual corneal incision. The application of femtosecond laser in AK provides accuracy and precision in corneal incision manufacturing and thus improves significantly in uncorrected and best-corrected visual acuity. The femtosecond laser can also be applied in case of high degrees of post-keratoplasty astigmatism. The incisions within the graft button present precise geometry and reliable depth of incision. Specific complications associated with femtosecond laser-assisted AK such as self-healing micro-corneal perforations and low-grade inflammation at the incision site appeared [18, 29].

3.7. Intracorneal ring segments

Intracorneal ring segments were small and curved when first proposed in 1978. Clear ring segments made of polymethylmethacrylate are implanted in the deep corneal stroma with the aim of generating modifications of corneal curvature and refractive changes. Peripheral intracorneal implantation has been permitted to correct low to moderate astigmatism and myopia and keratoconus by Food and Drug Administration (FDA). Complications such as incomplete tunnel formation, corneal perforation, endothelial perforation, corneal melting, and uneven implant placement may occur with the traditional technique. The femtosecond laser can be programmed to create corneal channels at a specific depth and orientation with high predictability and precision to allow safer insertion of Intacs segments. In patients with keratoconus, the femtosecond laser can be programmed to cut tunnels for the implantation of intracorneal ring segments, and it results in better safety owing to greater consistency of depth and uniformity [29, 30].

3.8. Penetrating keratoplasty (PKP)

Penetrating keratoplasty (PKP) developed rapidly after first being introduced in the early 1900s. The surgical outcomes rely on a centered and perpendicular cut of cornea, a well-matched donor button, and a recipient bed [18]. Manual PKP requires a long learning curve and a lengthy procedure time, which can be optimize using the femtosecond laser. The femtosecond laser can also achieve a higher precision in surgical steps, such as the donor cornea cutting. Moreover, the choice of shapes and diameters in femtosecond laser-assisted PKP is dependent on individualized clinical requirement. It enables advanced shaped corneal cuts creation, eliminates manual dissection, thus minimizing misalignments, and increases the stability of the wound. Some pattern of incisions that are not compassable with conventional technique can be achieved by femtosecond laser. However, postoperative regular and irregular astigmatism remain a major challenge in full-thickness keratoplasty.

3.9. Anterior lamellar keratoplasty (ALK)

Anterior lamellar keratoplasty (ALK) is a partial thickness corneal transplantation indicated for management of anterior corneal dystrophies degenerations, ulcers, and scars. The advantages of ALK over PKP include being less invasive and having a decreased rate of rejection. Femtosecond laser-assisted suture less anterior lamellar keratoplasty (FALK), first described in 2008, has been reported to be safe, effective, and stable. The femtosecond laser has reduced irregular astigmatism and accelerated visual recovery by its precision of pre-programmed corneal dissections at a variety of depths and orientations. As the corneal incision is well shaped, it can be converted to full-thickness keratoplasty in case of Descemet membrane perforation. The donor and recipient tissue are better positioned because of highly precise cuts assisted by the femtosecond laser, and thus sutures are typically removed earlier.

Limitations of ALK include the high cost and the slow growth of the epithelium over the graft [18, 30].

3.10. Femtosecond laser-assisted endothelial keratoplasty

3.10.1. Descemet's stripping endothelial keratoplasty (DSEK)

DSEK has been the preferred approach for treating corneal endothelial diseases, such as bullous keratopathy, Fuchs dystrophy, congenital hereditary endothelial dystrophy, and endothelium failure in previous penetrating keratoplasty. It removes the diseased endothelium and leaves the posterior cornea intact. However, technical challenges always afflict corneal surgeons. The application of the femtosecond laser alleviates certain difficulties. Significant advantages over manual dissection are that the femtosecond laser allows for increased automation and standardization in donor tissue preparation. In addition, lamellar interface preparation can be performed up to 3 weeks before the surgery, making it feasible for conversion to PKP owing to complications during preparation of the donor disk. It also creates a smoother donor-recipient interface to minimize induced refractive astigmatism. However, whether the femtosecond laser has an effect on endothelial cell loss and visual acuity after donor tissue preparation has yet to be determined [18, 30].

3.10.2. Descemet's membrane endothelial keratoplasty (DMEK)

Descemet's membrane endothelial keratoplasty (DMEK) has been proven to result in faster visual recovery, fewer higher order aberrations, and lower rejection rates by abundant evidence. The crucial technique is selective endothelial transplantation. The femtosecond laser offers a precise and predictable means to create the uniform thickness of the posterior stromal rim and control big bubble expansion that cannot be achieved by manual operation. The femtosecond laser not only avoids energy-associated damage but also results in smooth stromal interface by closer spots, line separations, and a low energy level. Thus, the femtosecond laser can be a novel approach for the donor grafts preparation of DMEK, which may reduce intraoperative graft manipulation and postoperative detachments [29, 31].

3.11. Cataract surgery

At 2008, the femtosecond laser was first performed in Europe in cataract surgery and was approved by the FDA in 2010. It was applied in the steps of corneal incision, arcuate corneal incisions, capsulorhexis, lens fragmentation, and liquefaction. The short pulses duration (10^{-15} s) make it a promising tool in cataract surgery [32].

3.11.1. Corneal incisions

Clear self-sealing corneal incision is of most importance in cataract surgery. In terms of length and tunnel structure, manual incision is difficult to control. In addition, bacteria have a chance to enter at low intraocular pressure because of the instability of manual incision and thus lead to endophthalmitis. Compared with manual incisions, corneal incisions made using the femtosecond laser are more precise in width, depth, and length. More consistency in the architecture is also achieved, which leads to better incision sealing without stromal hydration at the end of the surgery. The stability of the wound makes it more resistant to deformation and leakage. The femtosecond laser is used to create a clear corneal incision according to pre-programming, which requires a large amount of patient data to confirm, and definitive results. In cases of corneal astigmatism, an arcuate or a relaxing incision can be created, which has been reported to provide more stable and accurate long-term outcomes compared with toric intraocular lenses (IOLs). The femtosecond laser shows less damage by virtue of its construction and reduced mechanical stress during surgery, which may decrease corneal swelling after surgery.

3.11.2. Capsulotomy

A precise size and centration capsulorhexis are essential to optimize the IOL position. The IOL's longitudinal displacement per millimeter will lead to approximately 1.25-D refractive change, inducing myopia for an anterior shift and hyperopia for a posterior displacement. Capsulorhexis size is correlated with effective lens position. An insufficient overlap of the IOL, results in decentration, oblique astigmatism, and increased higher order aberration. A small capsulorhexis has been associated with anterior capsule fibrosis. Manual continuous curvilinear capsulorhexis (CCC) relies on the technique and proficiency of the surgeon. The femtosecond laser can be used to create a more precise, better-sized, and centered opening of the anterior capsule compared with the conventional CCC by dissecting it with a spiral laser pattern. Compared with the manual CCC group, the laser CCC group showed more accuracy and stability in anteroposterior and central IOL positioning. It is also more predictable in the refractive outcome, which is more important to patients with high expectation [33, 34].

3.11.3. Lens fragmentation and liquefaction

The increased energy for lens fragmentation and liquefaction delivered from the phacoemulsification probe to the eye can result in energy-associated capsule complications and corneal endothelial cell injury in manual phacoemulsification. The reduction in ultrasound energy can decrease the risk of such complications. It is reported that the femtosecond laser reduced the ultrasonic energy delivered during phacoemulsification significantly. However, its effect on endothelial damage is still unknown.

3.11.4. Limitation and complications

Although the femtosecond laser showed excellent advantages over manual operation in cataract surgery, there are still some inevitable limitations and complications.

1.   Additional operating room shifting time

     Patients need to be shifted under the operating microscope after application of laser treatment. This logistical issue results in increased time spent with each patient, which could lead to overall delay.

2. Applicability

The femtosecond laser relies on good anterior chamber imaging. Patients with poor eyelid opening, nystagmus, poor pupillary dilatation, corneal opacities, and ocular surface disease are poor candidates. It is also not suitable for patients with tremors or dementia in the initial docking system. The femtosecond laser is not available for grade 4 cataract according to the Lens Opacities Classification System III (LOCS III).

3. Complications

Capsular blockage syndrome: Large diameter hydrodissection cannula with high-speed fluid may inhibit a gas bubble that is formed from leaving the nucleus. Pressure elevation between the capsule leads to the rupture of the posterior capsule, and the lens may drop into the vitreous cavity. This is a learning curve-related complication, which can be avoided by a more cautious and skilled surgeon

Pupillary constriction: Bubble formation and suction force can trigger pupillary constriction by releasing small amounts of free radicals. In addition, a delay between femtolaser pretreatment and cataract surgery may result in pupil diameter changes (5–10 min is recommended).

Corneal incision sizing and positioning the initial docking of the laser ring is crucial to the accuracy of the intended femtolaser-created incision. Imperfect interface positioning causes inaccuracy of corneal incision sizing and positioning, which results in surgically induced astigmatism and complication in the manual final procedure [18, 35].

4. Summary

The surgery has gradually matured with the development of the modern excimer laser equipment and technology. Upgrading thermal control and ablation centration will achieve better corneal biomechanical results. We have reason to believe that a single excimer laser surgery would be the next breakthrough in refractive surgery, the next step toward the "perfection" of refractive surgery. Femtosecond laser advancements over the past two decades have brought revolutionary change in ophthalmic practice. It not only improves the safety and efficacy in corneal refractive surgery but also achieves remarkable advancement in the field of cataract surgery. Despite having certain limitations and complications, femtosecond lasers are promising in ophthalmology. Both excimer and femtosecond laser technology will serve for better human health as long as we apply it appropriately.

Author details

Liang Hu*, Yiqing Huang and Meng Lin

*Address all correspondence to: liang_hu@live.cn

School of Ophthalmology and Optometry, Wenzhou Medical University, Wenzhou, China

References

[1] Krueger RR, Rabinowitz YS,Binder PS. The 25th anniversary of excimer lasers in refractive surgery: historical review. J Refract Surg. 2010, 26(10): 749–760. 10.3928/1081597X-20100921-01

[2] Wagoner MD, Wickard JC, Wandling GR, Milder LC, Rauen MP, Kitzmann AS, Sutphin JE,Goins KM. Initial resident refractive surgical experience: outcomes of PRK and LASIK for myopia. J Refract Surg. 2011, 27(3): 181–188. 10.3928/1081597X-20100521-02

[3] O'Brart DPS. Excimer laser surface ablation: a review of recent literature. Clin Exp Optom. 2014, 97(1): 12–17. 10.1111/cxo.12061

[4] Pallikaris IG, Kalyvianaki MI, Katsanevaki VJ,Ginis HS. Epi-LASIK: Preliminary clinical results of an alternative surface ablation procedure. J Cataract Refract Surg. 2005, 31(5): 879–885. DOI 10.1016/j.jcrs.2004.09.052

[5] El Bahrawy M,Alio JL. Excimer laser 6(th) generation: state of the art and refractive surgical outcomes. Eye Vis (Lond). 2015, 2: 6. 10.1186/s40662-015-0015-5

[6] Wang DM, Du Y, Chen GS, Tang LS,He JF. Transepithelial photorefractive keratectomy mode using SCHWIND-ESIRIS excimer laser: initial clinical results. Int J Ophthalmol. 2012, 5(3): 334–337. 10.3980/j.issn.2222-3959.2012.03.16

[7] Kaluzny BJ, Szkulmowski M, Bukowska DM, Wojtkowski M. Spectral OCT with speckle contrast reduction for evaluation of the healing process after PRK and transepithelial PRK. Biomed Opt Express. 2014, 5(4): 1089–1098. 10.1364/BOE.5.001089

[8] Aslanides IM, Padroni S, Arba Mosquera S, Ioannides A,Mukherjee A. Comparison of single-step reverse transepithelial all-surface laser ablation (ASLA) to alcohol-assisted photorefractive keratectomy. Clin Ophthalmol. 2012, 6: 973–980. 10.2147/OPTH.S32374

[9] Fadlallah A, Fahed D, Khalil K, Dunia I, Menassa J, El Rami H, Chlela E,Fahed S. Transepithelial photorefractive keratectomy: clinical results. J Cataract Refract Surg. 2011, 37(10): 1852–1857. 10.1016/j.jcrs.2011.04.029

[10] Celik U, Bozkurt E, Celik B, Demirok A,Yilmaz OF. Pain, wound healing and refractive comparison of mechanical and transepithelial debridement in photorefractive keratectomy for myopia: results of 1 year follow-up. Contact Lens Anterior Eye. 2014, 37(6): 420–426. 10.1016/j.clae.2014.07.001

[11] Abdulaal MR, Wehbe HA,Awwad ST. One-step transepithelial photorefractive keratectomy with mitomycin C as an early treatment for LASIK flap buttonhole formation. J Refract Surg. 2015, 31(1): 48–52. 10.3928/1081597X-20141104-01

[12] Kanitkar KD, Camp J, Humble H, Shen DJ,Wang MX. Pain after epithelial removal by ethanol-assisted mechanical versus transepithelial excimer laser debridement. J Refract Surg. 2000, 16(5): 519–522.

[13] Pallikaris IG,Panagopoulou SI. PresbyLASIK approach for the correction of presbyopia. Curr Opin Ophthalmol. 2015, 26(4): 265–272. 10.1097/Icu.0000000000000162

[14] Alio JL, Amparo F, Ortiz D,Moreno L. Corneal multifocality with excimer laser for presbyopia correction. Curr Opin Ophthalmol. 2009, 20(4): 264–271. 10.1097/ICU.0b013e32832a7ded

[15] Epstein RL,Gurgos MA. Presbyopia treatment by monocular peripheral PresbyLASIK. J Refract Surg. 2009, 25(6): 516–523. 10.3928/1081597X-20090512-05

[16] Soong HK,Malta JB. Femtosecond lasers in ophthalmology. Am J Ophthalmol. 2009, 147(2): 189–197 e182. 10.1016/j.ajo.2008.08.026

[17] Lubatschowski H, Maatz G, Heisterkamp A, Hetzel U, Drommer W, Welling H,Ertmer W. Application of ultrashort laser pulses for intrastromal refractive surgery. Graefes Arch Clin Exp Ophthalmol. 2000, 238(1): 33–39.

[18] Aristeidou A, Taniguchi EV, Tsatsos M, Muller R, McAlinden C, Pineda R, Paschalis EI. The evolution of corneal and refractive surgery with the femtosecond laser. Eye Vis (Lond). 2015, 2: 12. 10.1186/s40662-015-0022-6

[19] Stonecipher KG, Dishler JG, Ignacio TS,Binder PS. Transient light sensitivity after femtosecond laser flap creation: clinical findings and management. J Cataract Refract Surg. 2006, 32(1): 91–94. 10.1016/j.jcrs.2005.11.015

[20] Liang Hu, Wenjia Xie, Lei Tang, Jia Chen, Dong Zhang, Peng Yu, Jia Qu. Corneal subbasal nerve density changes after laser in situ keratomileusis with mechanical microkeratome and femtosecond laser. Chin J Ophalmol, 2015, 51(1):39–44. 10.3760/cma.j.issn.0412-4081.2015.01.010

[21] Xie W, Zhang D, Chen J, Liu J, Yu Y,Hu L. Tear menisci after laser in situ keratomileusis with mechanical microkeratome and femtosecond laser. Invest Ophthalmol Vis Sci. 2014, 55(9): 5806–5812. 10.1167/iovs.13-13669

[22] Shah R,Shah S. Effect of scanning patterns on the results of femtosecond laser lenticule extraction refractive surgery. J Cataract Refract Surg. 2011, 37(9): 1636–1647. 10.1016/j.jcrs.2011.03.056

[23] Reinstein DZ, Archer TJ,Gobbe M. Small incision lenticule extraction (SMILE) history, fundamentals of a new refractive surgery technique and clinical outcomes. Eye Vis (Lond). 2014, 1: 3. 10.1186/s40662-014-0003-1

[24] Zhang Y, Chen YG,Xia YJ. Comparison of corneal flap morphology using AS-OCT in LASIK with the WaveLight FS200 femtosecond laser versus a mechanical microkeratome. J Refract Surg. 2013, 29(5): 320–324. 10.3928/1081597X-20130415-03

[25] Demirok A, Ozgurhan EB, Agca A, Kara N, Bozkurt E, Cankaya KI,Yilmaz OF. Corneal sensation after corneal refractive surgery with small incision lenticule extraction. Optom Vis Sci. 2013, 90(10): 1040–1047. 10.1097/OPX.0b013e31829d9926

[26] Denoyer A, Landman E, Trinh L, Faure JF, Auclin F,Baudouin C. Dry eye disease after refractive surgery: comparative outcomes of small incision lenticule extraction versus LASIK. Ophthalmology. 2015, 122(4): 669–676. 10.1016/j.ophtha.2014.10.004

[27] Dong Z,Zhou X. Irregular astigmatism after femtosecond laser refractive lenticule extraction. J Cataract Refract Surg. 2013, 39(6): 952–954. 10.1016/j.jcrs.2013.04.016

[28] Charman WN. Developments in the correction of presbyopia II: surgical approaches. Ophthalmic Physiol Opt. 2014, 34(4): 397–426. 10.1111/opo.12129

[29] Liu HH, Hu Y,Cui HP. Femtosecond laser in refractive and cataract surgeries. Int J Ophthalmol. 2015, 8(2): 419–426. 10.3980/j.issn.2222-3959.2015.02.36

[30] Mian SI,Shtein RM. Femtosecond laser-assisted corneal surgery. Curr Opin Ophthalmol. 2007, 18(4): 295–299. 10.1097/ICU.0b013e3281a4776c

[31] Jardine GJ, Holiman JD, Galloway JD, Stoeger CG, Chamberlain WD. Eye bank-prepared femtosecond laser-assisted automated descemet membrane endothelial grafts. Cornea. 2015, 34(7): 838–843. 10.1097/ICO.0000000000000453

[32] Ali MH, Javaid M, Jamal S,Butt NH. Femtosecond laser assisted cataract surgery, beginning of a new era in cataract surgery. Oman J Ophthalmol. 2015, 8(3): 141–146. 10.4103/0974-620X.169892

[33] Friedman NJ, Palanker DV, Schuele G, Andersen D, Marcellino G, Seibel BS, Batlle J, Feliz R, Talamo JH, Blumenkranz MS,Culbertson WW. Femtosecond laser capsulotomy. J Cataract Refract Surg. 2011, 37(7): 1189–1198. 10.1016/j.jcrs.2011.04.022

[34] Abouzeid H,Ferrini W. Femtosecond-laser assisted cataract surgery: a review. Acta Ophthalmol. 2014, 92(7): 597–603. 10.1111/aos.12416

[35] Nagy ZZ. New technology update: femtosecond laser in cataract surgery. Clin Ophthalmol. 2014, 8: 1157–1167. 10.2147/OPTH.S3

Chapter 17

Diode Laser-Based Sensors for Extreme Harsh Environment Data Acquisition

Chayan Mitra and Rachit Sharma

Additional information is available at the end of the chapter

http://dx.doi.org/10.5772/63971

Abstract

The world has witnessed several step changes in living standards, productivity, growth, and innovation. We are currently witnessing a convergence of intelligent devices, intelligent networks, and intelligent decision making. Obtaining long-term accurate, in situ, and real time data from the machines is necessary for enabling the industrial Internet. This relies heavily on sensor systems. Development of robust sensors that can operate reliably in extreme environments will make it possible to gather data from previously inaccessible locations in the equipment. This will enable machine operators to monitor and optimize the performance of their machines. Diode laser-based diagnostics technology has found applications in a variety of areas and a versatile range of operating conditions. It has proven to be a strong and reliable technique for remote measurements of concentrations and temperatures in harsh environments. Some of the major challenges for implementation of these sensors in real world are machine vibrations, window clogging, cooling, etc. In this chapter, the authors discuss about the application details and specific technologies suitable for the applications. Few case studies are considered, and the theoretical approach, algorithm development, and experimental validation are also discussed.

Keywords: lasers, sensor, harsh environment, tunable diode laser, industrial Internet, optical sensor

1. Introduction

Prior to late eighteenth and early nineteenth century, the human society was primarily agrarian and rural. Then, a step change of innovation occurred: the Industrial Revolution, which gave us a shift from hand tools to steam engine, the internal combustion engine,

telegraph, telephone, and electricity. Productivity and economic growth accelerated sharply [1]. A key upside of the feature characterizing this period was that it harnessed the efficiencies of hierarchical structures, with centralization of controls. It led to reduction in cost as the machines and fleets got larger and production volumes increased. A major downside was that it was more resource intensive. Much of the incremental innovation at later stages was focused on improving efficiency, reducing waste, and enhancing the working environment. The second wave of Industrial Revolution started with the advent of the electronics, computers, and Internet. Here, the key feature was the design of standards and protocols to permit incompatible machines in diverse locations to connect and exchange information. The explosive growth was a result of the combination of speed and volume. Deeper integration, flexible operation, and distributed intelligence led to the creation of new platforms for commerce and social exchange by driving down the cost. Ability for rapid exchange of information and decentralized decision-making process led to open innovation and knowledge-intensive growth. Today, we are witnessing another transformation by melding of the global industrial system with open computing and communication system. The industrial Internet is enabled by the coming together of intelligent devices, intelligent systems, and intelligent decision-making systems [1]. The architecture consists of three technology elements: brilliant machines, advanced analytics, and people at work. A brilliant machine is self-aware of its performance, health condition, and capability. This enables the machine to operate close to its performance boundary. The machine communicates with other machines and operators or service personnel through the Internet.

The first step in this revolution is the generation of data from the machines assuring the real-time health condition information of the machine. Widespread instrumentation of the machines is a necessary factor in this case. A suite of sensors on each machine will enable performance monitoring on a real-time basis and help the operators make the most of their assets [2]. The challenge comes from the increasing technical complexity of the assets in service. Performance data from a sensor located in an unmonitored location in a machine along with powerful software analytics and visualization tools will enable the operator to diagnose the problem with greater confidence. **Table 1** provides a breakup of the value opportunity of industrial Internet for various industry segments [3].

Industry segment	Global base	Capacity	Labor-hours/Year	Value
Power	56,620 (Gas + Steam Turbines)	4156 GW	52M	$ 7B
Aviation	21,500 Commercial Jet Aircraft	43,000 Jet engines	205M	$ 10B
Rail	120,000 Freight	7M People + 9.6T Freight tonne-km	52M	$ 3B
Healthcare	105,000 (CT + MRI Machines)		4M	$ 250M

Table 1. Value opportunity of industrial Internet for various industry segments [3].

Similarly, in the oil and gas sector, reduction in asset downtime (asset performance management) and operations optimization through predictive analytics and condition-based maintenance can result in substantial cost savings for the oil well owners. For example, the cost for surfacing a blowout preventer (BOP) from the seabed is around $10–$16 million and unplanned downtime costs a mid-size LNG facility $150 million per year [4]. Most of the downtime in deep sea drilling rig is caused by BOP-related problems, and it costs an oil company more than $ 1 million per day in lost productivity [5].

Therefore, nowadays when major industrial products, such as gas and steam turbines, aircraft engines, turbomachinery equipment, power transformers, and locomotives, are involved, the primary challenge is to keep the systems operating at peak performance to avoid unwarranted shutdowns. Continuous operation at peak performance not only demands high-fidelity system architecture and design, but also requires optimized operation and maintenance practices. This in-turn necessitates the usage of online sensor systems that can perform desired measurements for continuous monitoring of operational performance and overall system health. The idea is that measurements using multiple sensors in combination with environmental, operational, and performance-related parameters can provide a more accurate system health status. The sensor data can also be used along with statistical pattern recognition and machine-learning techniques to detect changes in machine parameter data, isolate faults, and estimate the remaining useful life (RUL) of the machines [6–9]. This approach assumes a product's loading and operating conditions, geometry, material properties, and failure mechanisms as the parameters to estimate RUL. The sensor systems are used to monitor these parameters for anomalies, faults, and failure predictions [10].

In an example provided in GE Report [11], the operators of Whitegate Power Station near Cork in Ireland placed more than 140 sensors throughout the plant. This allowed the operators to run the plant reliably and efficiently through round-the-clock monitoring and diagnostics of the plant. The sensors digitized the critical plant operating parameters (vibration, temperature, pressure, fuel mix, ambient temperature, and load) and helped to create a virtual dynamic model of the asset in the cloud, which is a mirror of the real asset. The model in conjunction with the sensor data gives us the ability to predict the plant's performance, understand trade-offs (adjust hedging strategies to manage fuel cost volatility), and enhance efficiency. The modeling approach discussed above requires directly sensed parameters, design parameters, and operating condition uncertainties, as well as inspection and historical reliability data. Several techniques, such as stochastic models (which take into account randomness of the operating profiles, extreme operating events), physics-based models, neural networks, or real-time probabilistic models, are used for this purpose. To a large extent, the integrity of the measured parameters determines the fidelity of the models used [9].

It is obvious from the above discussion that robust and reliable online sensors that can accurately measure the desired system parameters are crucial toward optimizing asset performance and maximizing its lifetime. Most of the above-mentioned industrial assets, such as gas turbines or aircraft engines, involve extreme harsh environments such as high temperatures, high pressures, vibrations, shocks, dust and soot load, reacting flow, and thermal transients. Several industrial challenges and applications can be addressed through sensing of parameters under these harsh conditions. Conventional sensors do not work reliably here (or even fail to perform) because these harsh conditions often lie outside the operational envelope of traditional techniques. Therefore, development of new and advanced harsh environment sensors is becoming increasingly important because such sensors can enable the industrial community to get enhanced value out of their assets. When classified broadly, harsh environment sensors serve the following key application areas: process optimization (or controls), prognostics/health management, better machine design, and monitoring/diagnostics. For all these applications, it is beneficial to have on-line/real-time, accurate, selective, and direct measurement of the harsh environment parameters. In addition, high measurement repeatability and ease of installation and maintenance are extremely desirable. As described in this chapter, optical harsh environment sensors can provide these major advantages over conventional techniques for a host of industrial applications.

Technology	Laser type	Detectors	Measured sample	Information
Laser-induced fluorescence	CW/pulsed, low power	Camera	Fluorescent molecules	Species concentration
Laser-induced incandescence	Pulsed, high power	Camera	Soot particles	Soot distribution
Laser Doppler anemometry	CW, low power	Photodiodes	Particles and droplets	Local flow velocity
Particle imaging velocimetry	Pulsed/modulated, high power	Camera	Seeding particles, fuel droplets	Flow field
Phase Doppler anemometry	CW, low power	Photodiodes	Droplets/Particles	Droplet/particle size
Laser-induced breakdown spectroscopy	Pulsed, high power	Spectrometer	Molecules	Elemental composition
Raman scattering	CW/pulsed, high power	Photodiodes/ spectrometer	Molecules	Concentration, temperature
Laser absorption spectroscopy	CW/pulsed, low power	Photodiode	Molecules	Concentration, temperature

Table 2. Overview of common optical techniques for industrial-sensing applications [12].

Laser-based optical sensors provide a unique method for measurement of fluid properties in industrial environments. Typical applications include flow or velocity measurement through techniques such as particle imaging velocimetry (PIV) or laser Doppler anemometry (LDA) and particle size measurement using phase Doppler anemometry (PDA). Laser-induced breakdown spectroscopy (LIBS) is commonly used where elemental composition measurement is required. Furthermore, for applications requiring concentration measurements, laser-induced fluorescence (LIF), Raman scattering, or laser absorption spectroscopy (LAS) is used depending on the type of sample. For fluorescing samples, LIF is preferred. On the other hand, Raman scattering is beneficial for measurements on di-atomics (H_2, N_2) or for analyzing gases at high pressures. Finally, LAS is used for applications where high sensitivity and selectivity are crucial. A brief snapshot of the various laser-based industrial-sensing techniques is provided in **Table 2**, and the reader is encouraged to refer the literature [12] for more details on these techniques. To remain within scope, this chapter will remain focused on absorption-based techniques, especially diode laser sensing, which is most promising for harsh environment applications.

Common industrial applications of conventional absorption-based optical sensors include gas monitoring, sensing, and analysis through techniques, such as Fourier transform infrared (FTIR), ultraviolet spectroscopy (UV), nondispersive infrared (NDIR), and photo-acoustic spectroscopy (PAS) [13, 14]. For harsh environment applications, such as power generation and energy systems, these techniques typically require gas extraction to condition the sampled gas. This often leads to unwanted lag in the measurement and requires frequent maintenance of the sampling system [15]. For applications requiring fast response time, high resolution, and good selectivity, such as industrial controls and process optimization, LAS has immense potential that is why it is an active area of research across the industrial and academic community. Tunable diode laser absorption spectroscopy (TDLAS) [16] and quantum cascade laser absorption spectroscopy (QCLAS) [17] are the two most common modalities in which LAS-based optical solutions can be implemented in harsh environments. This is mainly because these techniques, as discussed later in the chapter, can be implemented using economical, robust, and compact diode lasers/quantum cascade lasers that are specifically designed for the required application. In addition, diode lasers/quantum cascade lasers can be mass produced, require minimal maintenance, and have long operation lifetimes (>20,000 h) [18]. In the past decade, the industrial community has been increasingly adopting novel technology solutions based on these lasers. This is acting as the driving force behind the rapid advancement of the diode laser manufacturing industry toward lower cost, higher performance, and increased reliability. As this trend continues, these lasers will become even better and cheaper in the future, which will open new avenues toward novel and affordable optical solutions to today's unsolved challenges. Therefore, diode laser-based sensing techniques, such as TDLAS and QCLAS, are of utmost importance to the industrial community.

TDLAS/QCLAS-based sensors have immense potential and advantages for in situ measurements of concentration of constituents, temperature, and other wide varieties of gas parameters in challenging real-world environments [19–21]. In most of these applications, light emitted from a tunable diode laser system is passed through a gaseous medium to a detector. The transmitted radiation is then used to measure the gas temperature, species concentration, or pressure using spectral absorption models for the target absorbing species [22]. When implemented in line-of-sight [23, 24] or standoff configuration [25], these techniques can offer true in situ measurement capability in harsh environments with high temperatures or pressures. This is because the sensor can make reliable measurements through one or more transmitting windows while being completely decoupled from the harsh environment.

This chapter discusses the technology background of TDLAS from an applied experimental perspective. The two most common methodologies, that is, direct absorption spectroscopy (DAS) and wavelength modulation spectroscopy (WMS), will be covered. Subsequently, the key design philosophy, optomechanical architecture, and instrumentation of a TDLAS sensor will be presented. Finally, two harsh environment application examples will be provided to demonstrate the power of diode laser sensing toward solving complex real-world challenges. Please note that the implementation of QCLAS is essentially very similar to TDLAS, the only significant difference being that QCLAS uses mid-infrared quantum cascade lasers while TDLAS uses near-infrared tunable diode lasers. For the sake of simplicity, the term TDLAS is used in most places in this chapter but the concepts presented translate directly to QCLAS as well.

2. Technology background: tunable diode laser absorption spectroscopy

The fundamentals of molecular spectroscopy, including concepts such as vibration modes, absorption coefficient, line-shape, spectral width, and spectral broadening have been extensively studied and discussed in several books and articles [26, 27]. This section is presented from an applied experimental perspective and will use the aforementioned concepts assuming that the reader is equipped with basic understanding of molecular spectroscopy.

The absorption of optical radiation by gaseous medium is governed by the Beer–Lambert law. The law describes the optical transmission losses associated with a uniformly absorbing medium. When a narrow spectral radiation of frequency υ passes through a gaseous medium of length L [cm], the incident intensity I_0 and the transmitted intensity I_1 are related as,

$$I_1 = I_0 e^{(-k_\upsilon L)} \tag{1}$$

where k_υ [cm^{-1}] is the spectral absorption coefficient. It should be noted that for a mixture in pure gaseous phase (devoid of particulates, water droplets or condensed phases), the Beer–Lambert law assumes the optical scattering of the medium to be negligible. The spectral absorption coefficient k_υ is defined as below for an isolated (interference-free) vibrational transition,

$$k_\upsilon = PcS_i(T)\varphi_\upsilon \tag{2}$$

where P [atm] is the total gas pressure, c is the mole fraction of the species of interest, $S_i(T)$ [cm^{-2} atm^{-1}] is the temperature dependent line strength of the transition at temperature T [K], and φ_υ is the normalized line-shape function, such that

$$\int_\infty^{-\infty} \varphi_\upsilon d\upsilon = 1 \tag{3}$$

Using the above equations, the concentration c of the species of interest can be calculated.

The quantity $k_\upsilon L$, called absorbance, is of critical importance in deciding the capability and performance of a tunable diode laser-based gas sensor. For typical laser-based gas-sensing applications, trace concentrations need to be detected, and therefore, the absorbance to be detected is $\ll 1$. In such cases, the Eq. (1) reduces to

$$\frac{I_0 - I_1}{I_0} = \frac{\Delta I}{I_0} \sim k_\upsilon L \tag{4}$$

The quantity $\Delta I/I_0$ is a known as the fractional absorbance. The minimum fractional absorbance detectable by a TDLAS system is known as the minimum detectable absorbance (MDA). For a given TDLAS system, MDA is characteristically a function of the different system-level noises (such as laser excess noise, detector thermal noise, optical interference noise, or etalon noise) and does not depend on the species to be measured with that system. For a given path length, the MDA of the system can be used to calculate the minimum detectable concentration (or detection limit) for a measurable species using Eqs. (2) and (4). For most practical TDLAS applications, the MDA is around 10^{-5} to 10^{-6} and is often limited by the optical interference noise or etalon noise of the system [13, 28]. The two most common methodologies in which TDLAS is implemented are direct absorption spectroscopy (DAS) and wavelength modulation spectroscopy (WMS), which are discussed in the following subsections.

2.1. Direct laser absorption spectroscopy (DLAS/DAS)

A key requirement for both DAS and WMS is that the laser source must have a spectral width much narrower (at least 1–2 orders of magnitude) than the gas absorption feature to be measured. Distributed feedback diode lasers (DFB) in the near-infrared (NIR) and quantum cascade lasers (QCL) in the mid-infrared (MIR) can meet this requirement and serve as excellent sources for a majority of applications [18]. These lasers are generally available in both pulsed and continuous wave (cw) modes. The emitted wavelength of a diode laser is a function of the diode temperature and the injection current. Typically, a thermoelectric controller (TEC) is used to set (and control) the diode-operating temperature to a value where there desired wavelength can be reached at the desired injection current. For implementation of DAS [29], the injection current of the diode laser is scanned periodically in a sinusoidal, ramp, or sawtooth fashion. This leads to a related wavelength scanning of the laser. The scan current range has to be selected such that the resulting wavelength scan covers the absorption transition of interest. Typical scan frequencies are in the range of 5–200 Hz. It is highly advisable to use a wavelength-appropriate etalon to characterize the current–wavelength transfer function of the laser at the scan frequency [30].

The scanned laser is made to pass through an absorption cell where it interacts with the species of interest. The transmitted signal is measured using a photodiode (usually DC coupled). Most common detector types are indium gallium arsenide (InGaAs) for near-infrared and mercury cadmium telluride (MCT) for mid-infrared regions, respectively. The basic components and schematic of a direct absorption spectroscopy system are shown in **Figure 1**. In the absence of absorption, the detector signal will essentially represent the power vs current behavior of the laser. This behavior is typically linear when in small current ranges but could also be of higher order depending on the laser nonlinearity. In the presence of absorption, a typical DAS spectrum looks similar to the embedded graph in **Figure 1**, where a dip in transmission is observed as the laser wavelength scans through the absorption feature. The magnitudes of the photodiode signal at the absorption line center, with absorption and without absorption, are proportional to I_1 and I_0 in the Beer–Lambert law, respectively, and can be used to estimate the species concentration. A common way to calibrate a DAS system involves analyzing and plotting the photodiode signal as a function of the laser wavelength using the aforementioned current–wavelength transfer function. Subsequently, the integrated area under the absorption curve (which is directly proportional to the species number density) is used to correlate the DAS signal to species concentration [13].

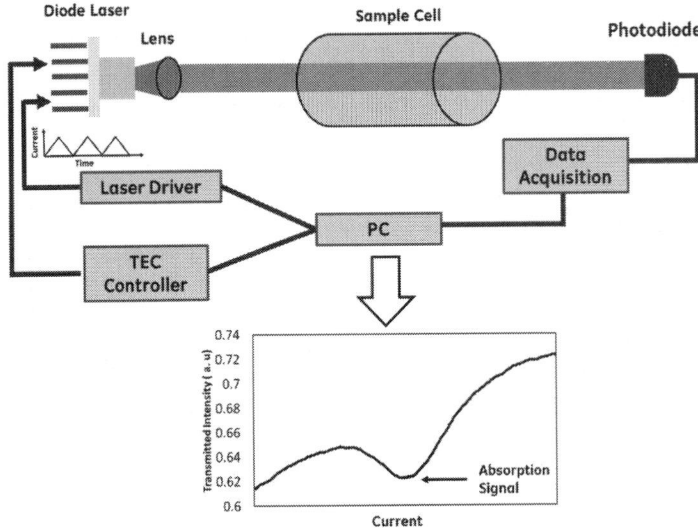

Figure 1. Schematic of a typical direct absorption spectroscopy system.

2.2. Wavelength modulation spectroscopy (WMS)

For TDLAS applications requiring high sensitivity, wavelength modulation spectroscopy (WMS) is a very effective technique [29, 31–33]. A typical WMS setup is shown in **Figure 2**. In addition to the laser scan (as in DAS), a fast current modulation (at a frequency f and amplitude a) is added to the injection current of the laser. The frequency f of the modulation signal is typically in the range 1–20 kHz. As in the case of direct absorption, the wavelength tuning properties of the laser, for both the scan and modulation frequencies, have to be well characterized with an appropriate etalon [30]. The transmitted signal measured by the photodiode is fed into a lock-in amplifier. As shown in **Figure 2**, it is critical that the reference signal of the lock-in amplifier is the same as the laser's modulation signal. It should be noted that this is a general requirement and applies differently depending on the type of lock-in amplifier. For example, in case the reference is internally generated in the lock-in, the reference signal from the lock-in can be scaled appropriately to modulate the laser. Or, if a software lock-in is used, then the PC generated reference can be used for this purpose. The lock-in analysis frequency is set to twice the modulation frequency of the laser (2f) and the spectrum is analyzed in a narrowband around the 2f frequency. This is known as second harmonic spectroscopy, and a typical second harmonic (2f) signal is shown in the inset of the **Figure 2**. As the modulated laser (at f) is scanned across a typical absorption line profile (Lorentzian, Gaussian, or Voigt), the transmitted on-absorption signal (at absorption line center) changes at a frequency of 2f, and the transmitted off-absorption signal changes at a frequency of 1f. Therefore, setting the lock-in band around 2f ensures that the system becomes more sensitive and selectively extracts the on-absorption signal. Also, choosing a higher f ensures a lower 1/f noise. Hence, the WMS technique is generally more suitable for high sensitivity applications compared to DAS. The typical absorbance limits achievable through WMS in the field are around 10^{-4} for standard WMS and 10^{-5}–10^{-6} for balanced detection-based WMS (compared to 10^{-2}–10^{-3} in typical DAS) [28].

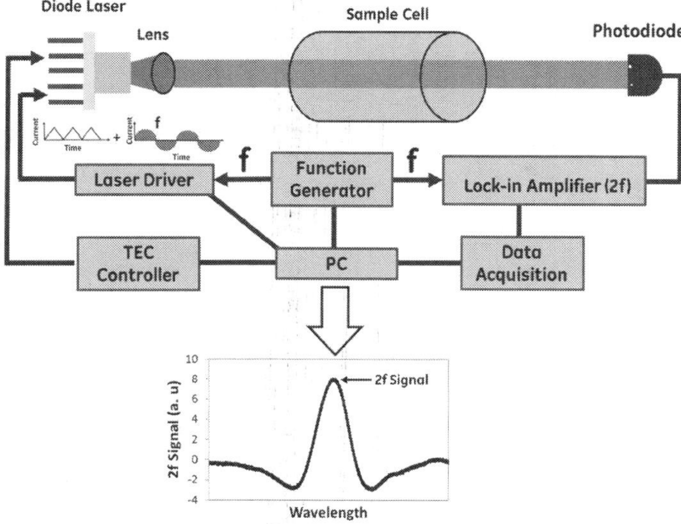

Figure 2. Schematic of a typical wavelength modulation spectroscopy system.

For WMS applications where the transmitted laser intensity fluctuates due to factors other than species concentration, an additional lock-in system is used to extract the 1f signal. The 2f and 1f signals are directly proportional to the transmitted laser intensity. Therefore, the 1f normalized 2f signal, also known as 2f/1f, is a good way to desensitize the WMS system to transmission intensity fluctuations caused by external influences [14]. This is a major advantage of WMS that makes it robust and field deployable in environments with high vibrations, dust load, fogged windows, and beam steering.

Another key advantage of WMS is that it is more sensitive to spectrally sharp absorption features. This is because the on-absorption 2f signal is stronger for larger absorption gradients around the absorption line center. The concept of modulation index (m), which is the ratio of the modulation amplitude (a) to the half-width half-maximum of the spectral line ($\Delta \upsilon$), is important to understand this. For a given absorption line, the optimized 2f and 1f signals are obtained at m = 2.2 [34]. For a single absorption line of interest, the appropriate m is chosen to meet this criterion. This makes the WMS system selectively more sensitive to spectral features of that particular width and the 2f contribution from broad spectral features and from molecules like water or heavy hydrocarbons that are significantly diminished. Hence, WMS is a powerful tool to overcome spectral interferences and to measure trace concentrations in complex gas mixtures.

The WMS signals are a strong function of the temperature and pressure of the sample gas, and therefore the calibration of such systems is a critical step toward ensuring accuracy and reliability. For sampling-based TDLAS measurements, sample is often filled in an absorption cell. These cells can be single pass, dual pass, or long path multipass cells [13]. The temperature and pressure of the absorption cell are controlled at fixed values, and calibration is performed at these conditions. For applications where temperature and pressure may vary, calibration is done for multiple operating conditions. Dynamic measurements of temperature and pressure combined with spectral models are then used to estimate the gas concentration as the conditions vary. For in situ applications with wide temperature and pressure fluctuations, the calibration-free WMS technique [35], pioneered by the Hanson group at Stanford University, has become widely acceptable. This technique involves thorough characterization of the instrumentation (lasers, detectors, amplifiers, etc.) and combining these details with the quantitative spectroscopy model. This semi-empirical model, where most of the real-world noise sources are accounted or corrected for, is then used to calculate the expected 2f/1f signal for an operating condition (known or measured temperature and pressure). The calculated signal is compared with the experimentally measured 2f/1f signal. The optimized concentration value in the model, which gives the best match between the two signals, is stated as the in situ concentration of the species.

3. Designing the TDLAS sensor

The design of a TDLAS system, especially for harsh environments, involves several critical steps. Details of the required detection limit, sample gas conditions, sample accessibility, installation methodology, and data reporting frequency are crucial toward designing a reliable, accurate, and robust sensor. The following subsections discuss these in more detail.

3.1. Wavelength selection

Selection of a suitable absorption line for a particular application is the first and the most important step in the sensor design process. The choice of the line is strongly dependent on the species of interest, the sample temperature and pressure, available path length, and background gas composition (background gas constitutes all the other species in the gas mixture apart from the species of interest). The HITRAN molecular spectroscopic database [36] is a fairly comprehensive database and is commonly used for simulations to assist in the spectral line selection process.

To begin, the selected spectral line should have sufficient absorbance to reach the required Lower detection limit (LDL) for the application path length, without considering interference. Absorbance values of 10^{-4}–10^{-5} are typically achievable in industrial TDLAS systems. Coarse spectral simulations, at the appropriate sample temperature and pressure, can be used to identify the potential candidate lines in the NIR and MIR regions that can meet the detection limit requirements. It should be noted that for very low detection limits (<1 ppm), the MIR region is more promising as it covers the intense fundamental absorption bands of most molecules. Subsequently, for each of the candidate lines, a rigorous background gas interference analysis needs to be done to investigate cross-sensitivity issues and potential LDL degradation. It is critical to perform the interference analysis for the full range of background gas composition variations. A specific case is discussed below to throw more light on the spectral line selection process.

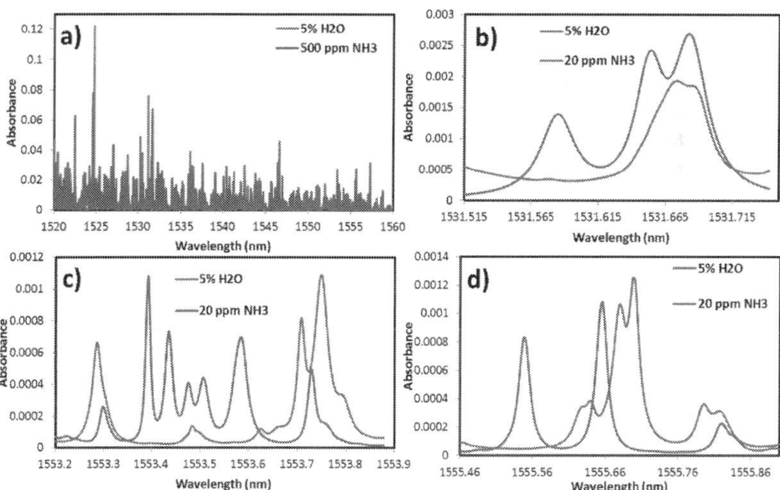

Figure 3. Spectroscopic absorbance simulations at 1.5 μm for pressure = 1 Bar, temperature = 500 K and path length = 10 m. (a) H_2O = 5%, NH_3 = 500 ppm. (b) H_2O = 5%, NH_3 = 20 ppm. (c) H_2O = 5%, NH_3 = 20 ppm. (d) H_2O = 5%, NH_3 = 20 ppm. Figures taken with permission from SPIE (From our published paper).

Assume a typical power plant emissions control application which requires measurement of ammonia (NH_3) with a sensitivity of <1 ppm at a path length of 10 m and sample gas temperature of 500 K. The average moisture level in the exhaust stream can be up to 5%. **Figure 3** shows the HITRAN spectral simulations for absorption line selection for this particular application. The spectral region around 1500 nm is good for ammonia detection because it covers strong absorption bands of ammonia. Also, this region allows for remote sensing as wavelengths around 1500 nm can be transmitted over fiber optics with low loss. **Figure 3a** shows a coarse simulation of ammonia lines (at the application conditions), with the appropriate moisture concentration. It is clear that many moisture lines are present which can potentially worsen the ammonia measurement fidelity. High-resolution simulations in shorter wavelength spans are conducted to identify three spectral lines at 1531.59, 1553.4, and 1555.56 nm where the spectral contribution of moisture is minimal. These are shown in **Figure 3b–d**, respectively. It should be noted that in all these regions, the moisture absorbance is nonzero but the spectra are relatively broader than the ammonia line. This is an acceptable solution when regions of zero background gas absorbance cannot be found. This is because, as explained in Section 2.2, WMS has the capability to be more sensitive toward sharp spectral features and can reject contributions from broad background gas features, like in this case. Once the absorption line is selected, the process of instrument development can be started as explained in the next subsection.

3.2. Optomechanical assembly and instrumentation

A well designed optomechanical assembly is crucial toward obtaining the optimum performance out of a TDLAS system and involves the following key considerations. All optical surfaces through which the laser passes, such as the laser window, the absorption cell window, and detector window, should have the appropriate antireflection (AR) coating for the laser wavelength. This ensures minimization of backscattering of photons into the laser which can severely increase the laser's excess noise. In case of fiber-coupled lasers, incorporation of an optical isolator in laser package (or right after the laser) is often a good idea to keep the excess noise in check. Furthermore, effort should be made to keep all windows wedged and tilted. This decreases the interference noise created by multiple reflections from parallel surfaces. In addition, in case of dual-pass or multipass systems, care should be taken to avoid the overlap of the different passes. This can also lead to significant interference noise in the system. Finally, all optical surfaces should be kept clean to the extent allowed by the application. It should be noted that some level of interference noise (also known as etalon noise [28]) will always be present in any practical TDLAS system as it cannot be completely avoided. However, if proper care is taken, as mentioned above, the absorbance noise level can typically be brought down to 10^{-4}–10^{-5}, which is sufficient for most industrial applications.

On the instrumentation, front, low noise laser current drivers are recommended. However, one must remember that the system noise level is often limited by the etalon noise. So, it is often sufficient if the stability of the current driver is enough to keep itself from becoming the dominant noise source. Typically, it is acceptable to have the noise induced by current drivers at about two orders of magnitude lower than etalon noise. Similar requirements apply to the detector as well where the thermal noise should be at least 10^{-7}–10^{-8} in equivalent absorbance or less.

The harsh environment implementation of the system depends on the application. Free space systems [23, 24] employ a transmitter (or pitch) to launch the laser into the harsh environment and a receiver (or catch) to collect the transmitted radiation. The design of the transmitter assembly can include refractive or reflective optics to ensure that the laser is launched with the required diameter and divergence. Similarly, the design of the receiver includes optics to collect the transmitted laser and direct it to the detector. Multiple lasers can be incorporated into the transmitter through optical multiplexing. On the receiver side, these lasers can be demultiplexed using dichroic mirrors or beam splitters or other similar optical elements. Other advanced techniques, such as time division multiplexing (TDM) and frequency division multiplexing (FDM), are also employed when demanded by the application [37]. In some free space configurations, the transmitter and receiver are packaged into a single assembly, and the laser is reflected back using a retroreflector located across the sample on the other side. This typically enables a dual pass system. While free space TDLAS systems are generally used to measure the line-of-sight average species concentration or temperature, some applications toward temperature profiling have also been reported [38, 39].

Applications where a line of sight or retroreflector is not possible, TDLAS in standoff mode [25] is often employed. This approach is similar to the retroreflector configuration described above except that the return signal is due to the backscattering from the sample gas. One expected challenge in the standoff mode is to get enough backscattered signal to do a meaningful analysis. Narrowband filters are often used in front of the detector to selectively separate the detection laser from the ambient radiation. Also, large area optics (typical diameter 2–4 inches) is often employed for optimum collection of the backscattered signal. Similar to free space systems, standoff systems also measure path average concentration or temperature value.

Sampling-based systems, which use multipass absorption cells, are also used for harsh environment applications. Herriott-type [40], White-type [41], and Chernin-type [42] cells are the most well-known ones. When absorption cells are used to analyze harsh environment gases, it is a key requirement to maintain the properties of the gas mixture (to the extent possible) during sampling and analysis. This is because the change in gas properties, such as cooling, can change the concentration of the species of interest. For example, if an exhaust gas sample is allowed to cool, then the water vapor would condense out taking with it a significant amount of exhaust gases. High-temperature multipass cells [43], in combination with a heated sampling line, serve as a good solution in such cases.

Since most TDLAS applications require fast response, a real-time data acquisition system can be highly beneficial. In WMS, the detector data are collected at a high sampling frequency, at least 20–50 times the modulation frequency (f) of the system. In case of DAS, as expected, this requirement is relaxed based on the scan frequency. In WMS, the acquired data can be processed using a hardware or software lock-in amplifier to generate the 2f and 1f signals. The choice of hardware vs software lock-in amplifier depends on the application. For applications with 1 or 2 lasers, a compact hardware lock-in amplifier can be optimal. However, for applications involving multiple lasers, the use of hardware lock-in amplifiers can be cumbersome and bulky. In such cases, an onboard PC/processor with software lock-in feature can be a much better solution. It should be noted that software lock-in features are commonly implemented in development environments such as National Instruments Labview and MATLAB. This concludes the basic overview of TDLAS sensor design and the following section will discuss some examples of how this technology is enabling real-world solutions to challenging industrial problems.

4. Harsh environment applications and examples

The applications of TDLAS and QCLAS techniques in harsh environment conditions are well known and widely discussed in the literature. The most common applications include measurement of trace gas concentrations [44–46], temperature [19, 47], combustion control [48], and plasma diagnostics [49]. It is beyond the scope of this chapter to discuss all these applications in detail. In the following subsections, two specific examples will be discussed which demonstrate the power of this technology in tackling real-world challenges where conventional techniques fail to perform well.

4.1. Steam quality sensor for steam turbine applications

This subsection is a summarized excerpt of the work published in reference [22]. Steam quality or steam wetness fraction is a critical operational parameter in steam turbines and is used in estimation of turbine efficiency and remaining life. It is a quantitative measure of the amount of water vapor and liquid water (usually microdroplets) present in the process steam. Steam quality or wetness fraction (X) is defined as follows

$$X = \frac{m_{vapor}}{m_{vapor} + m_{liquid}} \tag{5}$$

where m_{vapor} and m_{liquid} are the mass of vapor phase and mass of liquid phase, respectively.

The steam quality in a steam turbine needs to be closely monitored as an indicator of wear and tear of the machinery. The architecture of a steam turbine mainly includes three sections: HP (high pressure), IP (intermediate pressure), and LP (low pressure) respectively. The steam in HP and IP sections has little liquid water content, and therefore, the efficiency of these sections can be measured using standard temperature and pressure measurements. The LP section, however, can have steam with significant liquid water content (also known as wet steam), which can cause significant erosional damage to the rotating components [50, 51]. Therefore, the accurate measurement of steam wetness in the LP section becomes extremely critical. Most conventional methods are either based on steam sampling or based on calculations involving machine parameters (such as power output), both of which have low accuracy [22].

Therefore, a diode laser-based sensor capable of direct in situ measurement of steam quality in the hot and harsh LP section would be an ideal solution to this challenge.

As shown in Eq. (5), estimation of steam quality requires a quantitative measurement of water in liquid and vapor phases. **Figure 4** shows the schematic of a steam quality measurement system designed using two fixed wavelength broadband diode lasers. A laser at 945 nm (power 20 mW, spectral width ~2 nm) is used to probe the absorption of water vapor. Similarly, a laser at 1560 nm (20 mW, spectral width ~10 nm) is used for liquid water. Both lasers are passed through a steam chamber, which is capable of generating known vapor/liquid ratios. The path length of the steam chamber is predetermined. The intensities of both lasers before and after passage through the steam chamber are measured with Indium Gallium Arsenide (InGaAs) detectors. A comparison of the intensity loss is used to generate a quantitative measurement of the vapor and liquid concentrations respectively and hence, to estimate the steam quality, according to Eqs. (1) and (4).

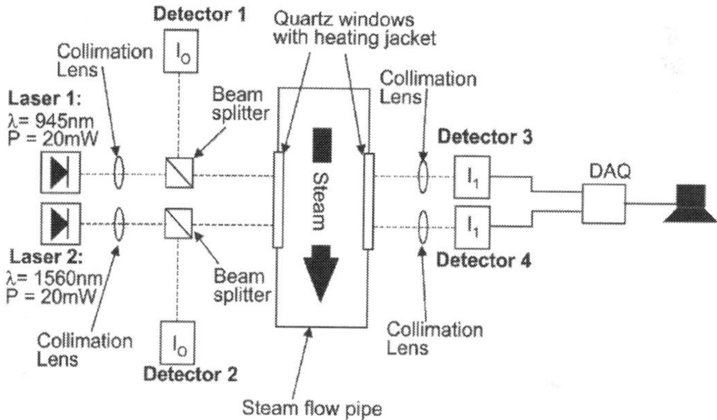

Figure 4. Schematic of the experimental setup for steam quality measurement. Figure taken with permission from IEEE (From our published paper).

Figure 5a and **b** shows a comparison of the laser-based steam quality measurement with the mass flow rate-based method under different water spray conditions. A reasonably good match is found between the two techniques and further improvements are possible through improvement of analysis algorithms for different temperature, pressure, and flow conditions. However, these results certainly demonstrate the power of laser-based techniques and the three main advantages offered in this particular application. First, this is a direct measurement of steam quality without any extraction or conditioning of the sample. The only inputs required in the method are temperature, pressure, and other standard operating parameters of the steam turbine. Second, the steam quality can be measured over a wide range (100–10%). This is a unique capability over conventional steam calorimeters which have a much smaller range of a (100–80%). Last but not the least, the measurement is real-time and can be used for better process optimization and steam turbine prognostics.

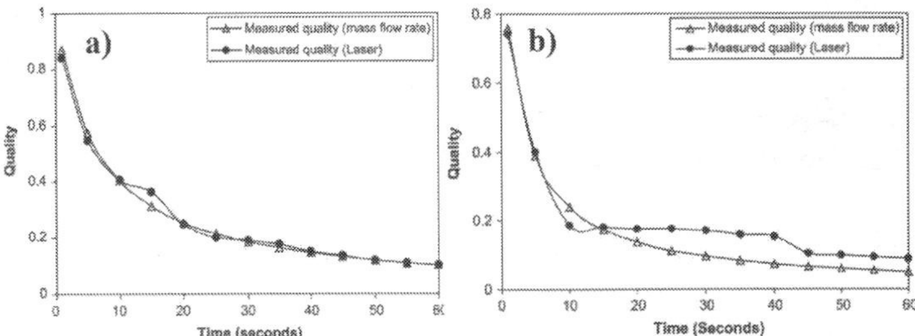

Figure 5. (a) Variation of steam quality with water spray in the in the steam flow pipe and comparison with calculation using mass flow rate. (b) Variation of steam quality with water spray in the steam flow pipe and comparison with calculation using mass flow rate at a different temperature, pressure, and flow rate condition. Figures taken with permission from IEEE (From our published paper).

4.2. Ammonia slip sensor for gas turbine applications

This subsection is a summarized excerpt of the work published in reference [14]. Gaseous emissions, such as NO_x, SO_x, and CO are strictly regulated by the environmental protection agency (EPA) in the United States and similar agencies around the world. To minimize NO_x emissions, a selective catalytic reduction (SCR) unit is commonly introduced in the exhaust gas path [52]. The gas temperatures in the harsh SCR exhaust environment are typically of the order of 250–380°C. The functioning of an SCR unit includes injection of ammonia (NH_3) gas to cause chemical reactions leading to reduction of NO_x into N_2 and H_2O. This process is depicted in **Figure 6**.

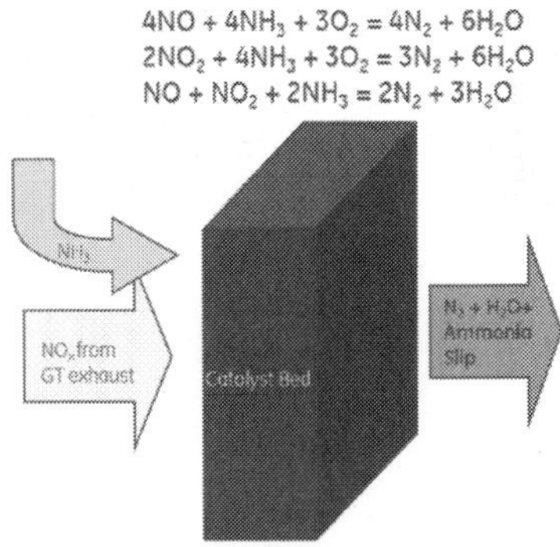

Figure 6. Schematic showing the function of a selective catalytic reduction unit in a power plant. (Figure taken with permission from Sharma et al. [14]).

The equations embedded in **Figure 6** suggest that the optimal performance of an SCR would require ammonia injection amount to vary with the NOx generated. Too much ammonia injection leads to incomplete reaction and too little leads to residual ammonia (ammonia slip) in the exhaust. The amount of ammonia injected is typically controlled by monitoring and minimizing the ammonia slip. The conventional sampling-based continuous emissions monitoring system (CEMS) [15] is the state-of-the art for this application and is often believed to be suboptimal due to sampling-related measurement lags of up to 2–3 min. The most efficient and desired way of controlling the SCR is to directly measure the ammonia slip in the harsh SCR environment (without sampling). An example of how a TDLAS-based sensor can address this highly challenging real-world problem follows here.

The TDLAS sensor discussed here is designed for direct operation in an SCR exhaust so that measurement lag time is minimized. The sensor is based on a free space line-of-sight methodology employing a laser transmitter (pitch) to emit the laser radiation and a receiver (catch) to collect the radiation after passage through the SCR exhaust. A diode laser around 1.5 μm is chosen to sense ammonia. Details on spectral analysis and line selection have been presented in the Section 3.1. In a combined cycle power plant, the SCR exhaust is usually into a heat recovery steam generator (HRSG). The environment inside a HRSG is fairly harsh due to high temperatures, engine vibrations, and floating dust or impurities. The challenges around beam steering caused by temperature gradients and transmission losses due to dust particles are addressed using the 2f/1f WMS methodology described in the Section 2.2. A major challenge specific to this application is the thermally induced misalignment of the beam since the typical line-of-sight path length in a HRSG can be up to 10 m. Therefore, maintaining alignment through all engine-operating conditions is critical toward achieving long-term reliability.

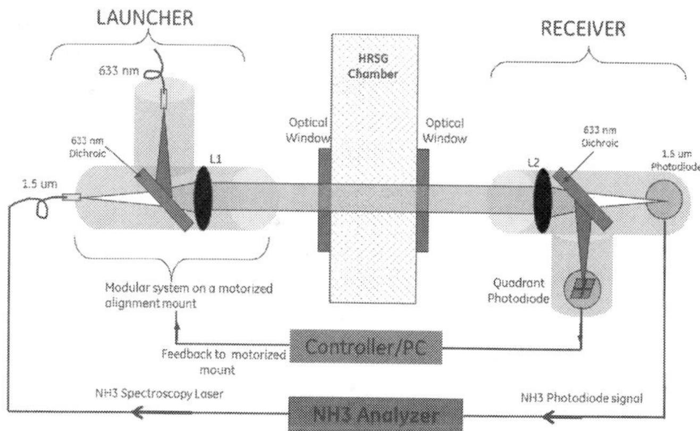

Figure 7. Schematic showing the hybrid ammonia slip sensor with alignment control and implementation in a HRSG. Figure taken with permission from SPIE (From our published paper).

The hybrid sensor, shown in **Figure 7**, is developed to address the misalignment challenge. A 633 nm fixed wavelength laser is multiplexed with the 1.5 μm ammonia spectroscopy laser at the transmitter (launcher) and demultiplexed at the receiver by using appropriate dichroic mirrors. The 633 nm laser wavelength is chosen because of two reasons: no absorption by exhaust gases and visibility to the human eye (helps in initial alignment of the system). As shown, the launcher optical assembly is mounted on a motion control stage. The demulti-plexed red laser is made incident on a quadrant photodiode which serves as a position sensitive detector. The error signal generated by the quadrant photodiode is used as a feedback signal to actively align the motion control stage to keep the red beam centered on the quadrant photodiode. As a consequence, the ammonia spectroscopy beam also remains centered on the 1.5 μm photodiode. This leads to overcoming of measurement errors induced by thermal misalignment and enables long-term reliable operation of the sensor.

Experimental results demonstrating the performance and value of the alignment control system are shown in **Figure 8**. **Figure 8a** shows the scenario with no vibrations/misalign-ments. As expected, the 2f/1f signal stability is the same with the alignment control system on or off. This is a key check to ensure that the alignment control system does not introduce any artifacts or errors by itself. **Figure 8b** shows the scenario under induced vibrations/misalign-ments. With the alignment control off, the total normalized transmitted power drops close to zero in a matter of seconds, hence indicating a severe misalignment of the system. However, with the alignment control on, the system is able to actively maintain alignment over time. About 20–25% power fluctuations are still observed which are experimentally found to be well within the correction capability of the 2f/1f WMS technique. For more details on the tests demonstrating performance of the hybrid sensor, the reader is advised to refer to reference [14].

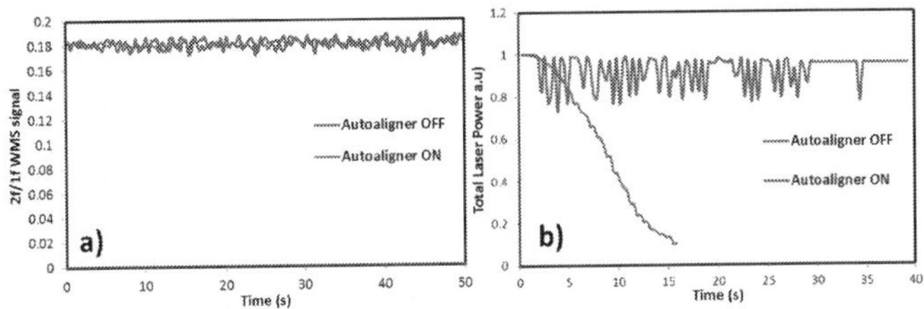

Figure 8. Experimental results showing the performance of the alignment control system. (a) 2f/1f WMS signals (with and without alignment control) as a function of time under no vibrations. (b) Transmitted spectroscopy laser power (with and without alignment control) as a function of time under induced vibrations. Figure taken with permission from SPIE (From our published paper).

5. Summary

In summary, this chapter presented the value and potential benefits of diode laser-based industrial harsh environment sensors. The chapter started with an overview of the industrial Internet and discussed the importance of sensors toward achieving enhanced output from industrial assets such as gas turbines, aircraft engines, and turbomachinery equipment . The discussion showed that real-time decision making through online sensors (that monitor the desired machine parameters) can enable optimized operation, performance enhancement, and extension of asset life. Subsequently, optical harsh environment sensors were introduced and the capabilities of diode laser-based techniques, that is TDLAS and QCLAS, were discussed.

The discussion was kept focused on these two most promising techniques (compared to other optical techniques) to remain within the scope of this chapter. The common implementation of TDLAS/QCLAS methodologies, that is direct absorption spectroscopy (DAS) and wavelength modulation spectroscopy (WMS), were discussed from an applied experimental perspective. The intention was to equip the reader with just the right level of information required to set up and carry out application-specific experiments. Appropriate references were provided for readers who wish to go deeper into these techniques. Next, the design philosophy and methodology behind a TDLAS sensor were discussed. Laser wavelength selection was presented as a crucial step where careful consideration to process temperature, pressure, and background gas interference needs to be given. Optomechanical configurations, namely line of sight, standoff, and extractive sampling, were then discussed followed by instrumentation and data acquisition basics. Finally, the power of diode laser technology in tackling real-world challenges was discussed using real-world examples. An in situ sensor for measurement of steam quality in the hot and harsh low pressure (LP) section of a steam turbine was presented. A solution based on two diode lasers, each targeted toward water in liquid phase and vapor phase, respectively, was discussed. Last but not the least, the chapter concluded with the discussion on a hybrid in situ ammonia slip sensor for power plant SCR control applications. A novel alignment control methodology in combination with a 2f/1f WMS scheme was shown to be an effective tool toward measuring gas concentrations in hot and harsh vibrating environments.

The field of optical and diode laser-based harsh environment sensing is rapidly evolving and is currently an area of active research in academia as well as industry. The various aspects of this technology space are being studied in details by different groups around the world and this chapter has, by no means, touched upon all of them. However, the authors do hope that this chapter has motivated the reader to think of new ideas and concepts that can advance the state-of-the art and application areas in this space and help the industrial community realize the full potential of their valuable assets.

Acknowledgements

The authors would like to acknowledge R&D support from the General Electric Company. The authors would also like to thank the Optics & Instrumentation team at GE Global Research, Bangalore for their contributions to the work presented in sections 4.1 and 4.2.

Author details

Chayan Mitra* and Rachit Sharma

*Address all correspondence to: chayan.mitra@ge.com

GE Global Research Center, Bangalore, India

References

[1] P. C. Evans and M. Anunziata, Industrial Internet: Pushing the Boundaries of Minds and Machines, 26 November 2012. [Online]. Available: www.ge.com/docs/chapters/Industrial_Internet.pdf. [Accessed 18 March 2016].

[2] J. Bruner, Industrial Internet – The Machines are Talking, O'Reilly Media, 2013 (Ebook ISBN:978-1-4493-6587-5).

[3] M. Annunziata and P. C. Evans, The Industrial Internet@Work, 2013. [Online]. Available: https://www.ge.com/sites/default/files/GE_IndustrialInternetatWork_WhitePaper_20131028.pdf. [Accessed 18 March 2016].

[4] M. Annunziata, Digital future of oil & gas & energy, [Online]. Available: http://gereports.cdnist.com/wp-content/uploads/2016/02/22094804/GE_Digital_Future_WP-02191611.pdf. [Accessed 18 March 2016].

[5] M. Egan, Deep Learning: New Subsea Service Model Helps Oil Drillers Limit Costs, 24 February 2016. [Online]. Available: http://www.gereports.com/deep-learning-new-subsea-service-model-helps-oil-drillers-limit-costs/ [Accessed 18 March 2016].

[6] B. Zhang, C. Sconyers, C. Byington, R. Patrick, M. Orchard and M. Vachtsevanos, Anomaly detection: a robust approach to detection of unanticipated faults, in: *International Conference on Prognostics and Health Management*, Denver, CO, 2008 (doi: 10.1109/PHM.2008.4711445).

[7] S. Mathew, D. Das, R. Rossenberger and M. Pecht, Failure mechanisms based prognostics, in: *International Conference on Prognostics and Health Management*, Denver, CO, 2008 (doi:10.1109/PHM.2008.4711438).

[8] A. Saxena, K. Goebel, D. Simon and N. Eklund, Damage propagation modeling for aircraft engine run-to-failure simulation, in: *International Conference on Prognostics and Health Management*, Denver, CO, 2008 (doi:10.1109/PHM.2008.4711414).

[9] M. J. Roemer and G. J. Kacprzynski, Advanced diagnostics and prognostics for gas turbine engine risk assessment, in: *IEEE Aerospace Conference Proceedings*, Big Sky, MT, 2000 (doi:10.1109/AERO.2000.877909).

[10] S. Cheng, M. H. Azarian and M. G. Pecht, Sensor systems for prognostics and health management, *Sensors*, vol. 10, p. 5774, 2010.

[11] T. Kellner, Here's How Digital Electricity Will Change The Power Industry All The Way To Your Home, 13 March 2016. [Online]. Available: http://www.gereports.com/heres-how-digital-electricity-will-change-the-power-industry-all-the-way-to-your-home/ [Accessed 18 March 2016].

[12] M. Lackner, Lasers in Chemistry, vol 1., Wiley, Weinheim, ISBN 978-3-527-31997-8, 2008.

[13] J. Hodgkinson and R. P. Tatam, Optical gas sensing: a review, *Measurement Science and Technology*, vol. 24, p. 012004, 2013.

[14] R. Sharma, S. Maity, A. Bekal, S. Vartak, A. K. Sridharan and C. Mitra, Optical sensors for harsh environment applications, in: *Proc. of SPIE*, vol. 9491, pp. 94910D-1, 2015.

[15] J. A. Jahnke, Continuous Emissions Monitoring, New York, USA, John Wiley & Sons, 2000.

[16] M. Lackner, Tunable diode laser absorption spectroscopy (TDLAS) in the process industries—a review, *Reviews in Chemical Engineering*, vol. 23.2, p. 65, 2007.

[17] L. Zhang, G. Tian, J. Li and B. Yu, Applications of absorption spectroscopy using quantum cascade lasers, *Applied Spectroscopy*, vol. 68, p. 1095, 2014.

[18] W. Zeller, L. Naehle, P. Fuchs, F. Gerschuetz, L. Hildebrandt and J. Koeth, DFB lasers between 760 nm and 16 μm for sensing applications, *Sensors*, vol. 10, p. 2492, 2010.

[19] R. K. Hanson and P. K. Falcone, Temperature measurement technique for high temperature gases using a tunable diode laser, *Applied Optics*, vol. 17, p. 2477, 1978.

[20] M. G. Allen, Diode laser absorption sensors for gas dynamic and combustion flows, *Measurement Science and Technology*, vol. 9, p. 545, 1998.

[21] V. Nagali and R. K. Hanson, Design of a diode laser sensor to monitor water vapor in high pressure combustion gases, *Applied Optics*, vol. 36, p. 3296, 1997.

[22] C. Mitra, S. Maity, A. Banerjee, A. Pandey, A. Behera and V. Jammu, Development of steam quality measurement and monitoring technique using absorption spectroscopy with diode lasers, *IEEE Sensors Journal*, vol. 11, p. 1214, 2011.

[23] S. T. Sanders, J. Wang, J. B. Jeffries and R. K. Hanson, Diode-laser absorption sensor for line-of-sight gas temperature distributions, *Applied Optics*, vol. 40, p. 4404, 2001.

[24] J. Wang, M. Maiorov, D. S. Baer, D. Z. Garbuzov, J. C. Connolly and R. K. Hanson, In situ combustion measurements of CO with diode-laser absorption near 2.3 μm, *Applied Optics*, vol. 39, p. 5579, 2000.

[25] M. B. Frish, R. T. Wainner, M. C. Laderer, B. Green and M. G. Allen, Standoff and miniature chemical vapor detectors based on tunable diode laser absorption spectroscopy, *IEEE Sensors Journal*, vol. 10, p. 639, 2010.

[26] C. N. Banwell and E. M. McCash, Fundamentals of Molecular Spectroscopy, 4th Edition, McGraw-Hill, London, 1994.

[27] J. D. Ingle and S. R. Crouch, Spectrochemical Analysis, Prentice Hall, London, 1988.

[28] J. Hodgkinson, D. Masiyano and R. P. Tatam, Gas cells for tunable diode laser absorption spectroscopy employing optical diffusers. Part 1: single and dual pass cells, *Applied Physics B*, vol. 100, p. 291, 2010.

[29] A. Klein, O. Witzel and V. Ebert, Rapid, time-division multiplexed, direct absorption- and wavelength modulation-spectroscopy, *Sensors,* vol. 14, p. 21497, 2014.

[30] X. Chao, J. B. Jeffries and R. K. Hanson, Wavelength-modulation-spectroscopy for real-time, in situ NO detection in combustion gases with a 5.2 µm quantum-cascade laser, *Applied Physics B,* vol. 106, p. 987, 2012.

[31] J. Reid and D. Labrie, Second-harmonic detection with tunable diode lasers—comparison of experiment and theory, *Applied Physics B,* vol. 26, p. 203, 1981.

[32] S. Schilt, L. Thevenaz and P. Robert, Wavelength modulation spectroscopy: combined frequency and intensity laser modulation, *Applied Optics,* vol. 42, p. 6728, 2003.

[33] P. Kluczynski, J. Gustafsson, A. M. Lindberg and O. Axner, Wavelength modulation absorption spectrometry—an extensive scrutiny of the generation of signals, *Spectrochimica Acta Part B: Atomic Spectroscopy,* vol. 56, p. 1277, 2001.

[34] G. Hancock, J. H. van Helden, R. Peverall, G. A. D. Ritchie and R. J. Walker, Direct and wavelength modulation spectroscopy using a cw external cavity quantum cascade laser, *Applied Physics Letters,* vol. 94, p. 201110, 2009.

[35] G. B. Rieker, J. B. Jeffries and R. K. Hanson, Calibration-free wavelength-modulation spectroscopy for measurements of gas temperature and concentration in harsh environments, *Applied Optics,* vol. 48, p. 5546, 2009.

[36] L. S. Rothman et al., The HITRAN2012 molecular spectroscopic database, *Journal of Quantitative Spectroscopy and Radiative Transfer,* vol. 130, p. 4, 2013.

[37] R. J. Muecke, P. W. Werle, F. Slemr and W. Prettl, Comparison of time and frequency multiplexing techniques in multicomponent FM spectroscopy,in: *Proc. SPIE Measurement of Atmospheric Gases,* vol. 1433, p. 136, 1991.

[38] X. Liu, J. B. Jeffries and R. K. Hanson, Measurement of nonuniform temperature distributions using line-of-sight absorption spectroscopy, *AIAA Journal,* vol. 45, p. 411, 2007.

[39] X. Yu, F. Li, L. Chen and X. Zhang, Spatial resolved temperature measurement based on absorption spectroscopy using a single tunable diode laser, *Acta Mechanica Sinica,* vol. 26, p. 147, 2010.

[40] D. R. Herriott, H. Kogelnik and R. Kompfner, Off-axis paths in spherical mirror interferometers, *Applied Optics,* vol. 3, p. 523, 1964.

[41] J. U. White, Long optical paths of large aperture, *Journal of Optical Society of America,* vol. 32, p. 285, 1942.

[42] S. M. Chernin and E. G. Barskaya, Optical multipass matrix systems, *Applied Optics,* vol. 30, p. 51, 1991.

[43] R. Bartlome, M. Baer and M. W. Sigrist, High-temperature multipass cell for infrared spectroscopy of heated gases and vapors, *Review of Scientific Instruments,* vol. 78, p. 013110, 2007.

[44] X. Chao, J. B. Jeffries and R. K. Hanson, Absorption sensor for CO in combustion gases using 2.3 µm tunable diode lasers, *Measurement and Science Technology,* vol. 20, 2009.

[45] R. Sur, K. Sun, J. B. Jeffries, R. K. Hanson, R. J. Pummill, T. Waind, D. R. Wagner and K. J. Whitty, TDLAS-based sensors for in situ measurement of syngas composition in a pressurized, oxygen-blown, entrained flow coal gasifier, *Applied Physics B,* vol. 116, p. 33, 2013.

[46] J. C. Wormhoudt, M. S. Zahniser, D. D. Nelson, J. B. McManus, R. C. Miake-Lye and C. E. Kolb, Infrared tunable diode laser measurements of nitrogen oxide species in an aircraft engine exhaust, in: *Proc. SPIE, Optical Techniques in Fluid, Thermal, and Combustion Flow,* vol. 2546, p. 552, 1995.

[47] M. G. Allen and W. J. Kessler, Simultaneous water vapor concentration and temperature measurements using 1.31-micron diode lasers, *AIAA Journal,* vol. 34, p. 483, 1996.

[48] E. R. Furlong, D. S. Baer and R. K. Hanson, Real-time adaptive combustion control using diode-laser absorption sensors, in: *Symposium (International) on Combustion,* vol. 27, p. 103, 1998.

[49] J. Röpcke, G. Lombardi, A. Rousseau and P. B. Davies, Application of mid-infrared tuneable diode laser absorption spectroscopy to plasma diagnostics: a review, *Plasma Sources Science and Technology,* vol. 15, 2006.

[50] B. Stanisa and V. Ivusic, Erosional behavior and mechanisms for steam turbine rotor blades, *Wear,* Vols. 186–187, p. 395, 1995.

[51] B. Stanisa and Z. Schauperl, Erosional behavior of turbine blade in nuclear power plant, *Wear,* vol. 254, p. 735, 2003.

[52] P. Forzatti, Present status and perspectives in de-NOx SCR catalysis, *Applied Catalysis A: General,* vol. 222, p. 221, 2001.